Swainson's Handbook of Technical and Quality Management for the Food Manufacturing Sector

Related Titles

Hygienic Design of Food Factories
(ISBN 9781845695644)
Sensory Analysis for Food and Beverage Quality Control
(ISBN 9781845694760)
Food Processing Technology
(ISBN 9781845692162)

Woodhead Publishing Series in Food Science, Technology and Nutrition

Swainson's Handbook of Technical and Quality Management for the Food Manufacturing Sector

Mark Swainson

ELSEVIER

WP
WOODHEAD
PUBLISHING
An imprint of Elsevier

Woodhead Publishing is an imprint of Elsevier
The Officers' Mess Business Centre, Royston Road, Duxford, CB22 4QH, United Kingdom
50 Hampshire Street, 5th Floor, Cambridge, MA 02139, United States
The Boulevard, Langford Lane, Kidlington, OX5 1GB, United Kingdom

Notices
Knowledge and best practice in this field are constantly changing. As new research and experience broaden our understanding, changes in research methods, professional practices, or medical treatment may become necessary.

Practitioners and researchers must always rely on their own experience and knowledge in evaluating and using any information, methods, compounds, or experiments described herein. In using such information or methods they should be mindful of their own safety and the safety of others, including parties for whom they have a professional responsibility.

To the fullest extent of the law, neither the Publisher nor the authors, contributors, or editors, assume any liability for any injury and/or damage to persons or property as a matter of products liability, negligence or otherwise, or from any use or operation of any methods, products, instructions, or ideas contained in the material herein.

British Library Cataloguing-in-Publication Data
A catalogue record for this book is available from the British Library

Library of Congress Cataloging-in-Publication Data
A catalog record for this book is available from the Library of Congress

ISBN: 978-1-78242-275-4 (print)

ISBN: 978-1-78242-287-7 (online)

For information on all Woodhead Publishing publications
visit our website at https://www.elsevier.com/books-and-journals

Publisher: Charlotte Cockle
Acquisition Editor: Nina Rosa de Araujo Bandeira
Editorial Project Manager: Karen R. Miller
Production Project Manager: Joy Christel Neumarin Honest Thangiah
Cover Designer: Alan Studholme

Typeset by MPS Limited, Chennai, India

Contents

List of contributors

Andy Bowles Specialist Food Law Solicitor, ABC Food Law, Innovation Centre, Norwich Research Park, Norwich, Norfolk, England

Steve Clemenson Group QA & Regulatory Manager, Nedspice EMEA BV, Rotterdam, The Netherlands

Wojciech J. Florkowski Department of Agricultural and Applied Economics, University of Georgia, Griffin, GA, United States

Luanne Gounder Johannesburg, South Africa

Fiona Houston Independent Researcher, United Kingdom

Colin & Lindy Jeffery Johannesburg, South Africa

Richard F. Stier Consulting Food Scientist, Sonoma, CA, United States

Belinda Stuart-Moonlight Moonlight Environmental Ltd - Scientific & Legal Consultancy, Folkestone, Kent, United Kingdom

Mark Swainson National Centre for Food Manufacturing, University of Lincoln, Holbeach, Lincolnshire, United Kingdom

Kevin Swoffer NSF International Global Food Division, Hanborough Business Park, Long Hanborough, Oxon, United Kingdom

Colin W. Wrigley University of Queensland, Brisbane, QLD, Australia

Biography

Since the early 1990s Mark Swainson has worked extensively both within and in partnership with the food industry. His industrial food manufacturing sector expertise developed while holding key quality, technical and operations management positions in the high-risk chilled and frozen food sectors, producing a wide range of product categories including ready meals, soups, sauces, dips, dressings, risottos, salads and pasta snacks. This involved close engagement with all parts of the international food supply chain to assure food quality, safety and legality from supplier control of ingredients and packaging, through operational and technical oversight of food factory processing, storage and distribution, and finally liaison with the end customer including the major retailers and food service groups.

Highly experienced and well qualified in industrial food technology, food processing and packaging operations, and in their underpinning technical standards and audit systems, his industry engagement continues, currently leading the research and higher education agendas of the UK National Centre for Food Manufacturing (NCFM), University of Lincoln.

Since joining the University of Lincoln Mark has worked extensively on the development and delivery of a range of food sector employee training and education programmes (both undergraduate and postgraduate), making a significant contribution to the skills and staff development needs of the industry. He has also structured the successful NCFM 'industry-focused' approach to research and development, working in partnership with the food sector to progress multidisciplinary projects and initiatives which address specific food sector challenges across food safety, quality and productivity agendas.

Preface

This book is focused on the expansive and highly demanding world of food industry technical and quality management. Food production is vital to all populations, and the largest manufacturing sector globally. It is complex and multifaceted, linking 'farm to fork'/'plough to plate' via food supply chains that combine the activities of farms, primary producers, food processors, storage and distribution networks, through to retailers, food service operations and markets, ultimately to satisfy the nutrition, energy and well-being needs of the end consumer.

With the benefit of a unique perspective which I have gained through working extensively in quality, technical and operations management roles at all levels within the food industry, and routinely working with the sector to develop employee focused education programmes and research services, I recognised the need for a book which provides insights and guidance on the 'applied practice' of industrial quality and technical management, written from the perspective of the industry practitioner.

I joined the industry in the early 1990s and I reflect that not a year has gone by without serious food scares and product recalls both nationally and internationally. This demonstrates the persistent challenge of controlling the major food hazard groups – physical, chemical (including allergens), and microbiological. More recently, these key risks have been compounded by the increasing criminal threats of food fraud, adulteration and intentional contamination.

Government and trade organisations often respond to issues and incidents with increasing legislation and guidelines. Society becomes increasingly litigious, and companies in the supply chain respond with added detail and focus on specifications, contracts, codes of practice, testing criteria, audits, supplier assurance programmes, compliance and certification standards, all often coupled with major fines and other penalties for any 'failures' encountered.

As a result, compared with past decades there is now a wealth of information in the food sector with regard to expected controls and practices to assure the quality, safety and legality of food production. Yet there appears to be limited coverage of the very diverse remits and particular challenges of those working within quality and technical roles focused on the control of product quality, safety and legality. Knowing what to do is half of the challenge, but being able to then make it happen is crucial and often less well considered in sector information resources.

There is increasing recognition of the need for more holistic approaches to the assurance of food safety, quality and legal compliance. These relatively new management approaches endeavour to combine the consideration of key factors including people, processes and products, in preference to viewing each quality management factor in isolation. Knowledge management and leadership are increasingly recognised as vital to the control of food safety and quality. Such rounded approaches are also being recognised in an emergent food safety and quality culture.

This book has therefore been written with a diverse reader profile in mind, including those holding technical and quality-related responsibilities in the food industry, together with readers who wish to gain an insight and greater understanding of the activities and demands of the sector when working to assure food product quality, safety and legality. The latter may include undergraduate and postgraduate study, researchers engaged with the food sector, certification and training organisations, industry managers, government enforcement bodies and policy decision-makers.

I have written the first part of this book to cover the more generic aspects of food quality and technical management activities. For the second part I am very grateful to the guest authors who have generously shared their expertise on a range of specialist topics, adding significant breadth and depth to the content covered.

Acknowledgements

This book has benefited greatly from the wisdom and experience of all of the contributing authors, who have kindly shared in-depth insights and invaluable knowledge of their subject areas. It has been a pleasure to correspond with you all.

I would like to express my added thanks to Lindy Jeffery who following the loss of Colin kindly agreed to maintain the inclusion of Colin's invaluable guest chapter on baked products.

The Elsevier publishing team have been of great support at all stages of the process, always so highly professional and enthusiastic. Particular thanks to Editorial Project Manager Karen Miller who has worked with me through the majority of the writing process including the key completion phases, and to my initial Commissioning Editor Nell Holden, who helped me to structure the book at the outset.

Past and present food industry colleagues, tutors and mentors have taught me a great deal about, not only the craft of 'food science and technology', but also the value of teamwork, resilience and integrity when working to assure food safety, quality and legality in the fast-moving and highly demanding food sector.

I am also very grateful to my family and friends for their great support and encouragement throughout the writing process. Thank you all.

How to use this book

This book is designed to be used in two ways. Those wishing to gain full insight into quality and technical applied management practices may read the book in its entirety. However, readers who regularly work in food safety and quality assurance processes are also advised to pick up and read chapters/sections whenever they require added perspective and information on a particular initiative or challenge.

To support on the intermittent use of this book, the "Contents" section provides an overview of the topics covered in each chapter and the "Index" contains a range of specific words and terms used within the text. Useful sources of further information ("References" and "Further Reading") are also included at the end of each chapter in order to inform ongoing research into the subject areas.

Introduction

Technical and quality management personnel perform vital roles within a food business. These team members are expected to champion and oversee all aspects of the operation related to product safety, quality and legality. Ultimately the quality/technical manager's goal is to ensure that the business has the right systems and procedures embedded to deliver consistently safe, quality products which meet all legislative requirements. Their function is also to be a knowledge resource, leader and ambassador for the business on all such matters internally, and with a wide range of external stakeholders including business customers, suppliers, third-party certification schemes and enforcement authorities. The context of the food manufacturing operation will usually define the appropriate systems and approaches required to assure the quality, safety and legality of the product manufactured. To structure and implement an effective food safety and quality management system, the quality/technical manager must consider the generic requirements of the food industry sector, and then overlay with any product/process specifics relevant to their particular operation.

This book particularly seeks to provide 'applied management' insights across the many aspects of food quality, safety and legality. The content applies to food industry operations across the globe, while providing some focused national insights.

Effective food production and supply requires a diverse range of skills, knowledge, experience and management practices, spanning numerous disciplines from food science and technology through to social sciences such as business (e.g., economics, productivity, sales and marketing) and psychology (e.g., human behaviour). This 'food system' is directed on local, national and international scales via a framework formed from legislation, guidelines, codes of practice, compliance standards, certification, specifications and contracts.

To be effective, those working in quality, technical and related roles require insight into the underpinning 'food science' and 'food technology'. Equally important are robust management skills to influence activities and control situations. This text seeks to reflect the broad nature of typical quality and technical activities, providing helpful information and practical considerations within the context of a sector where no 'one size fits all'. Businesses and people are all individuals. Job titles and remits vary widely both nationally and globally, as do product types and quality expectations. Hence, the first part of this book is intended to be generic, applying to the majority of the industry. Chapter 1 begins by considering food sector challenges and the role of technical and quality management. This is followed by a review of influencing factors and approaches toward defining technical and quality standards (Chapter 2).

Chapter 3 is focused on the purpose and content of a business food safety and quality management system, including information on system implementation, monitoring, review and accreditation. Subsequent chapters provide advice and guidance on key aspects of site food safety and quality management systems, elucidating raw materials and packaging supplier control, site standards, product control and Hazard Analysis and Critical Control Point considerations, operations and process control, personnel control, audits, and the control of nonconformance, recall and crisis management.

The final chapter in Part I (Chapter 11) covers 'managing the technical department'. This provides detailed consideration of quality and technical staff management, and other key skills/activities including the provision of technical services, reporting, budgets and oversight of outsourced services.

Drawing on the expertise and experience of guest authors, Part II focuses on complementary specialist topics including hygiene management (Belinda Stuart-Moonlight), third-party audit schemes (Kevin Swoffer), supplying supermarkets: good technical and quality management practices (Fiona Houston), an enforcement authority perspective on the food manufacturing sector (Andy Bowles), fruit and vegetables (Wojciech J. Florkowski), herbs and spices (Steve Clemenson), cereals (Colin W. Wrigley), baked products (Colin Jeffery, Luanne Gounder and Lindy Jeffery) and canning (Richard Stier). The product category of 'cook-chill' ready meals, soups and sauces is also reviewed in this section.

Part I

Aspects of Food Quality and Technical Management Activities

Food sector challenges and the role of technical and quality management

1

Mark Swainson
National Centre for Food Manufacturing, University of Lincoln, Holbeach, Lincolnshire, United Kingdom

1.1 Food industry overview

As one of the largest industries in the world the food sector is dynamic, often complex and always multifaceted. To work effectively the 'food system' requires the application of science and technology, coupled with business and management skills, across the agri-food supply chain 'from farm to fork'.

The nature of the food manufacturing sector being operated in will usually define the appropriate science, technology, systems and approaches best placed to assure the quality, safety and legality of the product manufactured. When seeking to control such matters it is very helpful to reflect on the breadth and complexity of the global 'food system'.

Providing for one of the most basic needs of mankind, that of the requirement for safe and nutritious food, the food industry faces an ever-growing challenge to feed a rapidly rising global population, which is predicted to reach 8.6 billion in 2030, 9.8 billion in 2050 and 11.2 billion in 2100 (United Nations 2017) (Fig. 1.1).

Consequently, food production and supply is often one of the largest employers and manufacturing sectors in countries as they focus on the provision of safe and nutritious food to their populations. The rapid and continuing population growth across the globe means that this is likely to continue. There is also an additional increasing productivity requirement as due to population growth, agriculture and food systems will have to operate on land resources which are increasingly limited by population growth.

Key aspects of the agri-food supply chain or 'food system' chain include:

- agriculture/farming;
- transport and distribution;
- food processing/manufacturing;
- food sellers – e.g., retailers (shops and supermarkets), food service outlets (restaurants, canteens) and food markets; and
- consumers.

Swainson's Handbook of Technical and Quality Management for the Food Manufacturing Sector.
DOI: https://doi.org/10.1016/B978-1-78242-275-4.00001-0

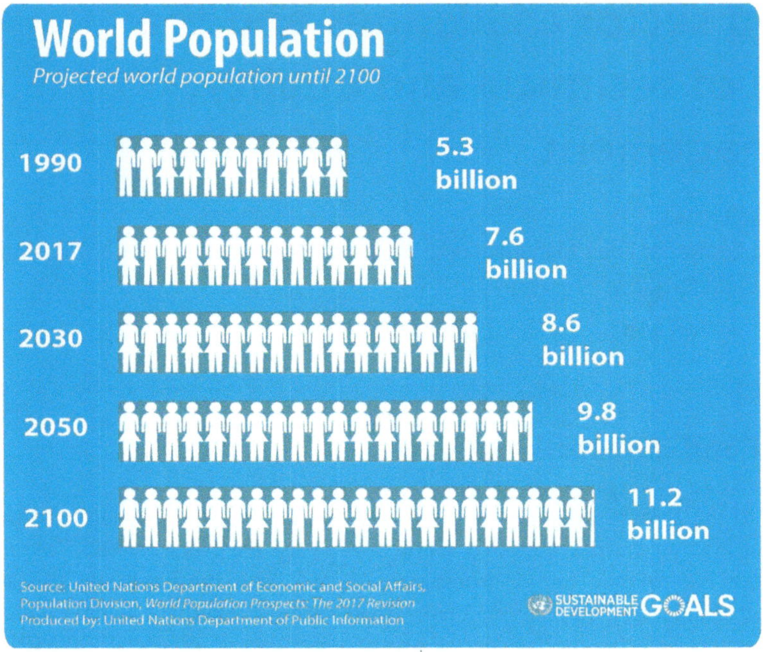

Figure 1.1 UN projected world population until 2100.
Source: From United Nations Department of Public Information, https://www.un.org/development/desa/publications/wp-content/uploads/sites/10/2017/06/Population_Graphic_R.jpg. © 2017 United Nations. Reprinted with the permission of the United Nations.

Alongside the food supply chain being often global and always critical to human survival, health and wellbeing, there are a significant amount of business interests engaged across the sector (Fig. 1.2).

All parts of the system require services such as:

- equipment manufacturers;
- technology providers;
- utilities – e.g., energy and water;
- packaging businesses;
- waste and recycling services; and
- transport and storage.

Food and drink manufacturers therefore have a very significant economic and social impact locally, nationally and globally. Food industry manufacturing businesses typically combine the produce from farms and primary production businesses, processing and 'adding value' to foodstuffs before the onward journey of the finished packaged product through to the end consumer, typically via retailers or food service groups.

The products manufactured in the sector range from base foods such as vegetables, fruits, cereals, raw meats, eggs and dairy products, oils, herbs, spices and

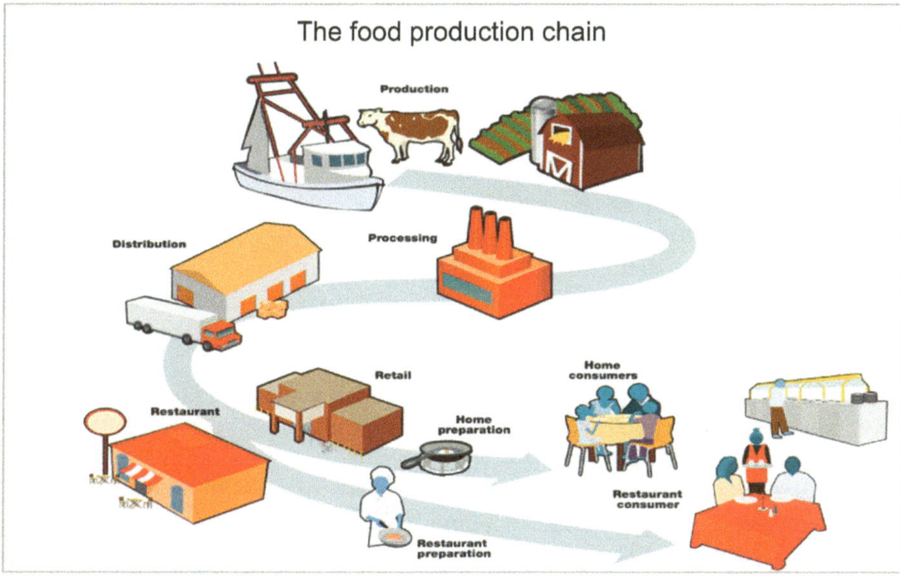

Figure 1.2 The food production chain.
Source: Centers for Disease Control and Prevention, 2017.

sugars, through to further processed foods which include ready-to-eat (RTE) foods such as sandwiches, wraps, salads and prepared fruits and ready-to-cook (RTC) foods such as ready peeled and washed vegetables and prepared meats, through to multicomponent products e.g., ready meals.

1.2 Typical challenges faced by the food production and supply system

1.2.1 Food safety: HACCP, TACCP and VACCP

HACCP: As foods are inevitably intended for consumption they have significant potential to cause harm if the food contains a particular hazard. Recognised hazard categories in the food industry include microbiological, chemical, physical and allergenic contaminants. These hazards and the approaches to their control form a significant part of the workload of the Quality or Technical Manager, working to identify the threats to food safety, form control strategies and then implement and manage appropriate systems/procedures which eliminate or reduce these hazards to an acceptable level. This approach is often termed the *hazard analysis and critical control point (HACCP)* system.

Adopted by the Codex Alimentarius Commissions the principles and practical application of this internationally recognised HACCP approach to controlling food safety is presented online by the United Nations Food and Agriculture Organisation (FAO) Corporate Document Repository (http://www.fao.org/docrep/005/y1579e/y1579e03.htm).

While HACCP is focused on controlling accidental contamination of foods, over time the potential for intentional contamination or other criminal activities with regard to food production has highlighted the need to develop a food safety assurance system focused on the prevention of deliberate food safety incidents or food fraud.

Threat assessment critical control point (TACCP). Due to a number of high profile food safety issues caused by criminal activity or malicious intent, TACCP has recently become recognised as a key aspect of food safety management. TACCP is defined by PAS 96:2014 as the 'systematic management of risk through the evaluation of threats, identification of vulnerabilities and implementation of controls to materials and products, purchasing, processes, premises, distribution networks and business systems by a knowledgeable and trusted team with the authority to implement changes to procedures'.

The threats considered and sought to be controlled by TACCP include: economically motivated adulteration (EMA), malicious contamination, extortion, espionage, counterfeiting and cyber-crime.

A particular aspect of TACCP is the importance of seeking to understand the mindset, motives, capabilities and opportunities of the potential 'attacker' of the food production system. The potential attacker is defined by seven separate categories of individual: the extortionist, the opportunist, the extremist, the irrational individual, the disgruntled individual, the hacktivist and other cyber criminals, and finally the professional criminal.

TACCP system development is best progressed by a multidisciplinary team in order to ensure knowledgeable coverage of as many aspects as possible. Many of the multidisciplinary team may be the same team members who make up the business HACCP team, though care should be taken not to assume that this is the case, as the TACCP team members will need to possess a number of different skill sets compared to the standard HACCP team. The TACCP team will need to consider: who might want to conduct an attack? How might the attack be carried out? Where are the vulnerabilities in the current system? How can they be stopped? (These are very different questions to those considered by the HACCP processes.)

'Attacks' can take many forms, from the replacement/substitution of ingredients, false or misleading marketing/labelling (often for commercial gain), through to intentional contamination. Once the threats, vulnerabilities and risks have been assessed then critical controls can be developed to provide the level of assurance required. The TACCP team should then routinely review these food protection arrangements.

Publicly available specification PAS 96:2017 'Guide to protecting and defending food and drink from deliberate attack' provides further details and is available via this link: https://www.food.gov.uk/sites/default/files/media/document/pas962017.pdf

VACCP (vulnerability assessment and critical control point). TACCP and VACCP can be viewed as interrelated processes which combine to help control intentional product safety threats. The TACCP focus is to identify specific threats of criminal/malicious activity which seek to undermine food safety or food quality. The VACCP focus is to consider how vulnerable each point in the supply chain is to that threat of criminal/malicious activity.

VACCP involves scanning the operational horizon for possible indicators of adulteration, contamination or substitution and any other potential breeches of supply chain integrity. The VACCP team are typically encouraged to 'think like a criminal' when seeking to identify opportunities for illegal commercial gain, or 'consider malicious intent' if seeking to contaminate the food products under consideration.

Product adulteration, dilution, counterfeiting, unapproved adjustments to recipes, mislabelling (or misdescription), incorrect species and incorrect country of origin labelling are all examples of actions to be considered when reviewing where these issues could occur in the product supply chain and manufacturing processes.

Vulnerabilities could be identified based on historical data (e.g., scanning for news reports of issues which have happened related to products or supply chains similar to those of the business in question) and also on the VACCP team's own thoughts and opinions of the potential opportunities for criminal/malicious activity.

Once a range of possible 'vulnerabilities' have been identified, the next role of the VACCP team is to design and apply 'mitigation strategies' (establish appropriate preventative measures) which will serve to either reduce or eliminate each specific vulnerability.

Example mitigation measures may include:

Raw material specifications – to state the key criteria/parameters that define product authenticity.

Supplier approval processes – to ensure that only checked and trusted suppliers are used.

Analytical test regimes – routine screening for indicators that food authenticity may have been compromised.

Ensuring use of tamper evident packaging.

Reducing the number of steps in the product supply chain to reduce the potential for sources of risk.

These 'food safety assurance' and 'food defence' activities and the expectation placed upon food business operators to always maintain and demonstrate 'due diligence' will be discussed in subsequent chapters.

1.2.2 Nutrition

There is a basic requirement for the consumer to derive from their food intake sufficient nutrients to help ensure their health and wellbeing. Most commonly the nutritional content of foods is defined in the food 'nutrition information' as the levels of energy (calories), total fat, saturated fat, carbohydrate (stated as a total and also the proportion derived from sugars), fibre, protein and salt.

Many food operations have to significantly focus on the challenge of ensuring that their foods are formulated to meet required standards and are then produced in such a way as to be consistent in their nutritional content, especially where health claims are being made such as 'low fat', 'sugar free', 'reduced salt', 'high in fibre', or perhaps are being eaten as part of a calorie-controlled diet.

Meeting these expectations consistently requires a food manufacturing operation to have close control on the nutritional variability of any ingredients (via defined specifications and supplier testing/auditing practices) and also consistency of their food production processes including measurement and processing practices.

With regard to processed foods there is an increasing consumer demand to avoid the use of synthetic additives (e.g., colours and flavours) and preservatives, exemplified by 'clean dec' initiatives (removal and reformulation of recipes to no longer have such materials present in the product Ingredient Declaration). Consumers are tending to prefer foods to be more 'healthy' and 'nutritious' in their appearance and content. In many countries 'organic' foods have been seen to develop significant markets over recent years. Also, there have been trends toward 'minimally processed' foods which are more natural in their appearance and nutritional content.

Food manufacturers faced with these challenges often have to work hard to find natural replacers for the synthetic additives and preservatives that were previously used. Recipe reformulation and adjustment of processing parameters are also often required as part of such adjustments, and these steps can prove challenging to complete in a manner which does not add additional cost to the product.

In addition, the issue of whether genetically modified (GM) foods are acceptable to the consumer continues to divide opinion and consequently impacts upon food markets, legislation and labelling.

1.2.3 Organoleptic quality

The 'organoleptic' characteristics of a food product are a combination of appearance, aroma, taste and texture. Consumers are often attracted by the appearance of a food well before they have tasted it (e.g., the picture on the front of the food packaging) and tend to like a variety of choice with regard to the organoleptic properties of their foods. As a result, food production businesses tend to spend a great deal of time and money on new product development activities including market analysis, product creation/formulation, product sensory assessment and consumer taste panelling.

Market trends towards 'minimally processed' foods that are more natural in their appearance and nutritional content also routinely drive recipe reformulation initiatives and adjustments to the food processing parameters.

Reformulation and adjustment of processes can also be a challenge to the food business when seeking to 'cost reduce' products in order to make them more profitable. Often the challenge is ensuring that the product organoleptic performance is not reduced beyond acceptable limits, while changing the recipe/process to facilitate the use of less costly ingredients (or use a smaller amount of the more expensive ingredients).

1.2.4 Product shelf life

Product shelf life is often the most important factor in terms of whether a product will be commercially viable. If the product has too short a shelf life it may have insufficient time to be produced, distributed, sold and consumed, leading to either the product not being launched, or if the product has already been launched before this is realised, then the occurrence of unacceptably high wastage levels at varying points in the supply chain will ultimately result in the product being delisted.

Achievement of an acceptable shelf life is therefore a key issue faced by the food manufacturer, often requiring a combination of activities in order to address the challenge. Depending on the product type/category, the shelf life achieved will be linked to the physical, organoleptic and microbiological properties of the raw materials and end product, the processes and packaging systems applied and finally the control and consistency possible in the storage and distribution chain, including when relevant to the in-home storage of the end consumer.

Food science & technology and sensory sciences combine when businesses are seeking to develop an acceptable shelf life as often the end point of the shelf life will be defined by either microbiological or organoleptic performance.

1.2.5 Commercial viability

As mentioned above, product shelf life is one of the largest contributors to the commercial viability of food products.

Other drivers of commercial viability which are challenges faced by the food industry are with regard to the cost of production and supply. Businesses must be able to make and supply the product at a low enough cost point to be able to then add sufficient margin to achieve a profit.

Key cost drivers within food production systems include the cost of raw materials (including ingredients and packaging), business staff/labour costs, utilities costs (often energy and water supplies) and distribution costs. Food businesses are therefore continually under pressure to keep these input costs as low as possible, and where such costs are subject to increase then cost savings will often be sought on other aspects of the production costs in order to still ensure an adequate profit can be made.

Typically, in food production systems such 'cost rise mitigating actions' include:

- Reducing staff/labour and replacing with computerised systems, automation and robotics.
- 'Cost engineering': the product to have a reduced recipe cost or a reduced cost of production.
- Increased pressure placed upon the raw material supply chain to supply ingredients and packaging at more competitive process, often leading to global sourcing and increased movement among different suppliers.
- Added pressure to apply lean manufacturing techniques which can lead to pressure to reduce cleaning and changeover times and increase the potential for operators to rush, potentially leading to a greater potential for processes not to be thoroughly completed and for mistakes to be made.
- Increasing application of microprocessor systems which control and link food processing machinery and equipment to better coordinate product flows and operational controls. Whole production systems can now be completely automated and controlled via a series of sensor and microprocessor technologies.

All of these actions can place additional pressures on the quality and technical team members as the food manufacturing site seeks to still maintain product quality, safety and legality, while dealing with changing workforce numbers, the application of new technologies, adjustments in product formulation and associated changes in suppliers (which often will require quality/technical checks and approval) and changes in processing techniques/technologies, often becoming more complex/sophisticated and therefore providing additional challenges with regard to their operation.

1.2.6 Environment

Food businesses have significant scope to affect the environment. Such potential business-related impact points can include:

- sourcing of raw materials;
- energy use/efficiency;
- waste outputs;
- transport;
- pollution (e.g. business noise/aromas); and
- water use and effluent.

The food industry has become increasingly aware of its impacts on the environment, often due to increasing government influence and legislation. This has led to a wide range of initiatives including to reduce energy use, change refrigerant gasses to more environmentally friendly alternatives, encourage cleaner air emissions, increase awareness of the business carbon footprint and reduction of 'trade effluent' (food process waste) emissions into water courses.

Increased consumer awareness of the potential environmental impacts of the industrial sector has also led to significant pressure for food processors to address any shortfalls in their environmental standards. Environmental pressure groups have become very adept at highlighting issues and swiftly building pressure on food businesses, especially major brands, to address matters. One such example is the environmental impact of transport and logistics, especially when transporting fresh product from abroad via air freight, typically expressed as 'air miles'/'carbon footprint'.

In the developed world there is typically very well-established legislation that serves to compel businesses to operate in an 'environmentally responsible manner', whereas in the developing world a lack of such legislation enables environmental pollution/damage to occur with usually very limited consequences (if any) applied to the businesses responsible.

In the developed world it is a commonplace for businesses to have formally written and implemented their own 'Environmental Policy', which considers the key impacts that the business can have upon the environment and the steps that the business is taking in order to ensure that these 'impacts' are at best positive, and at the very least any adverse impacts must be minimised/managed.

International Organization for Standardization, ISO 14001, provides an environmental management standard/framework within which to consider and control such environmental matters. 'ISO 14001 is an internationally agreed standard that sets out the requirements for an environmental management system. It helps organisations improve their environmental performance through more efficient use of resources and reduction of waste, gaining a competitive advantage and the trust of stakeholders' (ISO 14001 Key Benefits 2015).

ISO 14001 Key Benefits (2015) PDF document. https://committee.iso.org/files/live/sites/tc207sc1/files/ISO%2014001%20-%20key%20benefits_EN.pdf

1.2.7 Corporate social responsibility (CSR)

'Corporate social responsibility' typically applied in the context of businesses and organisations could be considered as a general term which encompasses ethical practice, social responsibility and environmental awareness.

Corporate social responsibility as a concept is not new. Major global brands have recognised for many years the importance of being seen to be 'doing the right thing', and not being found to be guilty of poor practice, be it adversely affecting their direct workforces or those within their supply chains, or perhaps damaging the environment either directly via the presence of their production plants or indirectly via the sourcing of their raw materials or the impacts of their finished products.

Many companies progress certain aspects of CSR without even realising that such activities could be considered as (and attributed to) CSR. For example, for cost efficiency reasons many organisations will already have implemented energy and waste reduction initiatives which could be considered beneficial to the environment and natural resources (e.g., carbon emission reduction schemes) and hence contribute toward the business CSR agenda.

A current and rapidly developing theme is the range of individual factors which collectively are covered by the term CSR. New CSR expectations are being routinely generated and developed, often as a result of increased consumer expectation, sometimes led by businesses seeking to 'do the right thing', and also as a result of pressure being brought to bear either via governmental compulsion or the orchestrated media campaign work of pressure groups who have become particularly adept at progressing their objectives via the use of 'name and shame' style campaigns.

Generally accepted aspects of CSR currently include:

- sustainability;
- workers' rights;
- ethical trade;
- health and safety;
- community wellbeing/impacts; and
- environmental protection/impacts.

1.2.8 Health and safety

Formal legislation covering the health and safety responsibilities of businesses and their workers is well established within most developed nations. As an example, with particular regard to the United Kingdom there is a clear expectation placed upon employers to ensure 'so far as is reasonably practicable the health, safety and welfare at work' of all employees (Health and Safety at Work etc. Act, 1974).

To help accentuate business focus and commitment to health and safety standards this legislation in the United Kingdom expects businesses which have a total number of employees of 5 or more to draft and display a written; 'Health and Safety Policy' within the workplace. Interestingly with regard to a more global perspective this particular law also makes provision for the health and safety consideration of *nondirect* employees such as designers, manufacturers, importers and suppliers. These nondirect employees may well be based overseas within developing countries operating to health and safety standards which are far less stringent than those in the United Kingdom.

While much of the developed world applies a similar legislative approach to compel businesses to consider and control health and safety, some of the developing world remains in a position where businesses seeking to save money or time can operate without due care and attention paid to the health and safety of those employed, without fear of particularly punitive consequences.

1.3 Management approach: overview of technical-related and quality-related roles

This book reviews the very wide ranging and demanding world of 'technical and quality management' operating within the food sector. Such technical/quality roles are many, varied and often overlapping/interlinked, including technical and quality managers, raw materials and specification technologists, hygiene managers, process and product development technologists and product/process innovation roles.

The wide range of 'technical' and 'quality' roles often operating within food manufacturing teams combine to ensure that the foods produced are safe to consume, of the required quality and legally compliant. Some of these roles are outward facing, routinely liaising with the raw materials supply base, customers or other stakeholders such as audit/accreditation bodies and government officials/enforcement agencies. Other roles are more factory focused including quality assurance and quality control staff who routinely check the production systems and products to ensure compliance, and also audit the site to maintain high standards of good manufacturing practice (GMP) and uphold the food safety and quality management system (FSQMS) including particular focus on the critical control points (CCPs) of the manufacturing operation.

These roles often require a combination of food science and technology knowledge, coupled with a detailed appreciation of the particular food processing system, and an ability to communicate with a wide range of internal and external stakeholders. Usually it is only by working as a team that food manufacturing businesses can ensure compliance to technical and quality standards while also operating efficiently and effectively to ensure a profitable, 'commercially viable' business which meets all of its customer expectations.

The 'technical manager' and/or 'quality manager' are key roles within a food business as these managers are expected to champion and oversee all aspects of the operation related to product safety, quality and legality. Typically, the individual will oversee the quality and technical staff on site and will be a key member of the site operations management team, often reporting directly into the site managing director or operations director. The technical manager will sometimes also hold responsibility for a hygiene department and also a process/development department, due to the major impact which these two sections can have on the correctness and safety of the products.

Large businesses may have a number of managers specifically concerned with aspects of product safety, quality and legality, there may be a specifications manager, a systems manager, a quality manager, an internal audit manager, etc. However, typically in large food organisations these individuals will all ultimately report to a site technical manager (or sometimes a technical director).

Small businesses may not have the resources to split the technical function of the business into a number of different management roles and therefore often the 'technical manager' role operates in a way which also absorbs and delivers the duties which could be conducted by a quality manager, a specifications technologist, a hygiene manager, and a process development manager. One model does not fit all businesses.

Ultimately the technical manager's goal is to ensure that the business has the right systems and procedures embedded to deliver consistently safe, quality products which meet all legislative requirements. Their role is also to be an ambassador for the business on all such matters, internally within the business, with the business customers, Third-Party Certification Schemes and Enforcement Authorities including the Environmental Health and Trading Standards Departments.

1.4 Skills, knowledge and expertise requirements

To be able to operate effectively in the food sector technical/quality staff will need to utilise and express a range of capabilities linked to food science and technology, coupled with management acumen and other business skills.

Key food science and technology aspects of the roles include: food processing technologies; principles of food physics and chemistry; microbiology; health and nutrition; global food supply chains; ingredients technology (e.g., chemical, physical and microbiological attributes of common food groups including cereals, fruit, vegetables, meats, dairy produce, herbs and spices); allergens and food intolerance; food safety management systems (including hazard analysis and critical control point (HACCP), risk assessment and measures to counter food fraud); specifications; traceability systems; packaging technology and food packaging systems; new product design and development; hygienic design (processing facilities and equipment); corporate social responsibility; ethical trade; environmental awareness; labelling requirements and food legislation.

Management and skills include: people supervision/management; data collection, interpretation and analysis; process monitoring systems; project management; ability to collate and present technical information; experimental techniques; regular review and consideration of food sector and related scientific developments; decision making; commercial awareness; complaints management (including trend analysis and design and implementation of corrective actions); continuous improvement (CI) ethos; investigation and problem solving techniques (e.g., root cause analysis); audit skills (both to be able to conduct audits and also host and progress customer and other stakeholder audits); recall and crisis management; interpersonal skills; leadership and staff development capability.

Accumulating such knowledge and skills to best advance a career in the food industry is usually achieved via a combination of study (both long and short courses across basic, intermediate and advanced levels), personal research, learning from others in organisations, coupled with 'experience' of working within the sector in technical and operations related roles.

1.5 Conduct of technical and quality personnel

Technical and quality personnel in the food sector are required to display and encourage high standards at all times. Typical identifiers of strong, capable technical personnel include:

- readily takes responsibility for tasks/areas of monitoring and control;
- has pride in their work and seeks to be proactive whenever possible;
- exhibits high integrity and always shows respect for colleagues and other stakeholders;
- ability to set and work to clear timelines, milestones and goals;
- tendency for self-development, always looking to further understand the operation and its underpinning science and technology or perhaps improve decision-making skills and supervisory/management practices;
- good communication skills, both verbally and written;
- forms effective relationships and has the ability to influence others, particularly on the maintenance of standards related to food safety, quality and legality;
- capable of working both as part of a team and also equally comfortable working alone when required;
- flexible and responsive to unexpected issues or change within the organisation;
- tenacity and focus to achieve the required outcomes;
- keen to share knowledge/information; and
- commitment to continuous improvement, always looking for ways to improve themselves and the organisation.

1.6 Concluding remarks

Often the biggest challenges for those attempting to maintain appropriate quality/technical standards within a food manufacturing business occur during times of trouble or change, for example, when a business is encountering a number of product safety or quality issues, requiring constant focus on the manufacturing operation and regular dialogue with the customer technologists and local authorities. Or perhaps when new products are being created for which there may be no previous 'track history' to be drawn on when seeking to control and take decisions with regard to the correctness of the end product.

With typically a high number of new food and drink products launching every year, the managers can be assured of many challenges to the robustness of the quality/technical procedures on site, especially so when all too often businesses are rushing to hit launch deadlines and therefore sometimes sacrifice the provision of sufficient trial time, research and development of appropriate 'right first time' standards.

Being able to manage effectively in the challenging food manufacturing sector is therefore a core focus of this book.

Further reading

United Nations 2017 – World Population Prospects: The 2017 Revision, published by the UN Department of Economic and Social Affairs. A comprehensive review of global demographic trends and future prospects. https://esa.un.org/unpd/wpp/

A wide range of helpful sources of information with regard to Food Defense, TACCP and VACCP, including:

Guidance:
– UK Food and Drink Federation (FDF)
http://www.fdf.org.uk/corporate_pubs/Food-Authenticity-guide-2014.pdf
– TACCP (Threat Assessment and Critical Control Point): a practical guide 2014
(Campden BRI) http://www.campdenbri.co.uk/publications/pubDetails.php?pubsID = 4640

Self-assessment:
– US Food and Drug Administration (FDA)
Vulnerability Assessment Software – Food Defense Plan Builder
https://www.fda.gov/food/fooddefense/toolseducationalmaterials/ucm349888.htm

Alerts and databases:
– US Food and Drug Administration (FDA)
Recalls and Enforcement Reports
http://www.fda.gov/Safety/Recalls/default.htm
– US Food and Drug Administration (FDA)
Import Alerts and Refusals
http://www.fda.gov/ForIndustry/ImportProgram/default.htm
– European Union Rapid Alert System
http://ec.europa.eu/food/safety/rasff/portal/index_en.htm

– UK Food Standards Agency (FSA)
https://www.food.gov.uk/business-guidance/food-incidents
– UK Serious Fraud Office
http://www.sfo.gov.uk
– Australia and New Zealand
http://www.foodstandards.gov.au/science/surveillance/Pages/default.aspx
– The Department of Food Safety, Govt. of NCT of Delhi
http://www.delhi.gov.in/wps/wcm/connect/DoIT_PFA/pfa/home

Standards:
– BRC Global Standard Food Safety Issue 8
https://www.brcbookshop.com/

Supply chain security
– BS ISO 28000, Specification for security management systems for the supply chain
– BS ISO 28002, Security management systems for the supply chain – Development of resilience in the supply chain – Requirements with guidance for use

Sources of information and intelligence about emerging risks to food supply:
– Infosan
http://www.who.int/foodsafety/areas_work/infosan/en/
– FAO Early Warning Bulletin
http://www.fao.org/food-chain-crisis/early-warning-bulletin/en/
– Global Information and Early Warning System (GIEWS)
http://www.fao.org/giews/english/index.htm
– Centre for the Protection of National Infrastructure Holistic management of employment risk (HoMER). London: CPNI, 2012. http://www.cpni.gov.uk/advice/Personnel-security1/homer/
– US Michigan State University, Food Fraud: http://foodfraud.msu.edu/
ISO 14001 Key Benefits (2015) PDF Document.
https://committee.iso.org/files/live/sites/tc207sc1/files/ISO%2014001%20-%20key%20benefits_EN.pdf
International Organization for Standardization, ISBN 978-92-67-10647-2.

Food Industry Standards and Interest Groups:
GlobalGAP – www.globalgap.org
British Retail Consortium (BRC) – www.brcglobalstandards.com
International Featured Standards (IFS) – www.ifs-certification.com
Global Food Safety Initiative (GFSI) – www.mygfsi.com
International Organisation for Standardisation – www.iso.org
Food Safety System Certification (FSSC) 22000 – www.fssc22000.com
Codex Alimentarius Commission – http://www.codexalimentarius.org

Defining 'technical and quality standards'

2

Mark Swainson
National Centre for Food Manufacturing, University of Lincoln, Holbeach, Lincolnshire, United Kingdom

2.1 Introduction

A critical part of the technical manager's role within a food manufacturing operation is to ensure that the appropriate systems and procedures are applied to support the production of consistently safe and high-quality products. This role must ensure that the product meets all of the relevant legislative requirements and is also expected to define and support the attainment and maintenance of such 'technical standards'.

In addition technical management also involves being an ambassador for the business on all related matters, both internally and 'outwardly facing', for example, representing the operation when dealing with business customers, third-party certification schemes and government enforcement authorities.

Operating across quality and technical remits within a business, the technical manager is typically concerned with the assurance of product quality, safety and legality. This involves covering a vast range of requirements, the specific detail of which will vary from business to business. Some factors upon which variation might depend include:

- product type;
- processing methods utilised;
- supply chain (both the chain from suppliers and also through to end customers);
- staff numbers, nature of employment contracts (e.g., temporary/permanent) and working hours/shift patterns operated;
- business output volumes/size of the operation;
- country of production (this will determine the relevant legislation); and
- country of sale (e.g., legislation requirements may differ from the country of manufacture).

Many technical managers will be promoted to that position over time within a business or will take on the role when joining a new business. In either case, the manager is likely to inherit a functioning manufacturing process, which will already be seeking to comply with quality, safety and legislative requirements.

At this stage, the new technical manager may assume that because the operation is running and appears to be under control, the existing set of quality, safety and

Swainson's Handbook of Technical and Quality Management for the Food Manufacturing Sector.
DOI: https://doi.org/10.1016/B978-1-78242-275-4.00002-2

legal standards must be correct and appropriate. However, the manager must bear in mind that it is possible for a company to be operational, supplying large amounts of product to a wide customer base, but at the same time be unknowingly operating outside of certain quality, safety or legality standards. This could be due to a lack of awareness of the extent of the challenges faced by the operation, insufficient risk assessment or perhaps a result of changing customer/legislative expectations over time, for example:

Quality: Quality standards can change in a business due to the requirements of new customers or as a result of revised customer expectations. Therefore, if not regularly reviewed, product specifications and operational standards can become out of date. Another example could be an end product that gradually moves away from its specification limits over a period of time due to incremental process or ingredient change reasons. Such an issue may not be noticed by the operation or the customer until the point of a formal quality review against reference samples or against details drawn from the original product specification.

Safety: Processes believed to be sufficient to control product safety can later require updating, due to external factors such as new scientific research or problem issues being reported within the sector. This type of event would suggest that there is a potential for a particular hazard to occur that was originally thought to be sufficiently controlled by the standard practices applied. An example of such an event might be new research findings with regard to the growth potential of specific foodborne pathogens.

Legality: With a regular stream of new legislation being introduced to the food manufacturing sector, it is possible for a business to believe that it is fully compliant, only to find that increases in legislative requirements have resulted in the business operating illegally. For example, changes to product labelling legislation can trigger product withdrawals or recalls if product packaging is found to be noncompliant with the latest legislation.

Therefore, whether implementing standards at a new manufacturing facility or taking on an existing operation, the technical manager should always consider all of the quality, safety and legality expectations from first principles, without making any assumptions.

2.2 Food industry standards and certification schemes

The nature of the food manufacturing sector in which a business operates will significantly influence the appropriate systems and approaches required to assure product quality, safety and legality. The technical manager is therefore well advised at the outset of their role to consider the nature and specific demands of the food industry sector within which they will be seeking to develop and apply technical and quality standards.

Sector considerations linked to technical and quality standards can include:

- What is commonly viewed as 'standard practice' and what is viewed as 'best practice' with regard to quality and technical management within the sector?
- In addition to general food legislation, is there any legislation specific to the particular type of products being manufactured?
- Are there industry guidelines or other guidance documents written for the sector that need to be taken into account?
- Who are the 'gold standard' leading manufacturers within the sector? What practices make them outstanding?
- Within the sector, does the customer typically have a major influence on the operating standards of the food manufacturing sites? If so how are customer expectations defined and applied?
- Is there a need for products to be manufactured in adherence to specific rules or requirements, for example products which will be marketed as 'organic', 'kosher' or 'halal'?
- What accreditation schemes/certification programmes are appropriate to the sector?

Certification schemes and accreditation programmes developed by particular food industry sectors can often become necessary standards for a business to hold in order to assure their customer of the quality, safety and legality of the manufacturing operation. Some examples of such systems and related initiatives include:

- Global Good Agricultural Practice (GlobalG.A.P.) – started out as an independent certification system for GAP and has evolved to become a globally recognised farm assurance programme.
- British Retail Consortium (BRC) Global Standards – specify the requirements to be met by an operation in matters related to the assurance of product safety, quality and legality. These standards are sector-specific and include the Global Standard for Food Safety, the Global Standard for Packaging and Packaging Materials and the Global Standard for Storage and Distribution Operations.
- International Featured Standards (IFS) – seek to provide uniform standards for the food supply chain to aid compliance with food safety, quality and legality requirements.
- Global Food Safety Initiative (GFSI) – a business-led initiative which seeks continuous improvement of food safety management systems (FSMSs) via involvement in technical working groups and via benchmarking initiatives that seek to influence the content of food industry compliance schemes.
- International Organisation for Standardisation (ISO) – provides a wide range of voluntary standards for industry. Of particular relevance to the food manufacturing sector are the ISO 9000 quality management series, ISO 14000 environmental management series and ISO 22000 which specifies the requirements for FSMSs.
- Food Safety System Certification (FSSC) 22000 – utilises both the ISO 22000 Food Safety Management standard and the Publicly Available Specification (PAS) 220 standard. PAS 220 was written with the intention of being used in conjunction with ISO 22000 in that it specifies the requirements for prerequisite programmes which assist in the control of food safety within a manufacturing operation. The scheme also utilises ISO 22003 which defines the rules applicable for the audit and certification of a FSMS. FSSC 22000 therefore provides a scheme against which the food safety systems of food manufacturing operations can be assessed and certified.

(Note: Web addresses for these systems and related initiatives are listed in the 'Further Reading' section at the end of this chapter).

Certification schemes can help technical management to map the business range of food safety-, quality- and legality-related systems and practices onto a clear series of well-considered standard requirements. As part of such schemes, independent third-party auditing services (certification bodies) are typically utilised to routinely assess and certify that the system/approach taken by the operation is applied and working correctly. Independent audits provide assurance to the business and its customers that the operation adheres to all of the required points within the selected scheme/standard and is making further improvements if necessary. When visiting a food manufacturing operation to assess against a particular certification scheme/standard, a third-party compliance auditor is likely to review a typical range of business quality, safety and legality aspects. The auditor will usually expect to review tangible evidence of all food safety and quality systems, procedures and related records including:

- designated FSMS (e.g., hazard analysis and critical control point (HACCP) plan, including an appropriate scope, risk assessment and underpinning validation and review documents);
- senior management commitment (e.g., evidence of routinely reviewing business quality, safety and legal compliance factors, and ensuring that related initiatives are adequately resourced);
- quality manual (including detail on all quality assurance (QA)- and quality control (QC)-related activities);
- standard operating procedures (SOPs) for key aspects of the operation;
- internal audits (both "systems" and "compliance" audits);
- specifications (both for raw materials and finished products);
- complaint records;
- nonconformance and corrective action systems;
- maintenance records;
- calibration records;
- pest control records;
- foreign body controls;
- shelf life assessment and monitoring records;
- temperature records;
- personnel controls (including hygiene and training);
- supplier assurance;
- allergen management;
- specific handling requirements (e.g., allergen management, identity preserved materials, genetically modified organisms);
- traceability;
- product recall/withdrawal; and
- crisis management.

Any instances of nonconformance raised during the audit typically have to be appropriately addressed within the timescales dictated by the selected standard before the operation can be awarded certification. In some cases, noncompliance with the standard can be so substantial that certification cannot be granted and a reaudit is required once the operation has had time to reflect on and address the issues or challenges raised. For further detail on customer and third-party audits, including compliance schemes, refer to Chapter 9, Audits.

Certification standards often contain hundreds of clauses, each of which has to be addressed. Such wide-ranging areas of expectation usually require the business to operate extensive systems and procedures to assure on-going, consistent control. For some businesses the expectations of certain standards can exceed the resources available to achieve and maintain the standard. In such cases the technical manager will be expected to either select a more appropriate standard for the size and nature of the business or to develop resource proposals and progress plans to help the business to eventually achieve the standard required. The achievement and maintenance of standards can prove to be a very onerous process for a business, demanding time and resources. The technical manager should therefore ensure that each standard selected and progressed will add significant benefit and value to the organisation.

Industry guidelines specific to the particular product sector being operated within can be a valuable source of information for the technical manager on recommended industry good/best practices and links to relevant legislation. Guidelines and guidance documentation are often produced by relevant sector trade associations or government bodies seeking to raise industry awareness of appropriate good practice within a particular sector. It is usually beneficial to follow guidelines prepared by reputable sector stakeholders, especially when they are linked to government bodies. This is because such information on good and best practice is often informed by wider sector knowledge and as a result often eventually becomes incorporated into legislative requirements as food laws are further developed over time.

When operating in countries which do not have relevant industry guidelines available, the technical manager would be well advised to review guidelines available in other countries in order to gain valuable information on global good and best practice in the relevant sector.

2.3 Legislation

For all food manufacturing businesses, the very minimum requirement of food quality and safety-related practice is to ensure that all processes operate within the requirements of the law. Compliance with such expectations is inspected routinely by members of the relevant enforcement/government authorities. In many countries, alongside the maintenance of good manufacturing practice, food manufacturing operations are legally obliged to operate a HACCP system of food safety control. Such systems are typically based on the HACCP principles and practices as defined by the Codex Alimentarius Commission. When joining a food manufacturing operation, the technical manager should find and review the existing food safety and HACCP plans, to check that the business system for product safety assurance is up to date, relevant to the operation and legally compliant.

Quality- and technical-related legislation will often be in the form of expectations linked to product weight/volume/count, and also with regard to product labelling requirements including ingredient declarations, product safety warnings and storage/shelf life information. The new technical manager is well advised to check the current inspection status of the business with the relevant authorities. Usually inspections against legislative requirements will be documented and this is especially the case when recommendations or requirements (and timescales) for improvement have been made by the enforcement official. With good proactive management by the technical manager, such officials can prove to be a key resource in helping the business to remain up to date and compliant with regard to its legal obligations. Refer Chapter 6, Product control and HACCP considerations, for detail on the HACCP system of food safety control and Chapter 15, Enforcement authority perspective on the food manufacturing sector, for more detail on legislation and regulation.

The technical manager should be aware that legal compliance requirements will also be placed on the raw material suppliers to the business (e.g., ingredient and food packaging businesses). Provision should therefore be made to ensure that all suppliers and goods supplied are checked for compliance with the appropriate legislation. Such checks can be covered via supplier approval and audit programmes, raw material specifications and via the requirement for certificates of analysis or certificates of conformance. Refer Chapter 4, Raw materials and packaging supplier control for more detail.

2.4 Customer expectations

All customers expect the food manufacturing operations they deal with to supply safe, legally compliant products that meet their quality expectations. Customers will typically agree a product specification with the food manufacturer, the content of which will define all of the criteria relevant to the product quality, safety and legal compliance. Key information within the specification should include:

- product name and description;
- product weight/volume/count and related controls;
- product packaging (may also include pack information and artwork details);
- detailed organoleptic description of the product. (e.g., appearance, aroma, taste and texture);
- raw materials information (for ingredients and packaging);
- end product recipe and/or ingredient declaration;
- any reworked product percentage allowances;
- manufacturing process details and related food safety and quality control points;
- product microbiological standards (e.g., targets and maximum limits at start and end of shelf life); Selected microorganisms, testing methods and frequencies;
- shelf life information (e.g., total number of days from the date of manufacture) and date coding style (e.g., 'use by' or 'best before' date coding);
- product storage requirements/recommendations (e.g., chilled temperatures, frozen, ambient) while sealed and upon opening;
- end product preparation guidelines/cooking methods;

- hazard control statements (e.g., assurance of absence of physical contaminants);
- nutritional information;
- legal compliance statements;
- 'may contain' or 'free from' status (e.g., for allergens such as nuts);
- product warnings (e.g., 'not suitable for...' or 'while every care is taken some bones may remain'); and
- other product information statements (e.g., packaged in a protective atmosphere, suitable for vegetarians, vegans, Halal, Kosher).

In an attempt to ensure that their expectations related to manufacturing standards are clear, many retailers and food service business groups will provide their manufacturing suppliers with their own supplier codes of practice/operating standards with which they expect their suppliers to conform when manufacturing their products. The extent of such codes of practice can range from a small number of additional expectations (over and above the business' legal obligations) through to a very comprehensive set of expectations covering all aspects of the supplier operation including:

- FSMS expectations (e.g., HACCP);
- senior management commitment;
- product withdrawal, recall and crisis management procedures;
- specification control;
- ingredient and packaging controls;
- product development and scale-up controls;
- factory design and operating standards;
- nonconformance reporting, controls and corrective action systems;
- hygienic design and maintenance of process equipment;
- calibration of equipment;
- employment rights;
- operator personal protective equipment and hygiene controls;
- factory staff recruitment and assessment (including medical screening);
- staff training;
- process control;
- factory cleaning systems and hygiene monitoring;
- product coding and labelling considerations;
- traceability systems and procedures;
- product storage, transport and delivery systems;
- allergen controls;
- line/process validation;
- product microbiological control and monitoring;
- utilities management (e.g., water use and wastewater controls);
- product weight/volume/count controls;
- quality policy and quality management system (QMS) content;
- internal auditing (e.g., systems audits and compliance audits);
- documentation controls;
- customer service and complaints management;
- factory fabric and equipment maintenance plans;
- site pest control; and

- corporate social responsibility (e.g., environmental controls and ethical trade expectations).

Customers often choose to monitor their supplier's food manufacturing sites (to ensure that the sites comply with their expected standards) via routine audits and/or the requirement for the food manufacturer to routinely report (e.g., by monthly self-assessment) on the status of key performance indicators (KPIs). These might include specific details and trend analyses on the number and type of complaints received; number of microbiological nonconformances; environmental monitoring results; taste panel assessment results; site audit performance; product rejections, recalls or withdrawals; status of forthcoming product specifications (e.g., under review, approved); any raw material supplier performance issues.

A number of retail and food service businesses choose to assure themselves that their food manufacturing suppliers are operating in compliance with a wide range of controls and codes of practice by making the attainment of a recognised Food Manufacturing Standard a prerequisite of supply. Such standards could include IFS, the BRC Global Standard or other standards recognised by bodies such as the GFSI.

2.5 Business approach and quality ethos (quality assurance/quality control)

The approach to "doing business" taken by a food manufacturing operation will be significantly defined by its owners and/or directors. Businesses vary widely in terms of their views and positions on the following factors, which in turn can influence the site approach toward product quality, safety and legality:

- willingness or ability to invest in the business infrastructure;
- provision of sufficient staff resources;
- extent to which the business is focussed on innovation;
- does the business want or need to achieve rapid growth?
- focus on product quality and customer service; and
- is the business 'risk averse' or open to taking risks for commercial gain?

Businesses that fail to invest in the infrastructure of their organisation are likely to suffer from product faults associated with poor standards of both process machinery and operating conditions. In addition, failure to employ a sufficient number of appropriately trained food manufacturing personnel will typically lead to manufacturing faults linked to insufficient staff resources, including mistakes made by personnel who are trying to cover an inappropriately high number of tasks.

A focus on innovation and/or rapid growth will typically place far more pressure on the quality, safety and legality positions of the manufacturing operation. The quality and technical resources of such operations should therefore be sufficiently robust to maintain control. Many businesses take risks in the process of trying to gain a market advantage, perhaps seeking to rush product development and scale-up processes to get the product onto the market fast, or maybe reducing the level of product inspection required in order to save staff costs. In such operating environments, the quality/technical manager has to have very good negotiating and persuasion skills in order to maintain a sufficient level of control within the organisation.

Some businesses choose to embed quality principles and practices throughout their whole organisation by implementing specific sets of rules and practices such as the internationally recognised ISO 9000 quality management standard. This International Organisation for Standardisation ISO 9000 standard provides a written set of quality and business practices, which when embedded within the organisation are intended to assure the business and its customers that the operation will be able to consistently meet quality expectations. The requirements of the ISO 9000 standard are designed to be highly generic so that the standard can be applied globally to any organisation which provides a form of product or service.

International Organization for Standardization 'Quality Management Principles' document introduces the quality management principles that ISO 9000, ISO 9001 and other related quality management standards are based on. (https://www.iso.org/files/live/sites/isoorg/files/archive/pdf/en/pub100080.pdf)

Such sets of rules serve to form the basis of a business QMS, which will be further considered in Chapter 3, The food safety and quality management system (FS & QMS).

Many business operations will choose to define their approach to quality using the terms Quality Assurance (QA) and/or Quality Control (QC). QA incorporates all of the planned actions, systems and procedures within the business that seek to provide confidence that product quality requirements will be and are being consistently fulfilled by the operation. QA is a proactive approach to product quality, typically involving a range of product checks, conducted at key points throughout the manufacturing operation to confirm compliance of the product against required standards. This approach therefore raises any nonconformances as early as possible within the manufacturing process.

Scheduled QA checks, tests or activities at every stage within the operation collectively contribute to the product being 'right first time'. The key factor is that the system is assuring the quality of the product by checking that key points are correct at each stage of the process, from checking the quality of the ingredients being delivered right through to checking that the product at each stage of its manufacturing operation is correct. This system is viewed as efficient because the sequential checks should pick up any quality variation/nonconformance early on in the process; the issue will therefore have had limited cost and time impacts and may still be correctable.

QC measures within a food manufacturing operation are checks or other activities focused on the inspection of the end product to confirm that the product batch/run/lot correctly fulfils the expected quality requirements typically defined and documented within the end product specification. QC is often highlighted as being a 'reactive' approach to the attainment of product quality, whereas QA is seen by contrast as being relatively proactive. The main issue with QC is that any faults detected in the end product at the final stage of manufacture may mean that the matter is difficult or costly to correct, as all manufacturing processes and costs have already been incurred. A QA approach is far more likely to detect emerging problem issues sooner because it is comprised of a series of checks carried out throughout the operation, confirming product correctness at each manufacturing phase.

Many food sector businesses employ both QA and QC systems, rather than focusing exclusively on either one. Typically, businesses will apply a QA approach to the manufacturing process as this is usually a very cost aware approach, picking up quality faults as early as possible in order to reduce their impact. A form of QC may then be applied at the end of the production operation as well, for example, by requiring a representative sample of each production batch to pass a formal quality inspection (e.g., perhaps inspected at the end of the process line or in a taste panel format) before being released for despatch to the customer. As a result, the business gains the benefits of both approaches to the assurance and control of product quality.

2.6 The role of senior management in implementing standards

The senior management team within a food manufacturing operation is extremely influential in terms of both setting and maintaining the business standards with regard to food safety, quality and legality.

Key areas of senior management impact on business quality and technical-related initiatives include:

- ensuring routine senior management review of product safety, quality and legality KPIs, including trending of customer complaints and other customer feedback; review of internal, second and third-party audit results; monitoring of operational process/product nonconformance rates and related corrective/preventative actions;
- provision of sufficient resources (e.g., financial resources to provide the appropriate personnel and equipment) for the operation to maintain quality, safety and legality;
- ensure that trained and capable 'nominated deputy' resources are in place to cover the absence of key personnel;
- involvement in food safety steering committees such as multidisciplinary HACCP team process reviews;
- ensure appropriate business recall and crisis management plans are in place within the business;
- driving a commitment to continuous improvement within the operation via the development and implementation of plans deemed beneficial to product quality, safety and the assurance of legality (e.g., progression of complaint reduction initiatives, driving improvement in site audit scores and reduction in site process/product nonconformance rates); and
- ensuring business communication of product quality, safety and legality control expectations and provide staff with updates on business performance against such requirements.

Staff at all levels within a food manufacturing business look to the senior management team to set the standards for the operation and to lead by example. Senior management involvement in site audits (internal, second and third-party audits) is usually a very positive indicator of commitment to the business technical standards, especially when such involvement is seen to be proactive and keen to improve performance.

2.7 Conclusion

The wide range of requirements placed upon food manufacturing operations to assure food safety, quality and legality result in the need for a competent and professional technical team, typically led by a technical or quality manager.

It is vitally important to the food business operator that the technical team have the capacity and capability to highlight new and existing requirements raised by customers, legislative/enforcement bodies, third-party auditors and other stakeholders, and are able to assimilate these requirements into the business food safety and quality management system (FS & QMS). A well-informed FS & QMS will then form the blueprint/instruction plan, from which the food production system will operate.

The technical team should help maintain the FS & QMS via a range of activities including system/check implementation (including related staff training), auditing of the operation and supply chain (including supplier approval), validation and verification studies, and representing the business during site visits/audits completed by or on the behalf of customers, audit standard bodies or enforcement authorities.

Successful businesses will ensure that all requirements (e.g., legal obligations, food certification standard expectations) are clear, implemented/embedded and maintained within the operation. The more advanced food businesses also seek to combine the activities of their engineers, production and technical teams, in recognition of the fact that the requirements of all three parties need to be addressed in order to achieve a commercially viable and well controlled operation.

Further reading

British Retail Consortium (BRC) – www.brcglobalstandards.com.
Codex Alimentarius Commission – http://www.codexalimentarius.org.
Food Safety System Certification (FSSC) 22000 – www.fssc22000.com.
Global Food Safety Initiative (GFSI) – www.mygfsi.com.
GlobalGAP – www.globalgap.org.
International Featured Standards (IFS) – www.ifs-certification.com.
International Organisation for Standardisation – www.iso.org.
International Organization for Standardization 'Quality Management Principles' (2015). https://www.iso.org/files/live/sites/isoorg/files/archive/pdf/en/pub100080.pdf
ISBN 978-92-67-10650-2

The food safety and quality management system (FS & QMS)

3

Mark Swainson
National Centre For Food Manufacturing, University of Lincoln, Holbeach, Lincolnshire, United Kingdom

3.1 Introduction

The food safety and quality management system (FS & QMS) of a food production operation is of vital importance as it supports the manufacture of consistently safe products that are of the required quality, protecting the business brands and ensuring legal compliance. The FS & QMS will typically comprise all procedures, practices, methods and work instructions conducted by the business in the assurance of product safety, quality and legality.

Businesses may split the documentation of these procedures into separate parts, maintaining a QMS, which focuses on the assurance of product quality, and a Food Safety Management System (FSMS), which focuses on the control of product safety factors. Safety factors are typically physical, chemical or microbiological hazards. Allergens are also categorised as a separate hazard group in many business food safety systems, whereas some businesses choose to cover allergens under the 'chemical hazards' category. The FSMS often incorporates the Hazard Analysis Critical Control Point (HACCP) programme for the site, particularly in countries where HACCP is a legal requirement.

While some businesses separate the documentation of their food safety and quality systems other operations hold all control system information in one all-encompassing FS & QMS Manual. In such cases, the terms 'Quality Management System' (QMS) and 'Quality Manual' can be used to refer to both food quality and food safety systems. Typically, businesses will ensure that their FS & QMS reflects adherence to all appropriate legislation, the expectations of customers and the requirements of relevant certification schemes and standards. As a result, the FS & QMS will vary between sites and businesses in scope and complexity.

3.2 Food safety and quality management system purpose

Development and maintenance of a FS & QMS manual can benefit a food manufacturing operation in a number of ways:

Swainson's Handbook of Technical and Quality Management for the Food Manufacturing Sector.
DOI: https://doi.org/10.1016/B978-1-78242-275-4.00003-4

- Focuses the business on the planning, development and application of all key activities required to assure product safety, quality and legality.
- Provides clarity, serving as a site blueprint and reference document on FS & QMS-related matters. This can be particularly of benefit to site staff, internal audits, customers, third-party compliance auditors and enforcement officials.
- Provides a reference point against which to conduct gap analysis on whether policies or practices are required to be implemented or adjusted in response to changing circumstances such as additional legislative requirements, new customer requirements or increases in accreditation standard expectations.
- Supports continuity and improvement, as the relevant defined policies, procedures and records within the system can be regularly reviewed and adjusted or improved in response to internal and external feedback (e.g., in response to site quality performance trends and customer complaints data).
- The compilation and documentation of the site FS & QMS enables new members of the site management team to comprehensively review the site operational systems committed to the assurance of product safety and quality. This is particularly beneficial to the training of staff operating in quality and technical related roles.
- FS & QMS can also help provide legal defences such as the demonstration of 'Due Diligence' in the event of food quality or safety-related investigations, i.e. the business could use the documented systems and related records to demonstrate that it has taken a proactive approach to the control of food quality and safety.

3.3 Terminology/definitions

Useful definitions when considering QMS include the following provided by ISO9000:2015 QMS – Terms and Definitions, which can be reviewed at this website: https://www.iso.org/obp/ui/#iso:std:iso:9000:ed-4:v1:en

Quality – Degree to which a set of inherent characteristics of an object fulfils requirements.

System – A set of interrelated or interacting elements.

Management system – A set of interrelated or interacting elements of an organisation to establish policies and objectives, and processes to achieve those objectives.

Quality management – Quality management can include establishing quality policies and quality objectives, and processes to achieve these quality objectives through quality planning, quality assurance, quality control and quality improvement.

Quality policy – Generally the quality policy is consistent with the overall policy of the organisation, can be aligned with the organisation's vision and mission and provides a framework for the setting of quality objectives.

The QMS is the part of an organisation's management system that focuses on the achievement of the quality objectives of the operation.

In addition, when considering the specifics of Food Safety Management, definitions drawn from the ISO22000 Food Safety Management System can also be helpful including (Source: https://www.iso.org/obp/ui/#iso:std:iso:22000:ed-1:v1:en):

Food safety policy – Overall intentions and direction of an organisation related to food safety as formally expressed by top management.

FSMS is often one part of a larger management system and is a network of interrelated elements, which combine to ensure that food does not cause adverse human health effects. Elements of the FSMS include programs, plans, policies, procedures, practices, processes, goals, objectives, methods, controls, roles, responsibilities, relationships, documents, records and resources.

3.4 Food safety & quality manual – format and content

The business FS & QMS should be compiled in the form of a document which states the operational structure, responsibilities and procedures necessary to consistently ensure product safety, quality and legality.

Key aspects of the system will include:

1. Being formalised (i.e. making the business position on these matters official. Stating the operational intent to meet the expectations of all relevant stakeholders).
2. Clearly documented (i.e. detailed collation of related information in written form).
3. Defining the operational structure (i.e. business hierarchy, management and reporting systems).
4. Stating detailed responsibilities (i.e. clear role descriptions for those business members who have a significant influence on product quality and safety).
5. Documenting the procedures, methods and work instructions required to achieve effective food safety and quality management (i.e. the requirement to formalise all operational steps and control activities involved).
6. Linking expectations to relevant records of actions/tests/assessments.

The ISO9000 Quality Documentation Pyramid provides a helpful schematic of the information and document hierarchy of such systems. A basic version of the pyramid is shown in Fig. 3.1.

FS & QMS documentation within a food manufacturing business is often held in both electronic and printed formats. Usually the documentation will open with a

Figure 3.1 Quality documentation pyramid.

Food Safety & Quality policy statement that makes clear the business focus on the production of safe, legally compliant food products and meeting the requirements of its customers. For example:

Food Safety & Quality Policy Statement: 'Food Business Operation Ltd will endeavour to produce and supply foods in a manner which consistently meets the expectations of our customers whilst adhering to all relevant legislative standards. We will always operate in a safe and diligent manner and will ensure that a continuous improvement ethos is embedded throughout our operation'.

This statement should have the full commitment of the management team, and typically be displayed within the business reception area as a framed document signed by the person who holds the overall responsibility for the operation of the manufacturing site. Doing this is intended to demonstrate 'senior management commitment' to the policy. The 'quality policy' statement is typically expected to be communicated clearly to the workforce, often via induction training for new employees and via display on staff noticeboards. This helps ensure that all staff are aware and routinely reminded of the importance of meeting customer requirements and legislative obligations.

The FS & QMS should include a clearly defined organisational structure for the business operation, which defines the reporting lines of management authority and accountability on site. The system should also highlight all of the business roles that are key to the assurance of product quality, safety and legality, and define who deputises in these roles when the responsible individuals are unavailable (e.g., due to holidays or illness). This can cover key roles at all levels within an organisation, from senior management decision makers through to factory-based staff responsible for the monitoring of Critical Control Points (CCPs). It is good practice, and often a customer or accreditation standard requirement, for the FS & QMS manual to include at least an outline of each of the business policies and procedures related to the control of food safety, quality and legality. This meets the requirement to formalise all steps and activities involved in the attainment of these goals. The manual must also be available to all business employees who may need to refer to it due to the nature of their roles within the organisation.

If a business chooses to just provide an outline of the food safety, quality and legality-related procedures, methods and instructions in the quality manual, in-depth information on the measures or actions required will often be made available in separate site documents such as Standard Operating Procedures (SOPs) or the site/area training manuals. Detailed written procedures are of critical importance to assure the control of operational processes or activities. An effective procedure will clearly state what measures or actions are required, and will explain how these activities should be conducted. The procedure should also make it clear who is required to conduct each activity, when the action is required and how it should be recorded. If a business elects to use a solely electronic-based system, it will be necessary to ensure that there are sufficient backup systems in place for use in the event of a disruption to the electronic-based system. If the electronic information is not backed up, the business FS & QMS will run the significant risk of being compromised.

3.5 Records

The generation and maintenance of records of all actions/tests/assessments conducted for the assurance of product control is a vital part of the FS & QMS. This recording process enables the business to:

- monitor/confirm product control;
- analyse trends;
- meet audit requirements; and
- provide proof of 'due diligence'/legal compliance.

Records generation, review and storage requirements should therefore be linked to the relevant procedures within the QMS and the format/content of the individual test or record sheets should be included within the site document control system.

3.6 Food safety and quality management system (FS & QMS) design, implementation, monitoring and review

A significant amount of technical and quality management time within a food manufacturing business will be focussed on the design, implementation, monitoring and review of the FS & QMS-related policies, procedures and practices, which seek to ensure operational control of food safety, quality and legality.

3.6.1 FS & QMS design

To facilitate the design of an appropriate and effective FS & QMS, a business must have a comprehensive understanding of customer expectations and legislative requirements, combined with having clearly defined its own business ethos and expectations on such matters. Good Manufacturing Practice (GMP) provides the basic foundation on which to support the production of safe food of the required quality. Guidancc on what constitutes GMP can be gained by referral to industry-related legislative requirements and sector guidelines. Of specific benefit in the system design phase are the Codex Alimentarius Commission (CAC) 'Food Hygiene (Basic Texts)' which have widely contributed to governmental legislation and guidance on such matters.

The following explanation is taken from the CAC Food Hygiene (Basic Texts) 4th edition:

> *The Codex Alimentarius Commission is an intergovernmental body with more than 180 members, within the framework of the Joint Food Standards Programme established by the Food and Agriculture Organization of the United Nations (FAO) and the World Health Organization (WHO), with the purpose of protecting the health of consumers and ensuring fair practices in the food trade. The Commission also promotes coordination of all food standards work undertaken by international governmental and nongovernmental organizations.*

The Codex Alimentarius (Latin, meaning Food Law or Code) is the result of the Commission's work: a collection of internationally adopted food standards, guidelines, codes of practice and other recommendations.

The Codex basic texts on food hygiene promote understanding of how rules and regulations on food hygiene are developed and applied. The General Principles of food hygiene cover hygiene practices from primary production through to final consumption, highlighting the key hygiene controls at each stage. This publication also contains the most internationally used description of the Hazard Analysis and Critical Control Point (HACCP) system and guidelines for its application.

Source: CAC Food Hygiene 4th edition, http://www.fao.org/docrep/012/a1552e/a1552e00.pdf

The general principles highlight that prior to applying specific food safety systems within the operation (such as HACCP), there first needs to be a strong foundation of food safety, in the form of GMP and Prerequisite Programmes (PRPs). The CAC 'Recommended International Code of Practice – General Principles of Food Hygiene' within the CAC 'Food Hygiene Basic Texts (2009)' provides detailed guidance on the design, implementation and maintenance of practices related to GMP and PRPs, such as raw materials control, factory facilities, process equipment, staff control, personal hygiene, cleaning systems and the management of waste. The reader is encouraged to review all guidance provided by the Codex Alimentarius Food Hygiene Basic Texts (2009) (http://www.fao.org/docrep/012/a1552e/a1552e00.pdf).

PRPs tend to be closely interlinked with site GMP and legislative requirements and are usually independent systems or practices for the control of food hazards (e.g., microbiological, physical or chemical contamination), which are not specific to a particular production process, but instead are applicable to the wider manufacturing operation.

PRPs often are categorised into the following areas:

- establishment design and facilities;
- air quality;
- water quality;
- temperature control;
- personnel;
- preventative maintenance programme;
- calibration systems for measurement equipment essential to product safety;
- waste management;
- raw material controls;
- cleaning and disinfection;
- pest control;
- physical and chemical contamination;
- allergen controls;
- management of product information;
- traceability, recall, returned goods, rejections/nonconforming products;
- work instructions/SOPs; and
- storage and transport.

Building on the site GMP and PRP foundations, additional customer expectations to be incorporated into the site FS & QMS will typically be made clear via agreed product specifications and the provision of supplier guidelines or codes of practice. Any additional expectations for site technical approval will also be incorporated. Such expectations may stem from either the customer's own technical/auditing team or the stipulated requirement for the business to hold a specific third-party technical standard accreditation.

In addition, the legislative requirements that must be fulfilled by the operation with regard to the assurance of product quality and safety can be found by reviewing relevant governmental laws, often supported and elaborated on by various guidance documents and sector codes of practice. There is usually a specific regulatory or enforcement section of the government, which holds responsibility for overseeing the food manufacturing sector at both local and national levels. Such government departments and related officials support and inform the food manufacturing sector as to how best to meet its legal responsibilities. Government departments may also include roles related to the enforcement of requirements, such as conducting product sampling, site visits and audits. For more information on the roles and expectations of enforcement authorities, please see Chapter 15, Enforcement authority perspective on the food manufacturing sector (UK EHO).

Worldwide, FSMSs are often based on the HACCP system of product safety control. HACCP is used to proactively control the safety of foodstuffs via a systematic approach to the identification, evaluation and preventative/proactive control of those food manufacturing stages of critical importance to the assurance of food safety.

There are seven HACCP principles:

1. hazard analysis;
2. identification of CCPs;
3. establishment of critical limits;
4. establishment of monitoring systems;
5. establishment of corrective action plans;
6. establishment of verification procedures; and
7. establishment of record keeping and documentation.

Alongside the provision of guidance on GMP and PRP foundations for food manufacturing operations, the CAC provides specific guidance with regard to the implementation and maintenance of HACCP systems within food manufacturing operations. For further details please review the HACCP guidance provided by the CAC 'Food Hygiene Basic Texts' (2009) (http://www.fao.org/docrep/012/a1552e/a1552e00.pdf). HACCP is also further discussed in Chapter 6, Product control and HACCP considerations.

Design of a FS & QMS is usually best progressed by experienced personnel operating within the business quality and technical functions. In circumstances where the business does not have sufficiently experienced personnel for the task, staff should seek appropriate training and guidance. The FS & QMS will be the foundation of business quality, safety and legality control and therefore must not be placed at risk of being undermined due to lack of sufficient knowledge.

In circumstances where there are insufficient FS & QMS development skills or experience within the business team, it may be beneficial to consider the use of external technical consultant support. A consultant with the appropriate product and process experience, alongside a track record of successful FS & QMS implementation, can provide a great insight into the system requirements and the best ways in which to document and implement the required practices within the operation. Where consultants are utilised, the business would be well advised to ensure that the permanent technical and quality staff are closely linked to the system design and implementation work. Once the consultant has completed their contract it will be the permanent business staff who will need to understand and maintain the FS & QMS.

Many businesses choose to compile their Food Safety & Quality Manual in the same order as their most prevalent customer codes of practice or required technical accreditation standard. The FS & QMS will document all relevant sections in the order of the corresponding subject-specific sections, as stated in the supplier compliance standard or the certification scheme of most influence to the organisation.

For example, if the most critical operational accreditation standard to the business is the International Featured Standards (IFS) food standard, then the site FS & QMS documentation would be likely to include the following section headers drawn from and clearly linked to the requirements as stated in the IFS Food Standard (Version 6) (https://www.ifs-certification.com/index.php/en/standards/251-ifs-food-en):

1. *Senior management responsibility*: 1.1 Corporate policy/Corporate principles; 1.2 Corporate structure; 1.3 Customer focus; 1.4 Management review.
2. *Quality and food safety management system*: 2.1 Quality Management; 2.1.1 Documentation requirements; 2.1.2 Record keeping; 2.2 Food Safety Management; 2.2.1 HACCP system; 2.2.2 HACCP team; 2.2.3 HACCP analysis.
3. *Resource management*: 3.1 Human resources management; 3.2 Human resources; 3.2.1 Personnel hygiene; 3.2.2 Protective clothing for personnel, contractors and visitors; 3.2.3 Procedures applicable to infectious diseases; 3.3 Training and instruction; 3.4 Sanitary facilities, equipment for personnel hygiene and staff facilities.
4. *Planning and production process*: 4.1 Contract agreement; 4.2 Specifications and formulas; 4.2.1 Specifications; 4.2.2 Formula/recipes; 4.3. Product development/Product modification/Modification of production processes; 4.4 Purchasing; 4.4.1 General purchasing; 4.4.2 Trade of manufactured goods; 4.5 Product packaging; 4.6 Factory location; 4.7 Factory Exterior; 4.8 Plant layout and process flows; 4.9 Constructional requirements for production and storage areas; 4.9.1 Constructional requirements 4.9.2 Walls; 4.9.3 Floors; 4.9.4 Ceilings/Overheads; 4.9.5 Windows and other openings; 4.9.6 Doors and gates; 4.9.7 Lighting; 4.9.8 Air conditioning/Ventilation; 4.9.9 Water supply; 4.9.10 Compressed air; 4.10 Cleaning and disinfection; 4.11 Waste disposal; 4.12 Risk of foreign material, metal, broken glass and wood; 4.13 Pest monitoring/Pest control; 4.14 Receipt of goods and storage; 4.15 Transport; 4.16 Maintenance and repair; 4.17 Equipment; 4.18 Traceability (including GMOs and allergens); 4.19 Genetically modified organisms (GMOs); 4.20 Allergens and specific conditions of production.
5. *Measurements, analysis, improvements*: 5.1 Internal audits; 5.2 Site factory inspections; 5.3 Process validation and control; 5.4 Calibration, adjustment and checking of measuring

and monitoring devices; 5.5 Quantity checking (quantity control/filling quantities); 5.6 Product analysis; 5.7 Product quarantine (blocking/hold) and product release; 5.8 Management of complaints from authorities and customers; 5.9 Management of incidents, product withdrawal, product recall; 5.10 Management of non-conformities and non-conforming products; 5.11 Corrective actions.
6. *Food defence and external inspections*: 6.1 Defence assessment; 6.2 Site Security; 6.3 Personnel and Visitor Security; 6.4 External Inspections.

Source: IFS Food Standard (Version 6): https://www.ifs-certification.com/index.php/en/standards/251-ifs-food-en

As a food business develops and evolves over time there will be numerous customer codes of practice, accreditation standards and legislative requirements that will influence the content of the site FS & QMS. Table 3.1 provides examples of a range of general FS & QMS content and considerations.

Each policy or requirement within the FS & QMS should be clearly documented, including a clear and concise description of the programme and its purpose. For example:

- What issues or risks does the programme seek to control with regard to product safety, quality or legality?
- How is the programme structured and progressed on a routine basis?

The policy/programme should define:

- *What* is being measured/assessed/controlled?
- *Why* is this important?
- *Where* will the checks/control procedures be conducted?
- *How* is the measurement/assessment/control conducted?
- *When* are these practices to be conducted (e.g., schedule/frequency)?
- *Who* is involved, and specifically who is responsible, for each control/process?
- *What* are the decision-making and control processes to be followed in the event of a non-conforming result or issue arising?
- *How* are the check/control records and any associated corrective actions to be documented?
- *How* is the programme/policy routinely assessed to confirm that it remains under control?

The programme/policy documentation should also state the defined roles responsible for each aspect of the policy (both individually and on a departmental basis) and the arrangements in place to cover occasions when the designated responsible individual is unavailable. The level of qualification and training required to fulfil each key role will also need to be considered. The programme or policy should also define any measures that need to be taken to ensure that it is running smoothly. Such product/process checks should be logged on product check records or process monitoring forms.

Each programme/policy should bear a clearly defined reference code or number to provide a link to the particular overarching clause or section in the FS & QMS Manual. This should also tie in with the document control measures in place; i.e., each document should bear information stating the specific and unique document

Table 3.1 Example FS & QMS content and considerations

Summary FS & QMS document:	Example contents:
Food Safety/HACCP Manual:	• Operational overview and scope of the site food safety/ HACCP plan • Product details and related process flows • Site Good Manufacturing Practice (GMP) • Risk analysis (typical hazard categories considered to include potential physical, chemical, microbiological and allergen contaminants) • Prerequisite programmes (PRP's) • Critical Control Points/key food safety check points, including related monitoring and nonconformance procedures • Validation reports for all key processes and machinery related to the control of food safety • Methods of verifying site food safety control • Related SOPs (Standard Operating Procedures) and staff training • System review protocols • Threat Assessment Critical Control Point (TACCP): TACCP team, risk assessments and issues identified. Controls implemented linked to materials and products, purchasing, processes, premises, distribution networks and business systems • VACCP (Vulnerability Assessment and Critical Control Point): VACCP team, 'vulnerabilities' identified, 'mitigation strategies' (preventative measures)
Quality Manual:	• Operational overview • Quality policy statement with regard to the production of safe food whilst meeting all legislative and customer requirements • Site organisation chart (management and departmental organograms) • Site management overview, including clear authority and reporting structures facilitating the control of product quality, safety and legality; senior management commitment to continuous improvement and corrective action Note that much of the following contents could be considered under categories of 'Good manufacturing practices' (GMP) and 'Prerequisite programmes' (PRPs): • Site security and control of external areas • Design and construction of premises • Hygienic design of equipment • Employee recruitment, personal protective equipment (PPE), personal hygiene and facilities

(Continued)

Table 3.1 (Continued)

Summary FS & QMS document:	Example contents:
	• Control of temporary staff and indirectly employed/contractor staff • Staff medical screening and return to work procedures • Site hygiene control programme • Product development control • Supplier assurance programme • Raw material specifications • Raw material inspection and acceptance • Process controls (including control of any reworking of products) • Calibration of equipment critical to product safety, quality and legality • Site maintenance programme • Control of foreign bodies • Glass and hard plastic controls • Weight/volume/count control • 'Special material' related controls, for example, allergen management procedures • Pest control programme • Product labelling, coding and traceability • Product transport controls • Customer feedback/complaints management • Factory environmental microbiology monitoring programme • Product inspection/analysis • Nonconformance and corrective action systems • Record keeping procedures • Recall/withdrawal procedures • Crisis management procedures • Finished product specifications • Internal audit programme • Training programme (general induction, departmental and role specific) • Business environmental management practices and policies • Ethical trade and corporate social responsibility (CSR) agendas
Standard Operating Procedures (SOPs)	• Documents/manuals detailing the specific practices or work instructions for each policy or practice of relevance to the FS & QMS. This information is often held at specific operational department level and also can be held by the site training department • All such SOPs should be clearly cross-referenced back to the overarching policy/procedures stated within the site FS & QMS • Information on how staff training is recorded and timescales for review/refresher training

(*Continued*)

Table 3.1 **(Continued)**

Summary FS & QMS document:	Example contents:
Special measures/ additional standards	Documentation with regard to the specific additional requirements of certified standards for products certified as: • Halal • Kosher • Vegan • Allergen free • Gluten free • Organic Or wider standards covering the site position and practices linked to: • Environmental management programme • Ethical trading • Corporate Social Responsibility (CSR)

number, authorisation date, revision number and the name and position of the individual who has approved the documentation (Document control is covered further in the next section of this chapter).

It is usually wise to avoid documenting work instructions and procedures in fine detail within the overarching general programme or policy statements of the FS & QMS Manual. Instead, fine details should be covered in the individual SOPs written for each task or role, as these detailed work instructions can be subject to regular review and refinement as manufacturing processes and products are subject to adjustment over time. This approach enables the detail of the system to be easily reviewed and improved over time, without creating additional administrative burdens incurred via impacting on the particular focus and goals of the specific overarching policy/procedure.

3.6.2 FS & QMS Implementation

The commitment of senior management is critical to the implementation phase of all elements of the FS & QMS. This commitment includes ensuring that the business team members expected to operate and adhere to the FS & QMS are given appropriate training and sufficient resources. The senior management team should also ensure an appropriate level of supervisory monitoring and control over the operation, which must play a lead role in supporting the operational uptake of all new systems, procedures and practices. Management within an organisation must assess the capacity of the operation to take on new practices or meet new requirements and must ensure that implementation progresses at a rate that reflects staff capability and the resources available. Implementing too many new initiatives at once, or progressing at too fast a rate, can distract or overload staff and consequently risk losing control of the operation.

New customer or compliance requirements often impose deadlines for the uptake of a new FS & QMS or the implementation of new elements in an existing system. When this occurs, there may be added challenges to the implementation process which need to be overcome. If there is not sufficient time or resources available to meet all of the necessary deadlines, the senior management team should be expected to both prioritise internal requirements and also manage external customer and enforcement expectations. The senior management team should have the capacity to arrange additional resources as necessary to ensure that all requirements are met.

During the implementation phase, the business would be well advised to increase the level of monitoring and internal audits carried out on site, not only on the operational aspects subject to review (which are at higher risk of initial faults while the system is embedded) but also focussed on other controls already in place. This is due to the fact that implementing new procedures can sometimes lead to resources being drained away from other areas of the operation, which may then as a result start to fail. Close supervision and audit focus will help guard against such issues, or at the very least highlight any problem issues at an early stage.

Document control is a vital aspect of the FS & QMS, which should be in place from the outset. It helps ensure that the use of all related documents (e.g., systems, procedures, methods and records) is clearly controlled within the business, including formal authorisation, clarity of sequential version numbers and control procedures for the implementation of new or updated documents and the effective removal of any old versions from the business operation. The appropriate logging and authorisation of procedures and records is especially vital to ensure business operational control and full traceability of system adjustments over time. Document control information will typically be displayed at the base of each piece of controlled paperwork. For example, the base of all pages of a printed food safety check procedure could state the following:

Document name:	Version number:
Date issued:	Authorised by:

If not properly considered, document control on manufacturing sites can become complicated, leading to a lack of control within the manufacturing environment and the potential for instances of nonconformance to be raised during site audits. Food manufacturing businesses are advised to devise and maintain a simple, effective system.

3.6.3 FS & QMS Monitoring and review

In the event of a food quality or safety issue, the ability to swiftly review the business position can often be vital to regaining control and making a confident response to the issues raised. To enable such benefits to the organisation, all documentation and records linked to the FS & QMS must be retained for a timescale, which significantly surpasses the use-by or best before dates of the products manufactured (as there may be issues raised by customers who have frozen products which were originally short shelf life chilled products, or have consumed products that were well past their best before dates). When setting the standard term for records storage prior to disposal it is advisable to consult the relevant enforcement authorities and key customers to help ensure that timescales are set appropriately and to the satisfaction of all stakeholders.

The manner in which documents and records are stored should facilitate swift and easy access in the event of an issue being raised. Whether boxed, palletised, stored in a warehouse, stored electronically or held off site at a third-party records storage business, it is a good practice to ensure that all documents and records can be recovered and reviewed within a few hours at most. Many food service groups and retailers will set defined 'maximum timescale' requirements for record availability, and often will expect the food manufacturing business to test their response rates, usually during audits or mock recall/withdrawal test scenarios.

The security of these key records and documents should also be thoroughly considered and controlled. Are electronic records routinely backed up? In the event of a fire, what is the likelihood of the documents and records being salvageable? Do site staff know not to dispose of or move any key documentation without prior authorisation?

The FS & QMS of a business should also be subject to routine review. With regard to the frequency and extent of review necessary, this will differ between businesses dependent on a range of factors including:

- Product complexity;
- Nature of the ingredients and packaging utilised;
- Customer expectation;
- Complexity of the supply chain;
- Extent of legislative requirements;
- Customer complaints;
- Management time and resources available.

With regard to product/process check records generated by the operation, it is vital that these should be thoroughly completed to the required check frequencies, as well as being accurate and legible. It is often advisable for the manufacturing operation to routinely check all records in advance of storage to ensure that all staff completing these records are doing so correctly and consistently.

This approach will provide ongoing assurance that the operation is under control and that the key information required will be available in the event of an issue being raised. Often such record accuracy checks are conducted by an area supervisor or manager at the end of a production shift. Checks can also sometimes be conducted by an independent technical administration resource. A key factor here is to complete the checks as soon as possible in order that if a problem is found, then not only can the relevant operator quickly receive appropriate correction and guidance, but also any product potentially adversely affected will be more likely to still be within the control of the business (or at least easier to recall if necessary).

With regard to the food safety elements of the management system, the approach typically recommended for the review of HACCP systems could prove to be a useful guide; i.e., to always conduct a food safety system review in response to:

- internal feedback, e.g., out of specification food analysis results or operational failure of defined CCPs;
- external feedback, e.g., customer complaints;
- new scientific knowledge, e.g., new research suggesting a change in the growth limits of certain foodborne microorganisms, which may be of relevance to the product and its manufacturing process; and
- updates to industry guidelines and legislation relevant to product safety.

The review of the FSMS is often best conducted by a multidisciplinary team within the business to ensure that there is a full working knowledge of the product, the production process and extrinsic factors such as underpinning scientific knowledge and appreciation of customer and legislative expectations (refer Chapter 6: Product control and HACCP considerations for further information).

The wider QMS aspects provide the business with the foundations on which to build the production of food products that meet all customer and legislative expectations.

Internal audits are typically utilised for the routine review of the site FS & QMS. In order to make the review workload manageable and to ensure that all aspects of the system are reviewed at the required frequency, an annual auditing plan should be devised. This should list the documents pertaining to all relevant sections of the FS & QMS, and the months in which they should be reviewed. This approach also provides a useful tracker of progress (Table 3.2).

It is important to ensure that internal audits are conducted by appropriately trained personnel. In addition, in order to ensure complete independence and impartiality, areas must not be audited by a team member who is linked to or has responsibility for that particular area (refer Chapter 9: Audits for further information).

In addition to checking that the relevant systems and procedures are in place and being applied correctly, internal audits should regularly check that all FS & QMS-related procedures and check record documents in use are within the document control system and are up to date. The document control system should also record any adjustments or additions to the range of documentation in use, ensure appropriate authorisation of system updates and ensure the removal of all superseded documents from the operating environment once they have been replaced.

3.7 System accreditation

Businesses may choose to obtain external recognition or accreditation of their FS & QMS for a number of reasons.

- Gain independent confirmation that the systems operated comply with industry standards and are sufficiently robust.

Table 3.2 **Example of a site QMS annual review plan**

QMS section	Jan	Feb	March	April	May	June	etc.
1 Senior management responsibility							
1.1 Corporate policy/ principles	C						
1.2 Corporate structure	C						
1.3 Customer focus		C					
1.4 Management review		C					
2 Quality and food safety management system							
2.1 Quality management			S				
2.1.1 Documentation requirements			S				
2.1.2 Record keeping			S				
2.2 Food safety management				S			
2.2.1 HACCP system				S			
2.2.2 HACCP team				S			
2.2.3 HACCP analysis				S			
... Continue list (drawn from the business QMS sections required to be internally audited) ...							

S, Internal audit scheduled; C, Audit completed.

- Customer assurance.
- Serve as a continuous improvement initiative.
- Gain a competitive advantage and access to new markets.
- Demonstrate control and legal 'due diligence'.
- Benefit from related impacts of the improvement in general site efficiency.

Gaining accreditation of a recognised quality standard typically starts with the selection of the appropriate 'scheme/standard' for the manufacturing operation. Such standards include:

- GlobalG.A.P;
- IFS (International Featured Standards);
- BRC (British Retail Consortium);
- ISO9001 Food quality management standard (International Organisation for Standardisation);
- ISO22000 FSMS-Food Safety Management System; and
- Food Safety System Certification (FSSC) 22000.

Refer Chapter 2, Defining technical and quality standards, for further information with regard to the selection of appropriate food industry standards and certification schemes.

After the appropriate standard has been selected by the organisation, the standard will often contain hundreds of clauses requiring the business to operate extensive systems and procedures to assure ongoing and consistent control, all of which will have to be met before the site can achieve full approval/accreditation according to the standard. Therefore, a process cycle of gap analysis, procedure implementation and review can continue up to the point where the business feels sufficiently confident that all requirements of the chosen standard are met. At this point, the business should seek to secure the services of an accredited/authorised independent third-party audit body to assess the operation against the requirements of the standard.

Once the audit has been completed, and any nonconformances raised dealt with appropriately (and within the timescales required), then the operation can be awarded certification or approval against the selected standard.

Accreditation will often be on a graded basis and usually will be for a fixed period, after which another audit will be required to ensure that the operation continues to meet the requirements of the standard, including incorporating any updates or additions to the standard since the last site assessment.

For further detail on customer and third-party audits, including compliance schemes, please refer Chapter 9, Audits.

3.8 Conclusion

At its simplest, the development of a business FS & QMS could be considered as being the progression of the following steps:

1. *Say What You Do* – The business will typically define and document its product safety and quality checks and controls as a series of procedures and policies collected together in a quality manual.
2. *Do What You Say* – The business will ensure that all of the quality manual requirements and responsibilities related to food safety and quality are clearly presented to the relevant team members and that training is provided where required. Progression of the food safety and quality management control measures is then supported by ongoing internal audits and assessments to ensure that the necessary procedures/checks are carried out and conducted correctly.
3. *Prove It* – Businesses should not only document their procedures, but also maintain records of all training conducted and store the record sheets of all product safety and quality checks. A business may also regularly undergo customer and enforcement agency audits where such materials will serve as proof of working correctly to the appropriate standards.
4. *Improve It* – Reflecting on the findings of routine internal reviews or perhaps customer or enforcement agency feedback, the site should seek to constantly address any issues and update/improve its operation.

Food quality and safety-related check measures within a manufacturing operation often appear to be straightforward when viewed in isolation. It is only when a business considers the fact that it needs to operate a very high number of such procedures that the complexity of the quality and technical systems required is evident.

Checks and controls within an operation will often have been built up over time, and as a consequence a business can find itself in a position where its product quality and safety check systems are not necessarily cohesive and efficient. A common issue in many businesses is a situation in which site quality checks are routinely added to every time a new issue or requirement is raised, without thought of integrating and streamlining the new checks with the existing checks/procedures or ensuring that the staff dealing with the checks have enough resources to cope with more checks being added to the list.

As a busy technical or quality manager it can be easy in such 'cumulative circumstances' to lose sight of the key points and to run quality systems that deliver the basic requirements but take up a lot of unnecessary time and effort due to the perceived operational requirements to check absolutely everything. Taking the time to routinely review and streamline the business food safety and quality system to become more effective and efficient is a key challenge for the site technical management team.

References

Food Hygiene Basic Texts (2009). Codex Alimentarius Commission 'Recommended International Code of Practice – General Principles of Food Hygiene' within the CAC. Codex Alimentarius Commission. <http://www.codexalimentarius.org> CAC Food Hygiene 4th edition. <http://www.fao.org/docrep/012/a1552e/a1552e00.pdf>.

ISO9000:2015 Quality management systems – Fundamentals and vocabulary. International Organisation for Standardisation. www.iso.org

ISO22000 Food Safety Management System. International Organisation for Standardisation. www.iso.org.

Further reading

GlobalGAP – www.globalgap.org
British Retail Consortium (BRC) – www.brcglobalstandards.com
International Featured Standards (IFS) – www.ifs-certification.com
Global Food Safety Initiative (GFSI) – www.mygfsi.com
International Organisation for Standardisation – www.iso.org
Food Safety System Certification (FSSC) 22000 – www.fssc22000.com
Codex Alimentarius Commission – http://www.codexalimentarius.org

Raw materials and packaging supplier control

4

Mark Swainson
National Centre for Food Manufacturing, University of Lincoln, Holbeach, Lincolnshire, United Kingdom

4.1 Introduction

The major raw material groups used in food manufacturing include vegetables, fruits, cereal crops, meats, dairy, herbs and spices. Common cereal crops include rice, wheat, barley, oats, maize, sorghum and rye. To the list of product ingredients can be added numerous other ingredient groups including additives, flavourings and alcohols.

Most food ingredients are the product of farming systems. The typical 'farm to fork' supply chain commences with primary produce grown/reared from farms/agricultural systems. In the European Union 'primary production' is defined in Article 3(17) of Regulation (EC) No 178/2002 as '...the production, rearing or growing of primary products including harvesting, milking and farmed animal production prior to slaughter. It also includes hunting and fishing and the harvesting of wild products'.

Primary produce is often supplied to food manufacturing businesses where the ingredient can undergo numerous processes before ultimately being packaged. The product can then travel in the supply chain, sometimes via intermediaries/logistics/marketing businesses, either directly for sale to the end consumer or on to food manufacturers who will utilise the ingredient to form a final saleable product (e.g., multicomponent foods such as ready meals, soups, sauces, dips and dressings). Eventually the food will find its way to the point of consumption via 'food retailers' (e.g., supermarkets), 'food service' (e.g., restaurants/catering establishments) or 'markets'.

As most of the food manufacturing sector involves the procurement and processing of raw materials to form the final product, technical control of the product ingredients and packaging is vital to the safety, quality and legality of the food produced. By understanding the nature of the raw materials, and how they are produced and processed prior to arrival at the manufacturing site, the technical manager will gain a greater insight into the specific control points required to achieve a consistent end product.

The best way to gain a detailed insight into the ingredient supply of a manufacturing site is for the technical manager to take the time to visit key suppliers in each of the ingredient categories. This will facilitate a thorough review and

Swainson's Handbook of Technical and Quality Management for the Food Manufacturing Sector.
DOI: https://doi.org/10.1016/B978-1-78242-275-4.00004-6

discussion on the nature of the ingredients and related processing. If this is not possible then there is a wealth of literature and online resources which can be studied in order to gain the level of insight and understanding required.

For example, the US EPA (Environmental Protection Agency) website contains extensive information and notes on production processes including meat packing plants, meat rendering plants, poultry processing, dairy products, natural and processed cheese, preserved fruits and vegetables, canned fruits and vegetables, dehydrated fruits and vegetables, pickles, sauces and salad dressings, grain processing, cereal breakfast food, pasta manufacturing, bread baking, fish processing, snack chip (crisps, etc.), deep fat frying. *Source*: AP 42, Fifth Edition, Volume I, Chapter 9: Food and Agricultural Industries. https://www3.epa.gov/ttn/chief/ap42/ch09/index.html

Government departments, trade associations and other interest groups are also very good sources of information, usually operating on a national and sometimes international scale to represent specific product sectors. In the United Kingdom these include the Food and Drink Federation, Department for Environment, Food and Rural Affairs, Food Standards Agency, British Frozen Food Federation, Chilled Food Association, Institute of Grocery Distribution, Soil Association (Supporting the Organic Sector), Agriculture and Horticulture Development Board, Vegetarian Society, Halal Food Authority and the Kashrut Division of the London Beth Din (a leading UK authority on Jewish Dietary Laws).

4.2 Ingredient supplier management

As a consistently correct raw material supply is vital to underpin a controlled food manufacturing process the technical manager should always recognise the importance of developing and maintaining key supplier links. Knowing the key technical contact to call at the supplier site can save valuable time in the event of requiring raw material specific advice or when seeking to address a particular problem. Maintaining a supplier database or contacts list will help to ensure that key contact information is kept up to date. Such a database approach can also enable technical colleagues to be able to swiftly search for and review information related to each supplier (e.g., find contact records and a synopsis of past and current supplier position/any issues) in the event of other key team members being unavailable.

Control of raw materials involves a range of activities including:

- supplier approval processes (e.g., site audits or self audit questionnaires (SAQs));
- use of accreditation schemes (e.g., BRC (British Retail Consortium), International Featured Stand (IFS));
- raw material specifications;
- inspection procedures (e.g., certificates of conformance/analysis, delivery inspection/sampling);
- traceability control (supporting product quality assurance and swift response in the event of any major quality/safety issues).

It is often said that with the right raw materials a business will be at least half of the way toward making the right end product. The technical manager is therefore well advised to seek to ensure that sufficient technical/quality resource is applied to the control of raw material supply into the business. 'Sufficient' resource will be dependent on the size of the food manufacturing business and the range/complexity/volumes of raw materials being handled.

Key raw material related roles within the business technical team can include:

- Raw materials supervisors/managers overseeing raw material supplier ingredient specifications, compliance/audit positions and handling of non conformance issues.
- Incoming goods 'inspectors' overseeing quality checks upon intake, checking that all certificates of conformance (CoC)/analysis are present and correct (if required), progression of further sampling such as routine microbiological or chemical laboratory tests (e.g., checks on product safety, adherence to nutritional standards, speciation checks, etc.), and management of any non conforming goods including placing the goods on hold and communicating the matter internally and externally.
- Packaging technologist. Setting and overseeing packaging specifications to meet the requirements of the business, assessing proposed new packaging through the factory handling trials, shelf life testing and transit trials. This specialist role will also typically be involved in packaging supplier approval processes including the progression of audit requirements.

When considering the question of 'what makes a good supplier?', the raw material supplier has to meet a wide range of requirements. With regard to 'technical standards' the materials supplied must consistently meet the agreed specification, operating to the quality, safety and legal standards required.

Alongside these expectations, the supplier must meet the 'purchasing department requirements' of supplying at a competitive price, in the required quantities, delivered at the required time, with sufficient shelf life remaining on the material, and without insisting on excessive minimum order quantities (MOQs). In addition, a food business may place a range of 'new product development (NPD) expectations' on its supplier base including the need for selected suppliers to generate and share innovative areas for ingredient utilisation.

In the process of supply to a food manufacturing business, ingredient specifications are typically utilised to define the supplier's commitment and the food manufacturer's requirements. The volume requirements of the business are typically calculated manually or via a stock inventory system which looks at the forthcoming production plan and then calculates which ingredients need to be bought in and by when.

Orders are then placed with the suppliers for delivery. For speed and to provide a traceability trail these 'orders' can be made electronically, with deliveries ensured to be within the required lead times. 'Lead time' is the amount of time that the supplier requires from receipt of the order to the point of being able to deliver the ingredient to the manufacturing site. Typically, this lead time can be a few days but can be as long a number of weeks, particularly in the case of ingredients that are not held ready-made and on-stock.

During a typical cycle the food manufacturer dialogue with suppliers can include aspects such as order placement, feedback on supply cost/service, business links and longer-term outlook, feedback on quality/complaints and also possibly rejection of stock.

Feedback can be verbal and/or written. Photographs are also often a good way to communicate some specific extra details of ingredient supply issues. Good dialogue with a supplier, especially built during times when positive feedback/praise is given with regard to the supply, helps to build a good rapport that can benefit the food business operator in the event of needing extra service or help from the supplier.

4.2.1 Formal review of supplier performance

A routine 'supplier review meeting' held between the supplier and the customer buying/purchasing and quality/technical departments prompts the opportunity to discuss the business outlook, highlight any challenges that can be foreseen such as significant increases in volume of supply requirements (perhaps to support a particular promotional activity), and consider the consistency of supply or any other particular technical issues (e.g., quality issues/non conformance to specification/ rejections).

Businesses may wish to track their supplier's performance over time to give additional oversight on any trends that may be developing. A supplier scorecard for use at routine supplier review meetings may include the following metrics:

- percentage of orders delivered on-time and in-full;
- price stability;
- number of quality/technical issues encountered and
- number of rejections.

These reviews may be conducted with the whole supplier base, or if resources and time are limited then the reviews may be held with a select number of the suppliers that are seen to be high priority, perhaps based on the proportionately high financial amounts spent with them, or on their potential to adversely impact on the food manufacturer if the supply does not progress consistently and smoothly.

Food manufacturing businesses are often keen to develop new types/ranges of products and update existing ranges/recipes. In support of such initiatives a 'supplier development presentation' can be held with each key supplier and will typically also involve the food manufacturer's NPD and purchasing and technical departments.

The supplier development presentation can be used to review and discuss various development related aspects such as:

- Highlight and discuss any new products/product ranges that can be supplied which may add variety/value.
- Review and discuss whether the food manufacturer is utilising the supply of existing products to the best end product advantage.

- Planning considerations. The meeting can look ahead and discuss factors which may benefit from additional focus such as uplift in supply requirements due to a planned promotional activity or known seasonal sales trends.
- Horizon scanning. The meeting can discuss the current market conditions and outlook which may influence price and consistency of supply.

This approach works equally well for both ingredient and packaging suppliers.

4.2.2 Contingency planning (raw materials supply)

Food manufacturers are well advised to consider what their response and contingency plans would be in the event of a quality, safety or legal fault with an ingredient, or failure to supply. Making such plans in advance of matters actually happening will ensure that there is a readiness, clarity and speed of response, all of which can be defined in the business food safety and quality manual.

With regard to minor issues which are highlighted sufficiently early, the business response may be to just place the problem related or non conforming raw materials on hold/quarantine to await return to the supplier and placement of a new order.

If the food manufacturer does not have time to wait for a resupply of the non-conforming raw material, and the non conformance is deemed by the technical manager as not having a significant impact on end product quality, safety or legality, then the food manufacturer may choose to use the non conforming stock and may take extra steps to ensure that any issues are managed to ensure that they do not subsequently become a problem.

There may be occasions when a raw material problem/fault comes to light after the raw material has been utilised in an end product and the product has already been despatched into the food supply chain. In these circumstances the technical manager needs to lead the business in its review of the extent of the problem, which end products are affected, where are they in the supply chain, and what is the extent of necessity in terms of triggering a product withdrawal/recall? Depending on the issue and extent of the situation perhaps the business 'crisis management team' involvement will be required. Other matters to be attended to will be the placing on hold of any remaining affected raw material, work in progress and finished goods stocks, and liaison with the necessary customer contacts and enforcement authorities on matters that need to be formally reported.

On some occasions there is not a raw material fault issue, there is just a situation where the supplier cannot supply the required raw materials in time. Perhaps they have been given insufficient lead time, have run out of stock or have encountered a manufacturing issue in their operation.

With regard to the loss of supply the technical team will usually be involved in responding swiftly to a situation in one of a number of ways:

- Can the business substitute the missing raw material with another similar raw material? (The technical team will need to consider this from all aspects of quality, safety and legality.)

- Is the raw material available from another supplier? The technical team may be asked to discuss the required specification in-depth with the alternative supplier in order to ascertain whether supply is possible.
- Swift approval of an alternative supplier. If the alternative supplier isn't currently on the food manufacturers 'approved suppliers' list then the technical department may be asked to respond swiftly to seek to approve the new supplier in advance of the raw materials being required. This could be via SAQ review and/or physical on-site inspection.
- Additional checks may be required on the new raw material to add further assurance/due diligence to the food manufacturer business position.

Replacing suppliers: The above steps are also very relevant in the event of the food manufacturing business seeking to replace particular suppliers over time, perhaps as a result of poor performance, insufficient capacity or uncompetitive pricing. The business technical department will often be involved in finding a replacement supplier and this process is made easier if the food manufacturer holds clear specifications of the raw material required and has a working history of its supply from previous providers.

Key steps in this process will include confirmation that the proposed new supplier can match the specification required (including confirmation via the assessment of samples) and technical audit approval.

In the circumstances of seeking to replace a supplier it is sometimes prudent where possible to build an ingredient stock holding on site in advance of delisting one supplier and replacing with another. This will aid the continuity of supply if any 'commencement of supply' issues are encountered with the new supplier. Another approach is to 'dual supply' (run the old and new supplier together side by side) for a while until sufficient confidence is built in the new supplier.

4.3 Risk assessment

The term 'risk' relates to the potential for harm, damage or loss. The management of risk is a very wide ranging subject for the food industry as the processes of the food supply chain are complex and multifaceted, with many opportunities for 'issues' to arise unexpectedly.

With regard to the control of raw material supply, 'risk assessment' can be viewed as the process of systematically reviewing the nature of the raw materials and their associated supply chain in order to highlight the potential risks, and then proactively putting in place systems and procedures which reduce these identified risks to an acceptable/managed level. These steps will include sufficient monitoring processes designed to highlight any issue as early as possible.

Steps for proactive avoidance of risk include:

- Supplier assurance. Only dealing with suppliers who have been formally reviewed (either by the food manufacturer or via third-party audit/certification standards) to confirm that they appear competent in their supply of safe and legal raw material to the required quality standards.

- Expectation and confirmation of raw material supplier tests prior to delivery/use by the food manufacturer. Such testing can be confirmed to the food manufacturer via the supply of certificates of analysis (CoA)/certificates of conformity (CoC).
- Raw material inspection processes, for example, setting appropriate frequencies of physical and organoleptic inspection upon delivery and also during storage and at point of use. Microbiological and chemical assessment may also be relevant to confirm matters such as safety, shelf life, nutritional data, pesticide/veterinary reside checks, meat speciation checks etc.

The TACCP ('Threat Assessment Critical Control Point') approach is being increasingly adopted across a number of food supply chains. Documented in Publicly Available Specification *PAS 96:2017* 'Guide to protecting and defending food and drink from deliberate attack', the PAS document highlights that while hazard analysis critical control point (HACCP) principles are generally utilised on a global basis to help reduce the potential of accidental product contamination to occur, HACCP principles have not been used/designed to detect or mitigate deliberate attacks on a food supply chain related system or process (e.g., deliberate contamination or fraud). While the immediate focus is on avoiding the food safety implications of deliberate/malicious acts, such matters are recognised to also harm organisations in a wider range of ways including damages to business reputation or financial extortion.

Designed to help protect the food and drink supply chain from deliberate attack the document guides food businesses through processes designed to 'improve the resilience of supply chains to fraud or deliberate attack'. The TACCP methodology helps business to evaluate, document, understand and control a wide range of threats to their supply chain.

TACCP principles include:

- Identification of potential threats and emerging risks to food supply chains, such as economically motivated adulteration or other illegal practices.
- Consideration of which points in the supply chain are vulnerable to such threats. (e.g., whereabouts in the supply chain or manufacturing processes could adulteration most easily go unnoticed?)
- Implementation of control measures to mitigate the risk.

The PAS document guides the user through to topics of types of threat, understanding the attacker (e.g., extortionist, opportunist, extremist, irrational, disgruntled, cyber space 'hacktivists' and professional criminals), TACCP, assessing vulnerabilities, critical controls, response to an incident and a review of food protection arrangements.

Copies are available online searching on the term 'Publicly Available Specification PAS 96:2017. One such site for the 'Guide to protecting and defending food and drink from deliberate attack' is via this web address: https://www.food.gov.uk/sites/default/files/media/document/pas962017.pdf

TACCP and VACCP (Vulnerability Assessment and Critical Control Point) are further considered in Chapter 1, Food sector challenges and the role of technical and quality management.

4.4 Specific risk materials

'Specific risk materials' handled by a food manufacturing operation could be viewed as materials posing an additional threat to the safety, legality or quality of the foods being produced, or due to their nature requiring particular care in their handling and use. Potential threats to the end consumer are typically the primary focus and any risks to food industry employees in contact with the food must also be considered.

Specific risk materials/issues could include:

- allergens;
- genetically modified ingredients (e.g., identity preserved status, approval for use and labelling expectations);
- food additives/E-numbers and
- high-value items increasing the risk of theft or related food fraud activities.

4.4.1 Allergens

Allergens are typically proteins in foods which cause susceptible consumers to have an allergic reaction (immune response) to the allergen. Symptoms can include coughing, dry itchy throat and tongue, itchy rash, nausea, diarrhoea, vomiting, wheezing and shortness of breath, swelling of the lips and throat, runny/blocked nose, sore, red and itchy eyes. Anaphylaxis/anaphylactic reaction is a particularly acute reaction to the allergen and at its most severe can result in hospitalisation and death.

Populations in different regions of the world show varying susceptibility to particular allergens. It is therefore vitally important that food businesses control the allergens of particular concern for the markets being supplied.

In the EU there are (at the time of writing) 14 recognised food allergens:

- cereals containing gluten (e.g., wheat, rye, barley, oats);
- crustaceans (e.g., prawns, crabs, lobster, crayfish);
- eggs;
- fish;
- peanuts;
- soybeans;
- milk;
- nuts (almonds, hazelnuts, walnuts, cashews, pecan nuts, Brazil nuts, pistachio nuts, macadamia (or Queensland) nuts);
- celery (including celeriac);
- mustard;
- sesame;
- sulphur dioxide/sulphites (where added and at a level above 10 mg/kg in the finished product);
- lupin (includes lupin seeds and flour and can be found in types of bread, pastries and pasta);
- molluscs (e.g., clams, mussels, whelks, oysters, snails and squid).

European food labelling rules are designed to ensure that consumers are provided with comprehensive ingredient listing information. There are specific labelling expectations to highlight the presence of any of the EU recognised allergens, with the intention of making it possible for people with food allergies to identify the foods they need to avoid.

Consequently, one of the most regularly occurring reasons for product recall in the EU is where 'undeclared allergens' have been highlighted as being present within products. In these cases the allergy sufferer would not be able to avoid the allergen they are susceptible to and therefore the products affected must be withdrawn/recalled in order to protect the consumer. Many such circumstances of undeclared allergens occur because of mistakes made in the food packaging label generation process, products packed into the wrong packaging, cross-contamination in the manufacturing process, mistakes made in the recipe/production processes, or insufficient raw material/ingredient information being known at the time of generating the product labels, only for allergen presence to come to light later once the product has been launched, manufactured and distributed.

Businesses can seek to control allergens via a range of measures, including the following.

Focus on ingredient suppliers:

- Use of 'approved suppliers' who have been risk assessed and audited to confirm that sufficient technical controls are in place.
- Use of allergen data declaration systems which require suppliers to provide full details as to the presence of allergens for each ingredient/product supplied. This information can then inform and trigger relevant allergen controls at the manufacturing plant and can also inform the final end product allergen labelling.

Allergen segregation in the storage and processing phases:

- Use of physical/spatial segregation, clear labelling and different types/colours of containers/packaging are all common methods of allergen segregation in food manufacturing environments.
- Hygienic design of process equipment/lines to enable thorough cleaning and inspection.

Allergen handling procedures within the manufacturing operation:

- Such measures can include segregation of staff working on allergen and non allergen products/ingredients.
- Use of separate utensils/containers/process lines for allergen containing products.
- Production planning to avoid processing allergen containing ingredients at the same time as non allergen containing products.
- Hygiene procedures to be appropriately designed, validated and regularly monitored/verified.

Control of end product labelling/artwork information:

- Ensuring that the business has sufficient systems for gathering the allergen information across the raw material supplier base.

- Review of manufacturing site controls to either underpin the product label accuracy or trigger the use of 'may contain' statements where it is felt there is potential for cross-contamination with allergens to occur in the raw material supply or product processing phases.
- Product specification and label artwork/ingredient declaration check procedures to ensure that the end product packaging/label information is correct.

4.4.2 Genetically modified food ingredients

The use of genetically modified (GM) foods is accepted in some regions of the world and is restricted in others. It is therefore vitally important that food manufacturers and distributors apply controls/measures appropriate for the market/regions being supplied.

GM materials can be present or involved in a wide variety of foods, including:

- Direct consumption of cereals, vegetables, fruits, meats, etc. produced utilising GM technologies.
- Use of GM feed for animal production systems. While the animal feed may be GM, the animals resulting from these systems are not GM as the GM DNA is not transferred via the feed process.
- GM plant processing to form other foodstuffs/ingredients, for example, GM maize processed to form flour and starch products.
- Foods can be derived from GM ingredients/components, but have no GM material (DNA) present in the end product. Sometimes termed 'GM derived' such foods can include oils where the source crop was GM (e.g., soya) but the nature of the oil manufacturing process results in no GM DNA remaining in the final end product, effectively meaning the end product is physically and chemically no different to a product made with non-GM material.
- Some food manufacturing processes can use GM materials as a 'process aid'. One common example of this is some vegetarian rennet utilised in cheese production which are created via the use of GMOs (genetically modified organisms). The rennet triggers the cheese curdling process but the GM material is not classed as forming part of the end product, hence the term 'process aid' rather than 'ingredient'.

The matter of GM material utilisation in food systems is complex and widely researched leading to a variety of views on its acceptance. Food businesses are therefore well advised to ensure that they are clear on the expectations of their markets and to ensure that their production systems sufficiently reflect and control these requirements.

Where 'GM use' restrictions/expectations are in place, typical control measures include:

- Ingredient supplier focus. Including the use of 'approved suppliers' who have been risk assessed and audited to confirm that sufficient technical controls are in place.
- Use of GM material declaration systems which require suppliers to provide full details as to any presence of GM material for each ingredient/product supplied. If necessary, this information can then inform and trigger relevant GM controls at the manufacturing plant and can also inform the final end product GM labelling.

- GM material segregation in the storage and processing phases. Use of physical/spatial segregation, clear labelling and different types/colours of containers/packaging are all common methods of GM material segregation in food manufacturing environments.
- GM material handling procedures within the manufacturing operation. Such measures can include segregation of staff working with GM products/ingredients. Use of separate utensils/containers/process lines for GM containing products. Production planning to avoid processing GM containing ingredients at the same time as non-GM containing products. Hygiene procedures to be appropriately designed, validated and regularly monitored/verified.
- Control of end product labelling/artwork information. Ensuring that the business has sufficient systems for gathering the GM information across the raw material suppliers. Product specification and label artwork/ingredient declaration check procedures to ensure that the end product packaging/label information is correct and meets the expectations of the market being supplied.

4.4.3 Food additives/E-numbers

Food businesses may require to treat additives/E-numbers as 'risk materials' due to the fact that they often require particular/special attention to ensure their control of use in the manufacturing process. Depending on regional expectations, food additives can have legislative or good practice guideline restrictions on the amounts in which they can be added to the foodstuffs. In addition, customers may have varying views on the acceptability of using additives in foods.

Typical controls for the handling of allergens/E-numbers in food manufacturing processes include:

- Ingredient supplier focus. Including the use of 'approved suppliers' which have been risk assessed and audited to confirm that sufficient technical controls are in place. Use of food additive/E-number declaration systems which require suppliers to provide full details on any presence of additives/E-numbers for each ingredient/product supplied. If necessary, this information can then inform and trigger relevant additive/E-number controls at the manufacturing plant and can also inform the final end product labelling where there is deemed to be sufficient amounts of such material remaining in the end product.
- Recipe control, including appropriate utilisation. Accurate calculation of allowed amounts in the product and accurate weighing/measurement into the product.
- Control of end product labelling/artwork information. Ensuring that the business has sufficient systems for gathering the additive/E-number presence information across the raw material suppliers. Product specification and label artwork/ingredient declaration check procedures to ensure that the end product packaging/label information is correct and meets the expectations of the market being supplied.

4.4.4 High-value items increasing the risk of theft or related food fraud activities

Food manufacturing businesses may wish to apply additional 'risk material controls' to certain foods/ingredients handled. Such circumstances may include:

- Particularly 'high-value' ingredients present on site leading to a high potential for theft. Example products include alcohols, caviar, truffles, chocolate, etc. In such circumstances businesses may choose to store such materials in locked areas or in areas which have additional security/surveillance measures in place.
- When handling foods that the food sector has experience of being susceptible to food fraud, for example the need for additional traceability and speciation checks to ensure that high-value fish/meat products are derived from the species expected.

4.5 Supplier approval

'Supplier approval' is a vital aspect of control in the food manufacturing sector. Without a technically sound raw material supply chain a food manufacturer is likely to have its control of food safety, quality and legality regularly undermined.

The term 'supplier approval' relates to the process of the food manufacturing business seeking to review, check and approve each provider of ingredients, materials or services prior to the commencement of supply. Thereby significantly increasing the chances of the required 'supply' arriving 'as required' and significantly reducing the potential for issues being encountered which could impact on the business operation and the safety, quality or legality of its products.

A typical approach is for food business operators to document their 'supplier approval' processes and their 'ongoing supply monitoring' procedures. Key focus points of the supplier approval procedure include seeking to ensure that suppliers are manufacturing their products under hygienic conditions, are effectively managing any risks to product quality and safety, and are aware of and fulfilling all of their legal and 'due diligence' obligations including the operation of effective traceability processes.

Supplier approval procedures are typically based on a combination of the following activities:

- Supplier audits conducted by the business utilising their own staff to visit, check and approve the proposed suppler.
- Third-party audits or certification utilised to confirm that the proposed supplier has been recently checked by a reputable auditor/audit body and found to be of an appropriate standard, for example, BRC, IFS, and ISO.
- Supplier self audit questionnaire (SAQ). Instead to physically visiting the supplier a series of documented questions are asked (via an e-mailed, online or posted document) which are designed to check the safety, quality and legality of the supplier's operation and its capability to supply consistently. The answers are then assessed by the food business to confirm that all is 'as required' before approval to supply is granted.

Ongoing monitoring of 'approved' status. All approaches to supplier approval will be expected to have a review frequency set. That is, how often does the food business commit to re-auditing their suppliers/reissuing the supplier questionnaires for review? How frequently are any third-party audits expected to take place?

It is prudent for a food business to formally place an expectation on its supplier to notify the food business if ever there is a significant occurrence that could affect its 'approved' status. For example, when using third-party audits as an assurance of supplier control, the supplier should be expected to inform the business of retained approval status or otherwise highlight immediately if certification/'audit approved' status has been lost. Other such expectations could be the requirement to inform the food business of any significant events such as product recalls, issues related to enforcement authority visits/audits or any legal actions being pursued against the supplier.

The particular supplier approval processes utilised and their frequencies of review will typically be set on the basis of 'risk assessment'. This approach can take the form of a supplier linked technical review to identify potential risks to product safety, quality and legality. In line with HACCP recognised categories of food hazards 'safety considerations' can include the potential for physical contamination with foreign bodies, presence of allergens, microbiological contamination and chemical contamination. Quality considerations can include the potential of a raw material to impact the quality of the final product. The results of this risk assessment can then inform the expectations for supplier approval and also ongoing monitoring/checking of the raw materials supplied.

As part of the risk assessment process many businesses will also choose to incorporate additional potential for 'business impact' into the assessment. For example, a supplier may be placed in a higher risk category (and therefore increased requirements for review and assurance) if they are a particularly high-value supplier, or if the business disruption would be very high in the event of a failure to supply.

There are a range of stakeholders in food businesses with regard to supplier approval and preferences as to which suppliers the business should particularly utilise/favour over time. Often such decisions will be influenced by a combination of technical compliance and commercial competitiveness.

Sections of the business involved in the approval of suppliers include:

- Quality/technical: Concerned with whether the supplier can and will supply materials to the required standards of quality, safety and legality (typically defined in agreed raw material specifications). Another related factor is whether the 'technical service' offered by the supplier is sufficient, as often food businesses require prompt and detailed feedback from their suppliers with regard to product information, audit status, complaints, queries, etc.
- Buying/purchasing: Will expect suppliers to meet the required materials specification at a competitive price. Other expectations will be for the supplier to meet the delivery timescales required to feed smoothly into the production schedule. Also, MOQs (minimum order quantities) will need to be at a level which does not risk the material running out of shelf life before it can all be used, or is so high that the food business cash flow would be adversely affected in tying up too much money in raw material stocks.
- Development/NPD: Require suppliers that can meet the quality requirements expected of the end product brief and are responsive to sample requests (where small amounts of new ingredients/packaging are required to develop new concepts/products which are yet to be

approved and manufactured on a commercial scale). Also, suppliers that can present innovation/new ideas to the food business will add significant value to the relationship.

The phases of new supplier approval will vary from business to business. A common process is for the development or purchasing team to find a new supplier who they feel may well be of benefit to the business. In these early stages dialogue with the potential supplier will be progressed by the purchasing or development team to confirm that the required quality of material can be supplied at costs, timescales and MOQs acceptable to the business. Once initial indications are positive the business development or quality/technical teams will seek to review samples and detailed material specifications to further confirm the potential to supply exactly what is required.

The technical team will then conduct a risk assessment via which the supplier approval 'audit requirements' (e.g., business audit, third-party audit, supplier questionnaire) will be set. The required audit approach will then be progressed. Upon approval being secured supply can then commence.

It is vital that each department is given sufficient time to conduct their checks thoroughly on prospective new suppliers. A hurried approach at this stage can easily lead to key issues being overlooked/missed. The temptation to rush can be in circumstances of the business not having an alternative 'already approved' supply, or having committed to a rapid launch schedule for a new product for which the raw materials have yet to be sourced.

Often with new suppliers it is advisable for the first deliveries to be subject to a high degree of review and checking against previously supplied samples to ensure that the supply is sound/secure. This also sends a clear message to the supplier that what has been agreed will be expected consistently from this point onwards.

More detail on the audit process can be found in Chapter 9, Audits.

4.5.1 Self audit questionnaires

SAQs enable a swift review of a supplier's capability to consistently supply the materials required. The SAQ can be posted, e-mailed or set up for online submission. It is vital that the SAQ is completed by a nominated representative of the supplier's business and that this individual is able to provide all of the knowledge/information required.

On return of the completed SAQ to the business it is very important that all answers/information provided are checked by an appropriate member of the technical team. This team member should have the knowledge and experience to be able to identify any answers/information provided which give rise to concern that the prospective supplier may have some issues with regard to being able to meet the expectations of the business. These questions/queries should then be formally followed up with the prospective supplier to clarify matters or request further assurance before approval to supply can be given.

Documenting all points of this process including the review of the SAQs and follow-up on any points will enable the business to show a good degree of due diligence with regard to the thorough review and approval of suppliers.

Typical SAQ sections and questions can include the following.

1. Supplier details and document sign-off:
   a. Business registered address and addresses of any other associated suppliers including sites holding stock/supplying on behalf of the business.
   b. Contact details including sales and technical representatives and 'out of hours' emergency contact details.
   c. Details of the business representative completing the SAQ including name and position held in the business.
   d. Document cover page signed and dated alongside a statement that the information given is accurate and that the business representative is authorised to provide the information on behalf of the business.
2. Current processes and operating position:
   a. Range of products manufactured and customers.
   b. Number of staff including total production staff and inspection staff.
   c. Factory site layout and process flows (including process flow diagrams).
   d. Audit approvals/third-party certification currently held.
   e. Details of product liability and recall insurance held by the business.
   f. Has the business ever been prosecuted for a food safety, food hygiene or other type of offence? If so details requested.
   g. How does the business remain up to date with legislation requirements and updates?
3. Buildings location, perimeter, construction and layout:
   a. Is smoking allowed anywhere on site? (If 'yes' then please state where.)
   b. Is provision made for staff to eat/drink while on site? If so are there designated areas?
   c. Is the business situated in an industrial area/urban area/rural area?
   d. Are there any 'high-risk' or 'high-care' designated operations on site? Please give details.
   e. Are these areas completely segregated from 'low-risk' operations?
   f. Do these areas have separate drainage from the rest of the factory?
   g. Are all windows and vents protected by insect screens?
   h. Are the factory floors and walls constructed from smooth, impervious and easy to clean material?
   i. Is the factory constructed in order to provide a 'one-way flow' of production so as to eliminate the potential for cross-contamination?
   j. Are all drainage systems adequately maintained and constructed so as to eliminate flow-back?
   k. Is flooring sloped so as to assist drainage?
   l. Is all glass on site covered?
   m. Does the business operate an air filtration system? Please state areas of operation and give details of system if 'yes'.
   n. Does the business operate air temperature control in production areas? Please state areas of operation and give details of system if 'yes'.
   o. Does the business perform factory structure and equipment condition audits? If so then please give frequencies, corrective action plans and time limits applied to rectification work.

4. Water supply and waste disposal:
   a. Is the site water supply drawn from the mains supply? If 'no' then please state nature of supply. Please state the name of the water company if 'yes'.
   b. Does the business perform chlorination of water on site? If 'yes' then please give details of the concentrations applied and detail how this system is monitored and controlled.
   c. Does the business operate a water filtration system? If 'yes' then please state areas of operation and give details of the system.
   d. Is sewage and/or effluent treatment carried out on site? Please give details if 'yes'.
   e. Is all food and packaging waste segregated?
   f. How is waste disposed of on-site?
   g. Does the business utilise an independent waste company to deal with waste?
   h. Please provide detail of any chemical or microbiological testing carried out on the water supply.
5. Site engineering/maintenance resources:
   a. Are engineers or any other maintenance staff required to change into protective clothing before entering production areas?
   b. Does the business employ an on-site maintenance person to control and repair factory structure, equipment and perimeter?
   c. Are all lubricants and fluids used food safe/food grade?
   d. Are all lubricants and fluids used free from nut and nut derivatives?
   e. Does the business have documented records of all tools issued and returned after each working day/shift?
   f. Does the business have engineering stores which control stocks of key spares/equipment?
   g. Does the business operate a planned preventative maintenance system and hold records of the work conducted and of any breakdowns/issues encountered?
6. Personnel and training:
   a. Are all employees formally trained with regard to food hygiene and their role within the business?
   b. Are canteen facilities provided which are segregated from the production areas?
   c. Does the business provide staff locker rooms and changing rooms?
   d. Does the business require that any employee returning from sickness must receive formal checks and clearance to return to work following cases of stomach and bowel infections/related illnesses?
   e. Does the business operate a recognised system for the reporting of illnesses and diseases?
   f. Are plasters blue and metal detectable?
   g. Does the business employ temporary/agency staff? If 'yes' how are these staff formally trained on the requirements of the business.
   h. Do all new employees receive a formal induction to the business? If 'yes' please give brief description of contents/approach.
   i. Are all employees required to pass a satisfactory medical? If 'yes' please give details.
   j. Please state the business policy with regard to the wearing of make-up and perfume by food handling staff and production personnel.
   k. Please state the business policy with regard to the wearing of jewellery.
   l. Please list all factory clothing and protective equipment supplied to employees for wearing in production areas.

- **m.** Please state boundaries where factory clothing and protective equipment must not be worn outside of and list all areas where factory clothing and protective equipment is not permitted (e.g., canteens, outside).
- **n.** Please define the business changing procedures for wearing and removing factory clothing and protective equipment.
- **o.** State who is responsible for laundering of factory clothing and protective equipment and the frequency of changing overalls/related work-wear.
- **p.** Please state the controls operated with regard to ensuring that site visitors and contractors adhere to all business procedures and policies.

7. Cleaning and hygiene:
- **a.** Does the business have documented and controlled hygiene schedules and procedures?
- **b.** Does the business have lockable segregated storage areas for cleaning chemicals?
- **c.** Does the business have 'control of substances hazardous to health' and health and safety data available for all chemicals used on site?
- **d.** Is this information easily accessible and readily available?
- **e.** Are GMP (good manufacturing practice) and CAYG (clean as you go) policies standard practice?
- **f.** Does the business have a separate clean-down/procedures for nut containing products?
- **g.** Does the business employ an independent hygiene company/cleaning chemical supplier? If 'yes' please state name.
- **h.** Does the business perform microbial swabbing in order to monitor and control the cleanliness of the factory environment? If 'yes' please give details including frequencies and documentation (e.g., *Listeria* sampling plan).

8. Control of raw materials (ingredients and packaging):
- **a.** Are all raw materials inspected in order to ensure correctness upon delivery?
- **b.** Does the business document all rejections and quality issues?
- **c.** Are all deliveries recorded upon intake for traceability purposes?
- **d.** Are all delivery vehicles checked on arrival for temperature and hygiene?
- **e.** Does the business perform microbiological analysis upon ingredients?
- **f.** Does the business employ stock rotation and control?
- **g.** Does the business audit stored ingredients for use-by control?
- **h.** Does the business hold ingredient specifications for each individual ingredient?
- **i.** Does the business employ any of the following inspection methods upon ingredients during preparation and/or on use in production? Visual inspection – X-ray metal detection – sieving?
- **j.** Does the business handle allergens? If so please state which allergens are handled and whether they are subject to specific handling and storage procedures.
- **k.** What is the business position on being able to supply 'allergen-free' end products?
- **l.** Does the business perform any chemical testing upon ingredients?
- **m.** Does the business perform ingredient supplier auditing? If 'yes' please give details including frequencies and documentation. If 'no' does the business employ a third party to perform raw material supplier audits? If 'yes' please give details including frequencies and documentation. If 'no' please give details of how the business controls and assesses raw material suppliers.
- **n.** Please state the minimum and maximum tolerances for temperature control of raw materials upon intake.

9. Storage and warehousing:
   a. Does the business have segregated storage areas for raw ingredients/packaging/finished product?
   b. Does the business have segregated storage areas for chilled/ambient/frozen/dry ingredients/allergens (e.g., nuts)?
   c. If these ingredients are handled, are the following storage areas segregated from each other? Meat/dairy/vegetables?
   d. Does the business have off-site (third-party) temperature monitoring and control of storage areas?
   e. Are all raw materials stored off the floor and away from walls to aid pest control and hygiene inspections?
   f. Does the business use wooden or plastic pallets? (If so, in which areas are they utilised?)
   g. Does the business ensure removal of all primary packaging before the ingredient is transferred to preparation or production?
   h. Are storage and warehouse areas covered by pest control procedures?
   i. Are intake personnel required to change into factory clothing when moving between storage areas/warehousing and delivery wagons?
   j. Are intake personnel required to change when entering preparation or production areas?
   k. Are finished product storage areas completely segregated from the raw material storage areas?
   l. Does the business employ a third-party off-site to store any goods, raw or finished? If 'yes' please state name and whether or not they are audited and at what frequency.
   m. Does the business employ a third party to further process any goods, raw or finished? If 'yes' please state name and whether or not they are audited and at what frequency:
10. Transportation:
   a. Does the business allow the same forklifts and reach trucks to move between outside, warehouse and production areas?
   b. Does the business operate temperature controlled distribution?
   c. Does the business perform temperature monitoring on distribution wagons?
   d. Does the business document the above checks and keep records?
   e. Does the business use wooden or plastic pallets for distribution of finished products?
   f. Does the business utilise a third-party haulier? If 'yes' please state name and whether or not they are audited and if so at what frequency are they audited?
   g. Please state the business minimum and maximum temperature tolerances for transportation of goods.
   h. Please describe the type of wagons that are used by the business, for example, open trailer, enclosed trailer, curtain sided, solid, temperature controlled, ambient, other (please state).
11. Technical resources and systems:
   a. Does the business have an operational quality/technical team on site? If 'yes' please provide the full structure in family-tree style layout.
   b. Are the technical team members formally qualified in their disciplines? Please give details.
   c. Please give details of any ongoing training and development that the technical team receive.
   d. Does the business operate a 'HACCP' system of product safety control?

If 'yes':

- **a.** Does the business have an HACCP team?
- **b.** Are members of the HACCP team formally trained in HACCP and its processes?
- **c.** What is the frequency of the HACCP team meetings?
- **d.** Are all faults and corrective actions documented?
- **e.** How often is the HACCP system formally reviewed?
- **f.** Please list the main process 'CCPs' (critical control points), measurement approach and tolerances.
- **g.** Does the business operate any other systems for the assurance of product safety, quality and legality? Please provide details.
- **h.** Please detail all finished product inspections conducted.
- **i.** Does the business perform microbiological sampling and testing on all products manufactured? Please provide details.
- **j.** Does the business have an on-site microbiology laboratory? Please give details of any accreditation which this microbiology laboratory holds and date achieved.
- **k.** Does the business maintain finished product specifications for everything manufactured/sold?
- **l.** Is the business able to supply certificates of conformance (CoC) or certificates of analysis (CoA)?
- **m.** Does the business have an operational glass policy in place? If 'yes' please state frequency of audits/checks.
- **n.** Does the business have a product recall system implemented?
- **o.** Does the business have a traceability system implemented?
- **p.** Does the business have a quarantine/isolation system in place for non conforming raw materials?
- **q.** Does the business have a quarantine/isolation system in place for non conforming finished products?
- **r.** Are process records archived for later reference/review?
- **s.** Are the following recorded and filed?
  - **i.** Customer complaints?
  - **ii.** Intake rejections?
  - **iii.** Product placed on quarantine/isolation/hold?
  - **iv.** Product removed from quarantine/isolation/hold?
- **t.** Does the business conduct daily/weekly taste panels? If 'yes' please state typical attendees and at what frequency the panels are carried out.
- **u.** With regard to foreign body detection, please state the systems utilised (e.g., metal detector, X-ray, sieving, etc.), the position in the production process and the sensitivities/tolerances within which the product is checked. Please also provide details as to the frequency of checks performed to ensure that the detection system/s are working correctly.
- **v.** With regard to weight control please state the type of check-weighing system utilised (e.g., manual scales, automatic check-weigher) and the position in the production process. Please also state whether documented records are kept for all production runs and please give details as to frequencies of calibration and accuracy checks performed.
- **w.** Does the business hold reference samples of each batch/production run manufactured?
- **x.** Please provide a copy of the most recent government/enforcement authority visit reports.

Other aspects of supply that can be reviewed via an SAQ approach include:

- ethical trading position,
- GM foods presence/policy,
- specific allergen controls and
- environmental policies.

4.6 Specifications

In most regions of the world ingredient suppliers will be expected by law to ensure that their products meet particular quality and safety standards. Usually a food manufacturer will also wish to stipulate far more detailed expectations of the supplier in order to ensure that the materials provided will be as required. The ingredient specification is therefore an ideal document within which to state and agree these requirements.

Ingredient specifications seek to define the key quality and safety criteria of the material, typically including physical, chemical and microbiological standards. Legislative requirements are also often noted directly or indirectly via the required standards stated.

As food businesses use specifications to define to their supplier exactly what is expected of the material being supplied, the specifications can be directly linked to formal contracts of supply. Use of a specified standard highlights the importance of consistency and guards against the potential for misunderstanding and error.

Raw material specifications provide key information on which to base a range for incoming goods checks. The specifications also provide information for utilisation when forming the final end product specification that the raw material is to be incorporated into, for example, when defining the allergen status of the end product and also the overall ingredient declaration.

The content of information in a specification for each ingredient will vary significantly depending on the complexity of the ingredient and potential risk that the food may present.

There are a number of routes/sources of information available when seeking to define an appropriate specification for a raw material. The raw material supplier should supply their own specification and alongside this the business should consider whether there are any other characteristics of the raw material that they would particularly expect to have formally documented in order to ensure that all requirements are clear.

Codex Alimetarius official standards http://www.codexalimentarius.org are often an ideal starting point for the consideration and setting of appropriate raw material specification standards.

Legislative requirements can also be considered via review of relevant enforcement authority information. Other sources of information for incorporation into raw material specifications include relevant sector trade and research associations.

Typical specification content can include:

- supplier name and address (and any additional details with regard to specific manufacturing sites);
- supplier management/staff contact details;
- date of specification and who the document was written/authorised by;
- product name, description and legal name (to be used in an end product ingredient declaration);
- ingredient declaration of the raw material;
- country/countries of origin;
- packaging (including weights, labels and traceability codes);
- shelf life and storage instructions (including storage and shelf life upon opening);
- minimum expected shelf life remaining upon delivery;
- product physical/appearance characteristics, for example, shape, size and related tolerances;
- defects, for example, known potential presence of foreign matter (e.g., soil, insects) and extraneous matter (derived from the animal/plant, but not the part being purchased);
- sensory/organoleptic profile, for example, appearance, aroma, taste, texture and
- manufacturing process information, including key CCPs.

Microbiological criteria. Key organisms of concern and target/maximum limits. Important to also define whether the stated standards are at the start or end of the intended shelf life.

Chemical testing standards, for example: Nutritional information (e.g., energy, protein, carbohydrate, fat, fibre, salt, vitamins & minerals); Speciation tests (e.g., to confirm that the species of meat being purchased is as expected). With regard to meat fat contents a 'chemical lean' percentage can often be provided alongside a 'visual lean' statement.

Allergen information. It is important that this section is related to the relevant 'recognised allergens' in the country that the raw material is to be supplied to and sold within. While the initial focus is on which allergenic ingredients have been intentionally added to the raw material, it is important that the supplier also considers whether there is the potential for unintentional presence of allergens in the raw material (e.g., perhaps due to the nature of the manufacturing process or the source of their raw material supply).

Legislative standards and expectations. Many food businesses when setting raw material specifications will seek to ensure that the relevant legal standards of supply are stated. This is especially important if the raw material is having to be manufactured to a higher level of standard due to being supplied to a different country where standards are higher and therefore necessary to state. One such example may be the stating of maximum residue levels (MRLs) for pesticides, the expectations of which may vary between countries.

It is good practice to ensure that all specifications are subject to document control procedures (including the bearing of an issue number, signed by an authorised representative of the supplier and dated). Also, businesses will typically state that no changes should be made to the raw material specification without advance agreement in writing being received.

Once the raw material specification is in a final form, the specification should receive a final check and sign-off (as approved) by a technical/raw materials team member. This check should be dated/recorded as part of the food manufacturing business's due diligence processes. In addition, as part of these processes raw material specifications should be subject to routine review by the business to ensure that all points remain relevant and up to date.

4.7 Monitoring (including certificates of conformance, certificates of analysis and supplier reviews)

All suppliers to a food business require monitoring to help ensure that the materials/services required are being consistently and correctly delivered. Supplier monitoring, if designed correctly, will contribute to the proactive control of product quality, safety and legality and consequently also further enable the food operation to demonstrate 'due diligence' on such matters.

Raw materials should be checked/inspected to:

- Ensure that the material meets the standards required (as defined in the agreed product specification).
- Ensure that the ingredient does not pose a risk to the end product quality, safety or legality.
- Provide helpful feedback to the supplier.
- Demonstrate due diligence.

Many food business operations handling raw materials will be 'adding value' via their manufacturing processes. The costs incurred by the business related to a final 'end product' which is subsequently found to be spoilt due to containing a substandard ingredient will be significantly higher than the cost of the individual ingredient alone. There are therefore very sound commercial reasons for a business to invest in the proactive monitoring of raw material supplies.

Food business operators will typically seek to inspect all raw material deliveries. Some businesses may seek to risk assess their supply base and place certain 'high-risk' raw materials/suppliers on a higher frequency/extent of inspection than lower risk supplies. The extent and frequency of inspection (including sampling plans) can be influenced by the amount of quality/technical resource available. Businesses must be careful to ensure that this key part of the production process is well controlled as it is usually impossible to achieve the required end product if the raw material supply is flawed.

Incoming goods tests will often involve one or a combination of the following:

- Physical inspection of the raw material to confirm that its physical and organoleptic properties match the specified requirements.
- Review of CoCs/CoAs provided for the specific material production runs/batches delivered to confirm that they are 'as required'. A CoC (certificate of conformance) is a written statement provided by the supplier to confirm that the raw material batch delivered has met all of the expected quality/safety tests. A CoA (certificate of analysis) is a written statement provided by the supplier which details the actual results of the quality/safety

tests conducted upon the delivered raw material batch. Clearly in both cases it is of most importance that the pre agreed tests are selected to be the most appropriate for the assurance of product quality/safety.

- Routine sampling testing of the incoming raw materials, often to a predetermined review schedule, for example, taking microbiological and/or chemical test samples for analysis at a laboratory to monitor matters such as microbiological quality/safety, nutritional standards, speciation checks, allergen absence checks, MRL checks on potential pesticide presence, etc.

A food manufacturer must decide the extent of the incoming goods checks required before the operation able to utilise the raw materials. If the manufacturer requires to await certain samples being analysed and confirmed to be within their specified acceptable tolerances before the raw materials can be used, then a typical approach (to prevent accidental use) is to place the delivered raw materials on hold/ on quarantine to await 'positive release' which can be progressed once the acceptable results are received.

To enable such an approach the business purchasing systems must ensure that the raw material is ordered and delivered well in advance of the actual date that it is required for the production process. This is in order to allow the required checks to be conducted prior to its use. The business is also well advised to check that the ingredient shelf life remaining at point of delivery allows for such time delays while the raw material is being tested.

Raw material physical check criteria should reflect the agreed requirements defined in the product specification. Examples may include:

- Meat products: Delivery temperature, fat content, gristle definition and allowance, measurement of dice/slice/mince size, organoleptic (appearance, aroma, taste, texture, colour) testing of a cooked sample.
- Vegetable products: Delivery temperature, skin presence (if a peeled product), measurement of dice/slice size, amount of off cuts/'fines', organoleptic testing.
- Dairy products: Delivery temperature, organoleptic testing including colour, viscosity, flavour. Viscosity checks may utilise specialist pieces of measurement equipment such as Brookfield or Bostwick Consistometers (especially if the raw material has a significant impact on the end product viscosity). Colourimeters can also be utilised in raw material testing where there is a particularly high degree of sensitivity required to ensure that measurements of product colour are precise.
- Herb and spice products: Confirmation that the product is 'free flowing' (if defined as required). Checks on microbiological loading. Ash tests to guard against product adulteration (if such tests have not been conducted by the supplier and provided in the form of CoAs/CoCs).

Raw material inspection can also occur at earlier points in the supply cycle.

'Pre supply' is the time when the business is negotiating with the supplier with regard to grade/quality of raw material that is required/available, and at what price. At this stage it is important to assess initial samples from the supplier to help ensure that both parties are clear on the required standards/specification.

It is good practice to document this review and agreement process in order to be able to demonstrate due diligence with regard to the control of this important process. Reference samples of the 'approved ingredient' should then be kept for future comparison when the first main deliveries arrive (to allow for shelf life issues perhaps the samples will need to be frozen, or if this is not possible then at least clearly documented with test measurements and digital photographs).

Upon commencement of supply raw material checks/inspection can occur 'before the ingredient arrives' at the manufacturing facility. Checks for compliance to specified requirements can be arranged to be proactively conducted by the supplier in advance of delivery. The satisfactory completion of these checks can then be corresponded via the provision of CoAs/CoCs with (or in advance of) the raw materials arriving on site at the manufacturing operation. This approach sits well with a 'right first time' quality assurance principle as such checks if conducted correctly will help to avoid increased time and financial costs if rejection were to occur at the point of delivery.

Supplier assurance measures can help a food manufacturer to ensure that their supplier's systems of product control are sufficient to achieve a consistently correct supply. Such measures can include supplier audits (or SAQs), working to documented and agreed specifications, a policy of clear dialogue and communication with the supply base, all coupled with a business policy to only deal with reputable businesses.

'Pre supply' meetings and also on the occasion of a supplier audit are an ideal time to discuss and explain to the supplier the particular requirements of their raw material and which aspects are specifically important to the food manufacturer as their end customer. If these discussions are also held at the raw material supplier site then the relevant quality and safety checks can be reviewed and discussed first hand, further increasing the likelihood of the raw materials being supplied 'right first time'.

'Post delivery' checks. A food processing business may choose not to fully inspect certain ingredients upon delivery as the check process may compromise the protective barrier and related shelf life effects provided by the original sealed packaging, for example, canned goods, vacuum packaged goods and dry goods sealed in airtight containers. In such cases these goods will benefit from a quality assurance system which triggers a quality check at the point of opening rather than at point of delivery. For example, staff awareness to conduct quality assurance checks on opening a new batch of raw material, or a computerised reminder system on the new lot codes/batch does being drawn from, can both be equally effective.

'Checks during storage'. From the point of delivery, an ingredient may be placed in the food manufacturing business stores awaiting use for a timescale anywhere between a number of minutes and a number of years. Typical ingredient shelf lives on delivery for fresh diced/sliced vegetables – 3 days chilled storage, fresh chilled meats – 3 days (or approximately 10 days if vacuum packed), frozen ingredients – typically 12–24 months, ambient ingredients (e.g., canned goods, dried powders/herbs and spices) – 12–24 months.

Significant product changes can occur over shelf life. Changes affecting food quality and safety performance can include:

- Microbiological spoilage and growth of any pathogens present (which can be accelerated by incorrect storage/contamination).

- Dehydration (e.g., frozen ingredients can suffer 'freezer burn').
- Moisture ingress (can lead to accelerated spoilage and powders ceasing to be 'free flowing').
- Organoleptic changes. Appearance, aroma, taste and texture could each vary over time, especially with regard to refrigerated fresh produce. Effectively the product structure is deteriorating during shelf life.

Incorrect or inappropriate storage can lead to the product being exposed to foreign body, microbiological or allergen contamination risks. Therefore there is good reason to routinely review the condition/specification compliance of the ingredients in storage/awaiting use within a food production operation. Such checks can often be a visual inspection by the production area or technical staff while also monitoring to ensure that ingredients have not exceeded their use-by/best before and therefore gone 'out of date'.

With regard to the frequency of stored ingredient checks, similar to the situation with incoming goods checks a 100% inspection/monitoring approach every day (perhaps even every shift) is labour intensive but will provide a high degree of assurance that all raw materials remain fit for purpose and are used within their shelf lives.

A stored ingredient check frequency based on a 'risk assessed' approach is likely to suggest that chilled, perishable ingredients (typically those bearing 'use-by' dates rather than 'best before' dates) are the most likely to encounter variation in performance during storage. In such cases a food manufacturing business may choose to assess the quality compliance of chilled perishable goods once per day, whereas for stable, longer shelf life ingredients, these may only be physically checked for example once per week.

Often businesses will incorporate these stored ingredient quality checks into a routine stock take for the ingredients stored within each area.

4.7.1 Raw material check records/documentation

When inspecting raw materials at the intake and storage stages, to ensure the consistency of the quality assessment approach taken, written procedures should be trained out to the relevant "checking staff" and should be available at the inspection point covering the frequency and method of inspection expected. To compile formal business due diligence/audit records all ingredient checks should be recorded on dedicated 'check forms'/'quality records' which should be document controlled and guide the user as to the expected range of checks to be progressed, and the acceptable tolerance limits of the check results.

If a particular check fails, the raw material should be clearly marked as 'on quarantine'/'on hold' to avoid accidental use while the appropriate rectification action is considered. The right action to take may be to reject/dispose of/sort/seek further assurance/use. The key point here is that such matters should be considered and decided upon by an appropriately qualified and experienced member of the business quality/technical staff who may also call on the expertise of the raw material supplier, related trade associations or other relevant information sources.

The raw material check records of a food business should be routinely audited to ensure that these inspections are being recorded clearly and that the required check frequencies are being adhered to. These records should be securely stored in date order as they will need to facilitate a rapid review in the event of an issue being raised.

4.7.2 End of shelf life/out-of-date raw material management

In many food operations, the cost of disposal of raw materials that have passed their stated shelf life and have not been used can be very high indeed. Ingredients typically go out of date due to one of the following reasons.

- The business has to order too much at one time due to high supplier MOQs.
- Predicted production volumes/orders are coming in lower than expected.
- The raw material has been delivered with less shelf life remaining than expected.
- Raw materials are being used in the incorrect order (e.g., perhaps 'FIFO' (first in first out) has not been adhered to).

All factors which result in out-of-date ingredients need to be focussed on by a business to reduce the volume/cost impact. Short shelf life ingredients are understandably the most prone to going out of date due to their smaller window of opportunity for use compared with longer life products which typically bear a 'best before' date (e.g., canned, dry or frozen goods).

Ingredients that are about to go out of date, if highlighted in time, can sometimes be addressed by the food manufacturer in the following ways to help avoid the write-off costs of disposal.

Out-of-date ingredient options include:

- Request a shelf life extension in writing from the supplier (perhaps the supplier knows that the raw material will actually last longer than the date provided on the label).
- Organoleptically assess and approve the ingredient for use internally.
- Microbiologically assess and approve for use.
- Freeze stock in advance of going out of date, thereby extending shelf life.

All of the above options need to be managed carefully as they can add complications from a process control or audit compliance/traceability viewpoint. Use of clear procedures which demonstrate due diligence and operational control are vital to underpin such cost saving approaches.

4.7.3 Calibration of inspection equipment

Incoming goods checks such as temperature, pH and viscosity rely on analytical pieces of equipment. Over time such equipment can fail to record accurately and therefore may mislead the operator into believing that a check is 'within specification' when actually it is not and vice versa. It is therefore vital that all test equipment is maintained in good working order and regularly calibrated and serviced.

For example, hand probes utilised in operations where temperature control is vital. Typically, these may have their calibration/accuracy checked once a day against a thermometer or other temperature/signal source which is traceable back to a national standard (and checked against this standard once per year). Another assurance is to check that the probe/thermometer reads 100 degrees centigrade in boiling water (corrected for atmospheric pressure in certain areas) and 0 degrees centigrade, the temperature at which ice melts, in a mixture of water and ice.

4.7.4 Dealing with non conformances

With regard to the management of situations where a raw material is found not to conform to the required standards, in an ideal world the raw material should be returned to the supplier and a request should be made that the supplier replaces with material that does meet the required standards.

However, in a food manufacturing setting sometimes circumstances are more complex than this. Perhaps if the ingredient is rejected then the factory cannot produce as the raw material supplier requires some time to resupply. Perhaps the ingredient does not meet the business expectations, but does not actually breach the specific requirements stated in the agreed specification. Perhaps the ingredient was inspected and accepted at the point of delivery, and only at the point of use is it now discovered to not meet the standards required.

In all such situations a 'non conformance' decision needs to be made. The appropriate business quality/technical team member must judge each individual situation, considering factors such as: How soon can a replacement material be sourced? Can the material be substituted for by another similar material? (there may be matters such as ingredient declaration, nutritional and allergen issues to be considered here). If using the raw material what will be the likely impact on the end product? Can the production process be adjusted in order to compensate for the particular ingredient issue? What will be the financial impact to the business of using/not using the raw material? What is the likelihood of being able to recover compensatory costs from the raw material supplier?

4.7.5 Recall and damage limitation

Sometimes a problem with a raw material may only come to light once part or all of the affected batch/lot has been used in the manufacture of end product/s, which may in turn already have been despatched to the end customers. In these circumstances the appropriate quality/technical team member needs to assess the situation quickly in order to then take the appropriate decision which, dependent on the nature and extent of the problem, may entail conducting a trade withdrawal or full product recall. Swift and accurate traceability systems will be of great help in these circumstances, as will a clearly detailed and up-to-date product recall procedure.

In these circumstances, as there will be significant exposure and impact on the onward supply chain/end customers it is likely that the food manufacturer will require their customer account managers and their supply chain/logistics managers to be involved in the withdrawal/recall process to expedite the process effectively while also providing additional assurance that matters are in control and measures will be taken to prevent any recurrence.

4.8 Traceability

European Union (EU) Food Law (2002) defines 'traceability' as 'the ability to track any food, feed, food-producing animal or substance that will be used for consumption, through all stages of production, processing and distribution'.

Under EU General Food Law operating clear and effective 'traceability' systems is compulsory for all food and feed businesses. Specifically, the legislation requires that 'all food and feed operators implement special traceability systems. They must be able to identify where their products have come from and where they are going, and to rapidly provide this information to the competent authorities' (EU General Food Law 2002).

Traceability systems support the control of product safety and enable a rapid response in the event of any major issues of concern. Such systems ensure that when the food business or the relevant regulatory authorities identify a particular risk, they will be able to trace the issue back to its source, enabling swift isolation and investigation of the problem, and can take the necessary steps to prevent contaminated products from reaching consumers (e.g., clear identification of the affected batches/date codes will allow targeted product withdrawals/recalls via the provision of accurate information to the public and the relevant supply chain operators, thereby ensuring effective reduction of impact to public health and minimising the disruption to trade).

The assurance of traceability also has major quality assurance benefits. In the EU there is 'sector-specific' legislation applied to particular categories of food products including fruit and vegetables, beef, fish, honey and olive oil. In these cases, the additional legislative requirements are applied so that consumers can identify the origin and authenticity of the particular foodstuffs.

With particular regard to GMOs (genetically modified organisms), the EU apply specific traceability rules designed to ensure that any GM content of a food product can be fully traced and also requires GMO presence labelling on the product packaging in order that consumers can make an informed choice as to whether they wish to consume such materials.

Animal welfare is another aspect of EU Food Law that benefits from clear traceability systems. In the EU meat producers must 'tag' every animal with details of their origin, and these details must follow the animal through the slaughter process, typically via the traceability code stamp of the abattoir. The particular codes and methods of attachment/marking used to achieve such traceability may vary between manufacturing sites and countries; however, the end result of all systems must be the same in that the traceability information required is consistently achieved/recorded.

4.9 Specifics of food packaging supply

Food packaging is typically derived from metal, glass, paper and board or plastics, and sometimes a combination of these. It is a key component/raw material required for most manufactured food products, with the packaging serving the '3 P's' of packaging function to (1) protect, (2) promote, and (3) preserve.

To maintain food in the required condition the packaging serves a key role in protecting against physical (insect, metals, glass, stones, etc.), microbiological (pathogens and spoilage microorganisms) and chemical (toxic or tainting) contamination. The packaging can provide different types of 'barrier'. Many packed products require a barrier against gas, moisture, light and aroma ingress/transfer to maintain their quality throughout the shelf life. Gas and moisture barrier capabilities are typically measured as levels/rates of 'transmissibility'. Tamper evident seals can also be seen as a 'physical barrier' against malicious contamination.

When considering gas barriers, it is important to consider what function the packaging has to fulfil. In some cases the packaging will need to be permeable to allow gas exchange, for example when considering respiration and ripening of fresh fruit and vegetables.

With regard to physical formats, the packaging structure may sometimes need to be rigid to help maintain the form of the product during storage and transit, for example, delicate cream cakes.

Packaging must meet relevant legislation requirements. In the EU this legislation expects that food packaging components must not be significantly transferred into the food during normal shelf life. For example the EU Plastics in Contact with Foodstuffs Directive: EC Directive 2002/72/EC and its amendments lay out in great detail what is required to ensure that plastics are safe for use in contact with foods.

Packaging must often convey information which is also linked to 'legal requirements' including:

- clear and accurate name of product;
- ingredients declaration (in the EU 'QUID' – Quantitative Ingredient Declaration);
- relevant warning statements;
- contents number/weight/volume;
- storage requirements and
- durability (use-by/best before).

The management of packaging suppliers and the application of related specifications is therefore of relevance to both ingredient and food packaging suppliers. In some food businesses there will be the resource of a formal 'packaging technologist' setting and overseeing packaging specifications to meet the requirements of the business, conducting assessments of proposed new packaging through the factory handling trials, shelf life trials and transit trials, and administering the day-to-day supplier review practices required. In other businesses such duties are picked up by members of the general quality/technical team.

4.9.1 Food packaging specifications

Packaging specifications should confirm the exact expectations of the food manufacturer (the 'packer/filler') while documenting the processing and product standards of the packaging supplier (the 'converter'). The specification for each packaging material will also serve to provide a checklist of tests/requirements of the packaging manufacturing process and also form a list of criteria to be checked against upon delivery to the food processor.

Specifications also help to ensure consistency between 'dual sourced' materials (where two different suppliers are expected to supply the same end product), and also ensure a 'level playing field' in the event of the business tendering for a new packaging supplier. They form a contractual obligation between parties and serve to demonstrate due diligence with regard to quality, safety and legality control of supply.

A technical team member of a food processing business giving consideration to the required specification content for an item of food packaging would benefit from the following range of likely considerations:

- The packaging manufacturer's standard product specification. Reputable, established suppliers will usually already have a standard specification for the item of packaging under consideration. This specification may well contain all of the aspects required for the food business to be satisfied.
- Occasions where the standard supplier specification may not be sufficient include on matters where the specific end use of the packaging is different to that of other end users. For example, a polypropylene pot may have been supplied to cold-fill dips businesses for many years; however, the latest end user wishes to utilise the pot for a hot-filled product, or perhaps for a product that is going to need to be microwaved in the packaging before consumption. Specification considerations now need to cover the stability/performance of the plastic packaging at high temperatures. Such matters may also require the packaging to undergo more tests with regard to its suitability for food contact/migration testing, as at high temperatures many plastics will often react very differently with the end product.

It is helpful to define which factors are critical to the performance of the packaging through key 'phases' including:

- the intended 'manufacturing' (e.g., packing/filling) process;
- the intended movement, storage and distribution chains (before and after the food manufacturing phase) and
- storage and use by the end user/consumer.

Such factors can include the following.

The product needs to run 'correctly' through the packer/filler process:

- Perform consistently at the required food packing line process/running rates.
- Remain intact through the rigours of process movement, temperatures, pressures, etc.
- Provide a protective barrier to the product which will not be adversely affected by the product or the process.
- Achieve and maintain the desired aesthetic appearance (e.g., no scratches, dents or product drips/smudges evident post process/distribution).

With regard to the intended movement, storage and distribution chains, the packaging will be required to:

- Maintain the required performance for the duration of the shelf life.
- Be sufficiently robust to withstand the physical pressures of the storage and distribution processes.
- Maintain performance within the likely range of storage and distribution temperatures.

With regard to the end consumer, the packaging needs to:

- Optimise sales performance on retail shelves (e.g., be eye catching and provide good clear information).
- Make the product easy to open (while also achieving 'tamper evidence' can also be a challenge on such points).
- Make the product easy to use? (e.g., 'easy pour' features, microwavable packaging).

Specified features of the primary packaging could therefore include:

- specific material type/types;
- physical dimensions;
- weight of pack/materials used;
- colour/artwork/print/label requirements;
- tamper evidence features;
- defined 'barrier properties' (e.g., gas/moisture barrier. Transmissibility figures for oxygen, moisture vapour, etc.);
- robustness/stress tests including flexibility, puncture resistance and tensile strength, stretch/elongation values;
- suitable processing temperatures (e.g., pack filling, processing and distribution temperatures (both hot and cold), and
- end use (e.g., suitable for Microwave? Ovenable? Home freezing?).

4.9.2 Tolerances

Like all manufacturing processes, the packaging will have an element of variability and therefore, where appropriate, tolerances should be listed on the specified criteria. Caution needs to be taken when considering the acceptable targets and limits on such matters. For example, when two packaging components are being utilised to form one item of packaging, such as a 'base' and a 'lid'. In this case 'lid fit' may be a major issue if both the base and also the lid dimensions can vary. For example, in the case of a polypropylene pot 'lid' and 'base', the packaging manufacturer may have a specified size diameter of 100 mm $\pm$ 1 mm in their injection-moulding production process. Therefore, if by chance when producing the packaging the lid happens to be manufactured on the 'small side' of the acceptable tolerance limits and the base happens to be manufactured on the 'large side' of the acceptable limits, then there is the potential for the lid to be 99 mm in diameter and the base to be 101 mm. In these circumstances will the two pack components still fit together tightly if both ends of the tolerance range are encountered during a production run?

If tolerances are set too wide, processing/performance issues may occur. However, if tolerances are set too tight, the cost of the packaging may significantly increase due to the supplier operational costs incurred in achieving the requirements, and/or due to the number of potential capable suppliers being reduced.

4.9.3 Measurement of key specified criteria/values

To ensure clarity and consistency it is vital that 'units of measurement' and 'methods of measurement' are agreed between the packaging supplier and the packer/filler. Typically for many key points of measurement packaging manufacturers will use standard methods defined by organisations such as British Standards Institute (BSI), International Organization for Standardization (ISO), and American Society for Testing and Materials (ASTM). Where a business has defined its own unique measure for a specific point it is important that the test method and approach is very clearly defined to enable repeatability of the test over time.

The likely range of packaging-related specifications held by a food manufacturing business includes:

- primary packaging (e.g., food contact material; base and lid (closure));
- direct label/print specifications;
- extra promotional labels;
- label adhesive specifications;
- case (typically corrugated board and possibly perforated to be 'shelf ready packaging');
- palletisation (type, stack method, dimensions);
- stretch wrap and
- distributor specific requirements.

4.9.4 Achieving the required 'quality' of supply for food packaging

Securing supply of the required quality of packaging could be defined as seeking 'conformance to the stated requirements'. As previously covered in this section written specifications are an efficient way of communicating such 'requirements' between the packaging supplier and the food business.

Dealing with reputable packaging suppliers who have a track record of manufacturing the type of packaging required is likely to provide significant advantages. Accredited/certified suppliers to particular standards can also provide additional assurance, for example, ISO 9000 Quality Standard and the BRC Global Standard for Packaging.

4.9.5 New packaging acceptance testing and packaging QC/QA

'Acceptance testing' is conducted by a food business on newly supplied materials in advance of the food business then accepting and regularly using the new item of packaging. The purpose of this acceptance testing is to confirm that the packaging is 'fit for purpose', proactively ensuring that if any problems are encountered then at least their business impacts will be minimised.

When setting the criteria for acceptance testing it is helpful to review the main packaging functions:

- contain,
- protect,
- promote and
- preserve.

New packaging acceptance testing typically includes:

- supplier specification content review (including delivery format, pack dimensions, barrier properties and details of food contact approval/migration testing);
- end user/consumer testing (with regard to use/handling and appearance/information);
- sales department acceptance;
- retailer acceptance (if own label);
- process line trials (to ensure the product runs on the intended lines in standard operating conditions);
- transit testing;
- shelf life testing and
- artwork checks (approval of graphics/text).

To facilitate a check of the packaging performance against specified requirements there is the need for a defined packaging specification which documents all key aspects of the product. Packaging supplier quality control systems and procedures can also be drawn upon when seeking to define the range of appropriate checks.

Example checks that could be conducted by the manufacturer/proposed user include:

- pack integrity (e.g., confirming sound structure, thorough seals, effective tamper evidence);
- label adhesion (e.g., correct alignment and thoroughly affixed);
- extent of gas (e.g., oxygen, carbon dioxide) and water vapour barrier transmissibility (if the required specialist test equipment can be accessed);
- shelf life tests (which can involve incubators with programmable temperature and humidity cycling) and
- possible challenge testing with known pathogens.

Primary packaging tests could also include:

- burst strength;
- compression resistance;
- impact resistance;
- opening forces required (i.e., closure release torque);
- headspace analysis (i.e., measuring the gas composition in the headspace of the pack);
- production trials (i.e., running/testing the packaging through the production process);
- 'transit trials' (i.e., running/testing the packaging through the distribution process) and
- end product shelf life trials (both microbiological and organoleptic).

Trials should be conducted in actual production conditions, with 100% correct products/materials; otherwise there will be uncertainty as to whether the materials will meet the requirements of the specific process.

Only when all aspects of the packaging have been confirmed to be 'fit for purpose' should the product then be purchased for use.

4.9.6 Food business packaging delivery checks and performance monitoring

A business cannot create 'good quality' by inspecting the packaging product on intake, but it can ensure that it selects a reputable supplier with good operating systems, who is clear upon the specific packaging requirements (e.g., via a clear dialogue on business/supply requirements, underpinned by the utilisation of product specifications).

Ongoing packaging delivery checks need to cover the basic performance features required of the packaging and the approach will typically reflect the amount of relevant testing equipment available at the end user food business site. The extent and frequency of the ongoing delivery checks will also be dictated by the level of risk to the food business and its perceived likelihood of an issue occurring (i.e., the testing requirements should be set on a 'risk assessed' basis).

Incoming packaging inspection could include:

- confirm that the packaging is delivered on a hygienic and structurally sound vehicle;
- packaging securely contained within the required outer wraps/cases;
- correct dimensions (including micron thickness of the packaging);
- lid fit/seal check (plus ease of opening/reclosing);
- tamper evidence present and effective;
- volume/weight fill check (fill level visually acceptable);
- check for contamination/imperfections;
- hot-fill/cold-fill check (to ensure correct stability during standard processing temperatures);
- chilled/frozen storage;
- correct pack copy... graphic/artwork check + confirm that the bar code / QR code scans ok;
- scratch test of surface (to ensure print is sound and any varnish/coating is firm), and
- traceability (ensure via recording of data/codes that all packaging is traceable back to source).

Performance monitoring tests could include:

- Routine formal review that the packaging is running effectively down the food manufacturing production line at the rates required. (It is likely that the production team would highlight if there were packaging-related problems occurring in the process areas, but this is not always guaranteed).
- Application of a range of tests/equipment to facilitate both online and offline pack inspection systems during the use of the packaging, including checks for:
  - empty containers (e.g., missed the filling stage);
  - imperfections/mis-shapes (e.g., use of vision systems);
  - fill levels/weights (e.g., use of dosing/weighing systems);
  - metal/glass detection (e.g., foreign body control systems);
  - seal quality (e.g., vacuum achieved? Airtight?);

– label presence and location (e.g., scanner/vision system plus could be confirmed correct via bar code check), and
– tamper evidence present.

4.9.7 Online versus offline inspection

4.9.7.1 Online (continuous) automatic packaging check systems offer/considerations:

- 100% inspection (compared to statistical sampling offline which only reviews a smaller percentage).
- Instant results.
- Need to ensure good calibration/maintenance of test equipment to avoid reliability issues.
- Additional capital cost.
- May affect line efficiency (may have to slow the line to be able to inspect every pack; may have to routinely calibrate; can sometimes break down).
- Requires appropriate machine operator training. Operator experience and control is a vital factor. The operators must ensure that the machines are run at the correct rates and are constantly provided with replenished packaging, and that online inspection systems are correctly set up and monitored.

4.9.7.2 Offline inspection systems offer/considerations:

- More time for detailed inspection when the pack is taken offline.
- Can facilitate 'destructive testing' where the packaging will be damaged/destroyed by the necessary test.
- Lower cost (as just taking a smaller 'representative' sample).
- Taking and assessing samples offline is not 100% inspection, but is demonstrating some due diligence.

Offline tests can include:

- 'torque meter' to confirm caps are on to a sufficient degree of tightness and
- leak detector for MAP/'airtight' sealed packs (to check that the pack atmosphere created is not being lost due to faulty packaging/seals).

4.9.8 Packaging troubleshooting

When utilising packaging in food manufacturing operations it is likely that at some point a problem/issue will be encountered. In such circumstances it is helpful to first seek to clearly identify/define the problem as anecdotal evidence/information can be very misleading.

The food business is likely to be keen to keep the production line running while the matter is resolved if at all possible. To help this be the case considerations could include (1) reducing packing speed, (2) switching to alternative packaging and (3) switching to a different product.

The technical team should investigate the matter, collecting data and operator feedback. All of this information should then be analysed, looking for coincidences, timing, symptoms, people, etc. Could there be one person causing the problem with incorrect set-up? Does a particular batch of packaging or product type tend to always encounter an issue?

Online the factory operators will have extensive experience of the machine and materials. They may also have prejudices or make false assumptions, perhaps due to not liking a particular product/supplier.

It is often a good course of action to call in the packaging supplier as swiftly as possible to review the issues being encountered and to advise accordingly. The supplier will know whether they are also encountering similar issues elsewhere (or perhaps they have encountered such an issue in the past). They also should have an excellent insight into the packaging and its performance/tolerances.

When speaking to suppliers to resolve problem issues it is important that both parties approach the dialogue in an open and proactive manner. Seeking to quickly blame a supplier for an issue can make them immediately defensive and in such circumstances they may not readily share information that may be helpful in resolving the matter quickly. Also, it is important to watch out for 'buck-passing' where a supplier seeks to blame everyone else (the food business, the machinery, the product) before accepting that the packaging fault is their responsibility.

If ever struggling to resolve a packaging-related matter it can be beneficial to 'hire in' a short-term packaging consultant to impartially focus on the issues and guide all parties to a resolution.

4.9.9 Future packaging considerations

Over the coming years there are likely to be a number of trends with regard to the use of food packaging by food manufacturing businesses, many of which can be interrelated. These trends are likely to impact on food businesses as they have the potential to drive packaging changes which may require new machinery, test methods or running systems in order to facilitate.

- Ongoing drive to optimise and increase pack shelf life via packaging structure and/or modified atmosphere in the pack.
- Increasing environmental and waste management focus.
- Seeking to reduce the amount of packaging used and related energy use (including pack 'light weighting'* initiatives).
- Seeking biodegradable and recyclable packaging options.
- Regular presentation/information/graphics changes to keep consumers interested/engaged.
- Increased use of composite materials to give additional properties/benefits.
- Developing use of nanotechnology. Packaging that incorporates nanomaterials can be engineered to have specific properties. The vast potential includes response to environmental conditions; self-repair; increased barrier properties; improved heat/cold resistance; antimicrobial/antifungal properties; alert the consumer to contamination/excessive age/temperature abuse/presence of pathogens.

*Light weighting (redeveloping the product packaging currently used in order to use less material and consequently make the packaging lighter/cheaper) continues to be a major project across many businesses in the food industry in order to become more financially lean, while also showing environmental awareness.

4.10 Summary

There are many measures that can be taken in order to ensure that the raw material supply to a business consistently meets quality, safety and legality requirements. Food processing businesses need to ensure that they recognise where the potential ingredient risks are, and have systems and procedures clearly in place to monitor and control such issues.

Decisions on non conforming materials must always be taken by trained and competent personnel with safety and proof of due diligence always at the forefront of all decisions taken.

With regard to standard raw material controls in the food manufacturing sector over the next generation there are likely to be ever present and increasing 'pressures' on the supply base which may translate into a disrupted or flawed supply if not kept well controlled and monitored. These supply chain pressures include:

- Continued price pressures on suppliers driven by the retailer and food service groups as they respond to their profit drivers and pressure from the consumer seeking 'value' and willing to shop around to get the best deals.
- Continued increases in audit standards and food safety legislation driven by concerns over food safety incidents across the global food supply chain.
- Suppliers must continue to raise operational and food safety standards in order to maintain the required audit accreditations (e.g., BRC Global Standard for Food/Packaging Standard, Organic Standards, etc.).
- Generally increasing costs of energy and water (utilities).
- Drives for automation to reduce labour costs and increase efficiency.
- Reduced capital investment for new factory structures/equipment.
- Limiting building and process line maintenance budgets.
- Increased focus on 'lean manufacturing' techniques and ergonomic area designs.
- Environmental awareness. Pressure on the supply chain to reduce impacts such as carbon footprints and air-miles, may lead to reduced supply options or increased costs (as suppliers calculate and pass on the true costs of doing business).
- Fair trade/ethical trading. Increased awareness upon such topics will gradually see a re - costing of the supply chain with further finance finding its way to the primary producer/grower. This may well increase price pressures upon the raw materials supply to a food manufacturer.

Consequently, food businesses are likely to encounter a lot of change taking place across the operations of suppliers as they seek to ensure that they respond to these issues and remain competitive. Therefore, all such changes require monitoring to ensure that they do not adversely impact upon the service or quality of supply.

References

EU General Food Law. Reg EC No. 178/2002. https://ec.europa.eu/food/safety/general_food_law_en

EU Plastics in Contact With Foodstuffs Directive: EC Directive 2002/72/EC. http://eur-lex.europa.eu/legal-content/EN/TXT/?uri = LEGISSUM:l21301

EPA (Environmental Protection Agency) USA – Website has extensive information and notes on production processes including across Food and Agricultural Industries. https://www3.epa.gov/ttn/chief/ap42/ch09/index.html

Publicly Available Specification PAS 96:2017 Guide to protecting and defending food and drink from deliberate attack. https://www.food.gov.uk/sites/default/files/media/document/pas962017.pdf

Further reading

ASTM Standards. https://www.astm.org/

British Standards Institute. www.bsigroup.com

Codex Alimentarius – International Food Standards. http://www.fao.org/fao-who-codexalimentarius/en/

Food Packaging: An Introduction. Tim Hutton. Campden Publications (2003).

International Organization for Standardization. http://www.iso.org/iso/home.html

Process & Packaging Machinery Association. http://www.ppma.co.uk

Plastics and Flexible Packaging Group. http://www.bpf.co.uk/packaging/Default.aspx

Raw materials and ingredients in food processing. Key Topic 11. Campden Publications (2005).

Site standards

5

Mark Swainson
National Centre for Food Manufacturing, University of Lincoln, Holbeach, Lincolnshire, United Kingdom

5.1 Importance to technical management

In the food manufacturing sector the root cause of many food quality and product safety faults can often be linked in some way to 'site standards'.

For example:

- Damage to floors or walls can cause the formation of microbiological reservoirs, harbouring pathogens and food spoilage microorganisms and thereby posing a persistent product contamination threat within factory areas.
- Peeling paint, loose fabric or poorly maintained equipment can lead to foreign body ingress into the food products being manufactured.
- Food processing equipment could cause product consistency issues due to faults with key food safety control features such as sensors or timers.
- Even world-class food safety systems can still be undermined by a lack of sufficient staff control. In the fast moving food manufacturing sector one small mistake by a staff member can sometimes escalate into a full-scale product recall incident. Poor factory standards can often cause poor staff morale which in turn can lead to a lack of attention to detail and consequentially raise the potential for mistakes to be made.

The technical manager is therefore well advised to focus particularly closely on 'site standards' when seeking to fulfil the wide management remit of working to assure product safety, quality and legality.

From a technical viewpoint the factory process flows, fabric and equipment must all be conducive to staff being able to progress their roles in a controlled, competent and hygienic manner. It is very difficult to set effective business standards and systems if working from the weak foundations of a site in poor condition, or if using substandard equipment.

Factory fabric and working conditions are likely to impact on the 'Food Safety Management Culture' of the organisation, which relies on staff feeling positive about their work and having a sense of care and diligence as to how they conduct their duties. Poorly maintained or substandard processing equipment will lead to variable product quality and in many cases the increased potential for food safety issues to occur.

In business environments where food processing areas or equipment are of a continually poor standard it can also be very frustrating for technical and quality staff to be raising the same area shortfall points each time an audit is conducted. In such circumstances production staff will often also feel demoralised when seeing

Swainson's Handbook of Technical and Quality Management for the Food Manufacturing Sector.
DOI: https://doi.org/10.1016/B978-1-78242-275-4.00005-8

their area audit scores constantly reduced beyond their control due to poorly maintained processing area equipment and fabric.

A business may have commenced their operations with good factory conditions and well-maintained equipment and it is to be expected that over time materials and equipment will deteriorate. This is one reason why the senior management of the site must ensure investment in site refurbishment and provide equipment maintenance budgets alongside related systems of site review and repair and preventative maintenance programmes.

It is therefore no surprise that most food quality and safety accreditation systems place a high expectation on the requirement for 'senior management commitment'. Appreciating the major impacts which the senior management can have on the site food safety management systems and on the food safety/quality culture of the organisation.

In some circumstances factories and equipment can become utilised for products which they were not originally designed to process, leading to potential threats to product safety or quality. For example, perhaps a piece of packing equipment that was designed to only pack low-risk products may now be being used in a high-risk area to pack a very different product. As a result, the 'make-do machinery' may well not be sufficiently designed to be fully cleanable to an appropriate high-risk area standard, thereby potentially leading to product safety or quality issues.

Customers, audit standards and legislation also tend to increase their expectations over time, and with the documentation of such requirements becoming increasingly prescriptive there is little tolerance for any shortfalls in operating standards.

5.2 'Designing in' and maintaining appropriate standards

Technical managers are well advised to ensure that they are always involved at an early stage when the business is considering new site fabric, machinery or equipment (or significant revisions to the existing operation). To secure a high level of involvement will often mean working closely with the engineering and production management, taking the time to explain the importance of hygienic design and additional technical validation requirements linked to any new machinery.

The technical manager will often be expected to provide information on specific requirements such as legislative compliance on matters of product safety, quality and the control of weights and measures. They may also provide detail as to the expectations of any applicable customer codes of practice. It is always advisable to proactively incorporate all such requirements into the design and development process at an early stage.

The complexity of some machinery/equipment can often raise the potential for product safety or quality issues which are unforeseen and only come to light when in conditions of 'full production'. Therefore ideally such equipment will be pretested by the business in advance of full installation and commencement of use at the manufacturing site. Completing advanced testing while the machinery is still at the supplier site can make the progression of any necessary modifications far easier to achieve swiftly.

Prepurchase or preproduction tests of equipment and machinery can include:

- Hygienic design and 'cleanability' assessed by both visual and surface swab inspection (perhaps including allergen clean-down efficiency testing alongside microbiological or ATP swab testing).
- Running at the required throughput rates to confirm machinery capability to achieve consistent and correct product quality during standard and peak operating conditions.
- Check that any product safety parameters are consistently achievable and subject to full validation/assessment.

These tests can often be completed while production and engineering departments are also assessing factors such as machine running speeds or ease of use, and while the engineering department is also testing machinery build quality, serviceability and checking the availability/order lead times of necessary spare parts.

Technical managers should try to form a strong 'triangle of control' on such matters working alongside the business production and engineering functions. At the outset of a project each team member should communicate clearly what they require to achieve ('success criteria') and the intended timescales. On encountering any issues, the relevant project team member should swiftly inform their colleagues and agree how the matter is to be raised and rectified with the machinery/equipment supplier.

Learning from issues encountered by others is a valuable proactive strategy. Many potential problem issues can be foreseen by requesting feedback on the design and performance of the particular type of machinery/equipment from other food manufacturing businesses that are already using the particular product.

Food manufacturing sites usually have a site maintenance budget allocated to ensure that repairs are swiftly and continuously addressed. Use of internal and external audit data (and also sometimes product complaint data) can be very helpful in highlighting where site fabric or equipment requires improvement, and consequently can steer the business with regard to the prioritisation of where the site maintenance budgets require spending.

Maintaining appropriate factory standards can be encouraged via:

- application of a 'clean as you go' policy for all factory areas,
- development of hygiene procedures and cleaning schedules,
- preventative maintenance programmes, and
- site audit schedules and environmental monitoring programmes.

5.2.1 *Being tenacious...*

Quality managers and technical managers who have to routinely raise the same fabric/equipment/hygiene nonconformance points at site management review meetings, without any sign of the issues ever being addressed, should consider whether there is a more effective way of securing the action required.

Options could include:

- Raise the pressure for improvement by escalating 'long-term/repeat issue' points to higher-level senior management review/director review meetings.
- Arrange to routinely tour the factory site with the managing director, ideally while also accompanied by the engineering and production management team. Fabric/equipment issue points can then be discussed and acted on swiftly. In addition, such walk-rounds can also serve to demonstrate to staff that factory standards are valued by the business management team.

Being a good technical manager often requires much more than knowledge of (and decision making upon) quality, safety and legality rules and regulations. Technical management can be enhanced by having very good 'people skills', being a strong networker, a team player, building appreciation and goodwill with colleagues and other business departments over time. In addition, an effective technical manager will be able to highlight any difficult or awkward points clearly and urgently in order to prompt colleagues into action when food safety, quality or legality is under threat.

Carrying the respect of the management team will help the technical manager to get issues such as fabric and equipment points addressed swiftly by fellow team members who appreciate the importance because their technical manager, who they respect the views of, has particularly highlighted the necessity to them.

'Timing' is often very important to get right as a technical manager. If the business has just had a very good financial period or has recently secured a new supply contract, then this could be a good time to highlight to the site senior management the need to address some fabric/equipment repairs or make other improvements. Similarly, if a business is wishing to gain more supply contracts and has a potential new customer coming to review the operation, this may also be a good time to remind the site that certain improvements still need to be made which currently will reflect poorly if noted by potential new customers.

Businesses are typically very concerned with regard to the consequences of failing customer audits. Therefore, highlighting the list of current site faults in advance of a major site audit can be very effective in having such issues addressed, as no manager wants to be the one who did not address an issue which ultimately had a detrimental effect on the outcome of an important site review.

5.3 Factory location considerations

When deciding on the location for a new food production facility a very wide range of factors have to be taken into account, many of which can have a direct or indirect impact on-site technical standards.

Such factors include:

- Can the business secure the premises and land required for the industrial purpose intended? For example, a noise or significant odour emitting site would not operate in harmony with a residential area and therefore may be subject to operational restrictions.

- Does the site have sufficient utility resources for the intended operation? For example, electricity, water, effluent treatment capacity and gas supply?
- Will the factory have a sufficient pool of labour nearby?
- Is the site close to distribution routes for key customers and suppliers?
- If necessary is there room to expand the factory operation on-site in the future?
- Is there sufficient proximity to any 'off-site' extra storage required? For example, stores for raw materials or finished products.
- Does the location pose any specific threats to the operation? For example, natural threats: Perhaps located in an area susceptible to flooding. Perhaps pest control could be a problem if there are habitats nearby which would be a regular source of pests. Also "man-made threats": The presence of other businesses which could drain the available labour pool or the capacity of utilities within the area. Perhaps plans to build a residential area nearby would impact on and restrict the business activities in the future.
- There may be incentives provided to businesses to locate in a particular area such as regional grants and business start-up support.

As corporate social responsibility considerations continue to develop in the food sector, businesses are increasingly expected to carefully consider their impacts on the local population and the environment.

Such considerations include:

- Are there any foreseeable site constraints/issues? For example, protected buildings due to historical significance? Protected species of animals or plants? Sites of special scientific interest?
- Will factory noise pollution, effluent discharge, operational transport or potential odour emissions lead to objections from local residents or other businesses? These issues could impact on the ability of the business to operate normally.

Typically, local government will be expected to consider such impacts when the business applies for consent to operate in the area, and will usually have to balance their considerations with the considerable benefits that the business can bring to the area, such as employment opportunities for the local community.

Distilling 'location considerations' down to the aspects that will most typically involve the site quality and technical management, the food manufacturing operation will be expected to be able to demonstrate that it is operating in compliance with all local, regional and national regulations, including the holding of site planning consent and approval from the relevant enforcement bodies to operate food manufacturing processes.

5.4 Site security

Food manufacturing sites are increasingly expected by their customers and third-party audit standards to apply effective controls to maintain a highly secure operation.

Potential threats to site security can range from vandalism and theft through to the potential for particular interest groups seeking to gain access in order to disrupt the operation. Undercover journalism has increased in recent years with articles or television programmes created utilising information or footage gained via access to food manufacturing operations.

In addition, wherever there is food manufacturing, there has always been the potential for malicious contamination incidents to occur due to the actions of unbalanced or disgruntled individuals. The emerging threat of terrorism in many nations is also a strong reminder of why the food manufacturing and supply chain needs to remain vigilant to counter such threats.

Site security suggestions and expectations of food sector compliance standards can include:

- Use of employee identification methods such as employee ID cards which incorporate a picture of the employee.
- Site attendance control and monitoring procedures to cover all employees.
- Visitor and contractor procedures maintained which ensure that the identity of all visitors is checked on arrival and that while on-site visitors and contractors should be accompanied/overseen by a member of staff.
- Maintaining site staff controls and visitor/contractor site procedures to ensure that video and audio recording devices cannot be brought onto site and used without the express permission of the business.
- The requirement for the site to hold a clear and up-to-date site plan documenting the layout and access to all areas of the operation.
- The boundaries of the manufacturing site to be clearly defined and controlled, ideally by perimeter fencing.
- Requirement for the operation of security barriers or checkpoints on entry/departure from the site.
- Necessity for site buildings to be accessed via secured entry points which require authorisation codes/keys/swipe cards or video recognition to gain entry.
- Site and its perimeter to be monitored utilising CCTV cameras.
- The opening of site mail/packages may require special training for staff in the identification and handling of suspicious items.
- Screening procedures to be utilised in the recruitment of all employees to check their identity and background.
- Backing up of all site computer systems, procedures and records deemed critical to product safety control.
- Routine spot checks on the contents of containers such as toolkits, lunchboxes and storage lockers.
- Staff training on the importance of always challenging any unknown visitors present within their operational areas, and also being vigilant for and reporting any unusual activity by members of staff.
- Business crisis management procedures to include reaction to potential breaches of site or product security.

5.5 Perimeter control

In addition to the security considerations highlighted in the previous section, site perimeter control measures are usually also required to help maintain a hygienic and pest-free manufacturing site. Such control measures include the requirement to keep all external areas clean and free from foodstuffs which would otherwise attract pests, and ensure that materials (e.g., waste, transport containers, equipment, etc.) are not held outside where they can provide a harbourage to pests such as rodents.

Vegetation surrounding the manufacturing site can also be a habitat or source of harbourage for potential pests and therefore it is important that such surroundings are kept under control (e.g., regular grounds maintenance, trimming of vegetation and the use of pest monitoring and control systems).

Where the site perimeter is shared with another business it is important that the food manufacturing operation is not impacted by any reduction in perimeter control standards operated by the neighbouring business. This can be guarded against via regular contact between the food manufacturer and the other business, perhaps highlighting when issues such as overgrown vegetation or the holding of redundant/waste materials could cause a potential pest harbourage problem if not swiftly addressed.

External drainage can pose challenges to the maintenance of hygienic manufacturing site grounds. Inadequate drainage capacity can cause water build-up around the manufacturing site, making it difficult to keep external areas clean, increasing the potential for attracting pests and also presenting the risk of flooding internal food manufacturing areas. Additional external drainage systems should therefore be installed where it is evident that the current levels of drainage on-site are insufficient.

Where the manufacturing site is operating separate storage units or other buildings on-site alongside the food manufacturing units, for example, chiller/freezer facilities, storage containers/silos, waste handling/effluent plant facilities, the business should ensure that these peripheral units are also adequately protected against pest ingress and are maintained in a good hygienic condition.

Interface points between outer areas and the factory unit must be appropriately sealed to guard against ingress from pests and other potential contaminants. For example, control of gaps created in factory walls to enable materials to be transferred into the factory (perhaps via pipework) from external storage containers.

Often food manufacturing businesses will be handling a large amount of empty packaging linked to the recycling/return of raw material and cleaning chemical containers, and also related to the transport of end products (e.g., reusable trays/crates/containers). Ideally such containers should not be stored outside as they can both attract and provide harbourage to pests, thereby being open to many forms of contamination.

Where outside storage is unavoidable businesses will typically be expected to conduct risk assessments to ensure that the potential food safety threats of such practices are acceptably low. The business will also be expected to monitor and control the situation to further guard against the occurrence of any problem issues. Such controls may include ensuring that all stored materials are clean, covered and subject to routine inspections, including particularly thorough inspection at the point of being transferred into factory areas or at the point of return to other supplier operations.

As control of the factory peripheral areas/perimeter is a key first line of defence with regard to food safety, these areas should always be included in site audit systems.

Additional measures to ensure site security should also be applied during times when the manufacturing site is not operational (e.g., during national holidays and at night or at weekends if the operation is not a 24 hour, 7 days a week operation).

5.6 Staff facilities

While at work factory staff will typically require areas within which to:

- store personal belongings,
- get changed into and out of factory clothing,
- use the toilet and
- eat and drink.

Staff facilities should be subject to site hygiene procedures and monitoring audits to support the maintenance of a high standard of cleanliness. This is particularly important as staff facilities can have a major influence on defining the site 'food safety culture'.

Smoking facilities: On sites where smoking is permitted it should be done so in designated areas which are positioned to be well away from the food production and storage facilities. Appropriate shelter for staff should be provided, with these areas also being well ventilated to prevent smoke build-up on staff clothes.

Changing rooms, eating/rest areas and toilet facilities should be provided on a scale which reflects staff numbers and must be maintained in a clean and hygienic condition. Specific design and maintenance aspects of each area should include the following considerations.

5.6.1 Staff changing rooms and storage facilities

Typically, staff will be required to change into specific work clothing such as overalls, boots and hairnets. Therefore, businesses should provide an appropriately sized changing room which:

- is easily accessible to staff;
- contains (or is in close proximity to) washing facilities;
- has an appropriate source of factory clothing and footwear;
- has a means of hanging clothes, such as hooks or pegs;
- is sufficiently private and
- allows for the drying of clothes if possible.

Any storage areas intended for wet clothes should be well ventilated to assist the drying process.

Storage systems should prevent the personal clothing of staff coming into contact with work clothes.

Signage should highlight the required factory clothing to be worn and the correct procedure for getting changed.

Staff lockers should have sloping tops and also should be raised above the floor to ease cleaning and reduce the potential for debris build-up.

Lockers should have internal partitions to facilitate the separation of items where necessary, for example, clean versus dirty; indoor versus outdoor wear.

Locker signage should define what is acceptable/unacceptable for storage within lockers.

If showers are provided then these should be designed and maintained to protect against health and safety risks such as the potential to harbour *Legionella.*

Systems should be in place for the hygienic distribution of clean factory wear, and the collection of dirty/used factory clothing in readiness for the designated laundry system.

Systems for the storage of factory footwear should be controlled and easily understood.

5.6.2 Staff catering and rest areas

These areas should be hygienically designed to prevent the build-up of food waste, and should be monitored to ensure that the areas are consistently maintained in a clean and hygienic condition.

Signage should highlight the importance of cleanliness/waste disposal.

Any chilled food storage facilities provided to staff should be hygienic and capable of consistent and effective temperature control (which should be confirmed by regular monitoring checks).

Food and drink vending machines should be subject to hygiene and temperature check controls wherever relevant.

Catering facilities staff should be subject to formal staff medical screening and hygiene training.

Staff catering operations must maintain food safety control systems such as HACCP (hazard analysis and critical control point).

5.6.3 Toilet facilities

Employees must be provided with appropriate and adequate toilet facilities.

Such facilities should accommodate the following aspects:

- Be separate for male and female.
- Should be physically separated from food handling areas.
- Positioning should necessitate a change out of work based factory clothing and protective equipment.
- Be clean, well lit and ventilated.
- Contain toilet paper and sanitary disposal facilities.
- Have hand washing and drying facilities available, with signage making clear the appropriate hand washing requirements and technique.
- Have hot and cold running water and soap.

- Taps should be nonhand contact, for example, sensor or knee operated.
- Have a washbasin and a method of drying hands.
- Contain showers when the nature of the business makes showering necessary.

The amount of toilet facilities and wash stations required should reflect the amount of staff on-site and in particular countries may be stipulated in food sector legislation.

Toilet facilities for disabled people must be adequately designed to allow access and also provide features such as grip handles and accessible sinks/hand dryers.

The hot water provided to sinks/wash stations should be sufficiently warm to encourage hand washing and not so hot as to make the hand washing process uncomfortable.

All wash stations should be knee, elbow or sensor operated in order to avoid the need for hand contact which could cause the spreading of microorganisms between users at the contact point. The soap provided at wash stations should ideally be in liquid form and should be bactericidal.

Signage should highlight the necessity to wash and dry hands thoroughly.

Hand drying systems should be either disposable paper towels or hygienically designed hand driers. Where paper towels are used care should be taken to ensure that the paper towels purchased are sufficiently robust as not to easily tear/deteriorate during use, as such debris could pose a contamination threat to the foods being processed.

The airflow of any hand driers utilised should be designed and directed so as not to pose an aerosol contamination risk to the manufacturing process.

Toilet areas must be subject to a regular cleaning regime which utilises entirely separate cleaning equipment to that used within the food production areas.

5.7 Footwear and related washing facilities

If not properly controlled, footwear has the potential to spread various forms of physical and microbiological contamination across a food manufacturing site. The nature of the manufacturing operation will typically dictate the extent to which footwear requires to be changed and cleaned in advance of entry into food processing areas.

It is good practice to keep factory footwear within the manufacturing building and to restrict external use in order to avoid the potential for picking up contaminants outside of the factory.

Steel toe-capped safety shoes are typically utilised in warehousing areas.

In food processing areas wellington boots are commonly utilised due to their cleanability and also their potential to have nonslip soles which can be of major health and safety benefit.

In advance of entry into food manufacturing areas the footwear should be washed and sanitised thoroughly to help avoid the movement of potential contaminants into the food processing environment. This can be achieved via manual brushing approaches (e.g., use of a bucket of detergent/sanitiser and a handheld brush) or via automated boot washing systems.

Boot washing systems should typically be sited outside of the production areas as the brushing action can often create aerosols/spray which could pose a contamination threat to the products being processed.

The design of automated boot washing systems should facilitate easy access to enable the equipment to be regularly and effectively cleaned. Otherwise such equipment can become microbiological traps more likely to spread contamination onto the boots.

Automated boot washing systems will often automatically draw on detergent/sanitiser solutions, therefore there should always be routine checks of such systems to ensure the consistent supply of the required chemicals and that these chemicals are being dosed at the correct concentrations.

Footbaths (e.g., 'walk-through' troughs containing sanitiser solution) are often considered to not be as effective as manual or automated footwear scrubbing systems due to their lack of physical abrasion of the boot surfaces and the potential for the solutions in the walk-through baths to become ineffective due to the build-up of food debris within the wash system over time. Where such systems are utilised, they should be subject to regular review to ensure that they are maintained in a clean and effective condition.

The adequacy of boot washing systems should be subject to regular monitoring and review (by visual inspection) both by area supervisors/management and via incorporation into area GMP (good manufacturing practice) audits. Boots should also be occasionally subject to microbiological swabbing in order to monitor that such footwear is being maintained in a microbiologically sound condition.

It is important that visitors and contractors adhere to the same footwear rules as the standard site personnel; therefore it may be beneficial to maintain sets of footwear designated for visitor/contractor use. Where a business decides that it wishes to avoid such processes via the use of plastic 'overshoes'/'shoe covers' the alternative foot covering must be sufficiently robust so as not to tear/become damaged while walking around the factory areas, and must be designed so as to provide thorough coverage of any external footwear being worn underneath.

5.8 Visitor and contractor control

Site visitors and contractors operating on-site require close control as otherwise they can pose a range of threats to operational food safety and quality.

Such threats include:

- Visitors to site may be carrying certain health risks which could compromise the safety of the business staff or the food products being manufactured.
- Individuals with malicious intent may seek to gain access to the manufacturing operation by presenting themselves as an approved visitor/contractor working on the site.
- Visitors and contractors may breach food safety controls if they are unaware of site rules and regulations. This is particularly the case with regard to site hygiene and allergen controls.

The use of checkpoints on entry to the manufacturing site provides an opportunity to check the identity of the visitor/contractor and to ensure that they are fully aware of the site rules and regulations with specific regard to the control of site security, food safety, and health and safety controls.

Visitors and contractors should typically also be screened for any particular health issues which may pose a threat to product safety, including requests to declare whether they are suffering from, have recently (within the past 28 days) suffered from or have been in contact with anyone suffering from any of the following: Typhoid or paratyphoid; diarrhoea/vomiting/stomach upsets; recurrent boils or septic fingers; infected or discharging ears; skin conditions, for example, dermatitis, psoriasis, eczema, etc. or other septic infections including recurrent infection of the nose, throat or eyes; illness of any kind while on holiday abroad; any potential infectious illness.

The visitor/contractor 'health declaration' should then be checked by a member of staff capable of making a risk assessment of the answers given and providing further guidance on the necessary business position with regard to any issues raised, including exclusion from the manufacturing site if any issues are deemed to be a threat to food safety or employee health.

Awareness and acceptance of site rules is typically ensured via the use of a visitor briefing document and questionnaire, with the visitor being expected to answer all questions and sign to demonstrate that they understand and accept all site rules.

Site visitor/contractor control briefing documents will cover aspects such as:

- Site GMP and food safety rules.
- The need to be accompanied by a member of staff while on-site.
- A synopsis of site hygiene procedures/expectations, including hand washing and protective clothing requirements.
- Site jewellery policy, for example, 'all jewellery and watches must be removed before entry into production areas'.
- Cuts and abrasions, for example, all cuts and abrasions must be covered with appropriate factory issue materials (e.g., blue metal-detectable plasters) which can be provided on request.
- Where applicable highlighting the extent of site controls on potentially allergenic materials.
- The importance of adhering to the requirements highlighted by the host and to the content of any food safety or health and safety related signage.
- Controls with regard to the acceptability of carrying personal belongings on-site.
- Site rules with regard to the use of audio/video recording equipment on-site.
- Site fire alarm evacuation procedures.
- Other site evacuation procedures if applicable (e.g., ammonia leak).

Visitors and contractors who are present in order to complete work within food manufacturing areas should have the purposes and nature of their work risk assessed in advance of commencement by relevant members of the business management team including the quality/technical management. They should ensure that their actions will not pose a threat to product safety, and assessed by the health and safety manager who should ensure that their actions will not pose a threat to the safety of site employees.

It may be that in order to fully ensure the safety of the work being completed, the working area requires to be screened/physically isolated from the manufacturing process, or it may be that the work needs to be scheduled for a time when the manufacturing operation is not running. On completion of such work a full clean-down and inspection should be conducted to ensure that there are no potential contaminants remaining which could have adverse impacts on product or operator safety.

Many manufacturing sites issue contractors with their rules and regulations for review and sign-off in advance of their scheduled time on-site, thereby ensuring that contractors arrive already well briefed on the site rules and regulations and therefore ready to commence work.

Contractor rules and regulations for a manufacturing site will typically cover both health and safety and food safety control requirements.

For example, content may cover:

- General summary: Legislative and moral requirements to adhere to the specific site rules and regulations.
- Definitions: Any potentially complex terminology used within the 'contractor requirements' documentation.
- Insurance: Requirements for the contractor to ensure that they are appropriately insured and clarifying which (if any) aspects of their work are insured by the food manufacturing site.
- Conduct: Expectation that the contractor must not progress any work which may cause risk to the health and safety of employees or the safety of the food products. Also, the requirement for the contractor to take all reasonable precautions and approve all work in advance with the manufacturing site.
- General site safety controls related to: Commencement of work; safe systems of work; special permissions/permits to work; designation of areas of contractor ownership while on-site; procedures to cover movement of staff and equipment on/off-site; access to and use of site materials and equipment including company vehicles (e.g., forklift trucks); use of site welfare facilities; first aid procedures; accident reporting requirements.
- Food safety controls including: Hygiene rules and regulations; health screening requirements/questionnaire; site dress code and jewellery policy; PPE (personal protective equipment); site smoking policy; site glass control procedures.
- Fire controls and related procedures.
- Discipline: Expectations for contractor staff and any subcontractors to operate responsibly and in full adherence to all site rules and regulations at all times.
- Site health and safety rules: Safe operating practices and related procedures; definition of legislative obligations and business rules/regulations; 'near-miss', dangerous occurrence and accident reporting procedures; display of safety signs/notices where necessary; use of safety barriers where appropriate; work conducted in a manner which does not endanger members of staff, site visitors, the public and other contractors; zero tolerance of drugs/alcohol on-site, or of staff under the influence of such substances; strict prohibition of any play, fighting or other unsafe acts; the importance of all staff setting excellent personal examples of safe and diligent working practices.

Secured site entry points which require authorisation codes/keys/swipe cards or video recognition to gain entry serve to ensure that visitors and contractors are only able to access key business areas with the express permission and presence of members of staff. In addition, the use of site CCTV cameras provides a useful reminder to all employees and visitors of the importance to act responsibly at all times.

The operation of a visitor/contractor permits/badges system helps ensure that business employees are aware of their status on-site, and staff training can raise awareness as to the importance of always challenging any unknown visitors present within their operational areas, and also being vigilant for and reporting any unusual activity.

5.9 Process flows

When designing a food manufacturing operation, implementing the right 'process flow' can be vital to both food safety and production efficiency.

Food manufacturing operations should avoid the potential for cross-contamination caused by processed foods coming into contact with unprocessed foods or waste materials. In addition, 'linear'/'sequential' food production will typically support the ergonomic flow of food manufacturing operations and avoid the inefficiencies and risks associated with more complex approaches.

A common process flow example of businesses seeking to avoid cross-contamination of the end product is the operation of high-care or high-risk areas within the manufacturing operation. The design and construction of such areas is covered in more detail later in this chapter.

When building a new facility, a business should proactively plan to help ensure that the production flow is logical, efficient and avoids bottlenecks. It is always important to also plan for 'waste flows' too. Waste could include waste ingredients, end product and packaging, the movement and disposal of which is sometimes overlooked by businesses until production commences and the inherent waste flows start to cause hygiene issues or inefficiencies.

5.9.1 Considerations...

Thorough advanced consideration of the processing sequences and space allocation required for each of the 'operations' to be carried out, and envisaging likely future requirements, is a process best progressed by a multidisciplinary team to ensure that as many factors as possible are reviewed.

Further approaches to help ensure that process flow plans are effective include:

- Check the intended sequential production operation against HACCP flow charts for similar businesses/processes.
- Cross-check against the requirements of audit standards relevant to the operation (e.g., British Retail Consortium (BRC), International Featured Standards (IFS), International Organisation for Standardisation (ISO), etc.). Many audit standards make very detailed statements on requirements for certain factory design and equipment features. It is therefore important to check requirements in advance to help ensure compliance.

- Visualise the operation (e.g., mark out the intended areas and doorways on the floor and 'walk the process').
- Visit and review similar operations.
- Use experienced architects/planners.

Well thought through processes have a logical flow/sequence and consequently avoid process flow bottlenecks. Considerations have to take account of all structural aspects such as drains, floors, walls, ceilings, airflow systems, together with the provision of utilities to areas (e.g., electricity, water, gas). Businesses can often struggle for years with a badly thought through production flow so it is worth the additional time and investment at the outset in order to get it right first time.

5.10 Storage segregation

5.10.1 Ingredients

When handling meat ingredients in a multicomponent food production facility it is usually good practice to store such materials in a segregated meat store. Segregation from other types of ingredients benefits microbiological control and if applicable also supports the business control/assurance of manufacturing vegetarian/vegan status products.

A multiingredient site will typically operate the following ingredient storage areas:

- meat chill store;
- combined or separate (depending on space/capacity needs) vegetable and dairy produce chill store;
- ambient store for herbs/spices/powders, canned goods, etc. and
- frozen storage

Site freezers often tend to hold meat, vegetable and dairy produce due to the reduced risk of cross-contamination (assuming that all product is held sealed in its primary and secondary packaging in the freezer store areas).

Some food processing facilities may also require the handling of both preprocessed and raw ingredients. In cases such as these the technical manager should conduct a risk assessment (with particular consideration of the further processing yet to take place) to enable a decision to be taken as to whether separate raw and preprocessed storage areas are a necessity.

Allergen considerations: Due to the risk of any allergens handled on-site contaminating nonallergen containing products:

- 'Very high health risk' designated ingredients (e.g., in Europe such materials could include nuts and unrefined sesame seed) may tend to be stored in spatially separate and clearly marked storage areas within ambient, chilled and frozen stores.
- Other allergens classed as being of slightly lower risk classification (e.g., in Europe these could include milk, egg, celery and mustard) may tend to be maintained within their packaging and controlled via staff allergen handling procedures.

Allergenic status of ingredients varies geographically across the globe, with certain ingredients being particularly allergenic in one region, but then not classed as being of allergenic importance in other regions. Allergen control measures will vary from site to site depending on the nature of the ingredients handled, the end product claims made (e.g., 'nut free') and the legislative expectations in the region where the end product is to be marketed. Customer expectations and technical requirements will also be reflected in manufacturer site operating procedures with regard to allergen control.

5.10.2 Secure storage for glass, alcohol and high-value ingredients

Locked storage areas can be of great benefit with regard to both high value and fragile items.

As broken glass presents a particular food product hazard due to its potential to harm and easily be unseen in the end product, many operations will choose to avoid purchasing ingredients in glass packaging if at all possible. Where the purchase of ingredients in glass containers is a necessity, storing such ingredients in a secure area before use can help to reduce the potential for breakage/mishandling.

High-value ingredients such as champagne, alcohols, truffles, caviar, etc. can be a particular temptation to the casual thief and therefore typically such items will benefit from secure storage in a locked and secure area to prevent losses (Figs 5.1–5.5).

5.11 Multilevel factory units

Mezzanine floors, also known as 'raised floors', can be a cost-effective way of increasing production space in food factory. A 'mezzanine' is an architectural term for an intermediate floor between main floors of a building, as can be seen in Fig. 5.1.

Figure 5.1 Mezzanine floor.
Source: Hampshire Mezzanine Floors, United Kingdom.

When installing such structures, it is important ensure that the higher-level activity does not pose a contamination risk to the activities/products located below, as the raised nature of a mezzanine area can provide a platform from which to spread potential contaminants if not adequately controlled.

5.12 High-care and high-risk environments

High-care or high-risk areas within a food manufacturing process flow are typically used to increase the control of food safety at a process stage where the product may be particularly susceptible to contamination. For example, the assembly and packing phase for cooked or partly processed ready to eat (RTE) foods.

'High-care' factory areas are typically processing areas which are segregated from the remainder of the manufacturing site and are designed to **minimise** the potential for postprocess contamination of the product being handled.

'High care': This term is utilised when the RTE food being processed, due to its composition, is likely to already have a significant microbiological load, and therefore the intention is to avoid further adding to the microbiological levels.

'High-risk' factory areas are food handling areas which are segregated from the remainder of the manufacturing site and are designed to **prevent** the potential for contamination of the product being handled.

'High risk': This term is utilised when the food product being handled has been processed to effectively eliminate the majority of the foodborne microorganisms, and therefore the intention of the area is to prevent the potential for postprocess contamination (Table 5.1).

With regard to 'high-care' and 'high-risk' areas it is typically recommended in the food manufacturing sector that operations should separate these areas from 'low-risk' areas via a full physical wall barrier forming a 'floor to ceiling' divide. This reduces the likelihood of potential product contaminants gaining entry into the area from the low-risk areas (Table 5.2).

When seeking to achieve and maintain such physical segregation, any openings in the wall which divide high-care/high-risk and low-risk areas (such as hatches and doorways) should be kept to a minimum. Openings provide a particular opportunity for ingress by potential product contaminants and therefore should be carefully considered in terms of their necessity, design and ongoing monitoring/management.

Similarly, drainage channels for the high-care/high-risk areas should be separated from low-risk area drains to prevent ingress of microorganisms linked to low-risk area processes.

High-risk food manufacturing environments are typically expected to be maintained under a measured positive pressure system to guard against the ingress into the area of airborne/aerosol-borne microbiological contaminants. Positive pressure can be achieved by channelling air into the area via an air ducting/ventilation system. The air is pumped into the area at a rate which ensures that at all room doorway and wall aperture points the air flows *from* the high-risk area *into* the adjoining

Table 5.1 **High-care area expectations**

High care	
Expectation	**Achieved by**
Prevent the transfer of microorganisms from low-risk factory areas into the high-care area.	Full hand washing and factory clothing change (including footwear) on entry to the high-care area. Separate cleaning equipment and utensils used compared to the low-risk factory areas. Area should have separate drainage channels to avoid flows into the high-care area from low-risk areas.
Maintain a hygienic product handling environment.	Hand wash systems available and regularly utilised. Dedicated utensil and equipment washing systems. Monitoring/audit systems.
Staff to be aware of the importance of the area in terms of product risk and to operate in a manner which supports the control of food safety.	Appropriate staff selection. Thorough staff training systems. Effective supervision and management. Monitoring/audit systems.

Table 5.2 **High-risk area expectations**

High risk	
Expectation	**Achieved by**
All points summarised in Table 5.1 plus...	*All points summarised in Table 5.1 plus...*
Assure the microbiological quality of the air in the area by preventing the flow of potentially contaminated air into the high-risk area.	Use airflows/fans to maintain the high-risk zone under 'positive pressure'. Filtered air: All area air should be subject to filtration, the standard/grade of which should be selected to be appropriate for the nature of the operation.
Consider the application of additional air treatment systems within the area.	Ozone and UV light treatment are two examples of air treatment systems utilised within the food manufacturing sector.

low-risk area, thereby reducing the potential for contaminants to flow from low-risk into the high-risk area. As positive pressure air is being drawn from outside and actively channelled into the high-risk area, it is very important that the air is filtered to reduce the potential for contaminants (e.g., dust and pests) to access the sensitive high-risk processing areas.

Air pressure is measured in Pascal's and many food safety compliance standards provide a minimum air pressure which the high-risk processing areas and their airflow systems should be designed to achieve. Food safety systems also expect the positive air pressure to be routinely monitored, as reduced efficiency of the air pumping system, or increased numbers of apertures placed within the area walls over time, can lead to the air pressure achieved dropping to a point where it becomes insufficient for the purposes required.

Ventilation filter grades are usually based on their effectiveness at screening out particulates of a certain size. As with air pressure, many food safety compliance standards stipulate a particular grade expected, again usually expected to be set on a risk assessed basis which considers the potential for contaminants to be present within the airflow, and the sensitivity of the product within the high-risk area.

When seeking to achieve and maintain an appropriate standard of air control in high-risk environments, in addition to the required air pressure and filtration standards, it is also good practice to achieve a number of 'air changes per hour' to ensure that the air quality in the area is maintained to the required standard. The appropriate number of air changes per hour should be set by the technical manager on a risk assessed basis, reflecting on the air filtration quality, the nature of the manufacturing operation (e.g., how likely is the quality of the air in the area likely to deteriorate due to operational processes such as cleaning or perhaps due to air influx from large doorways being opened routinely into the area) and also reflecting on the sensitivity of the end product.

The microbiological integrity of high-care and high-risk environments should be routinely checked via the progression of an environmental monitoring system which may include microbiological swabbing of surfaces across all points in the area and also could include the use of 'air plates' to assess the microbiological quality of the air.

Air filtration systems are further considered in the 'ceiling-based lighting, ventilation and extraction systems' section later in this chapter.

Roller doors should be avoided in high-care/high-risk areas as they present a particular potential to drip floor-based moisture and debris onto goods passing through while they are open (e.g., of particular concern in this regard is the potential for microbiological contaminants such as *Listeria*).

5.13 Factory structure and fabric

In general, the outer shell of a food manufacturing factory serves to provide physical support to the structures inside, and should form a robust weatherproof and pest proof barrier to the external environment. The number and positioning of doors, windows and other entry points should be well considered in the design phase to support operational flows of goods and people, and also to minimise the potential for pests and nonfactory personnel to be able to gain entry.

5.13.1 Floors

Factory foundations will typically have been dug and filled first with hardcore, followed by covering over with concrete. The depths of these layers will be dictated by the terrain, the weight of the buildings and processes to be placed upon them and by standard building good practice including relevant building regulations/legislation.

The final top surface which forms the factory floor should be structured to ensure that the business can demonstrate adherence to good hygiene practice and relevant legislation. Typically, the expectation will be for factory floors to be 'smooth and impervious', making them nonabsorbent and easy to clean. The floor should also be robust enough to withstand 'wear and tear' caused by the factory manufacturing processes (which may include potential damage caused by moving equipment, operating temperatures and cleaning chemicals).

Another key hygiene control feature of floors is that they should 'fall to drain' with a gradient sufficient enough to avoid the formation of any water pools.

As slips, trips and falls are a major cause of accidents within the food manufacturing industry, 'antislip' floor design is often used via the incorporation of reduced slip surfaces. Such 'rougher' surfaces can make the surface more prone to accumulation of food debris and therefore harder to clean. For these reasons the technical manager should be involved in the floor design decision-making process so that such challenges can be considered and managed.

Epoxy and polyurethane-based floors: These are very common types of flooring used within the food manufacturing sector. Food factory areas often have significant 'traffic' and see frequent use of cleaning chemicals and other materials which may cause the deterioration of the floor area over time. There are a wide range of choices of floor material available and food businesses are well advised to take the time to match the materials to their specific area needs. For example, epoxy floors tend to be harder than polyurethane floors so are often chosen for areas where there is heavy traffic (e.g., forklifts) such as warehouses. Polyurethane floors tend to be more elastic and therefore more scratch resistant, and their increased flexibility often make them a good choice for frozen stores. Resistance to water and chemical attack also vary across flooring materials.

Other materials utilised within the food manufacturing industry include tiled surfaces and fully interlocking, compact flooring systems. As with all floor designs the technical manager should ensure that whatever surface is used it should be smooth and impervious to facilitate good hygiene control.

In flooring systems for frozen storage areas, the flooring will often benefit from the incorporation of underfloor heating. Such heating helps to ensure that the floor remains at a temperature which is not particularly prone to expansion and contraction that would eventually lead to the development of stress fractures and consequentially deterioration over time. Underfloor heating also supports the cleaning of the freezer store floors, a process usually also aided by the use of specialist frozen store cleaning materials.

Figure 5.2 Polyurethane floor with antimicrobial properties.
Source: Flowcrete UK Ltd.

5.13.2 Drainage and trade effluent

Food manufacturing site drains will vary in terms of the nature of the materials they will typically be carrying. The business should ensure that there are clear site drainage plans in place to help ensure clarity and control of which drains are utilised for effluent generated by the factory processes, which drains handle domestic waste/sewage, and which handle surface/rainwater.

Directions of flow and appropriate capacities are vital factors requiring planning and control to avoid any hygiene issues which could be created by inappropriate use or insufficient flow volume capacities. To help avoid increased potential for cross-contamination to occur, high-care/high-risk drains should always be separated from low-risk area drainage. Where there is the potential for flow back into high-care/high-risk areas then 'non-return' flow systems could be utilised to prevent such instances.

As flooring systems are the foundations of the manufacturing facility it is vitally important that area drainage is thoroughly considered and well designed from the outset of the factory build. Having to redig floors/drainage channels in order to correct initial design faults or as a result of redeveloping factory areas can cause significant disruption to factory operations.

Depending on the nature of the operation, factory areas can be designed to have open or closed drainage channels. Open drainage channels offer the additional potential to keep the channel clean and clear, whereas closed drainage channels reduce the amount of open food waste/debris within the manufacturing area. Whichever is utilised it is vital that the design implemented facilitates thorough and regular cleaning.

Where food particulates are being handled by the manufacturing operation drain traps can be used to ensure that such food debris is not allowed into the drainage flow system where it could accumulate to cause drain blockages and also could increase the wastewater treatment charges applied by the relevant water/waste authorities. Good planning at this stage can significantly reduce factory food effluent impacts as typically the more 'suspended solids' can be removed from the liquid drain waste, the lower the 'chemical oxygen demand' or 'biological oxygen demand' will be.

Businesses can also reduce their factory food effluent impacts via installation of physical 'screens' (designed to remove particulates from the water flow), 'fat traps' (which effectively remove fatty deposits from the effluent flow) and dissolved air floatation systems.

Screens or drain traps can also serve to help prevent pest entry into the food manufacturing environment through the drainage system.

5.13.3 Factory walls

External walls provide strength and protection to the factory unit and strongly influence the appearance of the building. Outer walls are often required to support the weight loads of the factory roof, internal structures and equipment.

Walls enable control of the factory environment, serving to divide the external and internal climates of a building. Walls can also make the factory building more resistant to:

- moisture (e.g., with regard to rainfall and relative humidity);
- fire;
- noise (e.g., serving as sound insulation);
- temperature (e.g., thermal insulation);
- pests, and
- intruders

Internal walls can serve similar purposes to the external walls. Interlocking wall panels are a popular choice for many internal factory structures due to their relative ease of construction. Build materials will typically be dictated by the functions served by these walls and the weight loads that they will be required to support. Material and design factors can also include the extent of temperature insulation required and also the amount of fireproofing required.

Some interlocking wall panel systems can be utilised both horizontally and vertically, if ceiling materials would benefit from being of the same composition as the walls, which can often be the case where the construction materials have a food grade finish to make the surface smooth, impervious and resistant to cleaning chemicals and related processes.

While interlocking wall panels may be suitable for food factory use it is also worth ensuring that the join areas between the panels are sufficiently sealed so as not to become a trap for food debris which can eventually lead to hygiene issues such as providing harbourage for microorganisms or attracting pests.

It is also good practice for the joining point between walls and floors to be 'coved' to avoid potential debris traps and facilitate effective cleaning.

As food manufacturing environments can involve a great deal of product/process movement and therefore likely 'wear and tear', it is advisable for walls to be protected against damage via the use of barriers where there is the significant potential for impact damage to occur (e.g., perhaps due to forklift trucks or material movements occasionally impacting on the walls).

Figure 5.3 Flex Impact® modular safety barrier system by Boplan®.
Source: Boplan UK Ltd.

In many food manufacturing factories metallic cladding is utilised as an effective way to make walls more resistant to 'wear and tear'. This often involves the application of stainless steel chequer plate riveted onto the walls. Lighter aluminium cladding can also be used though it is not as resistant to cleaning chemicals as stainless steel.

Where cladding is utilised businesses should be particularly careful to use a good sealant when sealing the plate edges to the walls. If this seal is not completed thoroughly then a common problem is the ingress of food debris and moisture accumulating between the plate covering and the wall surface over time which can lead to hygiene/microbiological issues which are then very hard to access and address without removing the cladding.

Removable wall sections can be very beneficial to facilitate the occasional movement of particularly large process machinery into and out of production processing areas. When using such systems, it is important to ensure that these removable sections are just as effective as the standard wall panels in terms of structural stability, hygienic design and in posing a significant barrier to airflows and potential pest ingress. These factors are especially heightened when seeking to utilise removable wall panels in high-care/high-risk manufacturing environments.

5.13.4 Hand wash stations

Staff hand hygiene is fundamental to food product safety; therefore the location and design of hand wash stations is very important. Typically, hand wash stations are located on entry to food handling areas and also inside or close to changing rooms.

Hand wash stations are often manufactured from stainless steel to prevent rusting and are usually operated by proximity sensors, or via knee/foot pressure mechanisms to avoid hands having to come into contact with surfaces which would be likely to harbour bacteria due to frequent handling (e.g., taps).

These stations require a controlled hot water supply which will often be provided via a factory hot water ring-main heated by the site boiler system. Common problems with hand wash stations are that they are sometimes located so far from the hot water tank/boiler system that it takes a long time for the hand wash water to get hot when running the water, especially if used infrequently. Systems can be installed to solve this problem by continually circulating hot water past all hot water utilisation points so that hot water is always only a few centimetres away from each usage point and therefore instantly ready for effective use.

Hand wash stations in factory areas should always be plumbed directly into the drainage system to avoid any flow of sink wash water onto walls and floors. Stations should also be set up in a manner and location which does not raise any issues linked to aerosol generation during the hand washing process.

5.13.5 Doorway considerations

When planning doors and doorways typically businesses will need to consider the nature of the business and the movement that will take place through these areas.

Relevant design considerations include:

- size requirements;
- frequency of door traffic;
- cleanability;
- ease of opening/closing (e.g., auto open/shut requirements);
- insulation requirements, and
- help keep cold air in/warm air out, or vice versa (e.g., rapid open/close mechanism, insulation with strip curtains, use of air-blow systems).

Where airflow and temperature control are particularly important the business may choose to operate an 'interlocking doors' system where one set of doors has to be closed before the next set will open, thereby creating a bridging area of interlocked doors which is effective at minimising the airflow impacts on the area inside. Such systems are often utilised in frozen/cold storage facilities where the impacts of temperature loss on both product quality control and running costs can be particularly high.

It is usually good practice to incorporate auto-close mechanisms on factory doors, especially on external doorways to the building. 'Kick-plates' on swing doors are also often recommended to ensure that the doors are as robust as possible to general 'wear and tear'.

Health and safety expectations will usually require a number of fire exit doors to be present to facilitate fast and efficient staff movement to the outside of the factory. Such doorways can be misused by staff as fast thoroughfares into and out of the building, thereby causing a number of potential hygiene and pest issues. To help guard against such issues it is usually considered good practice to place a seal on the fire door opening mechanism which easily breaks when opened, and therefore is 'tamper evident' to both discourage and highlight misuse. If using such tamper evident seals, it is clearly important to ensure that (1) they do not impede the easy opening of the door in the event of an emergency, and (2) when breaking they do not pose a potential physical contamination risk to the products being processed.

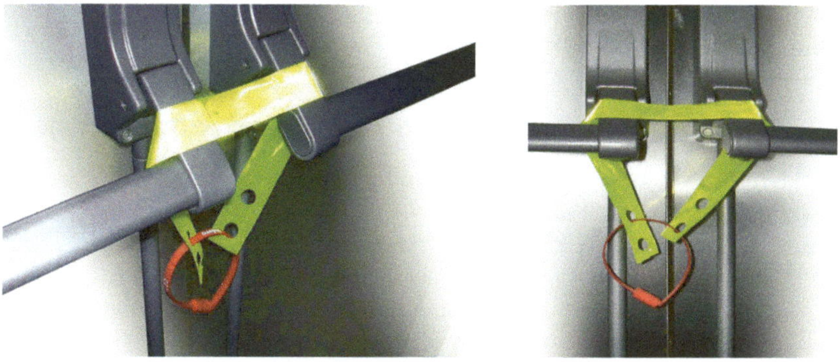

Figure 5.4 Emergency exit tamper evident seal.
Source: Unisto Ltd.

Another approach to discourage the misuse of fire doors is to place alarms on fire exit doors which sound upon opening. Door alarms can also be helpful in areas where the doors need to be kept shut. For example, doors into chilled and frozen storage areas can be fitted with timer alarms that sound after a preset amount of time if they have been left open.

External doors should always be kept closed when not in use. In the design phase consideration should be given to the size and frequency of traffic through the doorway. 'Fast doors' may facilitate swift opening and closing of doorways which are in frequent use.

For doorways where the door may be open for a significant amount of time the use of clear plastic strip curtains which are attached to and run for the full height of the doorway may be useful to help guard against pest access and temperature/airflow impacts while the doors are in use. When strip curtains are utilised, it is very important to ensure that the type installed are easily cleaned and sufficiently robust for the traffic that is going to be passing through them. Otherwise strip curtains can pose a hygiene issue as they trail over passing traffic and also pose a potential foreign body or pest control issue if they start to deteriorate. Some strip curtain systems are available which physically move apart when traffic pass through rather than trailing over the traffic.

5.13.6 Windows

Windows which are openable to the external environment should be proofed to prevent pest access to the factory or associated areas such as the site canteen, toilets and changing rooms. This proofing can take the form of fitted mesh screens which serve to prevent pest ingress when the window is open.

Windows within factory areas can provide communication, health and safety, and staff control benefits. As windows can pose a glass risk, windows which are necessary in factory areas should be made from clear flexible and shatterproof plastics or from toughened glass which is further protected from breakage via the application of clear plastic adhesive sheets.

5.13.7 Ceiling-based lighting, ventilation and extraction systems

Food manufacturing businesses often require their processing facility ceilings to have similar attributes to the walls in that they must be of hygienic design and suited to regular cleaning. Ceilings pose the additional challenge of being required to hold/incorporate lighting and often also accommodate ventilation/extraction systems to control airflow into or out of the area.

Ventilation/extraction grid apertures (the entry point into/out of the factory area) should be sealed flush to the surface to avoid forming potential debris traps and to facilitate effective cleaning. In addition, where there is a significant risk of unwanted dust/particulates (and potentially pests) flowing into the manufacturing area, filters should be positioned within the airflow ducting system.

Air filters are often utilised in positive air pressure systems where a high-care or high-risk food manufacturing environment is being maintained under positive pressure to further assure the microbiological, physical and pest control integrity of the production area ('high-care' and 'high-risk' environments are further discussed earlier in this chapter). The grade of air filter required should be selected by the technical manager with due consideration of the potential risks associated with the airflow into the production area and the sensitivity of the product/potential impacts. Many food safety certification systems would expect this 'consideration' to be documented in the form of a food safety risk assessment.

A number of standards also state separate levels of filtration grades dependent upon whether the food processing area is classified as high care or high risk. Filter grades are typically based on the efficiency of the filter to trap defined particle sizes.

Five to twenty-five air changes per hour can be sufficient to assure the appropriate air quality in many food processing operations; however, if the area poses additional challenges such as frequently opened doorways or hatches then far more frequent 'changes per hour' may be required.

The maintenance of air filtration systems is very important to the ongoing integrity of the airflows within the production area, and can often be overlooked as these systems are typically enclosed and placed in relatively inaccessible areas often within roof voids above the processing areas. The technical manager should therefore work closely with engineering/site maintenance colleagues to ensure that the filters are regularly checked to be working correctly and routinely changed to remain effective.

Use of airflow ducting can sometimes involve temperature differentials between the air being handled and the external environment, which in some cases can lead to the potential for condensate to build up. In addition, extraction systems are often handling very moist air (e.g., when being used to draw steam away from product heating systems). If excessive condensate is a potential issue then the ventilation/extraction system should be designed and positioned in such a way that any condensate is controlled in its flow and is not allowed to pose a contamination risk to the products in the manufacturing environment. Condensate is frequently linked with product microbiological contamination and therefore such systems must be carefully designed and monitored on an ongoing basis.

Lighting in the food manufacturing area will typically be inset into the ceiling and covered with clear plastic diffusers, or housed in sealed plastic units fixed to the ceiling. Such an approach is designed to reduce the potential glass risk associated with lighting units.

Good lighting levels in a food manufacturing environment are of vital importance to operational health and safety. Lighting aids operators in their work and supports effective visual inspections of products and area hygiene standards. In some countries there are legislative standards set for the levels of lighting required in manufacturing environments.

In food manufacturing areas, where the product is particularly subjected to visual quality inspection, the business should ensure that these checks are completed in an area which has very good lighting levels and may wish to also consider the use of 'natural light' systems (e.g., bulbs assured to emit the same wavelength as sunlight). Such lighting is often utilised in areas where some types of artificial lighting could affect the accuracy of visual assessment.

Lighting materials and designs must facilitate ease of regular cleaning and the avoidance of becoming debris traps. These lights must not pose a direct contamination risk to the food products. As lighting often necessitates the use of glass light bulbs/tubes, typically food safety standards will expect all glass materials to be covered/sealed within clear plastic diffuser units in order that the lighting is protected from breakage and will be contained in the event of the light shattering.

It is also possible to apply plastic shatterproof sleeves to light bulbs/strip lights which can provide a good degree of glass control/assurance.

When considering light diffusers, it should also be borne in mind that plastic materials which are relatively flexible and robust at a room temperature of 18°C can become very brittle and prone to impact damage at chilled or frozen temperatures. In addition, regular cleaning of such materials can cause the plastics to become brittle over time and therefore again more prone to damage or degradation which can pose a physical contamination risk to the food products being handled below.

For these reasons factory lighting benefits from being designed in a way that the changing of light bulbs/strip lights can be facilitated without the need to enter production areas. For example, 'inset' lighting systems where the bulbs are accessed and replaced via the roof void. This approach is especially useful as the changing of bulbs brings a heightened risk of glass breakage which usually therefore requires the stopping of production when lighting changes have to occur inside the factory area.

5.14 Area refrigeration systems

In food operations where areas are required to be maintained at chilled or frozen temperatures, refrigeration systems will be required. Such systems typically work via the use of evaporator units sited within the temperature controlled area. These units circulate the area air across a cooled surface area (e.g., plates cooled by a liquid coolant) which serves to reduce the temperature of the air via drawing the transferred heat away from the area. The heat taken away from the area is piped away within the system 'coolant' and to complete the refrigeration system cycle the coolant gas passes through a compressor/condenser unit which pressurises the gas back to a less active state while releasing its heat. The coolant is then ready to be recirculated through the system.

As heat is an output of the compressor process this phase is typically located in a factory engineering 'plant room' away from the manufacturing process. The plant room will usually have louvered vents to the outside of the building which are designed to facilitate good airflow of the heat away from the building (this assumes that the factory is not utilising a heat recovery system). These louvered vents can be covered with a mesh screen to help prevent pest ingress.

Condensate pipes are usually a feature of area evaporator units, serving to draw this moisture away from the unit. As condensate is often associated with the potential for microbiological harbourage and growth it is of vital importance that the condensate is efficiently collected from the unit catch trays and transferred to the drain system without any leakage. For these reasons evaporator units should not be positioned above areas where food products may be stored, or where process machines, utensils or packaging may be present. In addition, the condensate drain pipes should lead directly to the floor drain, ideally stopping a few centimetres above the floor drain point to provide a break in the pipework which helps prevent any microorganisms present in the floor drain from being able to transfer up the pipework and into the evaporator unit.

Ensuring that condensate pipes are designed to feature a 'U' bend or an 'S' bend will also help to ensure that any microorganisms which are able to become airborne from the floor drain area (e.g., perhaps via the creation of aerosols while cleaning the drains) will not be able to easily travel all of the way up the pipe into the refrigeration unit, as the bend serves as a water trap point. Such systems should however be avoided in storage areas maintained at 0 degrees centigrade or below, as the standing water in the pipe is likely to freeze and therefore cause a pipe blockage or breakage.

As evaporator units provide significant potential for the harbourage of microorganisms, including *Listeria*, their location and design should facilitate the routine treatment of the units with antibacterial compounds which once applied will provide a significant period of protection before requiring reapplication. Such treatments should be placed on the area cleaning schedules to help ensure that these hygiene aspects are regularly addressed and not overlooked.

Air socks are often utilised in chilled manufacturing environments to reduce the added wind-chill factor for factory staff due to cold airflows generated by refrigeration systems in the area. Air socks are usually attached to the ceiling and to the front face of the evaporator unit, the chilled air blows into the air sock, and slowly dissipates into the area through the mesh material, thereby eliminating area airflows and so reducing the wind-chill factor.

Figure 5.5 Area air socks.
Source: KE Fibertec NA, Inc.

As air socks are made of material it is important that they are cleaned regularly and sited in a position which avoids contact with food debris, otherwise the air socks can soon become sites for microbiological growth (particularly mould) which would then cause the air socks to become a significant distribution system for microorganisms in the area.

Factories may choose to operate air sock systems in both low-risk and high-care/high-risk environments. Where this is the case it is very important for the control of microorganisms to ensure that the low-risk and the high-care/high-risk air socks are washed separately and ensured to never be used in the incorrect area. Scheduling controls and use of different colour air socks for low-risk areas can both help in avoiding such issues.

5.15 Equipment (including hygienic design)

Food processing facility equipment is by its nature regularly in contact with or in close proximity to the food being processed and therefore has a significant potential to affect food quality and food safety.

Equipment can pose a range of food safety hazards including:

- physical (e.g. broken pieces; flaking paint);
- chemical (e.g., oil/grease/lubricants from machinery; cleaning chemical residues);
- microbiological (for example, dead-spots/micro-traps/harbourage points) and
- allergens (e.g., difficult to clean points within the equipment/product flow due to poor design or deterioration).

For such reasons all equipment utilised in food manufacturing areas should be hygienically designed and well maintained.

Hygienic design features can include	
Process areas	Use of smooth and impervious surfaces for floors and walls. Use of coving to make all wall/floor joins as cleanable as possible. Ensuring that areas are subject to adequate lighting. Avoidance of glass and use of protective covers where glass is unavoidable. Machinery/equipment must be positioned so as to enable ease of access for cleaning and inspection. Machinery and equipment must not be located in areas where the potential for contamination is increased, for example, close to hand wash sinks or waste disposal bins.
Machinery	Appropriate construction materials which are sufficiently robust, smooth and impervious. Must be designed to be easily cleaned with access to all food contact points. Where food contact is likely the materials utilised must be assured to be 'suitable for food contact'. Avoidance of painted or rusting surfaces which can flake into the product. Any materials which are susceptible to wear/breakage should be manufactured in a highly visible colour to aid detection if necessary, for example, blue conveyor belts. Ensure that all points are accessible and cleanable. Leads, cables and pipes to and from machinery must be designed to be cleanable and not to form food debris traps.
Utensils	Colour coded to avoid inappropriate use/cross-contamination. Where food contact is likely the materials utilised must be assured to be 'suitable for food contact'. Made from materials which are suitably robust, smooth and impervious. Well maintained or replaced regularly to avoid the build-up of scratches and pitting which can harbour food debris/allergens and make cleaning difficult.

It is good practice to subject proposed food operation equipment to thorough food safety risk assessments prior to purchase/installation. In addition, the layout/organisational flow of equipment also requires careful attention to avoid potential food safety issues.

Washing facilities, sinks and waste disposal systems should not be positioned close to production lines as such systems can generate food safety risks to the products nearby.

Equipment which can move between production areas must be ensured to be maintained in good hygienic condition, as often equipment that is 'shared' between areas can lead to confusion as to how frequently it is cleaned and by who.

Key points when selecting food processing equipment include:

- Confirm that the machinery can achieve the clearly defined business requirements (e.g., product quality, safety criteria, throughput rates, yields, etc.). Such points can be confirmed initially via supplier discussion, review of references from other business users and then specific trials on the food products intended to be processed.
- Ensure the correct 'build specification', for example, material types (e.g., stainless steel grade); IP rating (ingress protection against debris/moisture); user-friendly control interface; avoidance of glass where possible due to foreign body risk.
- Accessible for maintenance and easily cleanable.
- Reasonable recommended parts/spares inventory and avoidance of excessive lead times on replacement parts/repair/service.

With regard to what is a 'reasonable' recommended parts/spares inventory: While selling a machine the manufacturer should also provide a recommended spares list which the user should keep in stock as these parts are expected to wear out routinely over time. This list can be a good indicator as to how robust/'built to last' the machine is.

In the food manufacturing sector where products often go from being 'in demand' to being delisted as 'last year's fashion', many food production businesses choose to reduce their costs by purchasing second-hand equipment. There is often a wide range of food processing equipment available; however when developing its process, the business must always be careful to select the right types of equipment to create the desired end products, and avoid compromises which may arise if the piece of second-hand equipment available is not quite right for the production process requirements.

Food processing equipment can take many forms including:

- size/shape processing (e.g., dicers, slicers, mincers, grinders, choppers);
- mixers/blenders;
- heat processing (e.g., retorts, cooking vessels, heat exchanger units);
- dosing/depositing systems (e.g., pneumatic dosers and multihead weighers);
- packaging machines (e.g., vertical form fill seal, thermoform, flow-wrap, tray seal, rotary systems), and
- cooling/freezing systems (e.g., blast chillers, nitrogen tunnels, spiral chillers/freezers, 'individual quick freezing' systems).

5.15.1 Hygienic design

Hygienic design at the process specification and design stages is crucial to help avoid potential food safety problems. Hygienic design is often linked to ease of access and ease of cleaning and should therefore be designed into machinery/equipment from the outset.

The European Hygienic Engineering & Design Group (EHEDG) is a consortium of equipment manufacturers, food industries, research institutes as well as public health authorities. It has the specific aim to promote hygiene during the processing and packing of food products, and improving hygienic engineering and design in all aspects of food manufacture.

European legislation (EC Directive 2006/42/EC for machinery, EN 1672-2 and EN ISO 14159 hygiene requirement) requires that handling, preparation, processing and packaging of food is achieved hygienically using hygienic machinery and in hygienic premises.

Good practice in hygienic design includes: Materials and surfaces in contact with food must be 'fit for purpose' (e.g., resist corrosion, be impervious/nonabsorbent and easy to clean); such materials and surfaces must not transfer odour, taint or taste to the product, and must be made from nontoxic materials; fluids (both food based and derived from cleaning the machine), must be easily discharged from the machine.

Refer to the EHEDG website and the relevant European Legislation for further details on hygienic design considerations. The 'Further Reading' section at the end of this chapter provides the necessary contact details.

5.16 Washing facilities for utensils and equipment

Manual washing facilities for utensils and small pieces of equipment are often progressed via a segregated 'washing' area, utilising a two or three bath washing system. A typical approach is for the gross soiling/food debris to be rinsed off the equipment first, followed by immersion in the first bath containing a hot detergent wash with the equipment utensils being manually scrubbed clean using brushes/cloths. After the food debris has been removed via the detergent wash, immersion in the second bath containing a terminal sanitiser can prove effective against the persistence of bacteria. Finally, the use of a third bath containing a freshwater rinse, or 'air drying' direct from the second bath wash can both prove to be effective systems.

Food manufacturing operations are well advised to trial and develop washing systems to best suit their operations in terms of equipment and food types, coupled with the availability of washing/drying space and the throughput rates required.

The ideal type and concentration of the cleaning chemicals to be used and the optimum temperature of the wash water will vary between operations and is best confirmed by liaison with the hygiene chemical supplier and trialling within the factory.

Larger factory operations often operate an automated tray wash system where items to be cleaned are usually manually rinsed to remove gross soiling followed by being placed on the automated tray washer belt to then be drawn through a series of hot water jets which are dosed with concentrations of detergent followed by sanitiser/disinfectant, ending in a freshwater rinse to remove any remaining chemical.

Following washing, utensils, containers and other equipment can be placed on racks/shelves to facilitate the drying process. All items should be placed in a manner that ensures that the wash water runs effectively off the item and does not accumulate. Such drying areas are often best progressed with a ventilation system as the high moisture environment may otherwise restrict the rate of drying.

To help reduce the potential for microbiological and physical contamination, floor contact should be avoided throughout the washing and drying processes. In addition, equipment utilised for contact with floors (e.g., cleaning equipment) should not be cleaned through the same processes that are applied to clean food contact materials.

All utensils, equipment and containers should be stored in a clean and dry condition. On reuse, if a significant period of time has elapsed in storage then many food production operations would expect these items to be subject to a redisinfection prior to use.

Washing area staff/supervisors should regularly check to ensure that the right cleaning chemicals are being utilised and that the appropriate concentrations are being adhered to.

All washing systems, whether manual or automated, should be regularly reviewed to assure the effectiveness of the cleaning. Reviews can be via visual inspection and also occasionally (depending on the nature of the business) via the completion of swabbing/testing to confirm effectiveness against microorganisms or allergens.

5.17 Waste management

Waste is a necessary output of many food manufacturing operations. Ingredients, work in progress, end products and related packaging can all accumulate and if not handled/managed effectively such waste materials can pose a threat to food safety as well as impact negatively on process area flows and efficiencies.

Containers utilised for the collection and movement of waste within the food manufacturing operation must be dedicated to this sole purpose and must always be maintained in a clean and hygienic condition. Waste should not be allowed to accumulate and stand in factory areas for any significant amount of time as such issues can attract pests and encourage the growth of microorganisms.

External containers for the handling of waste materials must be fully proofed against pest ingress and must be kept covered at all times. These containers should be regularly cleaned and stored away from the production facility so as not to attract pests to the manufacturing operation.

Separation of waste streams can be important within waste handling operations, for example, when the business recycles food packaging materials. All such materials must be accumulated and stored in a manner which does not impact on the food safety of the manufacturing operation.

5.18 Utilities and services

Food manufacturing operations will usually require a consistent supply of utilities which may include:

- electricity,
- water and
- gas.

From the viewpoint of the technical manager, the utility which they are likely to have the most involvement with will be the site water supply, as water may be used on-site for:

- staff hygiene controls (e.g., hand washing);
- factory and equipment cleaning processes;
- product cleaning processes (e.g., washing of fresh produce), and
- use as an ingredient/component part of the end product.

Where the site water is likely to come into contact with the end product, it is important that the technical manager ensures that the water is of an appropriate potable (drinkable) water quality. This is typically achieved via dialogue with the water provider as the water provider should be conducting treatment processes and routine tests to ensure that the quality of the water supplied meets the expected (and often legislative) standards. The technical manager in these circumstances will usually request that the water provider supplies a routine Certificate of Analysis for the water provided which states the standards required and the results achieved. The technical manager should also ensure that the water provider clearly understands the standards required and the necessity to immediately inform the manufacturing site if ever there is doubt as to their ongoing ability to maintain the standards required.

Some manufacturing operations, often in response to the potential for water supply shortages, utilise their own borehole water. In such cases the technical manager should ensure that the water being drawn and utilised on-site is of the required standards, usually ensuring that samples are routinely sent for microbiological and chemical analysis, and liaising closely with colleagues in the engineering department who are often expected to control such systems.

The technical manager should be aware that the appropriate supply of water to the manufacturing site is only part of the challenge to be controlled. Once on-site the water will be transferred around the factory utilising pipework and sometimes also storage tanks, all of which have the potential to harbour microbiological contaminants if not subject to the appropriate controls. Cleaning of water storage tanks and pipework flow routes is therefore important to incorporate into cleaning procedures and schedules.

Some water on-site may also be passed through water softening systems, especially if being utilised for hot water. It is therefore important that the manufacturing site has full clarity on which water supply points in the factory are suitable for food use (i.e., potable) and which points should be utilised only for nonfood use.

The supply of compressed air to production machinery often also benefits from technical management consideration. Compressed airlines can be subject to condensate and also food debris, both of which can cause the accumulation of microbiological organisms which could pose a contamination threat to the foods being produced, especially when in close proximity due to typically being linked to the movement mechanisms of the process machinery.

Compressed air lines therefore require monitoring via both visual inspection and microbiological analysis (e.g., routine swab tests) to ensure that they are being hygienically maintained.

The supply lines/routes of utilities into factory production areas can also pose a significant hygiene risk if not properly designed and maintained. Cables and pipes of all sizes can form effective traps for the accumulation of food debris and moisture, and can be very difficult to gain access to and very difficult to clean thoroughly. The routing of such services within food manufacturing areas should consider how best to avoid the potential for contamination, and the materials utilised for cables/pipework should enable effective and routine cleaning.

Ducting and shielding can often be helpful in reducing the exposure of cables and pipework to potential water/food contaminants. When utilising such options, it is important to ensure that the ducts/shields can themselves be easily cleaned and do not form inaccessible debris traps/hygiene issues.

This is often an area of process operations that needs close control to ensure that both the health and safety requirements and the hygiene requirements sites can be effectively met.

5.19 Maintenance

It is vital that the manufacturing operation is well maintained in order to help ensure ongoing performance. Buildings maintenance and preventative maintenance programmes across process lines are key components of an effective engineering service to the business.

Such systems serve to catalogue and prioritise all preventative tasks and particular repairs associated with the maintenance of the operation and also aid the distribution of these duties across the engineering/maintenance team.

Many audit standards expect that process equipment which is defined as of major importance to product safety, quality or legality must be clearly covered by the site preventative maintenance programme. The use of computerised service/breakdown records can aid the analysis of process faults over time, highlighting where an aspect of the operation may be frequently found to be failing and therefore in need of additional focus/development.

Ideally engineering work should be completed on equipment outside of the food manufacturing areas. Engineering work within food processing areas should be subject to risk assessment in advance of commencement to ensure that product safety is not threatened by the tasks to be conducted and where necessary to preplan any particular additional precautionary controls to be implemented in support of such work.

Where significant work has been conducted on process areas/machinery the business should have food safety and hygiene systems in place which require a thorough clean and inspection of the area/machinery in advance of return to food production. Typically, the inspection and sign-off of approval to recommence production will be completed by a member of the quality/technical team alongside a member of the production team. Records of this process should be kept in order to provide evidence of appropriate control in the event of a subsequent issue.

The use of 'temporary repairs' across the site should be kept to a minimum (e.g., equipment being 'patched up' to keep working until a more permanent fix can be implemented). Such repairs must always ensure that they do not carry an additional risk to the safety of the food products being manufactured.

The engineering/maintenance team should always ensure that any materials used which have the potential for food contact must be assured to be food-safe. For example, lubricants utilised on the moving parts of food process machinery should be based on food oils and should be assured to be free from allergens.

Those charged with maintenance across the manufacturing operation are in a particular position of requiring control to avoid the potential for spreading contamination across processing areas/production lines. Engineers and other relevant technicians should therefore receive training as to the nature of the operation and the risk aspects associated with each element of the process.

Maintenance staff must be fully aware of and compliant with all staff changing and hand washing procedures when moving across the manufacturing site. Toolkits used by the maintenance staff should also be subject to hygiene expectations, being routinely cleaned and well maintained. When entering into high-care/high-risk manufacturing areas, any tools required to be taken into the area should be subject to cleaning and disinfection procedures on entry to the area. Training should incorporate relevant contamination threat categories of physical, chemical, microbiological and allergens.

Where contractors are utilised for maintenance and engineering aspects across the manufacturing site, these contractors must be formally briefed on the control requirements of the manufacturing site in advance of entry to the operation, and must confirm their acceptance to work within the hygiene control requirements of the operation.

5.20 Conclusion

The subject area of 'site standards' within a food manufacturing facility covers a very wide range of matters related to hygiene and process control. The key role of the technical manager is therefore to ensure that all technical requirements/expectations are clearly defined, that the senior management team provide sufficient funding for such matters, encourage effective staff training and also routinely review the control of all related site requirements.

Further reading

Campden BRI: https://www.campdenbri.co.uk

Guidelines on air quality standards for the food industry, No 12.

Guidelines for the hygienic design, construction and layout of food processing factories, No 39.

Chilled Foods Association (CFA). https://www.chilledfood.org CFA represents many of the United Kingdom's chilled food manufacturers. The CFA champion best practice across the industry and campaign actively on its behalf.

Codex Alimentarius Commission. http://www.codexalimentarius.org

The Codex Alimentarius, or 'Food Code' is a collection of standards, guidelines and codes of practice adopted by the Codex Alimentarius Commission.

EHEDG – European Hygienic Engineering & Design Group. https://www.ehedg.org A non-profit consortium of equipment manufacturers, food producers, suppliers to the food industry, research institutes and universities, public health authorities and governmental organisations.

Klipspringer Ltd: Cleaning equipment and temperature probes. http://www.klipspringer.co.uk

SYSPAL Ltd. Supplier of stainless steel products and services to the industry. https://www.syspal.com

Teknomek TM. Design and manufacture of stainless steel equipment. http://www.teknomek.co.uk

UK HSE (Health & Safety Executive) Information Source: The hygienic design of food machinery: Supply of Machinery (Safety) Regulations 2008 & Machinery Directive (2006/42/EC). http://www.hse.gov.uk/foi/internalops/ocs/500-599/520_22.htm

Product control and hazard analysis and critical control point (HACCP) considerations

6

Mark Swainson
National Centre for Food Manufacturing, University of Lincoln, Holbeach, Lincolnshire, United Kingdom

6.1 Introduction

Food production operations typically apply a range of systems and procedures to control product safety and legality. These activities usually combine to form the site hazard analysis and critical control point (HACCP) system and underpinning prerequisite programmes. This approach is common in the food sector which is to be expected as HACCP is a product safety control system recommended by the Codex Alimentarius Commission and the use of HACCP to assure product safety is a legal requirement for food manufacturing businesses operating in many countries including across the European Union (EU).

These systems typically involve the food operation seeking to control food safety and legality from the starting point of the supplier base (often covered within the prerequisite programmes under the term 'supplier assurance programme'), through HACCP system and prerequisite programme control of factory processing, storage, distribution and end use by the customer.

One of the most vital factors in food safety control is to ensure that each specific product is designed and developed with full consideration of safety, quality and legality at each stage. While this sounds obvious, all too often such key requirements can be overlooked during the product development process, leading to an intrinsically flawed end product which at best will take significant time and resources to correct prelaunch, and at worst will result in the sale of unsafe foods if the problem is not highlighted.

This chapter considers aspects of product safety, quality and legality control linked to product 'development', 'scale-up' and 'launch' operations in order to provide an insight into a range of key considerations required.

Swainson's Handbook of Technical and Quality Management for the Food Manufacturing Sector.
DOI: https://doi.org/10.1016/B978-1-78242-275-4.00006-X

6.2 Product control in design, development and launch

6.2.1 New product development

New product development (NPD) involves taking a 'new' product from generation of the initial 'idea'/'concept' through a series of steps/stages which if successful ultimately lead to the full-scale manufacture of the product and commercial supply into the food chain.

The term 'new' suggests that the particular product has not previously been manufactured and supplied by that manufacturer to the particular customer before. Product can be 'new' to the food market (innovation), or could just be 'new' to the company manufacturing the product. The product could also be 'new' by taking an existing product and changing it, perhaps to increase customer appeal or to improve its financial performance in other ways.

'Critical paths'/'project management': As there are a number of steps to be taken between the points of 'concept' and 'launch', with a great number of parties involved (including a range of business departments such as purchasing, development, technical and production, suppliers of ingredients, packaging, machinery and distribution services, artwork houses and enforcement authorities such as those involved with aspects of trading standards and environmental health), most businesses will apply project management techniques to each proposed NPD process in order to ensure that all steps are progressed by the right parties, in the correct sequence, and to the required timescales. Such planned approaches are sometimes called 'critical paths', where each 'critical' step/stage is defined and tracked through to completion.

As new products are a key driver for business success, to be efficient a business must seek to reduce its percentage of new product failures (as each failed product is effectively the loss of a significant financial investment).

Losses due to failed products include 'time and effort' (financial costs and also during the time committed the business could have been focused on more profitable activities), customer sensory evaluation panel and sales review costs (including trips to the customer for NPD-related reviews/meetings), packaging costs (including any 'minimum order quantities' committed to), ingredient costs, testing costs (including microbiological shelf life evaluation and nutritional), customer launch attendance costs, production line time and labour costs, etc.

Cost aware businesses are likely to track their percentage success rate for newly developed products and will take steps to help increase their 'success to failure' ratios. These steps can include thorough screening of initial ideas and market/consumer group analysis to help ensure that the product idea/concept has a good chance of success.

In view of the high business costs associated with a high rate of new product failure, it is vitally important that every stage of the NPD process is appropriately resourced and progressed as thoroughly and as professionally as possible to give the maximum chance of product success.

6.2.2 The key phases of new product development: overview

The commitment or instruction to develop a new product will typically have been initiated by the business requirement to:

- place a new product on the market (addressing a perceived sales opportunity);
- replace existing products in a range which has declining sales;
- add to an existing product range;
- use spare capacity within the factory to create saleable end products;
- use waste products generated by the current food manufacturing processes; and
- respond to market conditions. (e.g., making health claims, reducing price).

There are four phases in the NPD process:

6.2.3 New product strategy development

This first phase involves defining the intended product and perhaps also the 'product range' within which it might be placed. Development of the business strategy with regard to the intended new products/product ranges will include aspects of market analysis and business 'manufacturing and supply' cost estimation reviews. Once formed the plans for the intended new products should be documented for review and approval by marketing/sales and technical departments.

6.2.4 Product design and process development

This second phase involves the design of the intended product and development of its manufacturing process. Key aspects will include: design, review and optimisation of prototype products; confirm suitability to both commercial and technical requirements; market acceptance/consumer testing; develop a sales plan and volume predictions; analyse sales predictions and operating costs to confirm that the intended product is commercially viable.

Progressing trial production runs on the intended manufacturing lines is important to ensure that the trials will reflect full-scale production in due course. This typically involves running small-scale trial production volumes in order to create less waste, reduce cost and avoid interference with the current standard production operations. A key requirement at this stage is to manufacture the product to match the product samples that have been agreed with the customer and which up until this point may have been made only in the development kitchen.

Many businesses expect to run a minimum of three successful trials to provide assurance of process capability and to yield sufficient samples for photography, nutritional analysis, microbiological testing, confirmation of preparation/cooking instructions, transit/travel tests and organoleptic shelf life testing.

6.2.5 Product commercialisation

This third phase includes the progression of pack labelling and artwork designs: marketing activity; production trials and the setting-up of quality assurance procedures. Businesses moving into this phase will typically be incurring significant product development and launch costs. The marketing plan and sales predictions will be checked and used to define the planned first production volumes. A 'technical review' of the product should be completed by this point, and this review should serve to define and test all controls related to product safety, quality and legality. At this stage the manufacturing department should review the intended product volumes and production processes in order to form a production plan; by this stage the sales and production reviews will have yielded a great deal of information with which the business finance department will be able to complete a final review of manufacturing costs, sales prices, intended profits and any necessary investment requirements. All of which are intended to ensure that the business is likely to receive a sufficient 'return on investment' (ROI). Throughout this third phase there is likely to be regular correspondence with the intended sales outlets for the product.

6.2.6 Product launch and evaluation

This final phase of the NPD cycle involves the planning and organisation of the manufacturing department to produce the launch volumes required, launch the product and very importantly progress the completion of the 'postlaunch review'. These activities will include the organisation of full 'product roll-out' marketing activities; production and planning processes and also distribution of the initial launch and ongoing volumes to their required locations. If supplying a retailer branded product then this stage is also likely to involve a 'first production attendance' by the customer technologist and/or additional samples to be taken for quality review. This stage will also involve the completion of a postlaunch product and production process study for quality and efficiency review (which often requires the 'process technologist'/'process development team' to check and confirm that the product is running smoothly through the production system and then fine-tuning/trouble-shooting where necessary).

6.2.7 Business factors which impact on the new product development approach

In the food industry there is no one standard approach toward the stages of NPD. Business factors which impact on the NPD approach taken vary greatly, for example:

- Business strategy: Is the operation 'sales driven'? Sometimes a company will be less concerned as to the commercial viability of projects if its current strategy is to grow the total sales volume of the company or to meet customer requirements at any cost;

- Business size/resources: some businesses can afford structured and defined business teams, often leading to more formal processes. Whereas other businesses have fewer staff/resources leading to a smaller number of staff trying to cover everything;
- Business 'risk averse' vs 'risk takers'. Some businesses are very cautious and take a long time to develop and test new products/technologies. Whereas other businesses are willing to rush to market with limited resources made available for product testing or market research;
- Business customers: in some cases, especially when producing foods for a retailer 'own-brand', the specific customer (in this case the retailer/supermarket) expectations can define the key stages of the NPD process;
- Timescales to launch: some businesses may have the luxury of working up to a year in advance of market launches; however, most food manufacturing businesses are working 1–4 months ahead. There are also sometimes 'last-minute rush' products to carefully control (or guard against) when the business or customer wants to see the product launched as soon as is possible;
- Culture: communication and an inclusive team approach is critical to successful NPD. Anything less can cause serious issues/faults; and
- Strength, knowledge and experience of key departments (e.g., sales, development, technical/quality, production, engineering, purchasing and finance). All departments must be effective in order to best progress the NPD process. Any weak links can lead to mistakes being made or issues being overlooked.

There are many aspects to the development of a new product, a number of which are considered in this chapter. It should be noted that many businesses do not conduct each of these phases as separate and specific endeavours, instead many businesses form an approach that best suits their resources and time available.

Although NPD methods can vary greatly across businesses, a common factor is that a multidisciplinary team approach works best when progressing such a complex range of interconnected activities. Systems which ensure the focus and commitment of all departments usually ensures the best chance of success.

The food business operator needs to define the key NPD process stages/tasks to ensure that all requirements are covered. The use of critical paths/checklists significantly helps to control such matters.

6.3 The key product development phases in detail

6.3.1 Phase 1: Product strategy development

The NPD Phase 1 process of defining the new product includes: development of the business product strategy (e.g., new products/product ranges of interest) including consideration of the development options available; market analysis (including estimations of sales/profit to confirm whether the product will be 'commercially viable'); forming of draft specifications/blueprints for the intended new products which can then be reviewed and eventually approved by marketing/sales and technical departments.

Developing a new product strategy for a food business is often conducted by one or a number of product development and marketing/sales team members. The strategy is likely to be influenced by:

- 'Blue-sky' thinking: dreaming up new ideas from scratch. This requires a significant amount of creativity and perhaps also innovation of the process technologies to be utilised.
- Sector trends: (e.g., indulgence, healthy eating, sustainability, convenience, provenance and quality);
- Processing and packaging technologies available;
- Consumer group preferences. (e.g., how will the likely consumers feel about the product price and intended preparation processes); and
- Competitor review: define competitors, then review and react to their position (e.g., products/ranges, price points, pack/label claims, packaging and marketing approach).

Product formulation and packaging type are likely to be influenced by: intended consumer (including age and gender); target price (with consideration of consumer affluence, competitor product pricing and business operational costs/sales margin requirements): consumer panels/surveys; market analysis (possible sales volumes and growth potential).

Technical considerations at this stage may include:

- Is the new product 'seasonal' or 'all year available'? – if ingredients are required to be purchased 'fresh' then certain ingredients may have seasonal availability issues which could prevent them from being able to be used all year round. In such cases there may need to be technical arrangements made to source from other regions or switch to alternative sources (e.g., ambient stable or frozen formats);
- Intended pack 'portion size'. – this will usually be influenced by the target market and other products currently available in the sector. Where pack health claims are being made the portion control will also be required to support the claims being made and this will therefore necessitate close weight/volume control in the product packing phase.
- Product shelf life requirements – Longer shelf life requirements will affect the process technologies, product formulation and packaging to be used (e.g., use of appropriate processes, stabilizers, preservatives, use of acidity (pH) or water activity (aW) control to control microbiological activity and extend shelf life).

Once the new product ideas are taking shape the food production business will usually expect the development team to present the proposed product/range to key managers/staff in the areas of operations, technical/quality, purchasing and finance for consideration of any potential issues and check/comment on feasibility.

This wider business review of the NPD proposals can be achieved via the circulation of a product development summary brief containing:

- proposed product and (if relevant) product range: ideally including samples or pictures;
- sensory description: appearance, aroma, taste, texture, colour, viscosity, etc.
- intended sales price: with detail on product manufacturing costs and margin requirements.
- detail of any new equipment or other capital spend required to enable the product to be made;
- pack size and portion size;

- packaging type;
- Shelf life and storage requirements;
- distribution requirements;
- target customers;
- labelling claims required (e.g., 'free from', healthy claims, regional provenance).
- production/sales volume estimates;
- timelines for product trials and product launch date; and
- confirmation of whether the product will be 'seasonal' or 'all year round' supply.

'Technical feasibility': the review of whether a new product is likely to be feasible from a product safety/quality/legality viewpoint should be conducted by competent and experienced technical personnel and will include the following considerations:

- will the product be safe to consume over the intended shelf life? (e.g., what are the likely microbiological risks and necessary controls?);
- is the product likely to meet the sensory/quality requirements over the intended shelf life?
- are the proposed ingredients available and to the required specification? Are there likely to be any seasonal supply issues?
- can the product be produced utilising the existing production processes? And
- can any new process/technology required to be developed, installed and tested/validated within the required timescale?

These points are a general first review/consideration in order to flag up any likely concerns as early as possible. The more knowledge and experience the technical team has, the more reliable this review of 'technical feasibility' will be. There is a danger that a lack of sufficient knowledge may lead to issues being missed or to the technical team blocking ideas due to 'over-caution', rather than for 100% sound reasons.

'Financial feasibility': To be commercially viable businesses must avoid working to produce an attractive and safe product if it is not likely to be financially viable. The financial feasibility review should analyse and draw conclusions as to whether the product can be made to the required price point and deliver the required business and customer margin aspirations once all cost factors have been accounted for.

Considerations in this financial viability review will include: recipe cost, packaging cost, minimum order quantities, processing costs (e.g., labour, utilities, yields, throughput analysis), storage and distribution costs, and marketing costs (including any promotional activity).

Financial feasibility studies are often conducted by finance/management accounts team members and can be completed quickly if all relevant information is up to date and held on spreadsheets/databases and therefore easily applied to each new project.

Such new product feasibility review process can also be repeated with the intended end customer. For example, via presentation of samples at customer development/sales panels. Often businesses have already reviewed the feasibility of a project with the customer before gauging the views of the internal business team.

This can sometimes lead to problems when a customer has indicated a keenness to take the product and commitments have been given, only for the internal business reviews to then highlight that the product is not feasible/viable.

6.3.1.1 Project management and critical paths

Once a project is confirmed as technically feasible and financially viable, clear management is required to ensure that all necessary resources will be available/applied and that all teams are clear on their responsibilities and timescales required to hit the launch date.

Often businesses will use a checklist approach to the project management process and when combined with intended/required timescales such an approach is often called a 'critical path'.

The project critical path defines the specific tasks required to be completed, 'by who' and 'by when', in order to meet the project deadlines (and thereby ensure that the launch is kept on schedule).

Such specific tasks on the critical path of a new product launch can include the recipe confirmation date, product trialling and specific testing dates (e.g., shelf life assessment, nutritional testing, transit testing), packaging generation and printing dates, etc.

During the early phases of considering a new product, the business should reflect on whether it has the required resources to facilitate the development and launch processes.

Resources required may include:

- people (numbers available at the times required, training requirements, pay rates);
- management (extra people/shifts may necessitate extra supervision/management);
- CAPEX (capital expenditure): any new equipment required will often need to be justified against a business plan;
- storage space for increased raw materials/finished goods;
- segregated areas required for certain ingredient types (allergens) or packaging types (e.g., glass).
- new packaging design: in-house pack design capability (or sufficient finances to outsource); and
- market/consumer research.

6.3.2 Phase 2: Developing the product

Once Phase 1 'product strategy development (planning phase)' has been successfully completed (or at least sufficiently progressed), there is now focus on the requirement to fully develop the product.

This Phase 2 of NPD aims to: confirm the most appropriate method for product manufacture; confirm that the end product made will be able to be packed within the required packaging; confirm that the product can be made consistently to the required standards; confirm that the necessary product safety factors are consistently achievable; and confirm that the manufacturing methods and timescales will deliver a product which meets the required financial targets.

Typical stages of 'new product development' at this stage can include:

'*Bench work*' (typically small scale in a kitchen environment). This activity is called 'bench work' because most equipment used at this stage is very small scale (i.e., can fit on a table/bench). Here the business is seeking to confirm that the product quality is achievable and reproducible before tying up larger amounts of ingredients and process line costs. Samples produced at this stage can be held for reference against full factory trials and can also sometimes be used for preliminary pack artwork/nutritional/microbiological evaluation.

At this stage typically the development/process technologist and members of the technical team will be checking a range of points including: Are all proposed ingredients commercially available and to the required quality/specification (including: organic? vegetarian? allergen free? additive free? etc.); does the recipe meet all 'requirements' (e.g., customer requirements? expected nutritional standards? defined specific legal standards?); defining the best production method for the larger scale trials; conducting product assessment including 'organoleptic' and 'microbiological' shelf life reviews to confirm that the shelf life is safe, commercially viable and likely to be consistently achieved by the intended recipe and process; double-check product costing assumptions; confirm that the intended packaging is 'fit for purpose' (this can include a range of activities including specification of appropriate tolerances and checking of 'fitness for food contact', process line testing, shelf life and transit testing).

'*Pilot plant*' (small-scale equipment used to mimic the larger full-size factory equipment) can be utilised as an intermediate step between bench testing and full-scale factory trials. Product and process development work at this stage will give greater confidence that the product which is intended to be manufactured can be successfully produced on the full-scale production lines.

Typically, such trials are run on small-scale equipment which is situated out of the way of the standard production lines to avoid disruption. The smaller production volumes required by pilot lines serve to save time and cost while still achieving a trial product representative of the full-scale commercial production system.

At this pilot plant stage, there is the opportunity to recheck and confirm the range of product and process development points assessed at the bench testing stage including confirmation that: ingredients will be commercially available and will perform to the required quality/specification; product meets all manufacturer, customer and legislative requirements; the intended production methods are suitable; organoleptic and microbiological shelf life reviews are successful; product costing assumptions are accurate (including labour costs and run/throughput rates); packaging is fit for purpose.

'*Production scale trials*' are factory trials utilizing the standard production equipment/lines. Sometimes businesses will trial run the minimum batch size possible (to save cost/wastage). The use of the actual production equipment will give the best indication of any likely issues as full-scale production trials allow the assessment of the full standard production practices including routine product safety/quality monitoring , volume/throughput/labour and other costs.

Alongside checking ingredients, recipe, manufacturing method, product quality/safety over shelf life, costings and packaging fitness for purpose, these full production scale trials also enable a business to confirm/check a range of details including quality assurance points, product safety/HACCP, yields, throughput and packing rates, line hygiene methods and staff training requirements.

'*Large-scale consumer testing*' can also be facilitated by high-volume production trials, perhaps supplying product locally or to a limited number of stores to test the market. It should be noted that such consumer testing should only be conducted once the product and related processes have been confirmed to be safe, of the required quality and legally compliant.

6.3.3 Phase 3: Product commercialisation

'Product commercialisation' can be considered a key final phase of 'taking stock of the current position' before committing to the major expenditure of launching the product. In order to decide whether to progress to the full-scale launch the business management should review and consider a wide range of aspects including:

- Production/product feasibility, including consideration of any issues raised by the work conducted so far. Are both the product and process now clearly defined and performing consistently? Has the safety of the product/process been validated (if required)? Has shelf life testing confirmed that product safety and quality is as required over the entire shelf life of the product? Is the HACCP system confirmed to be set as required?
- Viability of the proposed market strategy including a review of predicted launch costs (including costs of advertising and promotional activity) and ongoing operating costs to enable an estimate of the likely financial returns/investment payback timescales.
- Availability of required resources. For example, any CAPEX (Capital Expenditure) required, equipment, staff, time, etc. Is the operations team ready to progress the launch with sufficient capacity and a clear production plan?
- Confirm packaging designs (including artwork and fitness for purpose).
- Process engineering (including focus on the successful implementation of the new product and any associated new machinery).
- Reconfirm 'strategic fit' with current business aims/priorities. It is surprising how many businesses get to this stage before reviewing whether launching the product is appropriate at this time. Sometimes this can be due to market conditions having changed significantly since the outset of the new product initiative.

This stage requires the engagement of a wide range of departments including development, production, sales, finance, technical and engineering. The stage therefore benefits from the use of a critical path approach with routine meetings to review progress and keep all stakeholders informed as to the current status and intended timelines.

6.3.4 Phase 4: Product launch (and 'postlaunch review')

The 'product launch' phase will typically involve coordinated action from a range of business departments, all focused on ensuring that the product is produced to the required standards and delivered into the supply chain 'on-time and in-full'.

Marketing/sales: Will inform (and sometimes provide training/training material) store/sales/restaurant staff operating at the point of sale to ensure that the product is handled and promoted correctly. The marketing/sales team will also confirm customer approval/acceptance of product and ensure that all plans and timescales for launch and ongoing supply are clearly communicated to the customer purchasing teams and also the manufacturer planning teams. Arrangement of promotional activity and advertising are often a key role of the marketing team as are the setting of targets for success and the progression of responsive strategies to be deployed if the product is not an immediate success.

Operations team: The 'operations team' of the manufacturing business will typically include the production, quality/technical and engineering departments. The production team with support from engineering will be expected to have the production lines operational and have sufficient capacity to manufacture the required products, in the required quantities, at the required time. Product safety/HACCP and quality assurance systems will be expected to be in place with all personnel trained on the correct manufacture/handling of the products and the associated tests/checks required.

Ingredient purchasing/buying department: These team members will be expected to have ordered the required amounts of ingredients and packaging stocks, and must have ensured that they are delivered within the required manufacturing timescale window.

Sales and technical teams will be expected to make provision for the customer (e.g., technologist/quality manager) to attend the first production runs/launch if required. Also, the manufacturer and perhaps the customer may require additional samples to be taken from the first production runs for further sampling/assessment to confirm that the required quality/safety/legality standards are being achieved.

The business may need to manufacture and build (accumulate) sufficient finished product stocks in readiness for the launch date. The business logistics/despatch departments will then be required to outload the product to the defined distribution route/destinations at the required time to meet the delivery schedule required by the customer.

6.3.4.1 Postlaunch review and evaluation

A key phase to help ensure the ongoing success of the launched product is that of 'postlaunch review and evaluation'. This phase includes the following checks and potential corrective actions:

- Review sales figures against original targets. If sales are lower than expected there may need to be action taken to help increase sales such as additional marketing activities. In addition, such information will be helpful in reviewing whether raw materials and packaging are over-ordered/over-stocked with corrective action to be taken if necessary. Similarly, if sales are progressing at significantly higher volumes than originally planned then this may require early correspondence to the purchasing and production planning teams in order that they can prepare accordingly.

- Review '*product performance*' at all key stages including raw materials/packaging supply, production line processing (e.g., are production labour costs, yields and throughput rates all in line with the costed standards?), storage (including whether the required product shelf life is being consistently achieved), distribution (are the transport arrangements working effectively and meeting the originally costed standards?). This review will provide an opportunity to proactively address any product issues including how efficiently the product is being manufactured and whether there are any quality/variability points evident which will benefit from being addressed. Most of the product performance review is best progressed by a member of the process development or production management team, as they will have access to both the product/process performance data and also have clear detail on the standards being operated to/required.
- 'Point of sale' review can be achieved via in-store (or in-restaurant) visits progressed by the sales/marketing team. Review of products at this point can include whether there is sufficient shelf space/position being made available to the product in-store and whether the product appearance on the shelf is attractive (e.g., product/packaging/label damage may be evident suggesting an issue occurring within the supply chain). Checks can also include whether there is sufficient and relevant sales/promotional materials available at the point of sale.
- Customer satisfaction/feedback can be gathered from direct store/outlet feedback and particularly from analysis of customer comments/complaints. This information will typically be handled by the business sales/marketing and technical/quality team members. Reacting quickly to customer panels and complaints feedback is vitally important in order to ensure that the products being manufactured are meeting all of the standards required and therefore have the best chance possible for commercial success.

As product design and product development practices have a significant effect on whether a new food product will meet all commercial and technical (food safety, quality and legality) expectations, the control ('product control') aspects of such activities are clearly defined in a range of available standards including ISO standards, compliance standards such as British Retail Consortium (BRC) and also many retailer standards.

When a business is seeking to supply into a particular food supply chain it is vital that with regard to aspects of 'product control' linked to NPD, the technical department conduct a thorough review of the relevant and required standards to be followed. These standards will be a combination of legal requirements, customer/supply chain standards and the manufacturing business's own internal operating standards.

6.4 Technical consideration of new product development procedures

The development of new products within a food business, from forming the initial concept through to recipe formulation and finally product scale-up trialling and assessment prelaunch, can be a very fast moving and demanding environment.

Such NPD operations are often complex due to the number of stakeholders involved, including the business:

- development department;
- sales department;
- process development (scale-up/factory trial technologists);
- quality/technical department;
- purchasing department;
- production department;
- engineering department;
- customer (often retail and food-service businesses will also have a number of their own departments involved including purchasing, sales, marketing, development and technical).

It is important to reflect at this stage that each of the stakeholder departments within the food business will have very different skill sets and goals to achieve. For example:

Sales/marketing – Are often creatively and commercially driven and very 'customer focused', sometimes to the extent of seeming to work more for the customer than for the manufacturing business that employs them. Getting this balance right is key to the long-term success of such roles as the sales team member must command the respect and appreciation of the customer and must also be effective for their employer. These roles often require and encourage innovation, 'unique selling points', tight product cost control and fast 'time to market' to help beat the competition.

Development – Are by their nature 'creative', and this can be expressed in wishing to utilise a diverse range of ingredients and processing technologies. Development key performance indicators (KPIs) within a business are often particularly focused on the product organoleptic performance criteria. Due to this fact development teams often wish to avoid being constrained by limited ingredient/packaging ranges and current factory processing considerations, seeing their role as to 'create' and for the process/production teams to then work out how to manufacture and commercialise the end product.

Production department – KPIs are usually focused on efficiency and throughput rates. Yields, processing rates and labour costs are paramount and therefore production teams prefer to run products that are familiar with and which run quickly and consistently down the existing production lines. It is therefore to be expected that these preferences can sometimes be in conflict with the development and sales team demands to produce new, varied and unique products.

Technical/quality – Are focused on setting the blueprint for and maintaining all three 'technical pillars' of quality, safety and legality. While product safety is paramount, there is no point in manufacturing a product that is also not of the required quality and legally compliant. Technical/quality departments will frequently serve as a bridge between development and the production/processing departments, often with the process technologist reporting into or working closely with technical, charged with scaling up new products from development and into the factory.

Process development (scale-up and factory trial technologists) – This is often the central role in taking products from the small-scale development kitchen through to full factory scale commercial production. The process development technologist has to have sufficient culinary appreciation to ensure focus on the organoleptic criteria while also having sufficient 'technical' knowledge of the underpinning food science and technology, combining this with a sound appreciation of the production processes within the business.

This combination of skills/expertise is utilised to achieve the commercial scale production of products which are 'as required' by the customer, running smoothly and efficiently along the production process and which meet all product safety, quality and legality requirements. This role is very demanding, progressing a lot of work directly while also coordinating many of the other departments through the scale-up process. For this reason, the process development technologist will often run and maintain the critical path for all new product launches.

With so many potentially conflicting KPIs and tight timescales within the NPD processes there is significant potential for matters to be overlooked and mistakes to be made. Sometimes new concept products which have already been shown/promised to a potential customer cannot in reality be achieved on existing processing lines. There are also many conflicts of interest which if not watched and managed carefully could impact significantly on product safety, quality, legality and on overall business performance.

Common issues include:

- Timescales committed to are unrealistic/insufficient for all departments to have completed their roles effectively.
- A required ingredient is not available from an approved supplier, requiring the business to deal with potentially unknown suppliers who have to be swiftly reviewed prior to supply for the imminent product launch.
- New packaging is on a long lead time, leaving insufficient time to effectively run all trials (including transit and shelf life testing) before launch.
- Production departments are so busy (or so efficiency focused) that finding sufficient 'line time' to run trials is difficult.
- To assure product safety/legality/shelf life, the product requires processing (e.g., time and temperature for cook-chilled foods) to such an extent that the quality deteriorates beyond acceptable limits. This is particularly a risk when initial development recipes have only been lightly processed at the low volume kitchen sample stage.
- Insufficient consumer testing before launch, leading to customer complaint trends after launch with regard to the quality, consistency or handleability/preparation of the product.

For such reasons it is vital that food businesses develop, implement and maintain a range of NPD control systems and procedures to best support the fast paced, demanding NPD scale-up and launch phases. Ensuring that throughout each phase of NPD all departments are focused on an appropriate range of key product quality, safety and legality factors (alongside the more operational/commercial factors).

6.4.1 Product design and development procedures

These systems and procedures seek to organise and control all stages of the NPD process to help deliver a product which:

- meets the customer's product quality expectations;
- launches on schedule;
- is correctly priced, including meeting the manufacturer and retailer margin aspirations;
- has sufficient shelf life to avoid wastage and encourage sales;
- is 'process friendly' (e.g., delivers a good yield, runs at a fast throughput rate and has competitive labour and utility costs);
- can be consistently made (e.g., no seasonal ingredient availability issues);
- is safe to consume; and
- meets all relevant legislative requirements.

When being audited, food manufacturing businesses will be expected to show particular focus and control on how the site NPD-related procedures help ensure product safety and legality.

6.5 Examples of how new product development (NPD)-related procedures incorporate and help achieve product safety and legality

6.5.1 Hazard analysis and critical control point study

All new products in their design and development phases benefit from being given full food safety consideration utilising HACCP principles. The appropriately qualified and experienced personnel in the business should consider: What hazard categories (physical, chemical (including allergens) and microbiological) will be particularly applicable to the product? Can these risks be controlled via application of the business's current/generic HACCP plan? Or will new/revised CCPs be required to ensure the food safety of the new product?

This is a point that can create problems within food businesses if there is an incorrect assumption that the new product will be covered by the current generic site HACCP plan. Without due formal HACCP team consideration products can have inherent food safety risks which go unrecognised through the development, scale-up and launch processes due to the lack of a formal hazard study being conducted.

It is therefore vital that no matter how fast moving the business NPD process is, there must always be a formal product-specific HACCP study incorporated at a sufficiently early stage within the site NPD procedures.

Within some businesses there is the additional challenge that, as an estimated 8 out of 10 proposed new products do not actually launch, how early in the product concept stages is it appropriate to instigate a product HACCP study if 80% of the time such a study will be wasted time?

There is no standard answer to this question. Each business needs to decide for itself at what point the HACCP study for a proposed new product should be progressed. Conduct the HACCP study too early and not all of the required details may be known (e.g., intended sources of ingredients, full definition of the production process required to achieve all of the organoleptic performance criteria). However, conduct the study too late and there will be insufficient time to correct any problems found before the launch date leading to awkward discussions with the customer on why their launch requirements have not been met, or worse still the business then deciding to launch the product while trying to also fix the problems found at the same time, thereby increasing product quality, safety and legality risks.

As the business sales and development departments are usually the earliest to work on new product projects and concepts, a way to help address the 'HACCP study timing issue' is for the technical department to describe to the development and sales departments a range of key 'early checkpoints' which can be proactively checked for in these initial stages, serving to provide an early alert to the technical team of potential issues:

These early checkpoints can include:

- Does the new product under consideration exhibit the typical properties, and take the standard production process, required to fit into the business generic HACCP plan? (e.g., For cold-blend dips perhaps the business has defined pH control as a key CCP. Perhaps for cook-chill products such as soups, sauces and ready meals the business generic HACCP plan requires all product components to receive a pasteurisation cook using set times and temperatures).
- The business policy on allergen presence, control and labelling (e.g., if the business is currently not handling particular allergens then introducing such materials onto site in new products is likely to require advanced consideration).
- Particular legislation requirements (e.g., for products which will be sold on the basis of health grounds the business will wish to review at an early phase that the necessary product nutritional claims are likely to be achieved).

By instilling some 'technically aware' product safety/legality knowledge within the departments that tend to first be involved with new products, it is possible to incorporate such 'quick checks' within the product development procedures which can then be used to highlight potential problems early to the technical/quality team for further consideration.

The full HACCP study for each new product can then be conducted later in the NPD process once all of the ingredients/suppliers have been secured, pack type/size and recipe formulation confirmed and the production process generally defined. This approach has the benefit of the general assurance that major hazards/issues are less likely to arise at the point of formal 'HACCP study' as there has already been a number of 'technical aware' prechecks earlier within the NPD process.

It is important that all such practices and procedures are formally documented in order to ensure consistency of approach and also demonstrate due diligence within the NPD process to both enforcement authorities and customer/third-party auditors.

This includes formally documenting the 'technical awareness training' provided to the development and sales team members.

6.5.2 Pathogen control and validation

The product HACCP plan within its scope should have highlighted the key pathogens of concern with regard to the product when considering the microbiological hazards involved. Such 'pathogens of concern' will tend to be highlighted via review of their likely presence in the ingredient supply, presence within the in-house process (including staff-related routes of contamination) and also the potential for postprocess contamination.

Product pathogens of concern must be addressed via appropriate product and process design, particularly around the aspects of:

- recipe/formulation;
- the manufacturing process parameters; and
- use of recognised preservation factors.

Use of ingredients which have been sourced to have good microbiological standards achieved by well-managed and consistent processes will provide a good starting point for the control of product pathogen risks.

Often one single processing point can provide the key CCP required for pathogen control, for example, sterilisation times and temperatures in canning processes. However, sometimes a combination of product and process factors are required to be utilised to achieve the degree of pathogen control required. The product formulation, manufacturing process, packaging and storage system can be utilised to address specific product pathogen risks by forming a set of 'hurdles' which when used in combination are sufficient to address the particular risk of pathogen growth/survival.

The combined hurdles cause the microorganisms of concern to become depleted of sufficient energy to eliminate, or reduce to an acceptable level, their risk to the end product.

The use of 'hurdle technology' to control product pathogen risk requires a good in-depth knowledge of microbiology and of the inhibitory measures which can be used to eliminate or sufficiently control pathogens.

Hurdles can include:

- time/temperature (e.g., use of recognised lethality rates to address target pathogens);
- product pH (i.e., use of product acidity levels);
- product water activity (the measurement of 'aW' is linked to the amount of water in the product that is available to support microbiological growth);
- availability of oxygen. (e.g., application of a modified atmosphere within the pack to reduce available oxygen down to levels which would prevent aerobic pathogen survival/growth) and
- use of preservatives.

When applying modified atmosphere packaging (MAP) technology businesses must ensure that while modifying the pack atmosphere to address aerobic microbiological challenges, it must avoid creating a packed atmosphere that would increase to an unacceptable extent the risk of pathogens that prefer to survive and grow in such low/no oxygen conditions, for example, anaerobes and facultative anaerobes.

With regard to the use of the temperature of the product storage/supply chain as a hurdle (e.g., at sufficiently low temperatures food pathogens cannot grow), businesses should reflect on the extent to which they can ensure that the required storage temperatures can be consistently achieved and maintained. This may be straightforward while the product is in the full control of the manufacturing site; however, it should be noted that sometimes storage temperature is not recognised as a sufficient hurdle for the control of pathogens as the consistency of the storage temperatures in the supply chain can sometimes not be assured (e.g., potential for in-transit, in-store and in-home temperature abuse).

By using expertise in understanding microbial growth and limits for growth/survival, hurdle technology can be utilised to preserve food using less severe processes than when relying on just one CCP.

6.5.3 Validation

The manufacturing process should be validated in order to confirm that the product process/hurdles to be applied are sufficiently effective against the target pathogens.

'Validation' has been defined as:

- 'Confirmation through the provision of objective evidence that the requirements for a specific intended use or application have been fulfilled' (ISO9000:2000).
- 'Obtaining evidence that the elements of the HACCP plan are effective' (Codex Alimentarius, 1997).
- 'That element of verification focused on collecting and evaluating scientific and technical information to determine whether the HACCP plan, when properly implemented, will effectively control the hazards' (NACMCF, 1997).

'Process validation is establishing documented evidence which provides a high degree of assurance that a specific process will consistently produce a product meeting its predetermined specifications and quality characteristics' (US FDA).

Predictive microbiology: microbiological modelling programmes such as 'ComBase' and the USDA 'Pathogen Modelling Programme' (PMP) can be used to assess the likely effect on product safety over shelf life when applying one or a combination of hurdles to address the product pathogen risk.

These programmes can also be invaluable in enabling a quick review of the potential product safety impacts of a revised product formulation or changes in postproduction storage/distribution (e.g., for chilled foods, the times and temperatures within the storage and distribution chain).

ComBase. https://www.combase.cc/index.php/en/

PMP, USDA. http://ars.usda.gov/services/docs.htm?docid = 6786

'Challenge testing' can also be utilised if the pathogen risk of survival/growth is not felt to be sufficiently understood. Challenge testing involves introducing the selected risk microorganisms into representative samples of the product, and then assessing the survival and growth potential of these pathogens when processed to the extent that the routine process will achieve, followed by holding within the typical production and distribution processes (e.g., storage and distribution times and temperatures for chilled foods).

The results from this challenge testing approach can be utilised to demonstrate 'due diligence' to customers, auditors and regulatory authorities, that the business has tested the effectiveness of its processes against specific pathogens to ensure food safety.

6.5.4 *Product trials (new product production/process trials required)*

New products will tend to have been developed on a small scale in a development kitchen/laboratory 'bench scale' environment, yet the processes that will finally manufacture the proposed products will typically be industrial scale processes, using much larger pieces of equipment and often taking far longer to produce the product due to the high volumes involved.

As the small development scale is unlikely to be sufficiently representative of the end product manufacturing process. With regard to product tests/trials to confirm that product safety, quality and legality can be achieved, such tests must be conducted on a factory scale, using the intended processing equipment and processing method.

It is important that these trials are recorded so that that a business has traceability back to the product formulation and processing methods in the subsequent event of a fault/problem. Also, such records can be used to prove to a customer/enforcement officer that due diligence has been maintained at each stage of the process.

The 'success criteria' for such full-scale production trials will include:

Safety factors:

- Confirm that the required safety-related parameters can be achieved (e.g., thermal process times and temperatures).
- Subsequent testing of product shelf life linked to product safety.

Quality factors:

- Product organoleptic performance confirmed as consistently 'as required' (e.g., appearance, aroma, taste and texture tests at the point of production and over the course of the product shelf life).

Legislative factors:

- Confirmed to meet all relevant legislative requirements (e.g., product weight statements and suitability of product packaging).

Commercial factors:

- Product price assumptions confirmed (e.g., process times and labour costs).
- Product is 'process friendly' (including good yields, throughput rates and labour performance).
- Can be consistently made (e.g., no seasonal ingredient availability issues).

6.5.5 Shelf life assessment

Product shelf life testing typically is required to assess microbiological spoilage rates and also to confirm that the product continues to meet the organoleptic and any chemical (e.g., nutritional) requirements throughout the duration of the shelf life.

To ensure that the results of this shelf life testing will be accurate, the product to be tested should have been manufactured on an appropriate industrial scale, through the intended manufacturing process, packed into the intended product packaging, and then stored and handled as it would be within the storage and distribution phases.

- Microbiological shelf life testing (e.g., product spoilage organisms);
- Chemical shelf life testing (e.g., product nutritional performance) and
- Organoleptic shelf life testing (e.g., appearance, aroma, taste and texture).

Organoleptic shelf life testing is typically conducted by the manufacturing business, routinely opening a representative number of appropriately stored packs at each point in the product's shelf life.

Chemical testing for product nutritional performance will tend to be conducted by a chemistry laboratory, owned either by the food manufacturer or more typically an independent contract laboratory. Samples will tend to be taken from the start, middle and end of the production run in order to also check for any product variations during the processing phase.

Microbiological shelf life testing should be conducted on a 'worst-case scenario' basis, for example, with respect to a 1 tonne batch of pasteurisation cooked soup which is then to be hot-filled into pots before cooling. If the product cooking temperature is classed as a CCP, and the batch of product is going to take in total over an hour to pack into the pots (which are then taken by conveyor into a blast chiller as they are packed), then the products that have been hot-fill packed first will have received the shortest thermal process, as later products will have been held at high temperatures while awaiting filling. It is therefore likely that the worst-case scenario for the product microbiological performance will be the packs from the start of the batch.

Conversely when considering organoleptic performance, due to hot holding at high temperatures while awaiting packing, the worst-case organoleptic performance would often likely be found at the end of the production run when the product has been held for the longest period of time at high temperatures and therefore is likely to show the most signs of deterioration due to holding times and temperatures.

Even if a new product is felt to be very similar in type/format to previously produced products, it is still advisable to conduct new product shelf life assessment checks. The smallest of changes in product formulation, product packaging, product ingredients or processing methods/postprocess handling/storage can be enough to make major changes in the product safety, shelf life or quality performance. Many third-party auditors and customer codes of practice/audit standards expect new products to all be shelf life assessed for these reasons.

As an additional monitoring method, shelf life testing is often also complemented with routine product 'end of life assessment/analysis'. Where, once the new product has been launched, routinely over time reference samples of production runs (either held within the business, or purchased back from the supply chain) are microbiologically, chemically and organoleptically assessed to confirm that the product is still performing acceptably at the end of the stated 'use by'/'best before' dates.

Aside from during the NPD 'scale-up and launch' phases and ongoing product monitoring, shelf life testing of a product is also appropriate when significant product/process changes have been made. For example:

- product recipe changes (e.g., perhaps due to the change the product has become less inhibitive to spoilage bacteria, for example, as a result of being made less acidic due to a recipe change);
- product packaging changes (e.g., perhaps a move to a less costly packaging has reduced the barrier properties of the packaging, increasing the potential for oxygen presence and therefore spoilage organism growth) and
- Product process changes (e.g., perhaps the product is now to be cooked at a slightly lower temperature in order to address organoleptic deterioration due to 'overcooking', thereby reducing the thermal process applied to likely product spoilage organisms and pathogens).

6.5.6 Allergen assessment

An 'Allergen risk assessment' should be made during the NPD phase as part of the general product food safety risk assessment (often deemed the HACCP 'hazard analysis' study).

To accentuate the importance of allergen control within the NPD process, it is worth reflecting that allergen-related product faults (e.g., 'unintended allergen presence' and 'allergen not declared on packaging') have been a very frequent cause of product withdrawal and recall across Europe for many years.

Allergen control-related checks can focus on awareness of allergen presence, control of allergens within the process and provision of consumer information with regard to the potential for allergen presence.

Related allergen control measures can include supplier 'allergy data' questionnaires to be held (and regularly reviewed) for each ingredient from each intended and approved supplier. Such information should be utilised to generate on-site awareness of allergen presence (as allergen presence is not always obvious from the ingredient name), application of process allergen controls and generation of appropriate allergen labelling upon the end product packaging.

The application of process allergen controls can include: allergen segregation; use of 'appropriate allergen use only' utensils/containers; staff practices and protective equipment to prevent cross-contamination of nonallergenic products;

production scheduling to move higher risk allergen products toward the end of the production plan; line hygiene procedures to confirm appropriate allergen cleaning has been conducted prior to moving process lines back to manufacturing nonallergenic products.

6.5.7 Confirm packaging will be 'fit for purpose'

Part of the product NPD controls to ensure product safety, quality and legality should be focused on the intended product packaging.

Checks required to confirm that the packaging is 'fit for purpose' will include:

- Purchase of the packaging to a specification which confirms that the packaging is suitable for food contact (e.g., 'food grade').
- Can the packaging supplier provide 'certificates of analysis'/'certificates of conformity' to prove that the packaging meets the relevant legislative requirements stipulated with regard to fitness to be in contact with food? (e.g., in the EU 'migration test' results are required to prove compliance with the 'plastics in contact with foods' regulations).
- Packaging performance testing. Changes in packaging properties may affect the end product pack performance (e.g., changes in the packaging 'moisture barrier' or 'oxygen barrier' properties may significantly affect the microbiological or organoleptic performance of the end product). This therefore requires checking during the NPD process to confirm that all packaging selected is appropriate to the end use.

6.5.8 Confirm that product labelling meets all relevant legislative requirements

At points in the NPD/scale-up processes the product labelling/artwork will require formal checks to ensure that it reflects the product specification and meets all legislative requirements (e.g., weight declaration, ingredient listing/declarations, allergen labelling requirements, modified atmosphere pack use statement, claims such as 'suitable for vegetarians').

Typically, businesses should have an artwork generation checkpoint at which the business ensures that the specification/artwork brief to be sent to the packaging printers is correct, and can then also conduct a full check of the proposed printed packaging when the artwork drafts are sent to the business for 'artwork approval' preprinting.

Due to the potential high cost and the product safety consequences of a pack label fault, such checks should always be conducted by appropriately trained and experienced members of the business team. Often the technical team will be required to check and approve all of the product safety/quality-related packaging points (e.g., ingredient declarations, allergen statements, warnings, weight statements) and the sales team will be expected to check and approve the commercial/'sales' aspects of the packaging (e.g., product 'serving suggestion', product description and picture).

As with all product checks it is necessary to record these packaging checks for both internal quality assurance/traceability and in order to demonstrate to an auditor that packaging/label control procedures are in place and being operated to.

A control system should also be in place to ensure that any product labels required for export will also meet the legislative and customer/market requirements for that particular country. Such a system may also require the utilisation of translation services.

Postlaunch a system should also be in place to smoothly administer any label changes necessitated by a change in customer requirements or legislative requirements.

6.5.9 System to control product formulation changes

Once launched there are a number of circumstances which may necessitate recipe/formulation changes. Such adjustments could affect product safety, quality or legality and therefore before implementation the proposed changes require formal review to check (perhaps via completion of a new product safety/HACCP study) that the product would not be adversely affected if the proposed changes went ahead.

Such product reformulation circumstances can include:

- customer request to address customer complaints/'quality feedback' via a recipe/process change.
- requirement to reduce product cost via 'cost engineering'/'value engineering' the product.
- customer request to make the product healthier (e.g., want to make additional nutritional claims such as 'now 25% less fat', 'reduced salt').
- customer request to remove/replace an ingredient (e.g., perhaps in response to supply chain/consumer pressure such as removal of allergens).

6.6 The importance of appropriate knowledge and management responsibility

This chapter seeks to highlight some of the wide range of factors involved in ensuring 'control' of new products from a safety/legality viewpoint.

In ensuring that new products are correctly designed to be safe, the key staff involved must work as a team and between them have a very sound appreciation of the food safety fundamentals of the HACCP system approach and prerequisite programmes (including good manufacturing practice (GMP)). There must be a clear understanding of all of the major food hazard groups (physical, chemical, allergens and microbiological) and how these groups are likely to impact on the product ingredient base and production process.

Any gaps within the business knowledge related to food safety/legality need to be addressed via the recruitment of new staff, current staff training and the use of specialists/consultants, as running without the required expertise will ultimately lead to product safety, quality and legality control points being missed.

6.6.1 Common fault areas

Focus on food safety, quality or legality can sometimes be lost within a business for these reasons:

- lack of clarity with regard to the safety, quality and legality standards required;
- lack of integration between product development skills, technical expertise, and the wider business teams.
- rushed development and scale-up processes.

Clear communication and firm adherence to business 'standard timescales' for product development and scale-up help to avoid such problems. The use of a new product 'critical path' will help ensure that all departments work in an integrated and coordinated way, routinely reviewing their progress during the development of each new product and communicating clearly and swiftly on any issues which are delaying or threatening progress.

6.7 Hazard analysis and critical control point considerations

6.7.1 Hazard analysis and critical control point and Codex Alimentarius

HACCP is a system which identifies, evaluates and controls hazards which are significant for food safety. The overall goal is the achievement of food safety, by identifying and preventing problems. The use of documented procedures and recorded evidence of compliance are also key aspects of a HACCP system to ensure consistency of approach and proof of application/performance.

'Hazard' – any biological, chemical (including allergens) or physical occurrence which may cause an unacceptable consumer risk.

'CCP' – typically a location, stage, operation or raw material which, if not adequately controlled, provides a threat to product safety. A CCP when adequately controlled, eliminates or reduces a hazard to an acceptable level.

In the food industry, especially in the EU where the application of a HACCP system to control food safety is required by law, most operations will formally include or refer to their HACCP system within their 'quality management system', most usually in the company 'quality manual' which is the document specifying the quality management system of an organisation (ISO9000:2000).

The role of Codex Alimentarius:

The Codex Alimentarius Commission was created in 1963 by Food and Agriculture Organization (FAO) of the United Nations and World Health Organization (WHO) to develop food standards, guidelines and related texts such as codes of practice under the Joint FAO/WHO Food Standards Programme. The main purpose of this programme is to protect the health of the consumers, ensure fair trade practices in the food trade, and promote coordination of all food standards work undertaken by international governmental and nongovernmental organisations.

In their basic texts on food hygiene, the Codex Alimentarius Commission defines contamination as 'the introduction or occurrence of a contaminant in food or food environment' (Codex Alimentarius 1997).

These basic Codex texts consider a range of measures that can be taken by governments, regulatory authorities, food businesses and food handlers to help assure food safety. The measures recommended within these texts are summarised in the 'Food hygiene (basic texts)' Fourth Edition (http://www.fao.org/docrep/012/a1552e/a1552e00.htm) and range from good site and equipment design through to the application of HACCP.

Section 5 'Control of Operation' in this Codex Code of Practice states the objectives of such operational controls are to 'produce food which is safe and suitable for human consumption by':

- formulating design requirements with respect to raw materials, composition, processing, distribution and consumer use to be met in the manufacture and handling of specific food items, and
- designing, implementing, monitoring and reviewing effective control systems.

The focus of HACCP is to develop and apply preventive measures to assure the safety and suitability of food at the appropriate stages of the operation.

There are many sources of information with regard to HACCP and its application. The following notes are intended to serve as a broad overview of the HACCP system approach to assuring food safety. It is highly recommended that readers further develop their knowledge via background reading, commencing with the Codex Alimetarius Commission Basic Texts of Food Hygiene. http://www.fao.org/docrep/012/a1552e/a1552e00.htm

6.7.2 Hazard analysis and critical control point history and approach

HACCP is a logical and systematic approach toward eliminating or minimising hazards and thereby controlling food safety. The HACCP technique seeks to recognise potential issues/threats to food safety and applies checks and methods to control these issues to help ensure the production of safe food.

HACCP was developed jointly in the early 1960s by the Pilsbury Corporation and National Aeronautics and Space Administration (NASA) in the United States. The system was originally developed for the control of microbiological safety and quality of foods to be consumed by astronauts. The intention was to develop a 'zero-defects' approach as astronauts contracting food poisoning in space could be catastrophic.

Controlling product safety via implementation and ongoing use of a HACCP system is a legal requirement for all food manufacturers within the EU, and the fact that this approach to food safety is recommended by Codex Alimentarius results in HACCP principles being expected to be applied in most food systems globally.

Consequently, any new processing line/factory will need to be subject to review and implementation of HACCP in advance of producing a saleable product.

The HACCP system approach involves the food manufacturing business forming a structured and detailed series of checks (and associated records) with the business resources being focused on those areas/aspects of most relevance. HACCP is a relatively low-cost and logical approach to the control of product safety, requiring production and technical staff to work together to assure food safety control.

6.7.3 Hazard groups (physical, chemical, allergens and microbiological)

The hazard groups which pose a contamination risk to foods and therefore require to be the subject of particular product contamination controls are:

- physical hazards;
- chemical hazards (including allergens) and
- microbiological hazards.

6.7.3.1 Physical hazards

'Physical hazards' are often referred to as foreign bodies/foreign material. The size, density and sharpness of the material will define the extent to which it is hazardous if consumed. The extent of the 'hazard' includes the potential for physical damage to the mouth, teeth, throat and digestive system, and/or the potential to form a choking hazard.

Physical hazards in foods can be literally anything. The common 'foreign bodies' encountered in the food sector include soft/hard plastic materials, stone, glass, metal, wood, soil and extraneous matter (physical matter derived from the ingredient, but not the part required, for example, a calyx present in chopped tomatoes, or bone present in minced beef).

Physical hazards can be introduced to the product in many ways: via the ingredients/packaging materials, factory staff, pests, processing equipment and factory fabric.

Food production operations seek to avoid and control the potential for such physical contamination via a range of checks, systems and procedures including:

- supplier assurance, including defined specifications and audits;
- ingredient and packaging inspection on delivery;
- sorting/washing/inspection of ingredients;
- sieving powdered ingredients;
- filtration of liquid ingredients;
- keeping the food products covered/enclosed as much as possible through the manufacturing process;
- preventative maintenance programmes for factory fabric/equipment;
- visual checks and/or camera inspection systems;
- metal detection (including metal detection machines and magnates) and
- X-ray detection.

6.7.3.2 *Chemical hazards*

Chemical hazards can take a wide range of forms and cause harm to the consumer or adversely affect the product in a variety of ways.

Chemical contaminant sources can include pesticides and veterinary residues used earlier in the ingredient supply chain, naturally occurring toxins in some ingredients, hygiene/cleaning chemicals, preservatives and additives, perfumes and scented deodorants, packaging and related inks/materials, processing/packing machinery oil and grease (especially when not sourced to be 'food safe').

Controls of chemical contaminants include:

- supplier assurance, including defined specifications and audits;
- ingredient testing for pesticide/veterinary residues;
- ingredient and packaging inspection on delivery;
- cleaning chemicals to be clearly labelled;
- use of auto-dosing equipment to avoid operator error when mixing cleaning chemical concentrations;
- hand soaps to be unscented;
- personnel rules not to wear perfume/scented products;
- machinery oils/grease/lubricants must be sourced to be suitable for food contact and
- utensils, containers and other food contact equipment must be 'food safe'/'food grade', that is suitable for food contact.

6.7.3.3 *Allergens*

An allergen is any material or substance that can cause an allergic reaction. Allergic/immunological responses may often affect a relatively small proportion of the total number food consumers; however, the potential severity of the reaction/response, including the worst-case scenarios of hospitalisation and death, means that this area of hazard control requires absolute attention to detail.

With regard to whether 'allergens' are best placed within the 'chemical hazards' or 'physical hazards' category, allergens tend to be classified as chemical hazards as although the product may be physically contaminated by the food allergen, the harm/hazard is actually caused by the chemical components of these materials.

Due to the major relevance of (and necessity for) allergen control within the food manufacturing sector, 'allergens' are often presented as a hazard category in their own right alongside physical, chemical and microbiological hazard categories. It is remarkable that over the past decades the topic of allergens in the food manufacturing sector has evolved from literally almost never being considered as a potential risk in the 1980s, through rising awareness in the 1990s where the subject became one of the first points on a technologist's risk assessment checklist when considering the potential to utilise new suppliers/manufacturing sites.

The key 'legally recognized' (and therefore expected to be subject to control/labelling) allergen groups within the EU include:

- milk;
- eggs;

- peanuts (also known as groundnuts or monkey nuts);
- nuts (almond, hazelnut, cashew, pecan, brazil, pistachio, macadamia);
- fish;
- shellfish (including mussels, crab and shrimps);
- soya;
- wheat/cereals containing gluten;
- sesame seeds;
- celery;
- lupin;
- molluscs;
- mustard and
- sulphur dioxide and sulphites.

Due to the legal requirement in the EU that the intentional use in the food product of any of these allergens must be highlighted on the food packaging label, the presence of allergenic material 'not highlighted' upon product labels has become one of the biggest categories of reasons for product withdrawal/recall. This is usually caused by the food operator either accidentally overlooking the labelling requirement/missing out the allergen from the pack label or accidentally adding the allergen to the product due to a recipe fault or lack of sufficient awareness as to the allergen presence within the ingredients utilised.

Severe allergic reactions cause:

- flushing of the skin;
- rashes/hives;
- swelling of the mouth and throat/difficulty swallowing or speaking;
- impact on heart rate;
- severe asthma;
- abdominal pain, nausea and/or vomiting;
- rapid feeling of weakness caused by a sudden drop in blood pressure;
- collapse and unconsciousness (anaphylactic shock) and
- loss of life.

Typical allergen controls applied in the food sector include:

- Supplier assurance, including expectations for site allergen controls. Specifications which state allergen status of each raw material. Supplier audits to check that the controls are in place and functioning.
- Ingredient testing for allergen presence.
- Ingredient and packaging inspection on delivery.
- Maintaining up-to-date allergen information on all raw materials on site.
- Allergen presence clearly labelled on all raw materials, work in progress and finished product packaging.
- Separate storage (segregation) of raw materials that may contain allergens.
- Use of separate processing equipment, utensils and production areas.
- Avoidance of staff contact in sensitive 'multiline' areas where cross-contamination may be a risk.
- Stock control and traceability systems to assure the correct picking and movement of raw materials, work in progress and finished product.

- Segregation and disposal procedures for materials adversely affected by allergen contamination/presence.
- Allergen awareness and associated control training for staff.
- Product withdrawal and recall procedures.
- Site visitor information and monitoring with regard to the site allergen status.

The anaphylaxis campaign promotes allergy awareness and the need for allergen control within food handling operations. More details on allergens and their control can be found at the anaphylaxis campaign website. www.anaphylaxis.org.uk

6.7.3.4 Microbiological hazards

Food poisoning is often the result of a loss of microbiological control in the supply chain or food manufacturing process.

Pathogenic (disease causing) bacteria cause by far the greatest number of cases of food poisoning. There are many different species of pathogenic bacteria, however, a significant amount of food poisoning cases are caused by a relatively small number of pathogens: *Campylobacter jejeuni*, *Salmonella* species, *Eschericha coli* 0157:H7, *Listeria monocytogenes* (often spelt 'Lysteria' in the United States), *Clostridium perfringens*, *Clostridium botulinum* and *Staphylococcus aureus*.

Food poisoning symptoms can include nausea, fatigue, headache, muscle pain, cold-chills or fever, stomach or abdominal pain, vomiting and diarrhoea.

Life-threatening conditions can arise as a consequence of food poisoning with the immunocompromised, infants and the elderly being particularly vulnerable.

Sources of microbiological contamination include:

- raw unprocessed food (e.g., cross-contamination potential);
- people (e.g. food processing staff can carry and transmit food pathogens);
- food processing equipment can serve as a reservoir for pathogens, passing this contamination on to the foods being manufactured;
- food waste, which while spoiling is increasing in its bacterial loading, including an escalating level of pathogen presence;
- pests;
- dust and soil;
- air and
- water.

General controls on microbiological contamination in the food manufacturing sector include:

- protection of the food from potential contamination sources. (e.g., the use of physical barriers/containers);
- cleaning of equipment/surfaces/environment to prevent presence and growth;
- bacterial reduction/elimination via processing methods (e.g., heat, acidity control, irradiation, antibacterial chemicals, irradiation, drying, ultraviolet, cold plasma, MAP) and
- in a worst-case scenario, a final control can be the supply chain withdrawal or product recall of products which have been despatched and subsequently found to contain unacceptable levels of food pathogens.

'*The Bad Bug Book*' is published by the Center for Food Safety and Applied Nutrition (CFSAN) of the Food and Drug Administration (FDA), US Department of Health and Human Services. The book is available online at http://www.fda.gov/Food/FoodborneIllnessContaminants/CausesOfIllnessBadBugBook/ and provides a very helpful reference guide to food microbiology, including information on organisms of specific concern to food groups.

6.7.4 The seven principles of HACCP

HACCP and its application within a food business is based on '*seven principles*'. These principles can be considered as key steps or stages in the development of a HACCP plan.

The seven principles of the HACCP system approach are:

1. assessment of hazards and risks;
2. determination of CCPs (using a CCP decision tree where beneficial);
3. establishment of critical limits, target levels and tolerances for the various CCPs;
4. setting of procedures to monitor each of the control points;
5. establishment of corrective actions when deviation from the set limits occurs;
6. development and maintenance of appropriate records and documentation and
7. establishment of verification procedures designed to confirm that what is said to be happening is actually happening within the production process.

A 'hazard' can be defined as any biological, chemical or physical occurrence that may cause an unacceptable consumer health risk (e.g., an unacceptable contamination by microorganisms, foreign bodies or chemicals/allergenic material). To progress the analysis of the hazards applicable to the food product the food business will need to progress the following:

- select the HACCP team (which should be multidisciplinary covering all aspects of the operation);
- define terms of reference of the HACCP system;
- describe the product, process and intended use;
- construct a flow diagram of the processes;
- verify the flow diagram on-site and
- list all hazards associated with each process step and list all measures which will control the hazards.

The HACCP team should be a multidisciplinary team (i.e., drawn from many different disciplines/departments) in order to give the best wide-ranging consideration and expertise to cover all possibilities/knowledge requirements during the HACCP process. Where the roles are present within a manufacturing company, the team should include the technical manager, production managers, engineers, microbiologists, quality assurance staff, specialist hygiene staff, research and development personnel, purchasing staff, line managers, supervisors and individual operatives with particular specialist knowledge (e.g., operatives on particular production or packing lines).

Where these named job roles are not specifically present within a food manufacturing business, the HACCP team must ensure that its team members include those individuals in the organisation who have this particular knowledge/ expertise in the various aspects of the business. Where there are any shortfalls in knowledge the business will be well advised to recruit to secure this expertise and in the short term utilise the required specialist skills on a consultancy basis.

When defining the 'terms of reference'/scope of the HACCP plan, the HACCP team will need to consider:

- What product, or group of products, are to be included in the plan?
- At which stages of the process will the HACCP plan start and end?
- Which class of hazard, or classes of hazards, are to be included? Physical, chemical, microbiological (and what range of microorganisms?), allergens.

6.7.5 Hazard analysis and critical control point terminology

- *HACCP* – A process by which the hazards and risks associated with the manufacture, storage and distribution of foods are identified and assessed, and appropriate controls which either eliminate or reduce the hazards to a manageable level are implemented at specific points.
- *Hazard* – Any biological, chemical or physical occurrence that may cause an unacceptable consumer health risk (e.g., an unacceptable contamination by microorganisms, chemicals/allergens or physical matter).
- *Contamination* – The transference of any unwanted and potentially harmful substance or material to the food.
- *Hazard analysis* – Is the collation and evaluation of information on hazards, including the conditions leading to their presence to decide which are significant to food safety and therefore must be addressed in the HACCP plan.
- *Risk* – The probability/likelihood that a hazard will occur, based on previous experience, sector-related data or expert opinion.
- *Severity* – Risk is usually coupled with the potential 'severity', for example, the probable number of consumers affected, and to what extent they will be ill/harmed by the issue.
- *CCP* – This is a location, stage or operation which, if not adequately controlled, provides a threat to consumer safety/product acceptability. A CCP should, when adequately controlled, eliminate or reduce a hazard to an acceptable level.
- *Critical limit* – This is defined as an absolute tolerance value which must be met for each control measure at a CCP, for example, a temperature, time, pH (critical limits separate acceptability from unacceptability).
- *Control measure* – Is any action or activity utilised to prevent or eliminate a food safety hazard, or reduce it to an acceptable level.
- *Monitoring* – Is the process/system of making observations or measurements of critical limits designed to ensure that CCPs are under control and therefore maintain product safety.
- *Corrective action* – This is the procedure to follow when a product falls outside of its critical limit.
- *Food safety management system (FSMS)* – Is a set of written procedures which define the range of actions taken by the food business operator to ensure that the food produce is safe to eat, of the required quality, and legally compliant.

- *Regulation (EC) No. 852/2004 on the hygiene of foodstuffs....* In the EU there is a duty placed on food business operators to put in place, implement and maintain a permanent procedure or procedures based on the Codex principles of HACCP.

6.7.6 The 12 steps to hazard analysis and critical control point implementation

The Codex Alimentarius HACCP system recommends 12 steps to its implementation (Fig. 6.1).

These 12 logical steps have been developed to assist not only with the development but also the implementation of the HACCP system. Essentially, the first five steps are the preliminary steps required before the HACCP study is developed and the last seven steps follow the principles of HACCP.

The first five steps are:

1. Assemble the HACCP team which should have multidisciplinary expertise (to achieve this external expertise may be required, especially for smaller businesses). The HACCP team is a group of people with appropriate expertise who develop, implement and regularly review the business HACCP system. HACCP team training is likely to be required and should be conducted by an appropriately qualified team member. Define the 'terms of reference' (stating which product or groups of products/processes are to be covered) and the scope of the HACCP study (stating which hazards are to be considered).
2. Describe the product/recipe/process. Relevant safety information will include: Product composition; water activity (aW); acidity (pH); processes such as heat treatment, freezing and brining; packaging; shelf life; storage conditions and method of distribution.
3. Identify the intended use of the product, that is, end user/consumer use of the product, for example, microwave reheating? Awareness of potential abuse, likely time outside of temperature control. Will the product be consumed by vulnerable groups, for example, babies, the immunocompromised or the elderly?
4. Construct a flow diagram – it is also usual to describe the process. The flow diagram must cover all steps/stages in the operation.
5. On-site confirmation of the flow diagram, via 'walking the line' to review first-hand all operational areas.

Then the final seven steps follow the 'seven principles' of the HACCP system approach, those being:

6. Conduct a hazard analysis and consider control measures (assessment of hazards and risks). Using the flow diagram, all the potential hazards at each step should be clearly listed and a hazard analysis conducted on each hazard. The presence of possible contaminants (e.g., microbiological food pathogens, chemicals, physical/foreign bodies and allergens), the multiplication or survival of pathogens and the production or presence in foods of toxins or the germination of spores must be considered.

   Codex states that 'hazard analysis involves evaluating information on the hazards and conditions leading to their presence to decide which are significant to food safety and therefore should be addressed by the HACCP plan. This will involve assessing the likelihood of the hazards (the 'risk') and the 'severity' of their adverse effects on health'.

   Within food operation HACCP plans the business is expected to review and develop a control strategy for all significant product hazards, and the best way to ensure that the

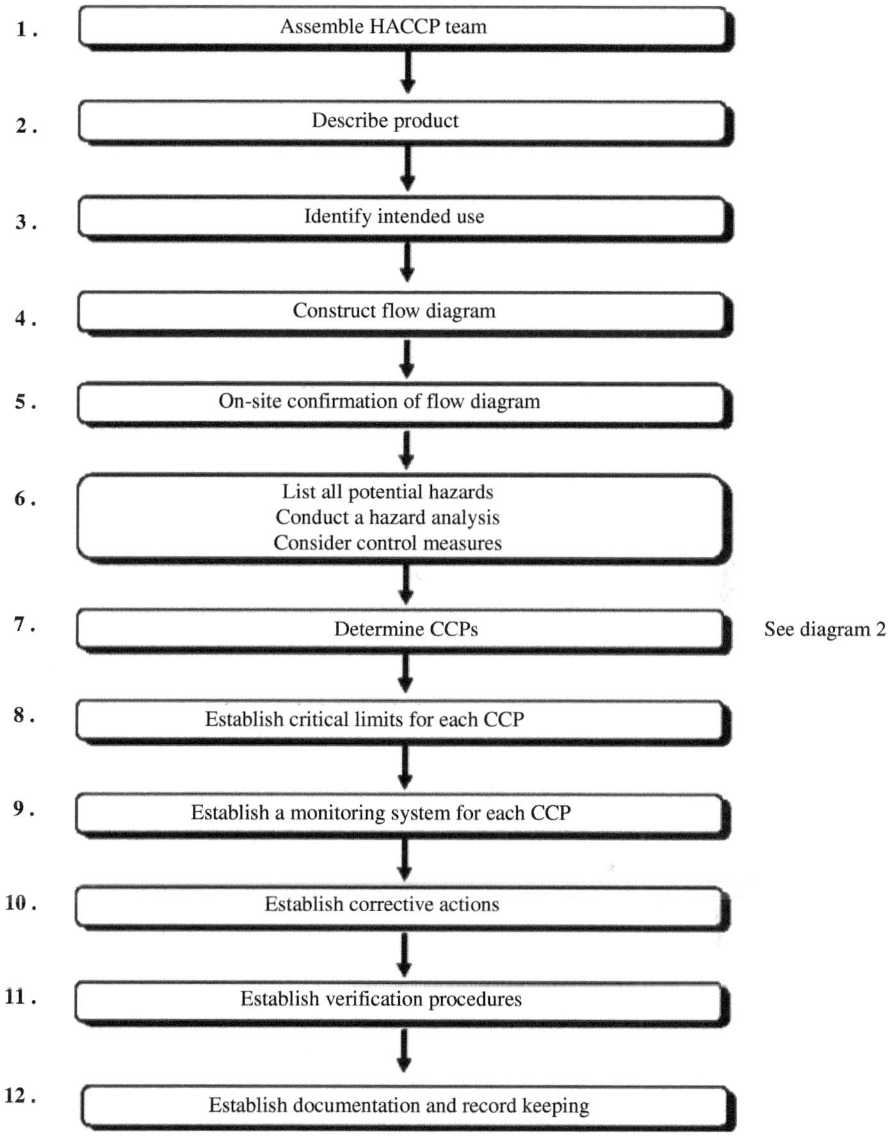

Figure 6.1 Logic Sequence For The Application of HACCP.
Note: 'Diagram 2' refers to the 'example of decision tree to identify CCPs' in Fig. 6.2.
Source: From Codex Alimentarius: Hazard Analysis And Critical Point (HACCP) System and Guidelines for its application. Food and Agriculture Organization of the United Nations and World Health Organization, Reproduced with permission. http://www.fao.org/docrep/006/y5307e/y5307e03.htm

controls for each possible hazard are appropriate and proportional to the level of concern is to assign a 'risk' status to each potential hazard.

'Risk' should be based on:

- severity of outcome (i.e., the health consequences of the hazard occurring due to loss of control).
- likelihood of occurrence;
- vulnerable groups of customers, and
- the number of those likely to be adversely affected by the issue.

7. Determination of CCPs (using a CCP decision tree where beneficial).

Control measures are 'actions required to prevent or eliminate a food safety hazard or reduce it to an acceptable level'. More than one control measure may be required to control a specific hazard and more than one hazard may be controlled by a specific control measure.

A CCP is a step in the process where control may be applied. Having identified a hazard at a particular step the following questions may be used to determine whether the step is a CCP.

Q1. Do control measures exist?

Q1a. Is control required at this step to ensure food safety?

Q2. Does the step eliminate the hazard or reduce it to an acceptable level? Is the step specifically designed to eliminate the hazard or reduce it to an acceptable level? Emphasis on the step, NOT the control measure at the step, for example, sterilisation of milk, chlorination of water, use of an X-ray machine to detect bones, rapid cooling of cooked meat to prevent spore germination.

Q3. Could contamination occur at unacceptable levels or could it increase to unacceptable levels? ('is it likely?' is a key point, NOT 'is it theoretically possible').

Q4. Will a subsequent step eliminate the hazard or reduce it to an acceptable level?

A decision tree can be used to identify CCPs (see Fig. 6.2).

8. Establishment of critical limits, target levels and tolerances for the various CCPs.

Critical limits need to be specified and, if at all possible, validated for each CCP. CCPs may include measurements of temperature, time, water activity (aW), acidity (pH), gas balance (if packing the product in a modified atmosphere).

The critical limits can be derived from mathematical modelling, challenge testing and also expert scientific opinion. Such critical limits are often determined by 'custom and practice' and experience, or from codes of practice, industry guides and legal standards.

9. Setting of procedures to monitor each of the control points. (Establish a monitoring procedure for each CCP).

10. Establishment of corrective actions to be taken when deviation from the set limits occurs.

11. Establish verification procedures, designed to confirm that what is said to be happening is actually happening within the operation/production process.

12. Establish the HACCP system documentation and record keeping.

Basically, the process of applying HACCP in a food system involves breaking down the production operation into a series of steps (often with the help of a flow diagram). Consideration of the hazards associated with each of these steps which may cause harm

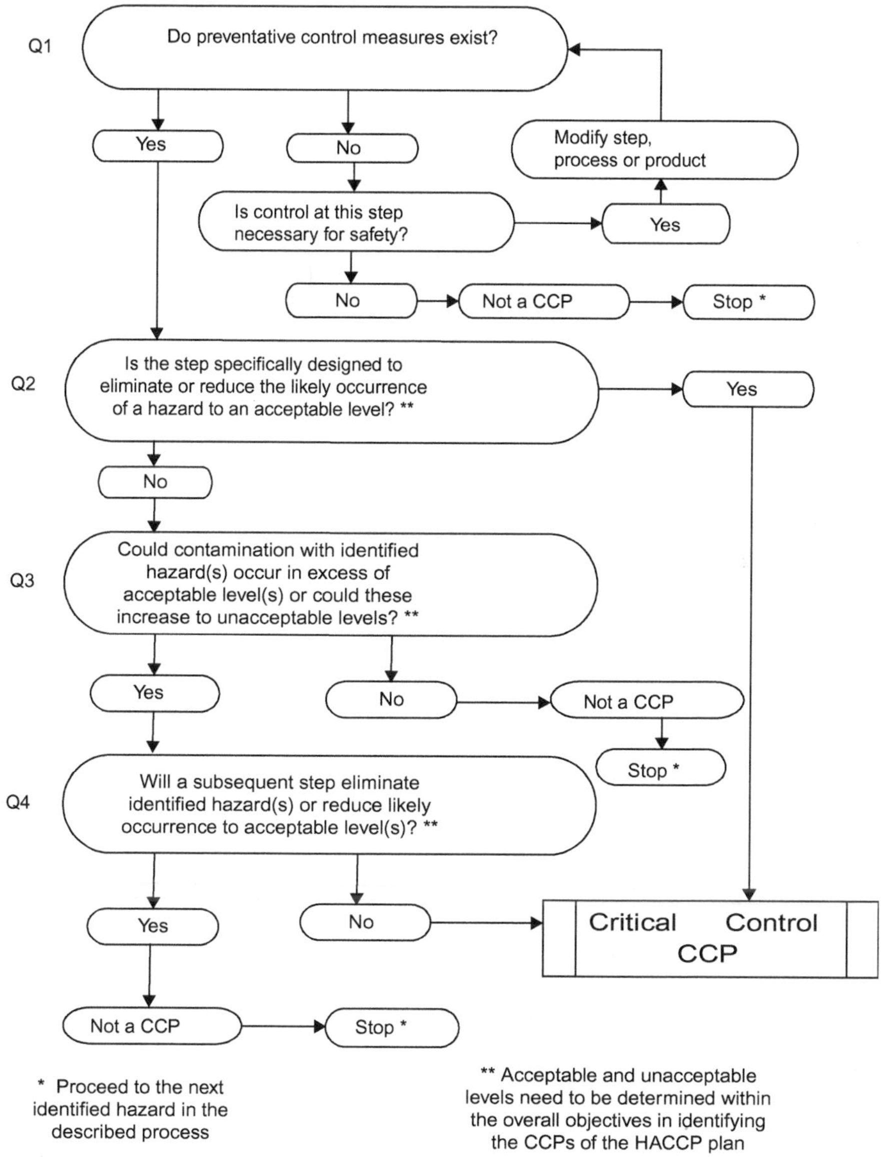

Figure 6.2 "Diagram 2". Example of Decision Tree to Identify CCPS (answer questions in sequence).
Source: From Codex Alimentarius: Hazard Analysis And Critical Point (HACCP) System and Guidelines for its application. Food and Agriculture Organization of the United Nations and World Health Organization, Reproduced with permission. http://www.fao.org/docrep/005/y1579e/y1579e03.htm

to the consumer. Followed by development and listing of appropriate control measures for the recognised hazards. For these 'control measures', the HACCP team must identify which steps are critical to control each of the recognised hazards. The team must then decide on the most appropriate way to check or monitor that the critical controls are working and must document each step (e.g., hazards identified, controls implemented, control procedures and records).

6.7.7 Hazard analysis and critical control point plan verification – specific points

The aim of 'verification' is to confirm that the HACCP plan is correct, effective and reflects what happens in practice. Verification is therefore defined as the application of methods, procedures, tests and other evaluations, in addition to monitoring to determine compliance with the HACCP plan.

Verification during HACCP development includes:

- Continuous checks/confirmations for accuracy/practicality, including 'walking the line'. 'Walking the line' is a key requirement of the HACCP team in order to ensure and continually confirm that the product and production lines are running in exactly the manner which the HACCP team envisages/expects. (This approach saves unforeseen surprises/problems at the HACCP implementation stage.)
- Involvement of production operatives/food handlers to ensure 'first-hand' knowledge of products and processes.
- Ensuring that the plan can be implemented in full (including setting realistic procedures and timescales).

Verification after HACCP implementation includes:

- requires the process to be audited against the HACCP plan;
- discrepancies to be noted and remedied;
- this may then result in modifications to the HACCP plan, product or process, and
- internal nonconformances, customer complaints and product microbiological results will all serve to provide HACCP plan 'verification' feedback.

6.7.8 Hazard analysis and critical control point plan validation – specific points

Validation can be defined as obtaining evidence that the elements of the HACCP plan are effective.

The validation process should be conducted during the factory, process and HACCP set-up phases. Validation seeks to ensure that the implemented HACCP plan is capable of consistently and effectively controlling the hazards identified.

Validation checks include:

- Ensuring that the plan is based on sound, up to date, scientific data. Does the study cover all hazards?
- Do the proposed CCPs control the hazards identified?
- Is the suggested corrective action effective?

- Does everyone understand their responsibility?
- Does everyone understand why the checks are CCPs?
- Is the site team committed to HACCP?

Key aspects of HACCP validation include the fact that validation must happen while the HACCP plan is being formed as progressing validation afterwards would risk the occurrence of unforeseen issues.

The most senior/experienced/knowledgeable site team members should form a 'validation team'. This validation team should include representatives from the original HACCP team plus production/process team members who will have ongoing responsibility for HACCP (this blend of team members will smoothly transfer system ownership from the team that created the plan to the team that will operate it).

The validation team must check that the hazards identified are correct and will be effectively controlled under the proposed plan. These validation team members and the work undertaken to validate the system must be clearly documented for future reference (e.g., for internal review, during audits and in the event of incidents when proof of 'due diligence' will be important to the defence of the operation).

There are two aspects to the work of the validation team. Check and confirm that:

1. the supporting evidence used in the HACCP study is correct
   and
2. the control measures, including monitoring and corrective actions, are correct.

6.7.8.1 Validation task 1: ensure that the 'supporting evidence' used in the HACCP study is correct

Confirm that evidence exists which (1) justifies the selection of the significant hazards and (2) confirms the effectiveness of the proposed control measures.

1. Evidence supporting the selection of significant hazards may come from the following:
   Scientific literature, professional bodies, trade associations, historical data, regulatory and legislative bodies, and company knowledge.
   This validation task should gather evidence to support both the inclusion and the exclusion of all relevant hazards considered during the hazard analysis.
2. Supporting evidence must be gathered which shows that the established target values and critical limits will adequately control the identified hazards to an acceptable and safe level.
   This may be achieved using the same sources that were used for the selection of the hazards and by testing.

Testing is the process by which the proposed control measures are positively tested for their effectiveness, for example, via deliberate contamination tests ('challenge tests'), heat distribution in thermal processing systems and penetration tests, 100% incubation or inspection of production lots, and mathematical modelling of microbial growth.

6.7.8.2 Validation task 2: the control measures, including monitoring and corrective actions

Validating the control measures involves cross-checking the HACCP plan and process flow diagram to confirm that the CCPs are correctly identified and that the established target values, critical limits and monitoring procedures are adequate.

The HACCP team should confirm that written procedures exist for all CCPs. Procedures must include check/test methods, frequency of assessment, acceptable values and actions to be taken if the readings/results are outside of the defined critical limits. All necessary calibration of measurement devices should also be covered.

It is vitally important to ensure that the corrective actions to be taken in the event of process deviations will definitely result in the adequate segregation of the nonconforming product. This is in order to ensure that this product does not reach the consumer.

Corrective action responsibilities and decisions on the future of the affected product must be documented to clearly define the mechanisms utilised to ensure that appropriate actions are taken to prevent any recurrence.

6.7.8.3 Trials of new processes/products

Conducting trials on key aspects defined within the HACCP plan will serve to ensure that safety points such as the following can be achieved consistently:

- desired raw material microbiological quality;
- heating and cooling times/temperatures;
- pH (where acid control is key);
- aW (where water activity control is required);
- metal detection/X-ray review of products for physical contamination control and
- required chill/frozen storage temperatures.

6.7.8.4 Shelf life evaluation of new products.

The total shelf life and its end point will usually need to be determined by microbiological and sensory (organoleptic) testing, using realistic or 'worst-case' product, storage and transport conditions.

It is usually the manufacturer's responsibility to determine the shelf life to be applied to the product. While there is little direct legislation upon how long foods should last, there is often shelf life guidance to be found when researching certain categories of foods.

Setting of product shelf life can be viewed as required to meet the most basic food law in that products must be 'fit for purpose' and not 'injurious to health'. Microbiological safety of foods is determined by the application of HACCP and predictive modelling/'challenge testing' where appropriate. Shelf life tests confirm how long a product remains within 'designated quality parameters' under normal processing and storage conditions.

Challenge tests confirm whether the nature of the product and its storage conditions will control the growth of pathogens during shelf life IF they were to be present in the product. (Challenge tests are usually completed if specific pathogen presence is a distinct possibility but not regular enough to be highlighted during standard product shelf life tests, for example, due to the nature of the raw materials or processing methods.)

6.7.9 Prerequisite programmes

Every HACCP system requires the foundation of a firm set of 'prerequisite programmes'/good manufacturing practice.

'Prerequisite programmes' are the business operating system requirements that need to be in place prior to the implementation of HACCP. Prerequisite programmes can be considered as 'generic controls' whereas the HACCP system then focuses on 'product-specific' CCPs.

Prerequisites are often essential to assure food safety but are excluded from the HACCP control chart in order to minimise the number of CCPs and avoid repetition. However, prerequisites should be considered in the verification of the total system.

The prerequisites for HACCP include the necessity for 'management commitment'/'adequate resources' to be available, plus a wide range of 'prerequisite programmes' including:

- approved supplier policy/selection;
- supplier quality assurance and performance monitoring;
- good factory/process design (including layout and process flow, grounds and security);
- equipment calibration.
- preventive maintenance programmes;
- personnel hygiene and competency;
- stock rotation;
- cleaning and disinfection;
- pest management;
- good housekeeping;
- waste management;
- staff facilities;
- categorisation of risk areas;
- control of glass and hard plastic;
- transport;
- training;
- quality management systems;
- labelling;
- traceability;
- contingency plans;
- incident management and
- product withdrawal/recall.

With regard to labelling and traceability, products should bear appropriate information to ensure that adequate and accessible information is available to the next person in the food chain to enable them to handle, store, process, prepare and display the product safely and correctly. Also the 'lot' or 'batch' can be easily identified and recalled if necessary.

No matter how well developed a HACCP system is, it will fail if it is fundamentally flawed due to poor or incorrect implementation of prerequisites and lack of good manufacturing practice within the business.

6.8 Concluding remarks

As demonstrated by this chapter and other chapters in this book, the control of product quality, safety and legality typically relies on a number of factors including the close control of NPD processes, the base foundations of the business prerequisite programmes and the application of HACCP-based principles for the specific control of food safety. Alongside all of this is the vital necessity to ensure that all relevant staff are fully aware of and adhering to the practices, systems, procedures and processes required to assure the safety, quality and legality of the foods being produced. This is typically achieved via training, supervision, monitoring programmes (including internal auditing) and increasingly an awareness of the importance of developing a strong 'food safety culture' within the organisation.

Further reading

Product Control in Design, Development and Launch...

BRC Global Standard Food Safety Issue 8 – http://www.brcbookshop.com

Creating New Foods, The Product Developers Guide. Earle M. D. and Earle R. (1999).

Food Quality Assurance, Principles and Practices. Inteaz Alli, 2004.

Guideline 46 on Shelf Life Evaluation provides an excellent review of shelf life test considerations. Campden BRI. https://www.campdenbri.co.uk

International Featured Standards (IFS) – www.ifs-certification.com

Product Development Guide for the Food and Drink Industry: Guideline No 8. Campden BRI. (1997).

The BRC global standard for food safety: a guide to a successful audit/Ron Kill Chichester : John Wiley & Sons, 2008.

HACCP Related Resources

Anaphylaxis Campaign promotes awareness and the need for allergen control. More details on allergens and their control can be found at the Anaphylaxis Campaign website – www.anaphylaxis.org.uk

British Retail Consortium (BRC) – https://www.brcglobalstandards.com/

Chilled Foods Association – http://www.chilledfood.org CFA was formed in 1989 to establish, continuously improve and promote best hygienic practice standards in the production of retailed chilled prepared food.

Codex Alimentarius Site Covering HACCP – http://www.fao.org/docrep/W8088E/w8088e05.htm

Food Hygiene (Basic Texts) Fourth Edition – Codex Alimentarius: http://www.fao.org/docrep/012/a1552e/a1552e00.htm

Food Safety Knowledge Network Basic Level Requirements: These materials were developed through a partnership between the Global Food Safety Initiative and Michigan State University to enhance the technical knowledge of individuals responsible for food safety: https://cnx.org/contents/nB_QF7rN@4.2:Vg7cYoGf@4/Food-Safety-Introduction-to-Pe

HACCP A Practical Approach. Sara Mortimore & Carol Wallace. 3rd Edition (2013). ISBN 0412570203

How to HACCP. Mike Dillon & Chris Griffith. 3rd Edition (2000). ISBN 1900134128

Hygiene for Management Highfield Publications. SPRENGER R. A. (2004).

Hygienic Design. EHEDG European Hygienic Engineering & Design Group. https://www.ehedg.org/guidelines/

'*The Bad Bug Book*' is published by the Center for Food Safety and Applied Nutrition (CFSAN) of the Food and Drug Administration (FDA), US Department of Health and Human Services. The book is available online at http://www.fda.gov/Food/FoodborneIllnessContaminants/CausesOfIllnessBadBugBook/

Validation and Verification of HACCP. ILSI Europe. (1999). http://ilsi.eu/publication/validation-and-verification-of-haccp/

Operations and process control

7

Mark Swainson
National Centre for Food Manufacturing, University of Lincoln, Holbeach, Lincolnshire, United Kingdom

7.1 Introduction

'Operations and process control' can be considered as the activities (e.g., operating procedures, monitoring, recording and follow-up actions) required to sufficiently control food processing. These practices could include the control of all raw materials (including ingredients and packaging), through all intermediary 'work-in-progress' phases and onto the review, approval and distribution of the end product. Ultimately the intention of operational controls is to ensure that only product which meets all quality, safety and legality requirements is supplied to the customer.

7.2 New process design and product flow

The design of a food manufacturing process will have a significant influence on the subsequent potential to consistently control the operation. When seeking to develop and implement a new manufacturing process, perhaps as an extension to existing business activities or as part of a brand-new food factory build project, it is vital to achieve an appropriate production process design and smooth production flow, as these factors can impact directly and indirectly on:

- food safety;
- food quality;
- product legality;
- efficiencies/production costs (e.g., yields, throughput rates and labour costs);
- health and safety; and
- area maintenance costs.

Ergonomic production flows can simplify and increase the speed of the manufacturing operation. Such ergonomic processes also help avoid operator health and wellbeing issues such as repetitive strain injury (RSI) and general tiredness.

Hygienic design of equipment will help ensure that food trap points are avoided, and that all parts are easily accessible, making cleaning far easier and thereby reducing the risks of end product contamination.

Use of appropriate equipment and machinery can help deliver the benefits of reducing downtime associated with both cleaning and maintenance. Yet when designing and installing new processes often there is a risk that a business

Swainson's Handbook of Technical and Quality Management for the Food Manufacturing Sector.
DOI: https://doi.org/10.1016/B978-1-78242-275-4.00007-1

requirement to save costs will lead to shortcuts being taken which have the potential to undermine one or more of the factors outlined above.

It is commercial reality that businesses will not always have a sufficient budget to achieve a perfect operation, and the assigned project managers will often be tasked with delivering the new process on schedule and at the same time meeting very tight cost constraints. Therefore, the technical manager needs not only to be able to define the requirements for the new process from a safety, quality and legality viewpoint, but also must be able to assess and guide the business on any 'adjustments away from the ideal', utilising knowledge of where the tolerances and limits are with regard to any technical 'leeway' or 'flexibility' requested.

Technical considerations when designing a new production process/product flow include the following:

1. Customer/consumer requirements:
   a. What is the end product specification? The equipment/process needs to be able to consistently produce the required end product across a range of criteria including: organoleptic performance (e.g., appearance, aroma, taste, and texture), microbiological levels, chemical/nutritional attributes, packaging size and required shelf life (use-by or best before date).
   b. Does the customer (or perhaps third-party certification standard required by the customer) have specific expectations with regard to the new process? (e.g., an expectation for certain production areas to be maintained at a particular air temperature, or specific layout requirements such as segregation of 'high-care' or 'high-risk' areas).
2. Financially sound planning: Is the proposed operation planned to realistic budgets that are based on accurate cost estimates? Is the new process underpinned by a well-considered business case? If not then there is an immediate risk that the operation will be underfunded from the start leading to potential shortcuts on build quality and process integrity which are then likely to impact at some point on product quality, safety, or legality.
3. Consider the production volumes/throughput rates required to be processed at each stage of the manufacturing operation: Review of 'peak rates' alongside 'standard rates' will ensure that areas and equipment types are designed to be of the appropriate size and help foresee and avoid (or prepare for) process 'bottlenecks' in the operation.
   a. The technical manager should first ensure that any process critical control points (CCPs) (e.g., CCPs such as product cooking and cooling times and temperatures) can be consistently achieved even during the 'worst-case' production flow scenarios. Such aspects can be checked via discussion and trials with suppliers of the key processing equipment, review of similar operations elsewhere, and sometimes can even be modelled through predictive software systems (e.g., thermodynamic modelling of cooling systems).
   b. While the production managers are checking that the proposed new production system can deliver the outputs required in terms of volumes, throughputs and labour rates, the technical team should also be checking that all production quality, safety and legality requirements can be consistently achieved at the settings proposed.
   c. It is worth being mindful that sometimes slow production flows/low production volumes also have the potential to cause inconsistency in the performance of processing operations. Equipment suppliers should be able to provide information on both the minimum and maximum tolerances of their processes, and line trials should be used to check minimum, average and maximum tolerances.

4. Planning of 'testing' resources: testing requirements must be considered, such as provision of appropriate equipment and sufficient dedicated space in which to conduct product quality, safety and legality tests/measurements. Area hygiene control (e.g., line hygiene test swabbing) should be planned into area/equipment design. Staff training related to new equipment and test procedures should also be planned for at this stage, especially as equipment suppliers can often provide tailored training as part of their sales package.
   a. Training of staff can easily be overlooked when businesses are developing new processes, only to find that when the new process is built and production needs to commence, the operators then struggle to run the line at anywhere near its optimum due to a lack of familiarity with the new process systems. Early involvement of the key production staff members and any dedicated site trainers can help make 'training and line familiarisation' a standard part of the installation and commissioning process rather than an afterthought 'bolted-on' at the end of the project.
5. Equipment requirements: Alongside assurance of having adequate capability and capacity, the build materials and quality of the intended process equipment need to be sufficiently sound so as not to pose a risk to the product over time. Ingress protection (IP) ratings should also receive good consideration as food manufacturing operations are often wet or dusty environments, and therefore if the equipment does not have a sufficiently high IP rating it can be at risk of failure caused by trapping unwanted materials over time. An insufficient IP rating can also lead to the equipment effectively being a debris trap and reservoir for microbiological growth.
   a. A matter that is often overlooked when considering the equipment requirements is that of 'hygienic design'. So often in the food industry product and area microbiological standards are undermined by the fact that key pieces of equipment just simply cannot be accessed and sufficiently cleaned within the hygiene 'down-time' made available in the production plan. If it is difficult to access and clean a piece of equipment it is very likely that production operators are not going to go to the trouble on a regular basis, thereby risking food quality and safety. Review of food industry sector 'hygienic design' standards [including those provided by interest groups such as the European Hygienic Engineering & Design Group (EHEDG)] can help the Technical Manager to work with equipment suppliers and colleagues to ensure that the equipment is sufficiently well designed to facilitate regular and effective cleaning.
6. Use of second-hand equipment: If any existing premises or equipment are intended to be included in the forthcoming new production/processing operations then they should be subject to formal review to confirm their ongoing suitability and fitness for purpose.
   a. Often, with a desire to limit the costs of the intended new processes, businesses seek to incorporate existing equipment into the new process. The Technical Manager should help ensure that such plans are checked to confirm that the older equipment/fabric is still in good condition, easily maintained in a hygienic state and capable of consistent performance at the intended processing rates.
   b. Awareness of 'hygienic design' has improved in the food manufacturing sector over recent years. Therefore, when considering the use of old food processing equipment in area development plans the Technical Manager should be particularly focused on checking that such equipment will not pose any food safety issues.
7. Size and layout requirements for any storage and preparation areas: While production teams may give general consideration to the likely volumes being processed and therefore the space required in the storage and preparation areas, the technical team should add to considerations by checking for aspects which may impact on the area design and scale such as:

   a. Requirements for ingredient/product segregation could be due to special materials handling requirements such as allergen control, genetically modified organism (GMO) control or protection of other statuses including organic, halal, kosher, vegetarian and vegan accreditations.
   b. Other area design requirements may include implementation of quality inspection or foreign body controls, for example perhaps in the form of inspection belts/tables, or the application of product sieving systems.
8. Timescales: The technical manager may need to plan line/equipment trials to validate the safety of the new production operation and confirm that the process is capable of consistently achieving the desired end product. The manager should therefore be aware of the intended lead times to the point of delivery and installation for all required materials and equipment, and be clear on the planned timescale before commencement of supply to the end customer. Such information will define the timelines within which all trials and training need to be conducted.
   a. For such trials the Technical Manager is well advised to make clear to the business that some of the results of the trials may not be known immediately, and that this time needs to be allowed for. Shelf-life trials may mean that the line is not confirmed to be working 'as required' for days, weeks, perhaps even months later, dependent on the intended shelf life of the product category.
   b. Sometimes there is no way around the fact that the shelf-life trials will take a significant amount of time. In these circumstances businesses simply have to wait for the results to be confirmed. Other times when timescales are tight and production for orders needs to commence, a business might consider starting supply, coding the product with a reduced shelf life which is set cautiously at a point up to which the product has been tested to and has secured good results.
   c. In some food product categories, the shelf-life testing of the product has the potential to be 'accelerated' via increasing the temperature or adjusting other storage conditions. However, such an approach should be taken forward cautiously and on the basis of a clearly defined and 'sector standard' approach, as such acceleration does not work in all food product categories. For example, elevating the storage temperature of chilled foods would then be selecting for the growth of certain microorganisms which may be far less likely to grow at the actual colder temperatures, running the risk of the shelf-life trial missing the growth of microorganisms that would be of more concern in the intended storage regime.
   d. During installation projects the timescales for completion can sometimes drift past the original plans. As the testing phase can often only be commenced once all of the process is installed, such circumstances can run the risk of the technical trials being rushed due to a reduced window of opportunity before being launched. The Technical Manager can guard against such circumstances by making clear from the outset exactly how much time is required for testing once the installation is complete, and that any delay early in the project will have a knock-on effect of then delaying the completion and results of the trials.
   e. A worst-case scenario at the end of a new process build is that the validation/confirmation trials do not yield the required end results, and as a consequence production cannot commence. The potential for this outcome can be significantly reduced by conducting preliminary trials designed to mimic the intended process, in advance of commencing the project. Such pilot studies could be conducted utilizing equipment/lines which have been borrowed or rented, just to help confirm in advance that the proposed operation is likely to yield an appropriate end product.

9. Planning sufficient access into and through the processing areas is important from a technical viewpoint as:
   a. Staff and visitors to the manufacturing operation can pose many different potential threats to the products if not sufficiently controlled. Area access locks/codes and the positioning of changing and hand-washing facilities require particular consideration.
   b. The movement of waste through a factory is often overlooked leading to disruption of process flows and, in worst-case scenarios; accumulations can pose a particular health and safety or even pest control risk. Waste hatches and dedicated throughfares can help ensure that the removal of waste from the processing operation is as smooth and as efficient as possible.
   c. If factory access points cannot accommodate the entry and installation of large pieces of machinery/equipment then the 'work-around' of cutting access holes into walls and ceilings can pose a significant product safety risk if not adequately controlled. For example, increased potential for foreign body and microbiological issues.
10. End product storage and distribution requirements:
    a. The appropriate storage and distribution of the end product is an important final phase of most operations. Some businesses manage all such operations on-site, maintaining end product storage areas and their own distribution vehicles. Other businesses select to place such operations in the hands of contracted suppliers with the choice often based on cost and convenience.
    b. The technical team will be expected to advise on the optimal conditions for end product storage and distribution. This could include recommended storage temperatures and for certain food types other environmental factors such as relative humidity or modified atmospheres.
    c. Product shelf life will be a key factor in ascertaining how much 'stock' can be held in storage. Shelf-life testing both microbiologically and organoleptically will typically be overseen by the technical department or associated 'process' departments. It is vitally important that the shelf-life trial protocols (i.e., the storage conditions and times tested over) reflect the true nature of the product storage and supply chain, with allowance for 'worst-case scenarios'.
    d. In order to maintain control and demonstrate due diligence, the storage and distribution processes will be required to record times and storage conditions (e.g., temperatures) and be appropriately alarmed in the event of such factors moving outside of acceptable tolerance range. The Technical team will be expected to ensure that these tolerances are set in line with industry good practice, customer expectations and legislative requirements.

The appropriate control of process design and product flow also involves a wider consideration of factors including:

- Food operations typically require a considerable amount of focus to enable them to run smoothly. Such matters commence with the application of good 'raw materials control', as without the appropriate ingredients and packaging, the required end product quality will be extremely difficult to achieve.
- All processes have the necessity for hygienic practice and the avoidance of cross-contamination with some requiring the installation of high care or high risk manufacturing environments (the management of which is considered later in this chapter and the structure/design of which is covered in Chapter 5: Site Standards – 'High Care and High Risk Environments').

- GMP, 'good manufacturing practice' and GHP 'good hygienic practice' are terms commonly utilised in the food manufacturing sector to sum up a wide range of operational and staff hygiene practices which serve to support the achievement of safe, legally compliant foods of the required quality.
- The application of a Hazard Analysis Critical Control Point (HACCP) system is also typically seen as a vital step in assuring food safety in manufacturing operations and in many countries the application of the HACCP system approach is also a legal requirement.
- Clearly defined, appropriate production processing times and treatments/tests are vital to the control of product safety and quality. The methods and parameters required to assure product safety should be derived from process validation studies and set to cope with both typical and worst-case product/process scenarios (see next section).

7.3 Process validation

'Validation' is a review and testing process utilised to confirm that any new (or revised/adjusted) equipment, product, process, or operational settings are capable of consistently achieving the required levels of product safety, quality and legality control.

Industry definitions of 'validation' include:

- Confirmation, through the provision of objective evidence, that the requirements for a specific intended use or application have been fulfilled (ISO 9000:2015).
- Validation: Obtaining evidence that a control measure or combination of control measures, if properly implemented, is capable of controlling the hazard to a specified outcome (Codex Alimentarius Commission 2008).
- That element of verification focused on collecting and evaluating scientific and technical information to determine if the HACCP plan, when properly implemented, will effectively control the hazards (National Advisory Committee on Microbiological Criteria for Food. NACMCF, 2014).
- Process validation is establishing documented evidence which provides a high degree of assurance that a specific process will consistently produce a product meeting its predetermined specifications and quality characteristics (US FDA 2011).

New pieces of machinery within food manufacturing operations, especially food processing equipment, can have a direct impact on the safety of the products being manufactured. Therefore prior to the public sale of products manufactured utilising new (or 'new to the process'), revised, or adjusted machinery, it is very advisable, and often forms part of the site HACCP plan requirements, to consider 'validation' of the 'new' manufacturing process to confirm that it is capable of producing consistently safe, correct quality, legally compliant products.

When considering which equipment and processes to validate, one way to consider such matters is to select those operations where there could be a risk to product safety if the equipment were not to work correctly or consistently. These considerations are best progressed by a multidisciplinary team (often the site HACCP team) to help ensure that all potential issues and scenarios are considered.

Food processing technologies requiring validation often include manufacturing stages which are designed to address particular food hazards. Such hazards can be biological, physical, or chemical depending on the particular piece of equipment/process, for example:

- Metal detectors, X-ray machines, or optical sorters. All are designed to detect and remove 'foreign material'/'physical contaminants' from the product being processed. Validation of these lines could involve adding a range of worst-case physical contaminants (or 'simulants' such as test sticks for metal detectors) to the product and then checking that the machinery both consistently detects the contaminant and efficiently removes the affected product/contaminant from the line via testing of the intended rejection mechanism.
- Cooking technology. Perhaps validation assessment is required for a cooking vessel applying a pasteurisation cook process. Or a retort cooking system being used to apply pasteurisation or sterelisation processes. All are intended to confer a degree of microbiological reduction/control to assure product safety and shelf life. In such cases often process validation will involve the use of equipment to monitor, at representative and worst-case points, the product processing time and temperatures, followed by microbiological testing (of samples taken before and after the process) to confirm that the microorganisms of concern have been adequately addressed.

In some cases where added assurance is required 'challenge testing' may be progressed which involves adding a known quantity of a target microorganism into the product (or simulant) followed by processing the product to the standard parameters intended, and then testing the resultant product to confirm that the process does adequately control the microorganism of concern. Such an approach is particularly useful where the microorganism of concern is a pathogen which is not regularly present within the raw materials for the product, but that its potential for occasional or increased presence is of sufficient concern to set a process which can adequately address the hazard, should it be present (see Section 7.3.3).

- Cooking can sometimes be a CCP for the control of chemical hazards. For example, cooking kidney beans to remove the natural plant toxin present in the raw form of the product. Use of time/temperature process sensors followed by product testing samples taken before and after the process to confirm that the levels of the toxin have been reduced to adequately safe levels will be involved in the validation of such processes.
- Cooling technology. Often where a product heating process has been involved there is a necessity to then cool the product to below a certain temperature within a set amount of time in order to avoid the growth of microorganisms. Validation in such circumstances could involve setting the required process equipment cooling times and operating temperatures and then sampling the temperatures of the product upon entry and exit of the cooling phase to confirm that the necessary cooling times and temperatures have been achieved. When cooling products there will often be a 'worst-case' area of the pack or batch where the product will be slowest to cool. It is important that enough trials are progressed to confirm where this point is and for this point to then be particularly monitored during the validation trials. Such a 'worst-case' approach should also be taken by running the trial with the maximum volume and highest rate of throughput possible in order that the validation results will be representative of what could happen during full-scale ongoing production runs. Another worst-case scenario may be in circumstances where the cooling system requires to sense the hot product loading before running the cooling

system at a high enough energy rate. In such cases, too low a product flow through the system may affect the effective operation of the cooler and therefore such a scenario should be checked/validated.

- MAP: modified atmosphere packaging; CAP: controlled atmosphere packaging. Close control of the specific gas levels within MAP/CAP food packaging is often utilised to achieve an extended product shelf life. Validation of these processes is likely to involve the use of a piece of equipment such as a gas analyser to confirm that the required atmosphere has been achieved within the pack and is subsequently maintained to a sufficient degree over the course of the intended product shelf life. Microbiological testing can then confirm that the microbiology of the product has remained within required limits and organoleptic testing can confirm that the product has remained of the required quality.
- Vegetable and fruit washing systems are often utilised to reduce the microbiological loading present on fresh produce in order to increase product safety assurance and/or extend product shelf lives. Validation studies may include confirming that the washing process equipment is capable of consistently maintaining the required level of produce washing chemical (e.g., free chlorine) and that this achieves the required amount of microbiological reduction.

7.3.1 Statistical significance

When conducting validation studies, it is important to ensure that the methods used are 'statistically significant'. That is, the sampling rates and amounts tested should be sufficient to highlight a problem if the process is not operating in a manner sufficiently robust to consistently control the particular hazard/s of concern. If sampling rates and amounts are too low it may be that a flawed process could go undetected with the worst-case scenario being that the business only realises that there is a problem when the customers start to complain.

'How much is enough' is the question and this will vary from process to process. It is wise to ensure that the sampling rates, amounts and methods selected meet requirements/considerations such as:

- typical industry 'standard practice' and 'best practice';
- conform to any specific customer/sector specifications/expectations; and
- meet all legal requirements (including confirmation of appropriate approach by engagement with relevant enforcement/government officials).

It is prudent to ensure that all potential stakeholders are involved at the outset of forming validation plans in order that if in the future an unexpected problem is found there will be very limited scope for lack of support from customers and enforcement authorities as the business showed a sufficient degree of 'due diligence' from the outset.

7.3.2 Validation records

In order to fulfil review, audit and other inspection requirements, and also to enable the demonstration of 'due diligence' in the event of an issue, it is vitally important that all validation studies are formally recorded, noting times, dates, background information, all matters reviewed, methodologies, results and conclusions drawn.

Such records should be held securely and reviewed regularly to ensure that the operation can always swiftly demonstrate a fully up-to-date approach to validation as part of the site food safety control system.

7.3.3 Pathogen control specifics

Within each product HACCP plan the scope should highlight the key pathogens of concern with regard to the product. These points are usually set when considering the microbiological hazards involved in the particular food system. Such 'pathogens of concern' will tend to be highlighted via review of their likely presence within the ingredient supply and potential for occurrence within the in-house process (including staff, factory and equipment routes of contamination and the potential for postprocess contamination).

The control of pathogens of potential concern can be approached via thorough product and process design. For example:

- Recipe/formulation can be designed to reduce the potential for pathogen presence and/or be inhibitory to the particular organisms of concern.
- The manufacturing process can be designed to eliminate (or reduce to a safe level) the pathogen loading which studies suggest may be present in the product formulation.

Use of ingredients which have been sourced to have good microbiological standards (often achieved by well-controlled processes) will provide a good starting point for the control of product pathogen risk.

The product formulation/manufacturing process can then be utilised to address pathogen risks via the application of 'hurdles' to pathogen survival and growth such as:

- Time/temperature (e.g., use of recognised thermal process lethality rates for target pathogens).
- Product pH (e.g., use of product acidity to eliminate/exclude specific pathogens).
- Product water activity (aw) sufficient to be inhibitory to the microorganisms of concern.
- MAP/CAP. Application of a modified or controlled atmosphere within the pack selected to reduce the growth/survival potential of certain key microorganisms.
- The temperature of the product storage, processing and supply chain phases [e.g., when stored at sufficiently low temperature food pathogens (if present) cannot grow].

The use of 'hurdle technology' to control product pathogen risk requires an in-depth knowledge of the measures which can be utilised to eliminate (or reduce to an acceptable level) the particular pathogens of concern.

Hurdles can be utilised in isolation or in combination in order to provide the required levels of safety/assurance, and product manufacturing processes should be validated in order to confirm that the hurdle/hurdles selected are sufficiently and consistently effective against the pathogens of concern.

Predictive microbiology: Microbiological modelling programmes such as ComBase and the USDA Pathogen Modelling Programme can be used to assess the likely effect on product safety over the course of the intended shelf life when applying recognised product treatments to address the pathogen risk to a product.

Such 'computational microbiology' approaches have been developed to utilise pathogen survival and growth data drawn from many scientific studies and can often prove invaluable in enabling a quick review of the potential product safety impacts of a revised product formulation or changes in postproduction storage/distribution (e.g., in the case of chilled foods, the times and temperatures within the storage and distribution chain).

PMP (Pathogen Modelling Programme), USDA.

https://pmp.errc.ars.usda.gov/PMPOnline.aspx

ComBase.

http://www.combase.cc/index.php/en/

Like all good tools this technology is only effective when used correctly. Users should always ensure that they receive full instruction and understand any limitations of such programmes prior to use.

'Challenge testing': Can be utilised if the issue of microorganism risk of survival/growth requires further assurance of control. Challenge testing involves introducing the selected risk organisms into representative samples of the product, and then assessing the growth potential of these microorganisms when processed and held within the typical and 'worst-case' production and distribution processes. Such testing is typically carried out in controlled laboratory conditions designed to closely reflect the actual full-scale product processing conditions.

When considering process validation techniques, the level of knowledge required with regard to the microbiology of such products and processes highlights the necessity for expertise in the particular microbiological hazards of concern and their relevance to the specific food product. Only then can informed decisions can be taken on how best to process the product and at the same time adequately address the identified product hazards, followed by completing process validation activities to further assure food safety.

'Worst-case scenarios': It is vitally important when running validation trials to always test the operation with the 'worst-case scenario' possible. That is, what circumstances are theoretically possible which would give the most challenge to the processing equipment or process being applied?

For example, in the circumstances of whole chickens being roasted via a continuous oven system, the cooking (roasting) of the whole birds will need to impart sufficient heat over an appropriate length of time to be able to effectively cook the chicken both from the viewpoint of microbiological safety and also organoleptic eating quality.

Potential 'process variables' to be considered when setting the criteria for such validation trials would include:

'Chicken variables': What are the range of temperatures that the raw whole chickens will be entering the cooker at? (incoming 'chilled chicken' temperatures may vary widely, e.g., from −2°C to +4°C); What are the weight/size tolerance and limits for the whole chickens to be processed? Worst-case scenarios would include the need to validate the intended system based on the largest bird size introduced to the cooker at the lowest possible temperatures.

'Equipment variables': Is the heat consistently applied by the oven system or could there be 'hot spots' and 'cold spots' within the oven? (in which case the business would need to check the processing temperatures of birds taken across the range of the belt, and in the case of using 'rotary' rather than 'belt'/'continuous' ovens it would be beneficial to check consistency across a wide range of rack locations); Does the oven work consistently when processing at low, medium and high volumes/throughput rates?

When considering the potential for such variables it is often advisable to utilise a multidisciplinary team to help ensure that all potential issues are raised and considered (this multidisciplinary approach would also be expected if conducting such activities as part of a HACCP programme).

7.3.4 Validation of allergen controls

Validation processes can be applied to the control of product allergens, often in the form of validating that the cleaning method used to prepare the line/equipment to produce an 'allergen-free' product is sufficient to remove to a safe level any allergenic material that could have been present in the previous product. For example, in circumstances when using the same equipment/processing operations to manufacture or process 'allergen-containing' and 'allergen-free' products.

When validating an allergen cleaning process, often allergen swabs will be used to confirm whether or not the target allergen/s have been removed to a sufficient extent following a 'worst-case' (applying the highest potential level of allergen content) production run. Some allergen test kits are manufactured to provide instant results, whereas other allergen presence tests may require to be sent to a specialist testing laboratory.

As with all validation processes, the extent of the assurance required by the validation study will be highly influenced by the inherent stability or variability of the process, and the likelihood and severity of impact linked to any potential issue.

Allergen control in the food manufacturing sector is a rapidly evolving subject area. Research is ongoing into the action of allergens and their threshold limits. This coupled with new technologies being developed to enable more accurate and sensitive test methods leads to regular revision of what is recognised as 'standard practice' on such matters. It is therefore advisable to always review the latest guidance and legislation on the matter of allergens and their control when considering the most appropriate approach to be taken for a particular operation/process.

7.4 Procedures and work instructions

Clear, unambiguous procedures and work instructions are vital to the control of all food manufacturing operations. Without such information staff are likely to operate in widely differing ways with inconsistency of approach leading to a high degree of product and process variability.

It is vital when writing procedures and work instructions to ensure that they will be easily understood by the intended user. The document should reflect the knowledge level of the intended staff member and also be written in a language or style that the user will be best able to understand. This can be of particular importance when the business has staff members for who the written language used may not be their own first language. Adding diagrams and pictures can often enhance the detail of work instructions and may also help them to be more easily understood.

Document control: The application of a 'document control system' will help ensure that all staff operate consistently to the most up-to-date site policies and procedures. This approach can be achieved by marking each work instruction/procedure document with the date that it was last updated and issued (e.g., trained-out or briefed-out) to the process operator/area, alongside a note of the 'version' or 'issue level' of the document. For example, if the work instruction has been updated four times since the original document was first issued/implemented then all versions present in the factory should now be 'Version 5'. Ensuring that all key work instructions are clearly marked with their version/issue level and their date of issue enables a business to be able to efficiently remove and replace older versions when they are superseded and be able to routinely check/audit that all paperwork in use on-site is up to date and being consistently worked to.

When a procedure or work instruction is changed, it is important to have within the document control system a robust procedure for the removal of the previous version which was in use and at the same time the system should issue the new version to all relevant staff. Such procedures may include a sign-off from all document holders confirming that they have updated their versions and have formally briefed/trained all staff affected on the new contents of the revised procedure/work instruction.

Storage and retrieval of records: The site document control system should also make provision for the secure storage and ease of retrieval of site documents and records. Consideration should be given to how long the records are required to be kept for. The minimum time required to hold records before disposal will be influenced by a number of factors including:

- the shelf life of the manufactured product to which the records relate (e.g., records are likely to be required for review as part of investigations into product issues/customer complaints);
- any specific legislative requirements;
- customer or third-party compliance scheme/audit requirements; and
- requirements of the business insurers.

When considering the timescales of record storage to adequately cover the shelf life of the product, allowance should also be made for any extension in life that may occur prior to consumption. Products typically stored chilled may be frozen by the customer to extend their shelf life, or ambient foods may be consumed well in excess of their 'best before' date. Therefore, access to records may be required for a lot longer than originally thought when just considering the standard product use-by/best before date.

The location of the records can be a key factor when considering the amount of time which it would take to access such records in an emergency (or traceability audit) situation. Where records are stored off-site, perhaps with a specialist document storage company, provision should be made in the storage contract to ensure that records can be accessed with minimal notice, 24 h/day, every day of the year.

Due to the requirement factors of storage space, cost and access time, food businesses are giving increasing consideration to 'paperless' systems, where records are held electronically on computer systems. Such systems facilitate rapid access and providing that log-in access is closely controlled these electronic records should be able to provide a good degree of traceability back to the individual employees processing the products and conducting the product safety and quality checks.

7.5 Control of risk materials (including allergens)

'Specific risk materials' within a food manufacturing operation could be considered as materials which pose an additional threat or risk to any of the products being manufactured, or perhaps due to the nature of the material, require particular care in their handling and use.

Such materials within a manufacturing operation which are considered as requiring particular care/specific handling requirements by food processing certification standards such as the BRC Global Standard for Food include:

- allergens and
- identity preserved materials.

The list of allergens required to be subject to particular control will vary dependent on the legislation and sector guidelines applicable in the country of manufacture/sale.

The identity preserved (IP) category may include products which have a specific identity such as organic produce, genetically modified (GM) materials, kosher foods and halal foods.

Due to their increased prevalence in recent years, and the fact that they are often a cause of food product recalls and withdrawals in many countries (e.g., recall categories such as 'unintended allergen presence' and 'allergen not declared on packaging'), the necessity for the control of food allergens tends to receive a high degree of attention with regard to the need for specific handling systems in food manufacturing operations.

Specific allergen handling measures can include:

- Supplier 'allergy data' information to be held (and regularly reviewed to ensure that the information remains up to date) for each ingredient in use, provided by each intended supplier. At the time of writing this 'data' in Europe would include a requirement for each supplier to list whether any of the 14 EU recognised allergens are present in the ingredient

which they supply. These listed EU allergens include: cereals containing gluten, crustaceans, eggs, fish, peanuts, soybeans, milk, nuts, celery, mustard, sesame seeds, sulphur dioxide and sulphites, lupin and molluscs.
- It is vital that this information is provided by each individual supplier as the information may vary between suppliers due to the nature of the supply chain or just due to the depth of knowledge that each individual business holds. Such allergen information should be utilised at the manufacturing operation to generate on-site awareness of allergen presence, where necessary trigger the application of allergen-related process controls, and also ensure the generation of appropriate 'contains allergen' labelling of the end product/packaging.
- A system of ingredient labelling to highlight allergen presence immediately on arrival at the manufacturing site is advisable as allergen presence is not always obvious from the ingredient name.
- Application of process allergen controls can include: Allergen segregation, use of 'appropriate allergen use only' utensils/containers, staff handling practices and protective equipment to prevent cross-contamination of nonallergenic products; production scheduling to move higher risk allergen products toward the end of the production plan; line hygiene procedures to confirm appropriate allergen cleaning has been conducted prior to moving process lines back to manufacturing nonallergenic products.
- The control of other 'risk materials' such as 'IP' materials can include some approaches similar to those utilised to control allergens (e.g., clear labelling and segregation). Caution must be taken not to confuse such materials in product handling systems on-site.

7.6 Packaging control

Food packaging has three main purposes: To protect, preserve and promote the product contained within.

'Is the packaging fit for purpose' is a question that should always be asked when considering new packaging or adjustments to existing food packaging systems. Considerations include:

- Suitability for food contact? If the packaging is to be in close contact with the food product has the packaging been tested and approved for 'food contact'?
- Can the packaging withstand the rigors of the production process? Processing methods, product formulation and variations in temperature can all place stresses on the packaging as it flows through the production process.
- Transit testing to confirm that the packaging is sufficiently robust. The supply chain from the manufacturing operation through to the end consumer can involve a number of potential damage points such as compression pressures due to stacking and also vibratory impacts caused by transit vehicles.
- Consumer trials. Is the packaging 'user friendly'? Can it be easily opened? Is 'tamper evident' packaging required? (a particular consideration if the product is going to be open to public access while on sale).

Specifications should be held for every packaging item clearly stipulating the product dimensions, composition, suitability for food contact (if applicable) together with any labelling requirements. Businesses may also wish to state on the specification that the packaging is intended for use within particular processing parameters. For example, 'freezer stable', 'suitable for processing/hot filling at temperatures up to XX degrees', and 'suitable for microwaving', etc.

Businesses are always well advised to trial new packaging or changes to existing packaging in advance of rolling out to full-scale distribution. Process trials should be organised to test the packaging performance through the standard and worst-case manufacturing conditions. These trial products can then be utilised for further 'acceptance testing' such as product shelf-life tests, supply chain 'transit testing' and final phase consumer/customer testing (including ease of opening and use/preparation checks).

7.7 Quantity control: requirements, systems and procedures

The quantity of product received for the price paid is usually a major deciding point in terms of whether a consumer wishes to purchase a product. In addition, some products are sold on the basis of having a particular portion control/nutritional content which could be made inaccurate via poor quantity control.

Due to the importance of such points most regions of the world have clear and specific laws related to quantity control. A review of products in many supermarkets today will highlight that food labels often do not just state the price of that particular pack, but also there will be other 'on-pack' or at 'point-of-sale' information regarding what the product price equates to in cost per 100 g or per kg of the product. This is in order to help the consumer to make an 'informed choice' on the comparative value of the product being considered against similar products.

Expectations for the control of product quantity will vary widely dependent on the particular product, customer requirements and relevant regional legislation.

Foods are often sold on the basis of reference to their 'weight' or 'volume'. For 'prepacked' foods (i.e., foods which are purchased in an already packaged format) the container will typically be marked with the correct 'net weight' or volume, and dependent on the country/legal expectations this quantity information will be stated in 'metric' (grams g/kilograms kg) or 'imperial' (pounds lb/ounces oz) units of measure.

'Net weight' refers to the quantity of the food contained within the packaging, and therefore excludes the weight of the packaging.

Measurement of product weight is also a key factor when selling 'loose products'. That is, products which at the point of sale are being drawn and sold from a bulk stock. For example, a grocery shop, cheese delicatessen, or a butcher's shop which has all of the produce on the counter in bulk format waiting for the customer to state how much they would like to buy. When selling bulk products on the basis of a measured weight, such businesses will usually have 'point of sale' displays stating the 'cost per quantity' of each product.

Most countries will have clear legislation stating the detailed requirements for quantity control of both prepacked and loose goods. Typically, such legislation relates to the use of weighing/measuring equipment including scales and weights and capacity measures (e.g., volumetric containers) which must be accurate and legally approved. Such 'legal approval' of measuring equipment/weights is typically administered by district government officials concerned with matters of 'trading standards'.

Quantity-related terms/definitions can include:

- Nominal weight: The intended pack filling weight.
- Gross pack weight: The combined weight of both the product and its packaging.
- Net pack weight: The weight of the product excluding the packaging material.
- Quantity declaration: A statement of the net amount of the product contained within the package.
- Tare weight: The weight of the product packaging material that is to be deducted from the gross weight to then obtain the net pack weight. (This term is often used when setting product weighing scales to ignore the weight of the packaging, i.e., when packing products to be sold by net weight and therefore the business only wants to measure the amount of product inside the pack.)
- Minimum weight: Typically this statement on the pack commits the supplier to ensuring that every pack will contain at the very least a minimum product weight of the amount stated.
- Average weight: Such a statement on the food packaging highlights to the consumer that 'on average' each pack will contain the stated amount. However, with this being an average the actual amount in that specific pack may be higher or lower than the stated amount and there will often be legal expectations on how widely this amount can vary.

'Minimum' versus 'average' weight packing: Most production processes will encounter less 'giveaway' if set-up to pack to 'average weight' rather than 'minimum weight'. This is because packing to minimum weight requires that every pack can only ever be at or above the stated 'minimum weight', whereas 'average weight' packed products provide the flexibility to pack an amount of packs at less than the stated weight, which enables the business to more finely control the overall amounts packed into each pack and thereby ensure that the overall batch average is very close to the average weight stated.

Therefore, for optimum cost control food packing businesses will often select 'average weight' as their approach to pack quantity measurement.

Minimum weight: With regard to factory control of 'minimum weight' packed batches, it is relatively easy for supervisors and managers to review batch packing records (e.g., packing weight batch summary records drawn from the packing area check-weighers), to check whether the packed batch was legally compliant. All that would be required to check is that:

1. The 'nominal weight' (the intended pack filling weight) had been set correctly.
2. The pack tare weight had been correctly allowed for/set.
3. No packs during the run had been passed and packed below the nominal weight.

The business would not want to see too many lightweight packs produced and rejected during the batch as they waste packaging and product due to being unsaleable. This is often the reason why minimum weight batches usually have an amount of 'giveaway' as operators run the filling weights a little higher in order to avoid 'underweight rejects'.

Average weight: In Europe, an 'Ô' mark on the food packaging weight statement indicates that the product complies with EU Average Weight legislation. In the United Kingdom, the current Weights and Measures (Packaged Goods) Regulations 2006 provide a regulatory framework for food packing businesses and they apply to the majority of packages intended for sale in pack sizes between 5 g (or 5 mL) and 25 kg (or 25 L).

These regulations expect that the 'e' marked food package will contain (on average) at least the quantity printed on the package weight statement.

With regard to these 'average weight' packing requirements in the United Kingdom and Europe the expectations can be summarised via what is known as the 'Three packer's rules',

'Three packers rules': Over the course of a production batch of packages:

1. The actual contents of the packages must not be less, on average, than the nominal quantity (Qn) which is the amount marked/stated on the pack label.
2. No more than one package in 40 (2.5% of the batch packs) may contain less than the Qn by more than a specified amount known as the tolerable negative error (TNE). This TNE figure varies and is set according to the Qn.
3. No package may be 'inadequate' in that it contains less than the Qn by more than twice the TNE.

One TNE value below the nominal weight is known as T1.

Two TNE values below the nominal weight are known as T2.

In the United Kingdom the acceptable TNE for packages, in relation to the Qn, is defined in the following table which also forms part of the Weights and Measures (Packaged Goods) Regulations 2006:

The TNE is the amount set out in the table below in relation to the Qn on the package:

Qn in grams or millilitres	**TNE**	
	As a % of Qn	**g or mL**
5–50	9	–
From 50 to 100	–	4.5
From 100 to 200	4.5	–
From 200 to 300	–	9
From 300 to 500	3	–
From 500 to 1,000	–	15
From 1,000 to 10,000	1.5	–
From 10,000 to 15,000	–	150
Above 15,000	1	–

Where the TNE is calculated as a percentage of the Qn, the amount must be rounded up to the nearest one-tenth of a gram or millilitre.

Businesses can use manual or automated weighing systems, and the type used is usually dictated by the throughput rates required, with automatic check-weighers typically being capable of far faster processing speeds.

When a business encounters packing weight control problems, these can sometimes be due to:

- Incorrect machine programming
  - Operator error, incorrectly setting the nominal weight or perhaps selecting the incorrect 'minimum'/'average' weight settings on the check-weigher.
  - Setting the incorrect packaging tare weight (setting too high a figure could quickly render a production batch illegal, whereas setting too low a tare figure would increase the batch 'giveaway' quantity).
- Poor line set up
  - Calibration of the check-weigher load cells using incorrect or inaccurate weights.
  - Check-weigher conveyor belt becoming worn or tracking badly leading to measurement inaccuracy.
  - Check-weigher load-cell damage.
  - Load-cell 'damping' caused by the build-up of food debris from product spills.

7.7.1 'Repeatability'

Sometimes weighing systems can have a less than perfect 'repeatability' of the weights being measured (i.e., repeatedly weighing the same test weight on the check-weigher can produce slightly different weight readings each time the weight is measured). Such variability can be due to a variety of factors including the sensitivity of the scales and the stability of the operating environment.

With regard to the stability of the operating environment, check-weighers can be affected by being positioned in the line of area air flows/vents, they can also be affected by area movements and equipment vibration among many other factors.

There are statistical measurement techniques that can be applied to assess whether the repeatability of weights on a check-weigher line are within an acceptable range (i.e., How accurately and consistently is the check-weigher working?). In the United Kingdom, this statistical assessment is called the ZOI test, the 'zone of indecision' with ZOI being 6 × standard deviation of a single pack passed over 100 times.

Where such tests are applied the ZOI has to be less than a quarter of the TNE, otherwise the business should make an allowance on the target to account for the variation in the repeatability of the instrument.

7.7.2 Keeping quantity control records

In order to be able to demonstrate legal compliance, clear records and traceability, it is vitally important that every batch of product packed across the check-weighers must at the end of the run have a print record from the check-weigher confirming the batch statistics. These statistics will include the nominal weight setting, the tare weight setting, whether the batch/run was packed to minimum or average weight legislation requirements, the total number of packs produced, total number of weight rejects and the average pack weight from the batch.

Also, if packing to average weight legislation the record would be required to show the T1 and T2 settings together with the pack quantities packed between these threshold figures.

As part of the site quality assurance system, this pack weight control information would typically be checked by an area supervisor at the end of each batch in order to confirm that all legislative requirements had been met in advance of the product being despatched to the customer.

Alternatives/Plan B's

- If the check-weigher printer failed a business could consider forming the batch records by writing down the required batch figures from the check-weigher screen together with a record of the operator name, time and date. For added proof a digital photograph could be taken of the check-weigher screen.
- If the check-weigher broke down then a manual weighing system could be utilised. It is always advisable to consult the local weights and measures enforcement body on such matters in order to ensure that the required alternative weight check frequency is set and followed (as just because a check-weigher can check every single pack, this 100% sampling rate is not necessarily a legal requirement).

When packing a product batch using a system of manually checking the pack weights (manually taking packs off the line as they are being packed and weighing them on a set of calibrated scales), the frequency of the pack weight checks should be in proportion to the amount of pack weight variability expected. In short, if the pack weights from one pack to the next are 'as required' and rarely vary, then the operation should be able to conduct pack weight checks during the batch far less frequently than if there is a great deal of variability from one pack to the next.

Every pack weight measured would need to be written on a manual pack weight record sheet, and at the end of the batch checks and calculations would have to be conducted on the weights recorded in order to confirm that the required legal rules for either 'minimum weight' or 'average weight' had been met.

7.8 Calibration

Routine calibration processes are important in quantity control to ensure that the measurement systems utilised are working correctly.

Scales and check-weighers used to check the end product pack sale weights within a manufacturing operation will need to be routinely checked and calibrated with a set of test weights which are appropriate to the range of pack weights being packed (e.g., manufacturers should not use a 5 kg weight to check the scales being used to assess the pack weights of a batch being packed to a 200 g pack weight).

Typically, food businesses will maintain a set of test weights which remain in the factory (and therefore are likely to encounter more 'wear and tear'), together with a set of annually calibrated weights which are used routinely to check the accuracy of the factory test weights, and which are kept outside of the factory areas to protect their accuracy over time.

7.9 Machinery design and acceptance testing

In the European Union machinery placed for sale in the European market must meet the requirements of the Machinery Directive 2006/42/EC and subsequent updates.

The 2006/42/EC directive highlights the need for:

- 'Safety by design'.
- Affixing a CE mark on machinery to confirm compliance (CE – 'Conformité Européene' is the French translation of 'European Conformity').
- Provide the end user with a 'declaration of conformity' and suitable instructions for use.

The Directive requires food machinery to have appropriate guarding and control systems to ensure the health and safety of operators and with specific regard to food safety the directive expects food machinery to be 'designed and constructed in such a way as to avoid any risk of infection, sickness or contagion'.

Compliant machines are expected to be designed and constructed in a manner which ensures that:

- The machine must be able to be cleaned before each use and be easily dismantled for cleaning where necessary.
- Any materials coming into contact with the food must satisfy certain conditions as to suitability and surfaces in contact with food must be smooth, have minimum edges, recesses, etc.
- Cleaning and disinfection chemicals must be completely discharged from the machine.
- Insects, organic matter, etc., must not be able to accumulate in areas that cannot be cleaned.
- Machine lubricants, etc., must not come into contact with foods.

There is also an expectation that the manufacturer provides instructions to the end user on the recommended cleaning products and methods for cleaning, disinfection and rinsing.

The European Commission Guide to the Machinery Directive 2006/42/EC 2nd Edition provides full details of the range of expectations: http://ec.europa.eu/DocsRoom/documents/9483/attachments/1/translations/en/renditions/native

There are a wide range of hygienic design considerations for machinery intended for use in the food and drink industry including:

- Review of the potential food safety 'hazards' [e.g., microbiological, chemical (including cleaning solutions and lubricants)] and physical/foreign body hazards.
- Is the machine going to be operating in 'low-risk' or 'high-care'/'high-risk' food processing areas? Is this likely to influence the design of the machinery? For example, the extent to which machine parts will be accessible for cleaning. Perhaps there will be increased expectation placed on machinery to be used in higher risk processing areas.
- Consideration of the physical operating environment. For example, Will the machine be exposed to high/low temperatures, chemicals (e.g., during cleaning), moisture?
- Will the products to be processed be consumed by vulnerable groups? (e.g., the elderly, infirm or children).
- What will be the required frequency of cleaning and inspection?

When involved in the sourcing and procurement of process machinery, it is vitally important that all of the business requirements of the machinery are defined in a written specification. The requirements specification can then be provided to the supplier (or prospective suppliers) in order to ensure that all design and end-use expectations are clearly agreed and covered.

General design considerations drawn from the UK HSE (Health and Safety Executive) information sheet on the hygienic design of machinery in the food and drink industries include:

- Food contact materials/surfaces must be smooth, impervious/nonabsorbent and easy to clean. They must also be suitably resistant to damage and corrosion.
- Relevant materials/surfaces must be suitable for food contact including being formed from nontoxic materials tested to ensure that they will not transfer to the product any odour, taste or taint.
  - Inner parts, including shafts and seals, must not present product trap points and should allow for routine cleaning and disinfection.
- Joints in the machinery must be smooth and impervious.
- Recesses in the machinery must be avoided.

EHEDG European Hygienic Engineering and Design Group provide a wide range of guidelines in support of hygienic design. https://www.ehedg.org/guidelines/

When purchasing food processing machinery, it is important to ensure that the business receives detailed operating instructions (and perhaps even specific support/training packages from the supplier). The operating instructions should cover the operation of the machine and recommended approaches to cleaning and maintenance (e.g., dismantling guidance, routine hygiene and deep cleaning instructions).

'Acceptance testing' is a term which covers the range of checks and assessments to be applied to food machinery in order to confirm that it is deemed to be 'fit for purpose'. Many of the requirements/considerations noted in this section would form a useful business checklist of points to be checked in advance of acceptance of new machinery.

Key considerations when 'acceptance testing' new process machinery include:

- Ideally do not take delivery of the new machinery until it has been confirmed to meet all acceptance testing criteria. Suppliers are usually more responsive and far better placed to address issues when the equipment is still on their site.
- If the machinery cannot be fully tested off-site at the supplier (or a third party) site then the business should take delivery but avoid paying the full price for the machinery until all acceptance testing has been successfully conducted. This way the supplier is encouraged to maintain a sense of urgency and commitment to the project.
- Machinery should be tested on the intended typical product range in conditions which reflect the standard production process. This may be a challenge to arrange but will avoid costly surprises later.
- A multidisciplinary team approach to machine assessment/validation trials tends to work best, serving to ensure that all aspects are covered. Experienced process, technical and engineering team members can be a very effective combination, with each team member considering the aspects of particular importance to their areas of expertise.

7.10 Product contamination hazard controls

A food 'hazard' can be considered as something with the potential to cause harm to the consumer. Typical hazard groups involved in product contamination can be biological, chemical or physical. All of which have the potential to render a product injurious to health and/or unfit for human consumption.

There are a vast range of potential sources of food contamination including employees, pests, raw materials/ingredients, process equipment and packaging materials.

The Codex Alimentarius Commission was created in 1963 by Food and Agriculture Organisation of the United Nations (FAO) and World Health Organisation (WHO) to develop food standards, guidelines and related texts such as codes of practice under the Joint FAO/WHO Food Standards Programme. The main purposes of this programme are protecting the health of consumers, ensuring fair trade practices in the food trade and promoting coordination of all food standards work undertaken by international governmental and nongovernmental organisations.

In their 'Basic Texts' on Food Hygiene the Codex Alimentarius Commission defines contamination as 'the introduction or occurrence of a contaminant in food or food environment' (Codex Alimentarius 2009). These Codex texts consider a range of measures that can be taken by governments, regulatory authorities, food businesses and food handlers to help assure food safety.

The measures recommended range from control of the primary production processes and supply chain, employee and environment control, good site and equipment design, through to the application of HACCP. To provide further insight, the relevant aspects covered are noted below and can be reviewed in detail at the Codex Alimentarius – 'Food Hygiene (Basic Texts) Fourth Edition' website: http://www.fao.org/docrep/012/a1552e/a1552e00.htm

"Primary Production" – Environmental hygiene; Hygienic production of food sources; Handling, storage and transport; Cleaning, maintenance and personnel hygiene.

"Establishment: Design and Facilities" – Location; Premises and rooms; Equipment; Facilities.

"Control of Operation" – Control of food hazards; Key aspects of hygiene control systems; Incoming material requirements; Packaging; Water; Management and supervision; Documentation and records; Recall procedures.

"Establishment: Maintenance and Sanitation" – Maintenance and cleaning; Cleaning programmes; Pest control systems; Waste management; Monitoring effectiveness.

"Personal Hygiene" – Health status; Illness and injuries; Personal cleanliness; Personal behaviour; Visitors.

"Transportation" – General; Requirements; Use and maintenance.

"Product Information and Consumer Awareness": Lot identification; Product information; Labelling; Consumer education.

"Training" – Awareness and responsibilities; Training programmes; Instruction and supervision; Refresher training.

"Hazard Analysis and Critical Control Point (HACCP) System and Guidelines for its Application" – Preamble; Definitions; Principles of the HACCP system; Guidelines for the application of the HACCP system.

Source: Food Hygiene (Basic Texts). fourth ed. Codex Alimentarius (2009).

At a food manufacturing facility, the Technical Manager could consider the following two phases to achieving the production of food which is safe, of the quality required, and legally compliant.

1. 'Process design' in ensuring the correct specification/requirements for all raw materials/ingredients, product development and formulation, processing, storage and distribution.
2. 'System design' in the development of clear and effective control systems, and their implementation, monitoring and routine review. With the rationale being the reduction of risk of unsafe/substandard food via the use of proactive/preventive 'hazard control measures' which assure the safety, quality and legality of the food at key stages in the operation.

In the European Union, the HACCP system of food safety control is a legal requirement for all food business operators. Under Regulation (EC) No. 852/2004 on the hygiene of foodstuffs there is a duty placed on food business operators to put in place, implement and maintain a permanent procedure or procedures based on the Codex Alimentarius principles of HACCP.

The process of identifying 'significant' hazards can be very challenging and, as with many aspects of HACCP, benefits from thorough research, sector knowledge, experience and a multidisciplinary team approach.

'Hazard analysis' involves the collection and review of information related to potential food hazards and the conditions which could lead to their occurrence/presence. The process must thoroughly evaluate the products, processes and business position in order to decide which of the points considered are significant to food safety and therefore must be addressed in the process HACCP plan.

Within food operation HACCP plans the business is expected to review and develop a control strategy for all significant product hazards. A useful way to ensure that the controls for each possible hazard are appropriate and proportional to the level of concern is to assign a 'risk' status to each potential hazard.

The level of 'Risk' should consider:

- Severity of outcome (e.g., health consequences of the hazard occurring due to loss of control).
- Likelihood of the issue occurring.
- Vulnerable groups of customers.
- Likely extent of the issue (e.g., number of consumers likely to be affected).
- When seeking to control food hazards the application of good manufacturing practices/good hygiene practices provides a vital foundation on which then to apply HACCP principles.

HACCP systems of food safety control often recognise three categories of food safety hazards:

- biological hazards (e.g., bacteria, viruses and parasites);

- chemical hazards (e.g., agricultural chemicals, cleaning chemicals, lubricants, naturally occurring compounds/toxins and allergens); and
- physical hazards (e.g., metal, glass, stone, plastic, wood and extraneous plant/animal matter).

Food hazards can sometimes develop within the foodstuffs over time, though more often hazards are introduced at particular stages of the food supply chain. The circumstances causing the contamination can occur at any point in the food manufacturing and supply chain, from the growth/rearing of primary produce, storage and distribution, further processing and finally the supply chain through to the end consumer. Many governments set particular standards for the maximum levels at which certain hazards can be present within foods and a common consensus is that the application of good agricultural practice, good manufacturing practice (GMP) and HACCP food safety control systems are the most effective way of reducing the threat from such hazards and thereby assuring food safety.

A 'preventive measure' or 'control' can be considered as 'a process or procedure to achieve conditions which will eliminate or control a hazard'.

For a food manufacturer utilising the supply of raw materials/primary produce to then manufacture an end product, typically within the business operational HACCP plan the food safety risks will be covered by either distinct CCPs, or via site 'prerequisite programmes'.

Effectively for each potential physical, chemical or biological contamination issue the business is seeking to eliminate (or reduce to an acceptable level) the risk of contamination. As part of this approach the site quality and technical management will be expected to develop strategies for the control of product contamination. These strategies will be embedded within the business HACCP/prerequisite programmes (which should be documented in the site Food Safety and Quality Manual) and will typically include themes such as:

- personnel hygiene expectations (including visitor and contractor controls);
- measures designed to avoid product contamination by/via raw materials and packaging;
- measures to prevent contamination by/via equipment and building fabrication;
- control measures to prevent pest-related contamination; and
- CCPs – measures applied at specific points in the manufacturing process to eliminate (or reduce to an acceptable level) defined product hazards.

The HACCP approach to food safety control and the associated importance of prerequisite programmes is also covered in Chapter 3: 'The Food Safety and Quality Management System (FS and QMS)' and Chapter 6: 'Product Control and HACCP Considerations'.

7.10.1 Physical hazards

Physical hazards in food (often referred to as foreign bodies/objects/matter) commonly include metal, glass, stone, plastic, wood and extraneous plant/animal matter. Dependent on their shape, size and the end consumer, such materials can pose a serious threat of injury via choking, cutting or damaged teeth when unexpectedly chewing on the hard object.

Babies and children are particularly susceptible to injury from physical hazards, often resulting in a higher degree of foreign material inspection being conducted at food companies producing for these vulnerable consumer groups.

The sources of physical contamination can be almost limitless, and often arise from raw materials/ingredients/packaging, employees/visitors, pests, factory equipment, machinery and the wider manufacturing environment.

Physical contamination controls in a food manufacturing operation include:

- ingredient/packaging inspection (on delivery, during storage and at point of use);
- washing and cleaning systems (e.g., fresh produce washing); and
- measures to prevent contamination by equipment and building fabrication:

For example, good design and ongoing monitoring of the factory structure (walls, floors, ceilings, doors, windows, etc.). Also, the food handling/processing equipment used must be of a sufficient design, nature and quality to prevent any risk of product contamination.

- Avoidance of 'open' processes (e.g., lidded process vessels).
- Preventative maintenance programs for equipment and processing areas.
- Sieves (e.g., for powders).
- Screens/filters (e.g., for solutions/liquids).
- Magnets (e.g., to remove metal debris in dried herb/spice processing).
- Metal detection (can detect nonferrous and ferrous metals including stainless steel).
- X-ray detection (work on density and therefore can detect a wide range of physical contaminants).
- Vision systems and visual inspection (e.g., while processing or packing the end product).

7.10.2 Product decontamination/dealing with 'contamination issues'

Primary producers and further processors often are dealing with product decontamination as part of their standard processing methods every day.

Due to their nature, farming and harvesting methods, vegetable, fruit, herb and spice processors typically apply a range of decontamination processes to help ensure that potential 'contaminants' such as extraneous matter (other parts of the plant rather than the part required, which could include thorns and woody stem material), soil, insects, stones and moulds/rots are removed before the product is progressed further along the food supply chain.

This decontamination is achieved via a wide range of process measures including washing, belt inspection, sieves, filters, vision/camera scanning detection, magnets, gravity-related measures such as flow through settlement tanks or air flow separators, metal detectors and X-ray detectors.

For further processors of multicomponent foods such as soups, sauces, dips, dressings and ready meals, this issue of decontamination is more complex. In such processes at the 'final product stage' if contamination is present then it is highly likely that 'something has gone wrong' earlier in the process or supply chain.

Perhaps the product was manufactured correctly, only to then find a contamination fault with one of the ingredients that had been used. Perhaps during the manufacturing process, a machinery breakage or area fabric issue caused contamination of the product with physical contaminants.

Clearly the list of 'possible contamination issues' is vast yet there are very few correction options available to the food manufacturer who has completed and packed a multicomponent product only to then find that a contamination issue has occurred.

Dependent on the nature of the contamination issue options to rectify, the matter might include

- Use of manual inspection (if the issue can be easily seen).
- Use of X-ray detection (if the issue can confidently be picked up by 'density difference' with the surrounding product).
- Split product/packs down, remove the food materials and reprocess/rework in a manner which resolves the contamination issue.

In the event that the product is spoilt/contaminated 'beyond repair' products should be inspected and samples saved to trace a contamination fault back to a particular process line problem, equipment/fabric breakage or supplier issue. While helping to find the root cause and thereby reduce the potential for future recurrence, such inspection work may also serve to support any cost compensation claims against the supplier or insurance claims against manufacturing/product insurance policies.

The issue of 'product liability insurance' should be considered by all food handling businesses at the outset of the operation and regularly reviewed over time to ensure that the extent of the cover held is appropriate. Considerations should include whether the insurance covers:

- Lost end product 'retail costs' or just the raw material/manufacturing costs?
- Trade withdrawal/recall costs with or without any associated customer fines?
- Prosecution/legal costs?
- Brand damage costs?

7.10.3 Chemical hazards

Typical sources of chemical contamination in the food supply chain include agricultural chemicals (e.g., pesticides, herbicides, veterinary treatments), naturally occurring compounds/toxins, cleaning chemicals, lubricants, food additives/preservatives, chemical reactions (e.g., acidic food corroding metallic surfaces), solvents and printing solutions and compounds associated with the food packaging (e.g., plasticisers and adhesives).

Chemical contamination controls include:

- Legislative requirements (e.g., approved chemicals and maximum residue levels for agricultural treatments).

- Presupply testing (certificates of conformance/analysis) to confirm absence or presence within acceptable levels.
- Supplier audit systems and procedures.
- Ensure food contact packaging is fit for purpose (e.g., 'food-grade' assurance via 'migration testing').
- Chemical handling procedures (e.g., instructions on dilution, correct application and appropriately labelled containers. Also, segregation from other reactive chemicals and food production areas).
- Maintenance programs/staff awareness (e.g., lubricants must be food-grade and any other chemical treatments used in the process area must avoid food contact).
- Use of 'suitable for food contact' utensils, containers and packaging.

To help avoid such food product contamination risks, raw materials and packaging should be purchased to an appropriate specification which covers all relevant expectations linked to microbiological, chemical and physical standards. The materials purchased should be routinely monitored for specification compliance via delivery inspection and/or the supply of certificates of conformance (CoC) or certificates of analysis (CoA).

It is also vital that all 'food handlers'/factory staff are thoroughly trained on good personal hygiene practices and must be well supervised to ensure ongoing compliance with the site standards.

7.10.4 Allergens

Allergens are often placed in the 'chemical hazards' category due to the fact that though a food may be physically contaminated with the allergen, the chemical composition of the allergen is the actual danger to those consumers who have an immunological allergic reaction to such materials.

Reactions to allergens can cause a range of issues including skin flushing and rashes, swollen throat (difficulty swallowing/breathing), heart rate changes, severe asthma, stomach pain, nausea, vomiting, sudden drop in blood pressure, collapse, unconsciousness (anaphylactic shock) and in the most extreme cases death.

The key food allergens within the UK/EU include

milk;
eggs;
peanuts (also known as groundnuts or monkey nuts);
nuts (e.g., almond, hazelnut, cashew, pecan, brazil, pistachio and macadamia);
fish;
shellfish (including mussels, crab and shrimps);
soya;
wheat/cereals containing gluten;
sesame seeds;
celery;
lupin;
molluscs;
mustard; and
sulphur dioxide and sulphites.

In the EU there is a legal requirement for the presence of such allergens to be clearly highlighted on the product packaging in order that the allergic consumer can be aware and avoid such materials.

Undeclared allergen presence within ingredients, manufacturers mistakenly failing to highlight allergen presence on food packaging, and accidental contamination of a nonallergen product with an allergen in the processing environment, are all potential allergen hazard threats to the safety of susceptible consumers.

Allergen contamination controls include:

- allergen control reviews and 'risk assessment' via raw material supplier audits;
- supplier approval processes and strict use of 'approved suppliers' only;
- request full allergen information on all raw materials in advance of use and regularly check to ensure that the information remains up to date;
- segregated storage for raw materials containing allergens;
- allergen handling measures during processing (e.g., separate cleaning and processing equipment, dedicated utensils and segregation of production areas);
- waste handling procedures for allergen-containing materials;
- staff training and awareness of allergy issues and necessary controls;
- visitor and contractor control procedures;
- full assurance of traceability for all raw materials and finished product;
- label allergen presence clearly on packaging; and
- effective product recall and crisis management procedures.

7.10.5 Biological hazards

Biological hazards in foods include bacteria, viruses and parasites. Pathogenic (illness causing) bacteria are a major challenge to the food sector. Foodborne pathogens are responsible for many cases of food poisoning including *Campylobacter jejeuni, Eschericha coli, Listeria monocytogenes, Staphylococcus aureus, Clostridium botulinum, Clostridium perfringens, Vibrio parahaemolyticus, Salmonella typhimurium and Salmonella enteritidis.*

Symptoms of food poisoning can include tiredness, nausea, headache, stomach pain, diarrhoea, vomiting, feeling cold and having a fever. Most food groups are capable of carrying pathogenic bacteria and food poisoning in the most extreme cases can be fatal, especially within particularly vulnerable groups such as the elderly, infants and the immunocompromised.

Microbiological contamination sources include:

- food handlers and other people (e.g., visitors and contractors);
- ingredient microbiological levels exceeding specified maximum and consequently leading to end product contamination;
- inappropriate selection of an inherently 'high microloading' ingredient for use in RTE 'ready to eat' end products;
- cross-contamination via 'RTE' food contact with raw food;
- contaminated equipment;
- waste food contact;

- air/airborne dust;
- soil;
- pests; and
- water.

General controls on microbiological contamination include

- physically guarding/protecting food from contamination;
- employee training to ensure that they are aware of the risks and their role in controlling them;
- reducing the potential for bacterial multiplication (e.g., via low storage temperatures); and
- removing/destroying bacteria (e.g., pasteurisation/sterelisation heat treatment, acidity (pH) control, chemical washes/dips/sprays, and perhaps increasingly in the coming years – irradiation).

'The Bad Bug Book – Foodborne Pathogenic Microorganisms and Natural Toxins Handbook' by the U.S. Food and Drug Administration is published online at http://www.fda.gov/Food/FoodborneIllnessContaminants/CausesOfIllnessBadBugBook/ This document provides a very helpful reference guide to food microbiology, including detailed information on organisms of specific concern to particular food groups.

7.10.6 Personal/personnel hygiene practices

Poor control and training of factory staff, especially those who are designated as food handlers, is likely to result in the contamination of the food products via poor practice and lack of care or awareness. Such contamination incidents could include the introduction of microbiological contamination due to poor/unhygienic handling, physical contamination by the introduction (usually not intentionally) of physical hazards, and chemical contamination via the lack of cleaning chemical measurement/control.

Food handler controls/associated good practice include:

- Hand-washing (to an appropriate method and frequency).
- Health/illness control (including exclusion from the workplace when appropriate and completion of a risk assessment review prior to returning to work).
- Control of cuts, boils and septic sores (e.g., coverage via blue metal detectable plasters).
- Initiatives to reduce 'staff – foreign body potential' (e.g., foreign body contamination awareness training).
- Use of appropriate protective clothing (e.g., clothing designed to protect the product from the employee).

Increasingly the importance of developing and maintaining a strong 'food safety culture' within food operations is being seen as fundamental to helping ensure that staff will adhere to the range of policies and procedures required for the control of product safety, quality and legality.

Further detail on personal/personnel hygiene control points is provided in Chapter 8: 'Personnel Control'. There are also many online resources available on the subject area such as 'Food Safety: Introduction to Personal Hygiene' provided by the Food Safety Knowledge Network at http://cnx.org/content/m33042/latest. This online material provides good consideration of control system development, hand-washing, personal cleanliness, illness management, personal behaviour and visitor procedures.

7.10.7 Pest control: Measures to prevent pest-related contamination

Pests can pose both a physical and microbiological contamination risk to food products being manufactured, stored and distributed. As a consequence the design and build of food handling sites must guard against pest ingress. For example, birds, rodents and insects must all be prevented from entering food processing environments by use of closed doors, strip curtains and covers (e.g., mesh) on holes/gaps within the fabric (where gaps are required for air extraction/chimneys, drain covers, etc.).

Third-party audit standards such as the BRC Global Standard for Food Safety typically require site pest control activities to be maintained by a competent pest control organisation or appropriately trained staff, expecting regular site inspections and treatments to help guard against pest issues from occurring. [Within the United Kingdom many customer codes of practice state that the food manufacturer should use a BPCA (British Pest Control Association) http://www.bpca.org.uk registered pest control contractor].

Trained and experienced pest control contractors/staff will support a business in implementing and maintaining the necessary pest control measures including:

- Ensuring that an effective pest control programme is in place with records maintained of all visits, up-to-date plans of site proofing, numbered pest control devices and baits/monitoring devices, site recommendations, pest activity information and fully detailed actions (and follow-up reviews) taken on any pest-related issues.
- The pest control programme will include regular review and inspection of site hygiene standards, fabric/proofing integrity and checks for pest activity.
- Fly killing devices (e.g., electric fly killer units) and pheromone traps must be correctly sited and operational. The business must avoid any product contamination risk of an insect being expelled from a killing device.
- Catch analysis from trapping devices should be conducted regularly to provide insights into levels of pest activity across the site.
- Use of toxic perimeter baits to reduce the risk of rodent ingress. Toxic perimeter baits are not permitted inside the factory areas where open product is present unless in use to treat an active infestation. All baits must be secured to prevent movement/risk to product.
- Application of pheromone traps in dry goods stores and roof voids to address the risk of moths.
- Use of crawling insect strip detectors to monitor such potential issues.
- Availability of resources necessary to enable a rapid response to any occurrence of pest control issues.
- Train site employees to recognise and report signs of pest activity.

- Ensure that all pest control activities are legally compliant (e.g., use of appropriate chemical treatments and awareness of protected species).
- Operate in line with good health and safety practice (e.g., safe handling and application of pesticides, including locked storage).
- Facilitate a routine (frequency based on risk – at least annual) in-depth pest control survey of the site by a qualified pest control specialist/expert to check for pest activity and to confirm that the current pest control plan/measures in place are appropriate. This is in addition to the more frequent routine site visits/reviews conducted by the pest control technician.

Alongside seeking to control pests once present on-site there are also a number of measures that can be taken to reduce the potential for such pests to be in the vicinity of the food facility, including:

- Check/audit that controls are in place at supplier sites to ensure that deliveries do not carry any pest risks. For example, shipping containers of bulk ingredients which at certain points may have been stored outside and accumulated pests before transport to the manufacturing site.
- Avoid provision of a pest habitat around the perimeter of the food site. Keep any grass/vegetation regularly mown and ideally have concreted/tarmac areas surrounding the factory.
- Do not allow food waste and previous food containing materials to accumulate/remain on the perimeter as these items will attract and retain pests due to the provision of food and shelter.

Chapter 11: 'Managing the Technical Department' further reviews 'pest control' activities when considering technical 'engagement with outsourced services' and also aspects of the 'pest control contract' in Chapter 11: Section covering 'Technical Budget Management'.

7.10.8 Malicious contamination/sabotage

Another way in which foods can become contaminated is via intentional addition/adulteration. Such circumstances can occur due to a multitude of different reasons including:

- Commercial advantage (e.g., herb and spice growers have been known to raise the weight of their containers of raw materials presale by adding heavy weights of rocks, stones and metal including iron filings. Another example of criminal adulteration for commercial gain is the 2008 melamine issue in China when the industrial chemical melamine was added to watered-down milk to hide the dilution by making it appear to have a higher protein content, leading to the injury and deaths of babies drinking the contaminated infant formula).
- Disgruntled employee (e.g., grievance with the business).
- 'Horseplay'/stupidity (e.g., staff member thinks it is amusing to pack assorted foreign bodies within packs).
- Extortion (e.g., member of the public threatens to contaminate products on the supermarket shelves).
- Mental instability of employee or member of the public.
- Terrorist threat.

Measures to counter such issues can include:

- Careful staff selection processes, including testing, interviewing and uptake of references.
- Training of staff to raise awareness of the consequences of product contamination.
- Strong and vigilant supervision of staff.
- Factory CCTV (closed circuit television monitoring use as a deterrent).
- Site security measures to prevent intruders.
- Batch traceability and records which ensure that any issues can be traced back to an individual (and then making this clear during staff training).
- Tamper evident packaging.
- Disciplinary/prosecution/prison consequences.

Future technology advances may also include sensors which can trigger 'contaminated' warnings to appear on packs that are not of the standard composition. (Already technology is available to highlight when a product has not been stored at the required temperature, or has been open for too long and may therefore be unsafe.)

7.10.9 Food law related to product contamination

Due to the primary need to protect public health the food law of most countries is usually clear on the expectation for food business operators to put in place procedures which manage food safety. As a consequence, such legislation will focus significantly on food contamination hazards and the control of potential issues.

In the United Kingdom, food safety legislation has used terms such as that the product must not be 'injurious to health' and must be 'of the nature and substance intended'. These clear expectations can prove invaluable to a quality/technical manager. For example, the manager could utilise these summary terms of 'expectation' in circumstances when dealing with business stakeholders who may have a less than appropriate level of 'risk awareness' on an issue of whether compromised foods are fit for sale. Also, such legal terms could be drawn upon when dealing with a raw material supplier who has supplied a contaminated ingredient and is then trying to absolve themselves of responsibility by blaming the issue on a 'natural occurrence' linked with the ingredient. In such a case, if the matter is a common/known issue then the supplier's processes and sorting systems should have been ensured to be far more rigorous at removing the known contaminants before then supplying the materials to their food processing customer where the issue would then cause even greater cost losses rendering 'unsaleable' large batches of multicomponent foods.

Food law/legislation is covered in detail in the Guest Chapter: 'Enforcement Authority Perspective on the Food Manufacturing Sector.'

7.11 Control of production startups, changeovers and shutdowns

Within a manufacturing and processing operation there will be operational phases during which there is an increased potential for product safety/quality/legality issues to occur. Such phases should therefore be subject to additional focus, confirmatory checks and monitoring to help guard against problem issues.

'Startups': Start up checks for process areas/lines should include consideration of the following:

- Health and safety: Is the line set-up correctly? Has the line been subjected to any maintenance since it was last used and is it now fit to return back into standard use? Is the area lighting appropriate? Are all safety systems (e.g., machine 'safety stop' devices) working correctly? Following clean-down are all electrics free from any water deposits?
- Hygiene: Is the area/line in an appropriate state of cleanliness to commence processing foods? Perhaps the area/line has been subject to a cleaning process and would benefit from a formal inspection to confirm that the cleaning has been conducted to the required standard. Have the process lines and equipment been returned to an operational state or are some set-up actions/checks yet to be completed before production/processing can recommence?

Product 'changeovers': When changing over from one product to the next on a processing line the challenge is often to reduce the 'down-time' to a minimum, while at the same time thoroughly ensuring that the last product has all passed safely through and the next product is thoroughly set-up and commenced correctly. The potential for rushing these processes to save processing time necessitates a great deal of focus to guard against the occurrence of problem issues. Factors that can be beneficial to consider at the point of product changeover include the following:

- At the point of completion of the last product were all key product safety checks still within their acceptable tolerance limits? For example, does a final check need to be conducted on any foreign body detection equipment? Perhaps checks on the metal detector are set to be conducted and recorded every 30 min, yet at 29 min after the last check the product run has completed. It clearly would be a weakness in the testing regime if 29 min of the production run could not be assured/confirmed to have been completed with a fully operational metal detection system. For such reasons, most product safety checks that are conducted intermittently will have a test frequency of 'Run checks on-start up, then every xx minutes during the process and finally check at the point of completion of the process'.
- What level of cleaning is required to take place between the product runs? Some products may be very similar in composition/recipe and therefore may not require as extensive a clean-down between products than if the recipes/products are completely different or if there perhaps are other matters to consider such as the potential for microbiological or allergen cross-contamination.
- When changing over to a new product it is important that any packaging related to the previous product is cleared from the machinery and area before the next product is commenced in order to avoid packing the next product into incorrect packaging.
- All production records associated with the last production batch should be checked by an appropriate team member (e.g., area supervisor or member of the quality department) to confirm that the batch has progressed correctly and met all check tolerances required. These records should then be signed off as checked, approved and removed for storage to make way for the online documentation and records of the next batch that is due to be processed. In the event of an issue being picked up on these checks, the fact that the

process has only just been completed should mean that the problem is still well within the control of the business with the affected stock still on-site.

'Shutdowns': When shutting down a line/area for a time after a product has been completed, it is usually important to ensure that the line/area is left in a clean and secure condition to then await the next time the area/line is required to be utilised. Shutdown considerations may include:

- Leaving any product debris behind can lead to microbiological problems building up in the area, as can leaving the area in a wet/dirty condition.
- It is helpful to leave the line/area in a condition that will make it easiest to recommence processing in due course. Having all equipment and related utensils cleaned and placed in their standard positions can be of great benefit when eventually looking to recommence production in the area.
- Perhaps it is standard practice to strip all of the machinery down to be deep-cleaned.
- It may be required to conduct 'chemical fogging' (the application of disinfection treatments as fogs/mists) in the area for additional microbiological control at the end of the production clean-down process.
- Some lines/areas may benefit from being covered once they have been cleaned.

7.11.1 Area/line checks

Often the best placed individuals to conduct and record these area and process line checks will be the area/process line staff themselves as they are likely to understand in detail how the area/process lines should look and therefore will be more likely to notice when something is not as it should be.

However, there may be times when an independent check is more appropriate. For example, perhaps site quality/technical staff should be utilised when deciding whether an area/process line has been sufficiently cleaned and is in an appropriate condition to commence processing the food products, as production staff may lack the detailed level of awareness/focus required for such specific checks.

Other circumstances where an independent assessment may be of benefit are when the area operators may be inclined to rush checks in order to recommence operational duties as quickly as possible. This can easily be the case in operations that place a high degree of pressure on employees to work as fast as possible to save time. While speed of work can increase volume output rates and reduce related labour costs, there needs to be an appropriate work rate set which also allows quality, safety and legality to also be thoroughly controlled. The technical manager's engagement with the operations team/production management should have a positive influence on agreeing appropriate running rates and check procedures.

7.12 Management of high-care and high-risk environments

The design and structure of high-care and high-risk areas is covered in Chapter 5: 'Site Standards'. With specific regard to the management of such areas within a manufacturing facility there are a number of key differences to be mindful of when compared to the operation of other process areas on-site. These differences are due to the fact that in high-care and high-risk areas the food product is usually in an open form, about to be packaged, and is not usually going to be subjected to any further decontamination processes before being consumed (e.g., the product is usually being sold classed as RTE–Ready To Eat). As a result, the microbiological loading on the product requires very close control in this area as the product loading will not usually be significantly reduced from this point onward, and in fact may actually increase during the shelf life of the product prior to consumption.

7.12.1 With regard to the terminology of 'high care' and 'high risk'

'High-care' factory areas are typically processing areas which are segregated from the remainder of the manufacturing site and are designed to *minimise* the potential for postprocess contamination of the product being handled. The term 'high care' is utilised when the RTE food being processed, due to its composition, is likely to already have a significant microbiological load, and therefore the intention is to avoid further adding to the microbiological levels (e.g., washed and prepared, RTE cut salad products).

'High-risk' factory areas are food handling areas which are segregated from the remainder of the manufacturing site and are designed to *prevent* the potential for contamination of the product being handled. The term 'high risk' is utilised when the food product being handled has been processed to effectively eliminate the majority of the foodborne microorganisms, and therefore the intention of the area is to prevent the potential for postprocess contamination (e.g., 'cook-chill' produced ready meals, soups and sauces which are classed as 'RTE' as the end consumer will only reheat the products rather than 'cook' the product, therefore microbiological reduction at this final heating stage before consumption cannot be assumed).

The elevated levels of risk management required in these areas can translate to a number of added controls being beneficial including:

- 'Higher level' staff training – Training staff operating in these areas to higher levels will increase their awareness and control when handling products, conducting checks on safety/quality/legality, and when progressing cleaning processes in the area.
- Low-risk to high-care/high-risk segregation – To prevent the spread of potential microbiological contaminants from the low-risk areas over to the high-care/high-risk area, the high-care/high-risk area will typically require full physical (walled-off) segregation from the low-risk areas (with only hatches to facilitate product flow and doors for entry/exit), separate drainage flowing out from the area, separate dedicated equipment, separate staff changing facilities with the requirement of a full change of factory clothing and personal protective equipment (PPE) into dedicated high-care/high-risk clothing/PPE. Engineers will often be expected to have dedicated toolkits for use in these areas to guard against cross-contamination carried from low-risk areas. Generally hand-washing will also be required during entry to the area and more frequent washing is likely to be required when working in the area.

- For added microbiological control, high-risk areas will also usually have the added requirement of the environment being maintained under positive pressure and utilising filtered air (such measures are also viewed as 'good practice' in high-care areas).
- Supervision – Due to the likely high impact of any issues arising from a lack of control in the high-care/high-risk area, these areas often benefit from a high degree of experienced supervision/management.
- Greater investment – As the product is in its most exposed form in this area, the quality of the equipment, machinery and any testing systems applied should be ensured to be of the highest level to best control and protect the product. Hygienic design of equipment to facilitate effective access for frequent cleaning and inspection is a key point to be ensured.
- Increased hygiene control – Due to the increased risk level the cleaning of the process lines and the surrounding environment must be optimised. This could be via the provision of good quality cleaning materials, specialist cleaning chemicals, validated hygiene procedures and the allowance of sufficient planned time for both routine cleaning/disinfection and for deep cleaning.
- Increased pest control – while all areas of the food processing facility should be effectively well proofed against the ingress of pests, due to the particularly exposed nature of the end product in high-care/high-risk areas this necessitates a particularly high degree of control/investment.

7.13 Control of maintenance activities

Site maintenance activities are typically conducted by engineering staff and focus on factory equipment, the building fabric and facilities. Such work is vital to maintaining production output and often has the potential to influence site Health and Safety and also product quality, safety and legality.

Technical expectations of site maintenance activities include the following:

- The need for a PPM – 'Planned Preventative Maintenance' approach to such activities in order to help ensure that all site equipment and fabric/facilities are checked and attended to at an appropriate frequency. This proactive approach will help avoid problem issues developing over time, and thereby contribute to a smooth-running site.
- The site maintenance system will benefit from a log of all maintenance, repair and installation work required in order to ensure that points can be prioritised (ideally by the operations team with technical input) and addressed within an appropriate timescale.
- The log of site work required should be routinely reviewed to ensure that any outstanding points are considered and adequately addressed (e.g., extra resources may be temporarily needed to address an issue, or perhaps a short-term solution has to be implemented while awaiting the required parts/expertise to arrive to facilitate a full repair).
- In advance of the maintenance activity, a risk assessment should be conducted to ensure that the work will not impact on product safety or quality. Where such a risk exists, it should be managed via a range of options including: physical removal of the work away from the production operation, time segregation where the work is completed outside of production hours, physical segregation, for example, via the use of screens.
- On outset and upon completion of line maintenance tasks, the equipment should be adequately cleaned.

- All maintenance activities must be completed by approved personnel, engineers or contractors to an appropriately high standard, and in a manner that is mindful of the potential impacts on the end product should any part of the repair work fail or be deemed unsatisfactory.
- Sufficiently robust materials should be utilised for all repairs. Use of ad hoc solutions such as tape and cardboard should be avoided.
- Records of all work undertaken should be maintained both to demonstrate that the repair and maintenance programme is effective and also to facilitate traceability/further investigation in the event of a product/process issue occurring which may have been affected by the maintenance work.
- Records of maintenance work should also incorporate or trigger an appropriate check to confirm that the work has been completed correctly and that any peripheral activities such as cleaning have been completed prior to the equipment/area being released back into a production mode.
- Engineering stores and workshops should be segregated from food manufacturing environments and maintained in a clean and tidy condition. Access should be restricted to approved personnel. It is important to monitor these areas via the site internal audit schedule to ensure that any potential contamination threats are adequately controlled.

7.13.1 Tools/toolkits

With specific regard to the site control of tools/toolkits, it is important that engineering personnel recognise the potential for product contamination posed by such materials. It is particularly important that they are:

- Kept in a good physical state and replaced when necessary.
- Maintained in a clean and hygienic condition.
- If kept in (or moving into) high-care/high-risk environments, this equipment must be subject to cleaning and disinfection regimes.
- Toolkits must be appropriately designed to help facilitate control of all pieces of equipment/tools (e.g., separate small compartments for smaller size items).

Tools and parts/spares should be kept on a register in order that they can be monitored. This is particularly important for such materials which are kept or used in food manufacturing areas, as any missing items could pose a contamination threat to the end product.

In addition, control can be further facilitated via the use of magnetic mats/trays to hold small parts when working on equipment/process lines. ‘Piece counts’ and checks can be utilised to ensure that all items are accounted for at the end of a task.

Lubricants are often required in food manufacturing environments where there are many moving parts on equipment/machinery. Any lubricants used should be ‘food grade’/‘food safe’ in order that if they do accidentally find their way into the end product then at least they will not present a food safety risk.

Glass control is a particular activity that requires close focus to ensure that broken glass cannot occur and become a contamination hazard. Engineering team members tasked with glass-related activities must be appropriately trained on the potential food impacts of broken glass and the importance therefore of ensuring that any glass (e.g., lights) cannot be changed in open food environments while there is food being processed. Screening and confirmation of all glass being removed from the area at the end of the task are vitally important to control glass risks.

7.14 Control of rework

'Rework' is a product which for a particular reason is not able to be packed and delivered into the supply chain, but is of a sufficient standard to be added back into the production process and 'reworked' to create an acceptable end product.

Control of rework processes can typically include:

- Securing end customer acceptance/approval that reworking practices can be progressed with regard to their specific products.
- Due consideration by appropriate staff on the acceptable percentage at which the reworked product can be added into a fresh production batch. This percentage could be set by progressing trials in advance, perhaps on a kitchen scale, to ensure that the maximum percentage set does not adversely affect the quality of the end product.
- Writing of process sheets/instructions to ensure that the rework is selected, made ready for reworking/addition, and added at the necessary stage of the manufacturing process. Such instructions may also expect additional processing and checks to facilitate the rework process.
- Ensuring that the batch traceability records account for the amount of rework added, and the related traceability records of that rework material.
- It is typically a good practice to ensure that there is a 'break' in the production batches containing rework over time, in order that there is no potential for food material in the system being 'unacceptably old' due to batches always containing reworked material from previous batches. Such break points will typically be set by the technical team, often in correspondence with the end customer.
- As reworking of product can sometimes involve removing the product from its packaging in readiness for being reworked, it is important that the staff opening the product exercise control and diligence to ensure that none of this packing becomes mixed in with the prepared rework. Otherwise there is a significant risk of the packaging becoming a physical contaminant when the rework is added to the fresh batch of product.

7.15 Waste management

Management of waste materials on a food manufacturing site must always take account of pest control and general food safety assurance via the use of dedicated and appropriately designed, lidded waste containers.

Waste flow out of a food processing facility should be designed to avoid contact with raw materials and the finished product. In addition, the waste containers should be placed on a regular and adequate cleaning cycle.

Where liquid leaks are likely to result from waste material, sufficient provision should be made including locating near to appropriate drainage and scheduling regular cleaning to avoid attracting pests and the potential for harbouring microbiological contamination issues.

Waste container design, size and movement schedules should be set to ensure that the waste is not allowed to dwell for significant amounts of time in food handling environments, and similarly is not able to reside on the manufacturing site for a length of time that may lead to the build-up of microbiological or pest control issues.

With regard to waste which is in the form of finished packaged end product, businesses typically have their own specific policies for the control of such material. Where the product is still fit for human consumption some businesses may seek to give this food away to charities or employees (perhaps via a staff shop). In such cases a business should be careful to ensure that the end recipient has clarity as to the use-by/best before status of the food, the expected storage conditions, appropriate handling/preparation instructions and any allergen presence/issues which may affect certain intolerant groups.

Where the food manufacturer packs food for a brand, the brand owner may object to such foods being given away or sold internally, or in such cases may wish for the brand, associated trademarks, their address and bar code detail on the packaging to be removed or redacted to help guard against any perceived brand damage or claims against the brand. It is always advisable for the technical management to have these 'expectation and allowance conversations' with the brand holder in advance, rather than waiting for an unexpected problem to occur and then having to move fast to decide what to do with the unsaleable end product.

7.16 Storage facilities

In food operations there are many technical considerations to be accommodated with regard to both raw material storage and finished product storage.

The construction materials of food stores should be 'food grade', typically smooth and impervious so that they can be easily and regularly cleaned and will not pose a microbiological harbourage, chemical taint or physical contamination threat to the stored foodstuffs.

Other 'food sector standard' area design aspects include:

- Antislip floors (as slips, trips and falls are a major accident category in the food sector).
- Area drains should be present and accessible to facilitate thorough cleaning.
- Floors should slope toward the area drain points to avoid pooling water.
- Coved wall-floor and wall-ceiling joints will ease cleaning and avoid debris/microbiological contamination trap points.
- All doorways should be thoroughly proofed to aid environment control and guard against pest ingress.
- Any strip curtains fitted in doorways should avoid trailing on the floor as this could easily lead to floor debris being transferred onto food products and employees during entry/exit.

- Lighting should provide an appropriate level of brightness with all glass bulbs/tubes adequately covered/shielded/sheathed to reduce the potential for the contamination threat of broken glass.
- Sufficient plastic pallets and/or racking should be provided to avoid any consideration of placing packages of foodstuffs directly onto the floor.

Access/doorways to the storage areas should be of sufficient size to allow the required type and flow of traffic in and out (e.g., staff, pallet forks, forklifts, etc.), with door mechanisms appropriately designed to swiftly open and close to help maintain the area operating conditions.

Stores must be designed to operate consistently within the required temperature range (and relative humidity where dry goods stores are concerned) to optimise and assure shelf life, safety and quality.

When setting the operating parameters (such as air temperature) for food stores, it is useful for the technical manager to consult legislation, good practice guides and customer/third-party audit standards in order to set appropriate limits. As food safety and quality can be impacted by incorrect settings or by area/equipment faults, or by doors being left open, it is important that stores are adequately monitored, ideally with alarms being raised when an issue/failure is evident. Placing of alarms on doors, which sound if the door is left open for too long, can be an additional way of ensuring that the inside environment remains protected.

Alongside audible alarms many storage sensor systems have the capacity to 'dial out'/send a message to mobile devices, which can prove invaluable in circumstances where the staff required to deal with a problem are not always on-site, and in some cases where there are no staff on-site at all to hear an audible alarm.

Records should also be maintained of ongoing store monitoring to be reviewed routinely to guard against an issue building up over time and also to prove to customers, third-party auditors and enforcement authorities that such aspects of the operation are clearly under control.

With regard to areas where there may be a significant difference between the outside air temperature and the in-store air temperature, fast doors, strip curtains, airlocks, etc., may be advisable to reduce the impact of regular movement into and out of the area (including reduction of the potential for temperature fluctuations and build-up of condensate). Such matters should be clearly considered and planned for at the design phase. Similar approaches can be relevant when considering doors from the food handling factory areas to the outer areas of the factory, where perhaps pest control is also a potential issue for consideration. For example, many businesses will operate pest-proofed doors (including tight seals and strip curtains) to the outside of the building.

Where refrigeration units/fans are operating within storage areas, their design should help ensure that condensate is not able to drop from the unit onto the floor below (and therefore potentially come into contact with food materials), as condensate can often hold microorganisms which could prove harmful if transferred to the food products. Condensate catch trays with the condensate caught and then transferred directly to a floor drain via pipework helps to reduce the potential for the spread of condensate contamination and many refrigeration system manufacturers will offer breaks and bends in the pipework, designed to prevent microorganisms from the floor drain being transferred up the condensate pipe into the refrigeration unit.

7.16.1 Additional storage facility requirements: 'Extra capacity'

Food manufacturing operations, usually as a result of growth or significant peaks in sales, often need to secure additional storage capacity for increases in handling raw materials or finished products. This section considers some of the challenges involved in such scenarios.

Storage for foodstuffs tends to fall into three general categories: frozen, chilled or ambient.

When seeking to increase storage capacity in any of these categories, the first question often asked is 'can the business free-up space in the current storage areas?' Sometimes the answer to this question is yes, when redundant stock lines are cleared and issues such as over-ordering and also holding excessive stock levels are taken into account and addressed. In addition, sometimes extra racking/stacking or use of rolling racking can also serve to increase the capacity of storage areas. However, business technical departments should always be involved in such considerations to ensure that stores are not allowed to become so overloaded that temperature control or stock control ceases to be effective.

Once a business has accepted that it requires extra food storage capacity above the current factory resources, there are a number of options to consider, all of which must conform to the same GMP requirements that would be expected of the current on-site storage areas. For example, the additional area should consider:

- Being well maintained in a hygienic and structurally sound condition.
- Being capable of consistently running at the required temperatures (and/or relative humidity if this aspect also requires controlling).
- Have sufficient monitoring and recording of running temperatures, with 'out-of-tolerance' temperature alarms if necessary.
- Have segregated storage areas for damaged/returns/quarantined stocks.
- Have segregated areas for the storage of allergenic foodstuffs.
- Area layout should facilitate (and not obstruct) good stock rotation practices.

There follows a range of options typically taken by food manufacturers to develop their raw material/finished product storage capability, all of which should be progressed with the above key GMP points in mind.

- Trailer storage: Trailer storage can be as simple as placing a truck trailer in the yard of the manufacturing business and utilising pallet trucks and forklifts to move products onto and off the trailer. Such storage can also involve connecting a trailer to an existing loading bay into the building where materials can then be moved in and out as required.
- Inflatable temperature control areas: Food manufacturing businesses can progress increases in storage capacity via investment in solutions on a temporary, permanent or semipermanent basis.

- 'Wide-span' temperature-controlled units: These are formed from bonded insulated wall panels with built-in refrigeration units. They can be built onto or inside an existing factory building.

The benefit of such flexible capacity solutions is that they can be increased/decreased in size over time in line with business fluctuations in volume output/requirements, making them particularly well suited to businesses which encounter variable sales rates (e.g., seasonal products).

While progressing such solutions it is vital that the technical department is involved to ensure that temperature control, temperature recording and hygiene requirements are all well considered and controlled over time to ultimately assure the safety and quality of the products to be stored within these locations.

7.16.2 Rental of 'off-site' storage

There are many ambient, chilled and frozen storage businesses that maintain large stores, within which storage space can be rented.

Considerations with regard to these options include location/distance from the manufacturing operation. Such sites should be close enough to enable practical operation/liaison with the main manufacturing site, and also if it is the end product that is to be stored at these locations then a decision needs to be made on whether to then pick orders and distribute direct from the rented/'off-site' store, or whether the stock has to be brought back to the main manufacturing site prior to distribution.

There are also additional technical considerations here with regard to the ease of sampling stocks, and ease of quarantine processes in the event of a problem batch/issue.

Every business will have its own preferences/requirements; however, a key technical point to bear in mind is that all movements of stock add complexity to the process, and with complexity can come mistakes if not tightly controlled. Such mistakes can include mispicks of stock (selecting the wrong product to fulfil an order) and loss of traceability, both of which can significantly impact on product safety, legality and quality.

7.16.3 Building or purchasing an additional storage location

Similar to the rental options outlined above, the proximity of an additional storage location to the main manufacturing site is a key factor in whether the arrangement is likely to be successful.

Renting or purchasing a storage site does give the added advantage of being able to staff the site with the business's own team members rather than having to utilise the staff of a particular business where just a small amount of storage space is being rented. In general, 'no one looks after your own business interests better than your own team members' is good advice to bear in mind in these circumstances.

When building/purchasing an additional storage location, the technical department should be closely involved from the start in order to ensure that running temperatures, records, temperature alarms, hygiene requirements, segregation requirements and traceability are all clearly defined and implemented.

When maintaining offsite storage a food manufacturer should also bear in mind the requirement for such stores to be audited alongside the main site when it comes to customer audits and third-party audit certification. It is therefore vitally important that these sites are aligned with the main site in terms of operating systems, traceability and internal audit standards, and thereby avoid any surprises or issues when customer/independent audits are conducted on the 'off-site storage facilities'.

Controlling product safety, quality and legality during times of business expansion can sometimes be the largest challenge that a technical team will face. New buildings/construction work, increasing capacity, changing operating systems, all while dealing with increased volume/customer pressures. As with so many technical aspects, a team approach working with other key members of the business operations team (production, engineering, logistics, etc.) will help ensure that all changes are clearly considered for pitfalls in advance and that new operating systems/facilities are well defined and correctly implemented. Thereby reducing risks to product safety, quality and legality and increasing the chances of getting everything 'right first time' (or at the very least spotting problems early while they can still be easily rectified).

7.17 Product transport and distribution

When considering the transport and distribution standards required for raw materials, packaging and finished product, focus should be on maintaining the product at the required storage temperature, in secure conditions, and ensuring a hygienic and pest-free environment.

Procedures should also guard against the potential for transfer of physical contamination, cross-contamination or taint to the materials being transported (e.g., via stipulation of what other materials are acceptable/unacceptable to be transported alongside the products, and via appropriate and thorough cleaning and hygiene procedures between journeys).

Threats/hazards could be in the form of:

- Other products transported alongside the food, for example highly aromatic/tainting chemicals or other pungent/tainting food products.
- Foreign body threats due to poor fabric/condition of the transport unit. e.g., Perhaps glass control of lighting may be an issue that needs to be guarded against.
- Pilfering/damage via transport staff mishandling or opening some of the product containers during transit. Such activity could also subsequently lead to pest ingress issues.

Additional factors in certain areas of the world may be the need to guard against very dry or high relative humidity conditions which may adversely affect the product or its packaging.

Where specific temperature control is required procedures and training should be in place to ensure that the driver knows and sets the required running temperatures and where appropriate the required alarm settings. Records of all running temperatures should be taken and stored to prove ongoing control and also to demonstrate 'due diligence' in the event of seeking to trace a shelf life, quality- or safety-related product fault.

For product safety/quality assurance and traceability purposes, it is important that the food manufacturer stipulates in advance to the transport company all requirements for data logging/recording the temperature during the journey. At this point it is vital to define whether what is being measured is the product temperature (by placing a probe within or between packs of product) or the air temperature within the unit [this is the most common and easiest option, and which can be seen to fluctuate significantly if there are many stops on route where the doors can be opened (perhaps to unload/load other products on route)].

It is usually vitally important that the food product is loaded onto the transport unit at the right temperature itself, as usually the temperature-controlled transport unit will be designed to maintain the required temperature but will not be designed to cope with further freezing/chilling of the product if it is loaded at too high a temperature. This is due to the sizing of the refrigeration system which has usually been produced to a specification only requiring to maintain air temperature and not requiring the extra capacity capability of actually lowering product temperatures which would require a far larger (and therefore typically more expensive) capacity system.

During the journey the chilled/frozen air within the transport unit must be able to circulate freely around the products. Too tightly stacking the product to the sides of the container can risk those contact points being maintained at an insufficient temperature (due to temperature conduction from the container walls). Additionally, during the transport journey there may be points on route where the unit doors are going to be opened, allowing ambient temperature air into the unit. If the chilled/frozen air cannot freely circulate within the unit (perhaps due to overloading, or stacking products too close together), such routine elevations in air temperature can have a detrimental effect on the quality/safety of the foodstuffs being transported.

Some products do require their transport units to be of a sufficient design/capacity to further cool the product during transit. One such example is fruit and vegetable products which due to continuing to respire following harvesting, can still be generating significant heat. Another example is newly harvested crops where the 'field heat' needs to be removed as swiftly as possible for the maximum benefit of product quality and shelf life. In such circumstances it is vital to have refrigeration/fans of the right capacity and products stacked/packed in a manner which allows the cooled air flow to freely circulate. Often this air circulation is achieved via fans/ducts forcing the cooled air through all points of the product.

All key operating aspects such as temperature control settings, data recording, alarm setting and actions to be taken in the event of a temperature fault or breakdown should be covered in formal written procedures, signed-off as trained against and understood by all relevant employees and subject to refresher training on an appropriate schedule.

In the circumstances where third party/contractors are utilised for transport and distribution, all expectations and requirements should be drawn up and signed off in the form of a service agreement/contract between the two businesses. These operations may also include the use of intermediary product stores in the distribution network, and these too should be subject to the same product quality and safety assurances. Service levels should be regularly reviewed and contractors should be subject to routine inspection visits/audits.

Contracts between transport businesses and food manufacturing businesses should clearly define:

- The temperature control settings/expectations and alarm requirements.
- Stipulations regarding driver training and expectations on holding and maintaining storage records.
- The timescales expected for the retrieval of the time/temperature records in the event of a query, complaint or audit.
- The length of time that the transport business is expected to store all records for.
- Expected actions to be taken in the event of the temperature-controlled unit developing a temperature control fault while in transit (break down procedures, who to call, what to do to and details of product loss insurance).
- What other products the foods can/cannot be transported alongside.

'Transit tests' should always be completed when seeking to take a new product and especially a new packaging type (or configuration) into the distribution network. There are many guidelines available on how to complete such tests effectively and often customers and audit standards will stipulate their own specific expectations. In general, the key aspects of transit tests are to:

- Ensure that the product and packaging are representative of what will eventually be placed into the food supply chain.
- Conduct the transit tests on a vehicle representative of what will actually be used routinely.
- Ensure that the trial time/journey length reflects the time that the product will typically spend in the standard food supply chain.
- A suggestion is that placing the transit test product over the rear axle of the distribution vehicle will provide the worst-case scenario test with regard to product movement/impact during the journey.
- Ensure that the product temperature and other environmental conditions match the ongoing standard settings for the supply chain.

Overland transport: Land-based transport is typically achieved via a range of different-sized vehicles selected to be most appropriate for the journey length and volume of product to be transported.

Articulated vehicles, trucks and vans are the most common methods of road transport. All of which can be fitted with freezer/refrigeration systems if required.

Multitemperature vehicles are versatile transport units, where different zones of the vehicle are running at frozen, chilled or ambient temperatures in order to accommodate mixed loads of foodstuffs (perhaps on route to the individual stores/restaurants from the distribution depot, or perhaps on route to the end consumer as part of a supermarket home delivery service).

Transport via sea freight: As approximately two-thirds of the world is covered by ocean it is no wonder that sea freight systems are well developed.

With particular regard to the sea transport of foodstuffs, technologies that help greatly are chilled/frozen freight containers which provide the required storage temperatures for the stocks contained within, and the development of modified atmosphere (MA) and controlled atmosphere (CA) storage systems.

MA/CA techniques have particularly enabled and optimised the transport of fruit and vegetables over long periods of time (often 30–50 days shipping within a container), where levels of CO_2 and O_2 have to be tightly controlled as the product respires in order to maintain shelf life and product quality.

With particular regard to meats, if not transporting in frozen conditions then the utilisation of vacuum packaging under chilled conditions (often 'deep chill' conditions of around $-1/-2°C$) can significantly increase product shelf life sufficient to facilitate a long sea journey. Timescales quoted in the industry include 12 weeks shelf life for beef products and 8 weeks shelf life for lamb products.

Transport via air freight: Transport via air is particularly costly and therefore typically only considered when no other transport route is able to deliver the products by the time at which they are required.

Fruits are a good example here. Given the choice on cost grounds most fruit importers would use the sea shipping system to transport the fruit from the country of origin to the UK market. However, such journeys usually take weeks and therefore have to be planned well in advance of the fruit stocks being required on the supermarket shelves. In the event of a stock shortage a final option becomes to 'air freight' the required product.

7.17.1 The Importance of packaging in product transportation

Aside from appropriate control of the transportation methods, it is of vital importance that the product packaging is designed to protect and preserve the product during the journey involved in reaching the customer.

Transit testing is the most common way of a business assuring itself that the product packaging will be fit for purpose in terms of physical protection of the product during distribution.

Shelf-life trials set at the appropriate storage conditions (typically temperature is the variable, and sometimes humidity) is the most common method utilised to confirm that the product within the intended packaging will have sufficient shelf life to still be saleable to the end consumer.

Packaging 'tamper evidence' is another important consideration. Usually for products to be sold within a retail outlet, where many end consumers could browse the product before purchase, tamper evident packaging is typically insisted upon. However, when supplying other food businesses ('business to business' B2B) or supplying into food service outlets, where only employees of the next business within the supply chain are going to be in contact with the product, the need for tamper evidence becomes a matter for discussion and agreement between the supplier and the customer.

Environmental considerations: As consumers become increasingly aware of the environmental impacts of their food choices, the method of food transport and its impacts on the environment are placed under increasing scrutiny to the point where it is now often able to affect decision making. For example, calculation of the 'carbon footprint' is increasingly being utilised to assess the environmental impacts of a food product, its packaging and its transport, and in some cases is being labelled on foods where the consumer may wish to make this part of their purchasing decision.

As a consequence, technical departments are increasingly being required to understand and perhaps also calculate the carbon footprint of the products being manufactured.

With retailers also now taking a lead on the issue of carbon footprint calculation and labelling this is likely to be an increasing subject area over the coming years and is likely to ultimately result in increasing targets to reduce environmental impacts.

Reducing the use of air freight is generally regarded as one potential significant area of lowering CO_2 emissions. As is ensuring that, when products are transported by road/sea, the content of the containers should be optimised to save the financial and environmental costs of transporting air (gaps) rather than product. Such measures may well impact on product packaging design in future and on the design of transport units/containers.

Transport via rail and by internal country waterways/canals may also see a resurgence in the future as the environment increasingly impacts on operational decisions.

Chapter 11: 'Managing the Technical Department' notes the business technical department interaction with 'product transport and distribution' in the section covering 'Engagement with Outsourced Services'.

Further reading

Allergens: The Anaphylaxis Campaign. www.anaphylaxis.org.uk

Audit/Inspection/Compliance:

British Retail Consortium (BRC) – www.brcglobalstandards.com

International Featured Standards (IFS) – www.ifs-certification.com

American Institute of Baking (AIB) – http://www.aibonline.org/aibOnline/en/

International Organization for Standardization (ISO) – https://www.iso.org/home.html

Global Food Safety Initiative (GFSI) – http://www.mygfsi.com/

Chilled Food Association (CFA). http://www.chilledfood.org The CFA was formed in 1989 to establish, continuously improve and promote best hygienic practice standards in the production of retailed chilled prepared food.

Codex Alimentarius Commission – http://www.codexalimentarius.org

EU Food Law:

https://ec.europa.eu/food/safety/biosafety/food_hygiene/legislation_en

https://ec.europa.eu/food/safety/general_food_law_en

Food Hygiene (Basic Texts). Fourth Edition. Codex Alimentarius (2009).
http://www.fao.org/docrep/012/a1552e/a1552e00.htm

Food Safety: Introduction to Personal Hygiene. http://cnx.org/content/m33042/latest

These materials were developed through a partnership between the Global Food Safety Initiative and Michigan State University. Food Safety Knowledge Network.

Guidelines for the Validation of Food Safety Control Measures. http://www.fao.org/input/download/standards/11022/CXG_069e.pdf

Codex Alimentarius Commission 2008.

Hygiene for Management. R.A.Sprenger. 19th Edition. Highfield International Ltd. (2017).

ISO 9000:2015 Quality management systems – Fundamentals and vocabulary. https://www.iso.org/obp/ui/#iso:std:iso:9000:ed-4:v1:en

HACCP:

Codex Alimentarius Site Covering HACCP http://www.fao.org/DOCREP/005/Y1579E/y1579e03.htm#TopOfPage

National Advisory Committee on Microbiological Criteria for Food (NACMCF). HACCP Principles & Application Guidelines 2014. https://www.fda.gov/food/guidanceregulation/haccp/ucm2006801.htm

Hygienic Design:

Campden BRI (UK). This research association runs courses on hygiene in engineering. It has also been accredited by BSI to test hygienic design conformity under the Machinery Directive. https://www.campdenbri.co.uk/

EHEDG: European Hygienic Engineering and Design Group was founded in 1989 as a nonprofit consortium of equipment manufacturers, food producers, suppliers to the food industry, research institutes and universities, public health authorities and governmental organisations. https://www.ehedg.org/

Machinery Expectations:

EU Law on new machinery: DIRECTIVE 2006/42/EC OF THE EUROPEAN PARLIAMENT AND OF THE COUNCIL of 17 May 2006 on machinery, and amending Directive 95/16/EC (recast). http://eur-lex.europa.eu/LexUriServ/LexUriServ.do?uri = OJ:L:2006:157:0024:0086:en:PDF

European Commission Guide to the Machinery Directive 2006/42/EC 2nd Edition. http://ec.europa.eu/DocsRoom/documents/9483/attachments/1/translations/en/renditions/native

UK Health & Safety Executive (The hygienic design of food machinery: Supply of Machinery (Safety) Regulations 2008 and Machinery Directive (2006/42/EC)).

Predictive Microbiology Tools:

PMP (Pathogen Modelling Programme), USDA. https://pmp.errc.ars.usda.gov/PMPOnline.aspx

ComBase. http://www.combase.cc/index.php/en/

The Bad Bug Book – Foodborne Pathogenic Microorganisms and Natural Toxins Handbook (Second Edition, 2012). U.S. Food and Drug Administration. Published online at
http://www.fda.gov/Food/FoodborneIllnessContaminants/CausesOfIllnessBadBugBook/

Example Government Authority: UK Food Standards Agency – www.food.gov.uk

'Food Handlers Fitness to Work' guidance.

'Safer Food Better Business' initiative.

General UK/EU Food law.

Food Safety Guidance.

Personnel control

Mark Swainson
National Centre for Food Manufacturing, University of Lincoln, Holbeach, Lincolnshire, United Kingdom

8

8.1 Staff impacts on quality and technical systems

The personnel of a food manufacturing operation are of critical importance to the control of food safety, quality and legality. Well-trained and motivated staff will be effective in the achievement and monitoring of product quality and safety control points and in the general maintenance of good manufacturing practice.

Food businesses can employ the best quality and technical management staff possible; they can build state-of-the-art manufacturing facilities and design superb products and processes which are comprehensively validated. Yet if the site staff are not selected correctly, are inadequately trained and/or poorly controlled, the business is likely to be unmanageable, with food safety, quality and legality in doubt at all times.

It is worth reflecting that a significant proportion of process operation nonconformances are caused by staff errors. It is therefore clear that staff training and supervision will always play a major part in operational performance and therefore should be a priority focus of the business senior management team who should seek to define the appropriate standards and facilitate their achievement by funding key roles within the business, and via provision of equipment and training.

A common challenge for food operations is that personnel will carry many microorganisms and therefore specific procedures and controls are required to help avoid these microorganisms contaminating the foods being processed. Microorganisms will be present on the external surface of the personnel member (e.g., on ears, eyes, nose, throat, hair and skin including hands), and also will be carried internally (e.g., present within the digestive system and excreted via faeces).

Personnel therefore have the potential to microbiologically contaminate foodstuffs either via direct contamination (e.g., direct contact such as touching/handling the food) or via indirect contamination where the personnel touch and contaminate surfaces/materials (e.g., utensils) which then in turn come into contact with and contaminate the food. Microorganisms carried in the nose, throat and mouth can also be spread to the food via coughing, sneezing, spitting or whistling.

The food and drink industry is very diverse with regard to its approach to factory staff. Some food manufacturing businesses may provide excellent rates of pay and full-time contracts, seeking to gain the operational control benefits of a consistent workforce with a low staff turnover.

Swainson's Handbook of Technical and Quality Management for the Food Manufacturing Sector.
DOI: https://doi.org/10.1016/B978-1-78242-275-4.00008-3

A number of food production sectors require the use of temporary staff to help manage fluctuations in production requirements, perhaps due to producing seasonal products or products where sales volumes are significantly influenced by the weather. Food production businesses can also often offer relatively low-paid employment, especially where roles are viewed as being 'unskilled'.

There is likely to be a large difference between the operational control systems of processors which utilise a high number of relatively unskilled factory staff, compared with those processors which operate far more 'high-tech' operations (e.g., controlling relatively complex and automated processes).

The impacts of a particular business 'staffing/labour approach' on the site technical and quality standards are included in Table 8.1.

The staff of a food business will typically have a huge impact on the consistency of product quality and safety that can be achieved. Well-trained, motivated and skilled staff are not only more efficient and thorough, they also tend to make fewer mistakes.

A point worth reflecting on is that the food and drink sector is a very competitive market, with the frequent potential for cheaper competitors and ongoing price reduction pressures from customers. The sector often has to operate on a continuous 'cost reduction' model, with the ongoing challenge of staff and their departments being asked to work more and more effectively and efficiently. Ultimately in such cases there is often increasing potential for mistakes to be made if not tightly controlled and monitored by the technical and quality standards operated by the business.

8.2 Staff selection

As food handling staff will have a direct and immediate impact on the quality, safety and legality of the products, it is vitally important to ensure that staff selection processes are robust and favour individuals who are clean and tidy in their appearance and have the right attitude toward the importance of food quality and safety.

For example, for 'food handlers' recruitment panels should seek and select:

- clean hands;
- tidy, trimmed (short) fingernails, with no evidence of nail-biting;
- clean skin which appears free from infection;
- appreciation of the need for food quality and safety (and awareness of their potential impacts on such matters); and
- potential new staff should be keen to learn, including attendance of food safety training sessions.

As it is always very difficult at the interview stage to be certain that the selected staff are going to be effective in their roles, it is important that all new staff have to satisfactorily complete a probationary period before securing a full contract of employment. During this probationary period staff should be particularly monitored to ensure that they are capable and competent in their duties particularly related to food quality, safety and legality.

Table 8.1 Impacts of a business 'staffing/labour approach' on the site technical and quality standards

Staff approach	Considerations
High staff numbers in relatively basic unskilled roles	Routine quality/safety training sessions will need to be frequently administered to staff, especially in 'high staff turnover' operations. Quality and safety check information will need to be written very clearly utilising terminology best suited to the relevant employees. In operations utilising a high percentage of migrant labour, the materials relating to technical/quality training and process control standards may benefit from being translated into the common languages of the workforce. Picture standards can often convey a great deal of information more easily than written materials, especially where additional language or literacy challenges may exist. Frequent audits of good manufacturing practice (GMP) and a focus on general staff adherence to operational standards may be particularly beneficial. A high degree of staff supervision and related supervisory structures will be required.
Low staff numbers in highly skilled roles	The depth and detail of technical- and quality-related training materials need to reflect the complexity of the operation. Tolerances and thresholds for quality, safety and legality parameters need to be clear and unambiguous, as often the highly skilled staff will be responsible for quality, safety and legality checks/decision making. Training systems should particularly incorporate scheduled refresher training as highly skilled staff in low numbers can often result in a relatively low staff turnover with team members having significant lengths of service in the business. Audits of key process areas should seek to particularly engage with the skilled staff, checking their specific knowledge of their roles and responsibilities via a 'question and answer' method as well as observing their actions/practice.

8.3 Medical screening (staff, contractors and visitors)

Food manufacturing operations must ensure that each member of staff is sufficiently fit and able to conduct their intended role and does not pose any increased risk to food safety or to the health of their coworkers. Such controls also need to be extended to cover any visitors to the manufacturing site including contractors, agency staff and guests.

Medical screening is typically utilised to help ensure that the health condition of anyone working within or visiting the manufacturing process will be ensured to not pose an unacceptable contamination threat to the foods being processed and/or to the employees of the organisation.

Screening can take a number of forms within a food manufacturing business:

- 'Preemployment medical screening' of proposed new staff.
- Staff 'return from abroad' screening. Checks made to ensure that staff returning from abroad (on holiday or on business) do not pose any increased risk to the manufacturing process.
- Staff 'return from illness' checks. Often in the form of a questionnaire completed with a line supervisor/manager or nominated health representative, with the intention of ensuring that the employee is sufficiently fit to return to work and does not pose any additional threat to the food products or to their fellow workers.
- 'Visitor health questionnaires'. Typically used to screen nonpermanent factory staff such as guests and contractors.

Businesses may also cover temporary/agency staff via their visitor health questionnaires or may choose to have a separate control process in place via the specific temporary staff recruitment agency used (which can save time as the temporary staff can be checked and approved before arriving on site).

Typically medical screening will take the form of a written questionnaire asking questions which the UK Food Standards Agency (FSA) suggests should include:

Name:
Unique identifying personnel number (e.g., payroll number):
Department (or business if visiting site):
At present, or in the last 7 days, are you suffering from

1. diarrhoea and/or vomiting? Yes/No
2. stomach pain, nausea or fever? Yes/No

At present, are you suffering from:

1. skin infections of the hands, arms or face
   (e.g., boils, styes, septic fingers, discharge from eye/ear/gums/mouth)? Yes/No
2. jaundice? Yes/No

Do you suffer from

1. recurring bowel disorder? Yes/No
2. recurring infections of the skin, ear or throat? Yes/No

Have you ever had typhoid or paratyphoid fever or are you now known to be a carrier of *Salmonella* typhi or paratyphi? Yes/No
Are you a carrier of any type of *Salmonella*? Yes/No
In the last 21 days have you had contact with anyone, at home or abroad, who may have been suffering from typhoid or paratyphoid? Yes/No
Countries visited in the last 6 weeks (please list):
Date:
Signature:

Record of any further action taken:
The UK FSA suggests that any 'Yes' answer should require an assessment of the individual's suitability to work/enter the factory environment. Such checks and any necessary follow-up action should take place in advance of the individual being able to access to the facility.

The answers to the questionnaire should be reviewed by a member of staff trained to consider the answers given and competently confirm that the individual does not pose a risk to the operation, or is capable of deciding on the further actions required (including exclusion) if an unacceptable level of risk is detected by the answers given.

Information provided in the UK FSA guidance document, 'food handlers fitness to work' (available via the UK FSA website www.food.gov.uk) can help in the consideration of such matters, as can health professionals and relevant enforcement authorities.

8.4 Illness reporting and return to work

Illness reporting: Staff who become ill could pose a contamination threat to the food products and/or fellow employees. It is therefore vital that staff report any illness immediately to their line supervisor/manager (or in some cases the site medical department or first aider) and be excluded from food handling duties and food handling areas until their contamination threat has passed.

All illnesses will have differing levels of potential for contamination; therefore it is important that the employee (and business) takes the advice of medical professionals with regard to the extent and length of exclusion required, and the setting of appropriate conditions which have to be met before the individual can return to work.

Return to work: When an employee has been ill or has visited an environment/location which may raise their risk of carrying potential pathogens/viruses, a food manufacturing operation is well advised to operate a 'return to work' procedure. This procedure must be designed to ensure that employees are only able to resume their work once it is confirmed that they do not pose an increased threat to the food products or to their coworkers.

Legislative expectations and industry standards vary around the world and therefore each food manufacturing operation should take the time to set their own 'return-to-work policy' and procedure via a process of discussion with their governing bodies, enforcement authorities, benchmarking against similar operations, and consideration of all customer and third-party accreditation standards relevant to the site.

In the United Kingdom, if an employee has been absent from work with vomiting or diarrhoea then there is usually an expectation that the employee should not return to work in food handling duties and related areas until they have been clear of any further instances of vomiting or diarrhoea for 48 hours. It is important to note that this 48 hours is expected to be achieved without the use of medication designed to combat such symptoms, as medication can mask the fact that the employee is still in an infectious condition.

If the individual is diagnosed with a specific infection then the recommended course of action may need to be designed to be more appropriate for the nature of the issue. Such 'higher impact requiring extra control' infections include *Salmonella* typhi and paratyphi A, B or C (enteric fever); verocytotoxin-producing *Escherichia coli* ; norovirus; hepatitis A; infected or injured skin; *Entamoeba histolytica* (amoebic dysentery); *Shigella dysenteriae*, flexneri and boydii; worms – threadworm and *Taenia solium*; *Vibrio cholera* 01 and 0139. As previously stated it is important that the advice of medical professionals is sought with regard to the extent and length of exclusion required, and the setting of appropriate conditions to be met in advance of returning to work.

Employees returning to work should be interviewed by an appropriate team member to confirm that they have met all of the requirements to allow their return; this may include receipt of medical documentation to prove that health/check/test requirements have been completed and approved. The returning employee should also be briefed to be extra thorough with their personal hygiene practices (particularly including handwashing), as they may still pose an increased risk for a while after their symptoms are no longer evident.

As infections are not always the cause of vomiting or diarrhoea, exclusion may not be deemed necessary if it is clear that there is a noninfective cause to the symptoms exhibited. Such noninfective causes could include response or reaction to medical treatments; inflammation of the bowel including diverticulitis, ulcerative colitis and Crohn's disease; irritable bowel syndrome (IBS); cancer of the bowel; malabsorption syndromes (e.g., coeliac disease and cystic fibrosis); pregnancy-related morning sickness; 'dietary indiscretion' (e.g., excessive consumption of highly spiced food or alcohol).

If there is uncertainty as to whether or not the cause of vomiting or diarrhoea is due to an infection then it is always best to exclude the individual until it is very evident that it is safe for them to return to work. If in doubt then always exclude from food handling/the food environment and seek professional advice.

Further useful detail on such matters is contained in the UK FSA guidance document, 'food handlers: fitness to work' (available via the UK FSA website www.food.gov.uk).

8.5 Personal medicine control

Most personnel should be able to keep their medicines outside of the food production areas, locked securely in their locker/personal belongings.

In the event that the personnel member must carry the medication on them during work, it is important that the employee is briefed on the potential product contamination dangers posed by the medication and its packaging. These threats must be managed by only carrying the amount of medication and associated packaging that is absolutely necessary, storing it in a safe and secure place (e.g., an inside zipped pocket) and if further assurance of control is needed then the materials could be 'signed in' and 'signed out' of the area by a supervisor/manager each work session.

Such procedures should be documented and highlighted during staff induction and also refresher training to ensure that a consistent approach is maintained across the site. Documentation (e.g., procedures, training records and control-check records) also ensures that adherence can be demonstrated during site reviews/audits.

8.6 Personnel hygiene standards and related facilities

The nature of food processing operations usually requires personnel to operate in close proximity to the foodstuffs being processed. It is therefore vitally important that staff follow strict rules with regard to hygiene so as not to pose a threat to product safety or quality. These 'rules' often take the form of a site 'personal hygiene policy'.

Food handling businesses vary widely in terms of the levels of risk and product susceptibility to contamination. Businesses are well advised to conduct their own risk assessments to highlight the range of potential impacts people can have on the products. This will in turn inform the process of setting appropriate rules and regulations related to personnel hygiene.

The following list is of 'personnel rules' commonly applied to help assure food quality and safety:

- No jewellery to be worn – exceptions sometimes are made for plain wedding rings and wedding wristbands. Staff wearing such items should be requested to ensure that the item fits well, is kept clean and is not allowed to impact on the ability to regularly wash hands and wrists.
- Watches must not be worn due to their high potential for hygiene and physical contamination risk.
- Piercings such as rings and studs should not be worn if visible (e.g., ears, eyebrows, nose and tongue).
- Fingernails should be kept short and clean. No nail paint, varnish or false nails.
- Perfumes/aftershave should not be worn – these can taint the food and also affect staff sensory assessment ability.

- Staff should arrive at work in a clean and hygienic condition.
- Cuts and grazes. Such skin wounds can pose a contamination risk to the food and therefore should be treated and covered before work can commence in food processing areas.

Typically, blue, detectable (e.g., can be picked up by metal detector/X-ray) plasters are recommended to be used in order that dressings can be easily visually identified or detected if all or part are accidentally dropped into the foods being processed. If the cut or graze is on the hand then covering with an additional blue plastic glove could be recommended to add further assurance on the security of the dressing.

Where detectable dressings/plasters are used, it is important to test each new consignment of materials to confirm that they will actually be picked up by the site foreign body detection systems. A single unit of each dressing/plaster type should be passed through the detection system with the product to confirm that the system is sufficiently sensitive to detect and reject the material on its standard settings. These routine checks should be recorded to demonstrate 'due diligence' which would prove particularly beneficial in the event of an issue/complaint related to product contamination by such materials.

Site facilities should be used to wash thoroughly during the process of changing into protective equipment. For this purpose, it is vitally important that the business provides appropriate facilities including changing rooms and handwashing stations for the washing and drying of hands. It is generally good practice to wash and sanitise hands on entry to all food production areas. Businesses may also set a required frequency of handwashing while staff are working in order to address the potential for build-up of food/organisms on the hands over time.

Soaps used should be bactericidal and should not be perfumed. Sanitisers should be quick drying, and there should be clear signage both to remind staff to wash their hands and also to instruct/guide on the correct method of handwashing.

Other site facilities which support the control of hygiene include sufficient and suitable toilet and shower facilities, the plentiful supply of clean factory clothing/personal protective equipment (PPE) and the provision of refrigerated storage in canteen/rest areas for staff food and drink.

Many businesses choose to make their manufacturing sites 'no smoking' sites. In circumstances where smoking is allowed it is good practice to have a well-ventilated designated smoking area, ideally located outside. It is important that such areas are kept tidy to ensure a good impression of the site and help avoid issues such as cigarette butts being walked into the factory via employee shoes.

Staff adherence to hygiene rules should be routinely monitored by colleagues, supervisors and management. It is also beneficial for these matters to be assessed during internal audits to help ensure that standards are not allowed to reduce over time.

The provision of appropriate facilities and factory clothing materials/PPE is covered in further detail in Chapter 5, Site standards, and also these matters are discussed in Section 8.12.

8.7 Personal protective clothing and equipment

Food businesses should always provide appropriate protective clothing and equipment to employees and visitors who require to visit food processing and storage areas. It is often worthwhile highlighting that the clothing is utilised as much to protect the food from the individual as it is to protect the individual from the food.

The extent and style of personal protective clothing and equipment will vary dependent on the nature of the food processing facility and the particular areas being entered. To define the appropriate materials to be worn in each area the business team should conduct a risk assessment of its facilities and personnel, and should also incorporate the requirements of any customer and third-party accreditation standards worked to, and any specific legislative requirements.

Personal protective materials for wearing in food handling roles/areas can include:

- coats/overalls
- safety shoes
- wellington boots
- aprons
- gloves
- hairnets/mob caps
- beard snoods
- hats

All staff should ideally be fully covered by their protective clothing and equipment.

Coats/overalls should always be ensured not to have external pockets or sewn on buttons to avoid easily dropping potential product contaminants. On sites where different factory product risk zones are operated it is important that clothing and equipment which are only to be used in specific areas must be marked to ensure that these materials are only ever utilised in their authorized areas (e.g., different coloured coats/boots for use in low-risk and high-care/high-risk environments). This will reduce the potential for cross-contamination by microorganisms that could be carried from low-risk into high-care/high-risk areas on clothing/equipment if separate materials were not required.

All materials should be present in sufficient amounts (and sizes) to ensure that there will always be enough clothing/equipment immediately available for use. If not then this situation will risk staff and visitors not following the rules required to protect the food products. Another issue that occurs when materials are not in plentiful supply is hoarding of clothing/equipment by staff which further compounds the shortage situation.

Protective clothing should either be 'single use' (disposable) or laundered via a facility audited and verified to be effective in its cleaning processes. Many businesses use contract cleaning companies and others set up and utilise their own in-house laundry services. Verification of effectiveness can vary depending on the risks associated with the particular food process, and is likely to involve visual inspection and microbiological testing/swabbing.

For added food safety control it is expected that any high-care/high-risk area clothing should be laundered separately from low-risk area materials. In addition, care should be taken to ensure that all detergents and conditioners used in the wash process must be nonfragranced to avoid potential taints being passed onto the foodstuffs and to ensure that aromas in the factory areas and product assessment processes are not overwhelmed by such fragrances.

It is usually not appropriate for staff to take home and launder their own protective clothing due to the potential for variability of cleaning effectiveness and the added potential for contamination.

Businesses should always ensure robust segregation practices between clean and dirty clothing/materials in order to help avoid the potential for cross-contamination.

It is important that food sites make clear to employees where protective clothing should and should not be worn via both training and signage. Prohibited areas are likely to include toilet areas, designated smoking zones and the canteen, all where there may be far too great a potential for contamination to occur.

Gloves for use in food manufacturing environments are often plastic and usually coloured blue in case any part becomes detached and contaminates the product.

Appropriate changing facilities should be provided for male and female staff. Where clothing/equipment is not 'single use', appropriate cleaning methods and change frequencies should be set on a risk-assessed basis.

Factory footwear/boots often have multiple purposes in a food factory environment. They typically need to be nonslip and made of nonabsorbent easy to clean materials. Wellington boots often fulfil the required criteria. Sufficient boot wash stations/automatic cleaners must always be provided in sufficient amounts and are often best placed close to the factory area entrance points. When operating high-care/high-risk environments it is also beneficial to colour code the boots to be used in these areas in order to distinguish and separate from footwear which is to be worn in low-risk areas, thereby helping to avoid the spread of microorganisms.

The order in which staff should change into and out of protective clothing should be defined so as to pose the minimum amount of potential for contamination to occur.

8.8 Capability and training

Often 'product quality faults'/'product safety issues' can be traced back to mistakes made by people. It therefore stands to reason that investing in and developing skilled, motivated staff will lead to improved product quality and consistency. While there is usually a financial outlay required to train employees, the cost to the business of not having competent employees will always be far greater in the long run.

8.8.1 Qualifications and experience

Qualifications and experience gained prior to employment at the current food business can be a very good indicator of whether the prospective employee is likely to understand and competently conduct their tasks related to food safety, quality and legality. Therefore companies are well advised to think carefully about the job advertisement and application screening process to help select for the candidates best positioned to be able to conduct the required role or to be most receptive to the training required to be completed on commencement of employment.

In more senior roles prior qualifications alongside experience may be the only factors that secure their employment with the food business, with little thought to any further formal training being required outside of an induction to the company. It is therefore important that a food business ensures that the qualifications and experience held by prospective new management staff are relevant to the food sector being operated within and that the knowledge gained via progression of their qualifications has been maintained and ideally built upon over time. This could be via demonstration of a commitment to the continuation of their professional development, perhaps via attendance on short courses, progression of further qualifications and via membership of relevant professional bodies.

8.8.2 Induction

When taking on new employees at any level in the organisation, it is vitally important that the new starter is swiftly introduced to the company structures and systems. The employee should receive particular guidance on all of the rules and regulations relevant to the business and their role within it.

In food manufacturing operations 'induction' is a vital stage at which to ensure that each employee understands the importance of food safety, quality and legality, and in doing so also clearly appreciates the actions that they will be expected to take in order to maintain control of such aspects of the operation.

For these reasons many businesses will combine basic food safety training with the general induction programme. Even if a new recruit already holds qualifications in basic or even more advanced food safety practices, induction training on these matters provides an invaluable opportunity for the new employee to reflect on such requirements within the context of the particular business operation.

At the end of this training it is important to confirm that the new employee understands what they have been taught and will therefore be capable of adhering to the site requirements when they start work. Such confirmation can be achieved via the new recruit sitting a test to be taken following the training. The test should have a required pass mark to be achieved before they will be allowed to commence work.

Use of a company 'induction manual'/'induction training pack' which lists all of the information required will help to ensure that each new employee receives consistent and correct information during the induction process. This should be delivered by an experienced employee who has been confirmed to be competent in their training approach.

Receiving a written induction pack will also enable the new employee to further review and refer to the information over time, as the initial days in a new food business can require a lot of information to be absorbed all at once.

It is important that inductions into the business are conducted as soon as possible on commencement of employment. Allowing time to pass between work commencement and formal induction risks the employee progressing work in the manufacturing environment without a full understanding of the key safety and quality risks in the operation, and the tasks involved in maintaining control. If a delay cannot be avoided then the new member of staff should be closely supervised to help avoid any issues arising.

8.8.3 Role-specific training

Many roles within a food manufacturing business will be based in a particular area of the operating environment and often require the use of specific pieces of equipment. Therefore while induction training can provide the new employee with sufficient information to commence work within the business, it is the role-specific training that will then enable the employee to become a productive member of the workforce.

The first step in being able to train an employee is to define the key tasks/duties required of their specific job role. Such listings (likely to be incorporated in job descriptions) can then be linked to area training procedures/training manuals.

Role-specific training can be approached in a number of different ways within a food operation as shown in Table 8.2.

All training methods have their benefits and drawbacks, therefore in practice businesses should utilise a combination of approaches in order to help them best address the training needs of their employees.

Businesses should always consider whether they are using the most appropriate approach for the learner and the subject matter. Some learners enjoy being trained via practical demonstration whereas others prefer a more formal teaching/study approach. Where migrant labour is used these employees may benefit from being taught in their particular language.

At the end of all staff training it is important to ensure that each staff member has understood the information presented. This can be achieved via a number of different approaches such as end of session tests or requesting a confirmation of knowledge via demonstration of a specific task.

Demonstration of understanding and competence is of particular importance when the task or subject knowledge is related to food safety 'critical control points' (CCPs) for the product or process. Such circumstances also accentuate the need for clear records of the training activity and confirmation of knowledge, as these records will be required during site audits and also potentially in order to demonstrate 'due diligence' in the event of a food safety issue.

Table 8.2 Comparison of approaches to training in a food operation

Training approach	Method	Potential advantages	Potential disadvantages
Buddy system	New employee is expected to 'watch and learn' from a more experienced employee in their team/area.	Quick to get started. New employee learns while working.	Training is only as good as the knowledge and communication ability of the selected 'Buddy'. The workplace environment could be distracting. Increased potential for any bad habits to be passed on to the new employee.
Company trainer	Trainer provides information and demonstration, ideally drawing from experience and a specific training manual.	Focused training approach.	Time available can be limited if site training is under resourced.
	Can be conducted either in the workplace or in a dedicated classroom style environment.	Consistency of information given.	Trainer may be an excellent all-rounder but may lack specific knowledge/ 'know –how' that can only be gained by experience.
Classroom-based training	Staff spend time in a dedicated study room to be formally taught by a trainer/tutor.	Easier for staff to focus on the information presented as they will not be distracted by the processing environment.	Learning may lack the 'demonstration opportunity' of the workplace.
Use of external courses	Employees sent to an external training provider to progress their studies, or the external training resource visits the business in order to deliver the training required.	Staff can be more focused on presentations from external trainers due to the newness of the situation or also due to an appreciation that the external trainer has wide-ranging expertise in the subject matter.	External trainers can sometimes not fully appreciate the specific work-based context of the employees being trained. Additional cost to the business engaging external training resources.

All business training records should list key information points including:

- employee name;
- payroll number (or another unique identifier);
- job role;
- key competencies required in role;
- type and content of training undertaken;
- date of training;
- name of the trainer;
- review/confirmation of competency date and assessor; and
- details of ongoing/refresher training required.

The training phase is a time when mistakes are far more likely to be made; therefore it is very important that all trainees are well supervised throughout the training period.

8.8.4 Ongoing monitoring

Good supervision and management should serve to continuously monitor the ongoing competency and actions of staff. In addition, the results/outcomes of the staff work are likely to highlight any particular areas for improvement. Such issues can then be addressed by retraining and reassessment.

Where staff performance is related to the control of CCPs, businesses will often incorporate checks on staff competency during routine area audits. For example, audit record sheets may require the internal auditor to request that a line employee demonstrates their competence on a particular CCP each time the specific area/line is reviewed.

Hand hygiene standards can be monitored using both routine and random microbiological swabbing. Typically, specific pathogens of concern to the site are selected for such assessment initiatives. A good example is *Staphylococcus aureus* which is commonly carried on the hands and therefore if present will be in a position from which to contaminate the foods being processed. A targeted swabbing programme will therefore enable carriers to be identified. These carriers can then receive training on the additional food contamination risks that they may present (dependent on the nature of the foods being manufactured and their role within it), and therefore the additional precautions required. These additional measures may include the requirement to always wear gloves or perhaps even the necessity for exclusion from certain high-risk areas of the business.

Over time employees can forget the training that they have been given or may develop bad habits/'short-cutting' during their routine activities. For such reasons 'refresher training' at set intervals should always be progressed for all employees. Clear training matrices and records will facilitate the planning of refresher training, the frequency of which should be set at a rate appropriate for the complexity and importance of the task.

8.9 Supervision

All staff should be consistently supervised to help plan and progress workloads and ensure that all team members always follow the required good manufacturing practices. Good supervisors will be organised, motivational, lead by example and be very good communicators.

Studies are increasingly recognizing that it is not enough for a manufacturing business to just have the right facilities and good staff training. The management approach/culture of an organisation also has to be right if such processes are going to be well controlled and consistent in their performance. Supervisory and management staff are vital to such agendas and should therefore be selected and developed very carefully.

8.10 Incentivisation

It is an all too common issue in the food manufacturing sector that key tasks with major food quality, safety or legality implications in factories are given to staff who are the least incentivised to 'get it right'. The factory team member being paid 'minimum wage' is often unlikely to go looking for his/her next piece of work once one task is completed, often seeking to take a break and wait to be instructed again, or may work at a slower pace to make the job last.

Such aspects have the potential to be highly inefficient. Worse still a lack of care and attention to detail due to poor pay can have huge impacts on product quality, safety or legality. Just consider the impact of mistakes in the form of factory recipe faults, batch mixing/processing faults, packaging selection faults, coding set up errors, etc.

Food businesses are well advised to ensure that their employees feel that they are paid a fair/appropriate rate for their work, and that each employee is clear that in-return there is an expectation for hard work and a conscientious attitude.

There are numerous options for further incentivisation of the workforce including bonuses to be paid upon reaching a particular work rate or quality. Whether such initiatives work well often relies on the initiative being clearly defined and felt to be both achievable and fair to all employees.

8.11 Temporary/agency staff

Particular challenges can often be associated with the utilisation of 'temporary'/ 'agency staff' in a food processing operation. When correctly managed and well supervised, temporary staff are a very efficient way for food businesses to balance staff requirement peaks and troughs linked to production output. Such an approach is especially useful for businesses working with seasonal products and products where the weather can cause major fluctuations in how many staff are required from one day to the next (e.g., the production of salads and sandwiches, the sales of which can be heavily affected by weather conditions).

Temporary staff can be called in to work at short notice and often do not need to be kept on site (and thereby the business saving the cost of their wages) for any period of time after the work required is complete. However, with the 'drop-in, dropout' nature of such temporary roles, when rushed/not correctly managed temporary staff can often have the associated risks of not being inducted properly (and therefore working while unaware of basic company procedures), and insufficiently trained to complete tasks requiring particular skills.

In these operational circumstances inexperienced or 'too busy' supervisors can lead temporary staff onto a processing line without sufficient operational knowledge and without experienced 'buddy' team members close by, leading to the highly increased likelihood of poor quality work or mistakes being made.

All staff, full-time, part-time and temporary/agency workers are vital to the assurance of product safety, quality and legality in food operations. No matter what circumstances, the basic requirements are to always ensure that every worker is well trained in the role/tasks expected of them and then supervised/managed to ensure that their performance is consistently 'as required'.

8.12 Visitor and contractor control

The presence of both visitors and contractors in the food business have equally as much potential for threat to the end product safety, quality and legality as do the site staff. In fact, it could be argued that they present more risk due to the fact that they are unfamiliar with the site and its day to day workings. For this reason, it is vitally important that the presence of visitors and contractors on site is very well controlled. The considerations required to achieve this are covered in Chapter 5 'Site standards'.

8.13 Food safety culture

A food business can have the most comprehensive systems and procedures covering food quality, safety and legality, yet if the staff do not feel that the business values themselves and their quality of work then it is highly likely that the business will struggle to achieve the control of product quality, safety and legality required.

Achieving a positive 'food safety culture' in an organisation is of vital importance. The term 'food safety culture' draws on areas of behavioural science and organisational theory, sometimes summarised as 'the way things are done around here'. For example, do staff feel compelled to follow the food hygiene rules and regulations because they know it is the right thing to do, or because they feel watched and compelled by their area supervisor/manager, and are therefore likely to ignore the rules when no-one is watching?

To help influence human behaviour, and in this case the behaviour of the food business workers, it is important that there is a clear commitment to food quality, safety and legality from the 'top down' in the organisation. Managers and supervisors should not be surprised to see staff members disregarding rules if they see that the higher echelons in the organisation do not adhere to the business rules or do not make the necessary arrangements for the staff to have all of the necessary resources required to progress their work properly and effectively.

Employees need to be confident that they know the right way of progressing their roles, and this confidence can be developed by the business operating effective systems of training and communication. Staff also need to consistently see that the supervisors and managers have high regard for and follow the same rules, and while doing so always treat everyone with respect.

Indicators of the level of 'food safety culture' in an organisation include: the extent to which staff highlight issues/problems; observation of staff behaviour (e.g., compliance with food hygiene rules); results of product quality assessments; site audit result outcomes; percentages of staff passing food safety compliance checks in the workplace.

The concept of 'food safety culture', its definition, how to measure it, and how to influence it over time, is the subject of many studies and the modern technical manager is well advised to follow the progress of this emerging area and give due consideration to how it relates to and could benefit their own operation.

8.14 Summary

A significant amount of the control challenge linked to personnel rules and regulations is ensuring that all employees are well selected, fully aware of what is expected of them, trained on how to achieve these requirements, and all then followed by ongoing development and monitoring to assure compliance.

For such reasons businesses need to ensure that they have clear systems and procedures, effectively administered by the site supervisors and management team, which seek to achieve compliance with all such matters on an ongoing basis. Comprehensive training programmes, refresher training and routine review of capabilities including audits to confirm ongoing adherence, are all important factors in ensuring that the business stays in control.

Site operations will always be most effective if the business can embed 'good practice' in the culture of the organisation. For this to occur it is vital that the site management both consistently reflect and personally adhere to all of the site rules and requirements, and also provide the appropriate resources/investment on site to enable all staff to be effective in their work to produce foods which consistently meet all expectations of safety, quality and legality.

Further reading

British Retail Consortium (BRC) – www.brcglobalstandards.com.

Codex Alimentarius Commission – http://www.codexalimentarius.org.

European Union Regulation (EC) 852/2004 on the hygiene of foodstuffs. http://eur-lex.europa.eu/LexUriServ/LexUriServ.do?uri = OJ:L:2004:139:0001:0054:en:PDF.

Food Handlers Fitness to Work Guidance – https://www.food.gov.uk/business-industry/guidancenotes/hygguid/foodhandlersguide.

Food Safety Culture – Creating a Behavior-Based Food Safety Management System. Frank Yianaas. Springer (2009).

Food Safety = Behavior. Frank Yianaas. Springer (2015).

Hygiene for Management. R.A. Sprenger. 19th Edition. Highfield Publishing (2017).

International Featured Standards (IFS) – www.ifs-certification.com.

Safer Food Better Business Initiative (UK Food Standards Agency) – https://www.food.gov.uk/business-industry/sfbb.

UK Food Standards Agency – www.food.gov.uk.

Audits

9

Mark Swainson
National Centre for Food Manufacturing, University of Lincoln, Holbeach, Lincolnshire, United Kingdom

9.1 Introduction

The International Organisation for Standardisation ISO 9000 Quality Management standard defines an 'audit' as a 'systematic, independent and documented process for obtaining audit evidence and evaluating it objectively to determine the extent to which audit criteria are fulfilled'.

'Audit criteria' can be thought of as the expected standards against which the business/area is to be judged (e.g., a compiled list of expectations and requirements forming the 'Technical Standard' to be audited against).

Over recent years rising public awareness of food quality and safety issues have prompted significantly increased levels of food legislation in the United Kingdom and EU, and a major review of the approach to food safety controls in the United States of America. In the early 1990s in the United Kingdom the traditional approach to food safety and quality control was one of food-related standards being administered and enforced by regulatory bodies based regionally such as the Environmental Health and Trading Standards departments, a system which continues to this day under the current direction of the UK Food Standards Agency (FSA).

Alongside these increases in control and supporting legislation during the 1990s the following audit-related initiatives developed in the United Kingdom and wider EU:

- As a result of increasing customer expectation and legislative standards, coupled with the necessity to demonstrate 'due diligence' (as a recognised legal defence in the event of a food safety issue), manufacturers started to hold a greater responsibility for the safety and quality of their products. This required significant development of the manufacturer's own 'internal audit' processes. Such audit activities can be classified as 'process audits' and 'systems audits'. Process audits seek to focus on whether particular check points [e.g. critical control points (CCPs)] in the process are working correctly with the test results falling consistently within the stated limits/tolerances. Systems audits concentrate on confirming that the required quality and safety 'systems' are in place and being clearly adhered to, which can involve checking that the relevant procedures are up to date, in use and generating the required check records.
- All audit activities should ideally combine to help the business demonstrate 'due diligence', where the manufacturer is able to show that they are taking 'all reasonable

Swainson's Handbook of Technical and Quality Management for the Food Manufacturing Sector.
DOI: https://doi.org/10.1016/B978-1-78242-275-4.00009-5

precautions' to help avoid a particular problem issue (e.g., a fault in food safety, quality or legality) from occurring.

- Businesses 'second-party auditing' their suppliers became increasingly common in order to both increase supply chain control and demonstrate 'due diligence'.
- 'Third-party' compliance schemes and associated audit systems were developed to provide additional independent assurance of aspects such as food safety, quality control and legislative compliance. There are now a wide range of compliance/certification schemes operating in the food sector.

The need for confirmation and control of operating standards, product safety, quality and legality has led to an ever-increasing amount of first-party, second-party and third-party audits across the food sector. As a consequence, staff skills and resources have had to be developed in the food supply chain to be able to support these increasing expectations/requirements. There is also an emerging area of auditing/assessment with regard to the measurement of 'food safety culture' within an organisation. This is also likely to become increasingly prevalent over the coming years.

9.2 'First-party' internal audit practices and systems

Internal audits are a key part of a food manufacturer's food safety and quality management system (FSQMS). These audits are conducted by members of the business on a scheduled basis to confirm that all aspects of the FSQMS are operating correctly.

The auditor conducting the particular internal audit should be independent of both the area and the procedures being audited against. This independence is to help ensure complete impartiality. For this reason businesses will ideally have a number of staff drawn from across the business and trained to operate as internal auditors when required. In some cases businesses instead opt to hire a contractor/consultant to carry out such audits. The key point here is that the hired auditor is working on behalf of the business and is not conducting an audit on behalf of a customer or compliance scheme, this approach therefore remains in the category of 'internal audit'.

Such audits should review the current practice in selected aspects of the operation against the expected procedures stated in the business FSQMS. By doing this on a scheduled basis the business can assure itself that it is operating correctly, demonstrating due diligence and also can proactively pick up any problems internally which can then be swiftly rectified in the event of any issue being found.

In manufacturing facilities these audits can be classified as 'systems audits' checking that the system/procedure in place is 'as required' and being adhered to, including use of the latest versions of written procedures and related check sheets.

'Process audit' approaches include area, process line and staff specific checks across a wide range of aspects from ensuring that CCPs are being monitored, are running within established limits and are being recorded correctly, through to checking equipment condition, area hygiene standards and checks on staff standards and knowledge/behaviour. For example, auditors might wish to use questions or request demonstrations to confirm that the staff members understand key aspects of their role such as how to set up and run metal detection checks, or perhaps check awareness and understanding of any area-specific allergen control requirements.

A business may also choose to conduct 'product audits' where the product itself is subject to formal audit review, such as taking a range of reference samples of a product from recent production runs and assessing the samples against the product specification to check whether all manufacturer and customer expectations are being met. For these checks it is often very useful to also purchase product samples from the end point in the supply chain, for example, the food service outlet or retail store, as this will confirm whether the product is maintaining its required properties right up to the key point of the end consumer.

9.2.1 Internal audits: A typical approach

9.2.1.1 Purpose of internal audits

The core purpose of internal audits is to ensure that product safety/quality/legality systems and procedures are confirmed to be in-place and operating effectively. Such an approach, if effective, should highlight any problem issues internally before the business encounters any related product safety, quality or legality issues, and certainly therefore in advance of such problem issues being identified by customer, third-party compliance or enforcement official audits.

9.2.1.2 What aspects of the operation are to be reviewed? Defining the scope

The scope of internal audits should be to focus on the business systems and procedures related to product safety, quality and legality. The business should take a similar approach as customer and third party compliance audits on the review of systems which are key to product safety, quality and legality, methodically reviewing first that there is a system in place and then confirming the business compliance to that system.

Typically, if a food manufacturing business has to comply with a particular customer/compliance scheme standard, then a usual approach is to map the scope of the internal audit system against the requirements of the standards to be complied with. For example, with regard to the BRC Global Standard for Food, a food manufacturer should be progressing routine internal audits to formally review compliance to the product safety/quality/legality aspects of the standard including:

- Senior management review/commitment;
- HACCP;
- quality management (e.g., quality policy statement);
- food safety and quality management system (FSQMS) manual;
- organisational structure;
- customer focus;
- document control;

- procedures;
- record keeping;
- specifications;
- internal audits;
- corrective actions;
- traceability;
- organoleptic assessment;
- management of incidents;
- complaints;
- purchasing, supplier approval and monitoring;
- resource management;
- location;
- perimeter and grounds;
- factory layout and product flow;
- fabrication;
- equipment;
- maintenance;
- staff facilities;
- physical and chemical contamination risk;
- hygiene and housekeeping;
- waste disposal;
- pest control;
- transport;
- product design and development;
- product packaging;
- product analysis and inspection;
- stock control;
- foreign body detection;
- product release;
- control of nonconforming product;
- control of operations;
- quantity control;
- equipment and process validation;
- calibration;
- specific handling requirements;
- personal hygiene;
- medical screening;
- protective clothing; and
- training.

Before the development of formal operating standards/compliance schemes within the United Kingdom, audits of 'area hygiene' and 'staff practices' tended to be the main internal audits conducted by a business. The ever-increasing focus on customer/third-party compliance scheme standards has driven a far wider awareness of the importance of routinely and systematically reviewing all of the business systems and procedures related to the control of product safety, quality and legality.

Internal audits seek to confirm and assure the food business that its systems and procedures related to product safety/quality/legality remain in place (i.e., the system/procedure has not lapsed) and continue to be operationally effective.

Auditor assessment of area/staff compliance with the designated policies, systems and procedures can be achieved via:

- interview of factory managers, supervisors and staff;
- observation of relevant operator practices and process operations;
- inspection/review of process and product records;
- completion of traceability studies while auditing can be used to ensure that the business operation/operating systems are clearly linked and cohesive (a point that will serve the business well during audit traceability studies and also in the event of a product problem/crisis/recall).

9.2.2 Internal audit frequency/schedule

A typical and often recommended business approach is to ensure that all points to be audited are reviewed at a frequency which reflects the likelihood of change/issues occurring and which also takes account of the severity of the outcome if a point does lapse. Such an assessment should also take account of the previous internal audit findings and also any particular issues/complaints/quarantined stocks linked to that particular system/procedure point.

All points should be audited at least annually, with many points being reviewed more frequently. The schedule of internal audits will ideally be displayed/mapped on an audit matrix/chart (the chart highlighting the audits to be conducted alongside the months from January to December, with a grid table enabling the user to check and sign off when each audit has been completed and the corrective actions confirmed, to help keep on track). To ensure clear focus and appropriate audit time allowance, internal audits should be spaced evenly across the year. Figure 9.1 provides an example of a 'live' internal audit schedule.

Internal auditor training: While typically most internal audits will be carried out by members of the quality/technical department, the food manufacturer is encouraged to develop a team of internal auditors drawn from across the many different departments of the business. In this way it can be ensured that:

- The audit workload is easier to manage as it is spread across a greater number of staff.
- The audits routinely benefit from fresh sets of eyes/different perspectives.
- Audits can always be conducted on an impartial basis as it can be ensured that a member of the internal audit team never has to audit their own area of the business.

The internal audits must only be conducted by site staff who have been formally trained and reviewed to confirm their competence. Each internal auditor should be trained via either attendance on an external auditor training course or via formal training carried out within the food business by an experienced and ideally formally qualified auditor (often the site quality or technical manager will conduct all supplier audits for the business and will hold a formal auditor qualification such as the ISO 'lead auditor' qualification).

Figure 9.1 Example of an internal audit schedule.

Company Internal audit schedule:

/	Audit completed: corrective actions complete and signed-off as confirmed by internal auditor.
–	Audit part completed: Follow-up of corrective actions underway and not yet reviewed by the internal auditor.
X	Audit pending: planned audit scheduled.

Section of food safety and quality management system	**Jan**	**Feb**	**Mar**	**Apr**	**May**	**June**	**July**	**Aug**	**Sep**	**Oct**	**Nov**	**Dec**
Senior management review	/			/			–			X		
HACCP		/						X				
Quality management			/								X	
Food Safety and Quality Management System manual				/								X
Organisational structure	/											
Customer focus					/							
Document control						/						
Procedures		/										
Record keeping							/					
Specifications			–						X			

This is a shortened example. A typical business schedule will state all key sections of the Food Safety and Quality Management System to help ensure a comprehensive range of reviews at the required frequency.

Auditor impartiality/independence: It is a vital factor within the internal audit system that each internal auditor must not be required to audit their own department or personal work. This helps ensure impartiality and avoids conflicts of interest which may otherwise lead to a 'problem issue' going undetected/not reported due to the personal interests of the auditor.

Small businesses tend to find that almost all of their internal audits are conducted by the site quality/technical manager. In such situations it is important to utilise at least one more internal auditor drawn from a different department. This team member can be used to independently audit the systems/procedures/actions of the site quality/technical department.

Audit reporting, records and follow-up: Each audit on an aspect of the business policies and procedures related to product safety/quality/legality needs to formally document 'evidence of compliance' to the required standard and record any evidence/findings of noncompliance to the requirements.

The audit records should document a range of points including:

- Date, time and auditor name;
- The area and topic/system/policy audited;
- How the audit was conducted together with findings [e.g., names of staff questioned/reviewed, work instructions/systems/procedures audited (including any document control reference numbers/codes), details of check/log records examined (including names, dates, times and document reference numbers)].

Internal audit results must be swiftly raised with the managers and supervisors responsible for the relevant business/process areas.

A nonconformance could be considered as any situation where there is a difference between documented requirements and what is actually observed in practice. Wherever possible, issues/nonconformances to the required standards should be graded in order to provide a further insight as to the level/seriousness of the issue. For example:

'Critical nonconformance' – the issue poses a serious and immediate risk to food safety/quality/legality.

'Major nonconformance' – the issue highlights a system failure or significant potential for a food safety risk/legality issue.

'Minor nonconformance' – isolated incident with no serious impact on food safety, quality or legality.

Auditors should also be given the flexibility to make 'recommendations'. Perhaps in the circumstances where an issue is close to being a nonconformance but deemed not quite deficient enough and can be easily corrected if a recommendation is followed.

The 'corrective action' required to address each specific nonconformance needs to be agreed between the auditor and the area representative, either at the time of the audit, or at a follow-up meeting. The timescales for any corrective actions also need to be clearly agreed/committed to.

Follow-up: After the agreed corrective action timescale, a follow-up audit/review should be conducted to confirm that all of the agreed corrective actions have been correctly progressed and appear to have had the desired 'compliance' end result.

In order to ensure that corrective actions taken remain in place it is advisable to place these points on an increased frequency of review. Such an approach ensures that the corrective actions are embedded within the practices/procedures, rather than swiftly lapsing as soon as the 'audit focus' has been taken away.

Report circulation: As part of business compliance to operating standards/-systems, internal audits and their results are expected to be reported back to routine management team meetings and also summaries are expected to be reviewed at senior management review meetings.

Such an approach helps ensure that a clear 'compliance' message is sent to the business teams and that management are provided with the opportunity to reflect on the current standards being operated to within the business and whether any further resources are required within the operation (e.g., adequate staffing in areas, staff training, etc.).

9.3 'Second-party audits' – supplier and customer audits

A second-party audit typically involves a supplier being audited by their customer. In the food manufacturing sector such audits will typically seek to review a wide range of supplier activities which can have an impact on product quality, safety and legality.

Supplier audits are typically carried out by an appropriately trained member of the technical function as part of a supplier assurance programme.

'Suppliers' requiring auditing can include

- ingredient suppliers;
- packaging suppliers;
- labour agencies;
- machinery/equipment suppliers;
- laboratory service providers; and
- storage, transport/logistics services.

Businesses will tend to develop their own audit approach, checklists and grading systems as different aspects will vary in importance from one business to another. Similarly, the extent and nature of auditor resources available will also vary.

The frequency at which a business will conduct supplier audits will tend to be dictated by:

- 'risk assessment' of the supplier base including the likelihood of supplier issues and the potential impact/severity of those issues;
- end customer expectations;
- degree of financial or legal liability in the event of an issue with the supplier; and
- the auditor resources available.

A common approach in the manufacturing sector is for the higher risk suppliers to receive physical 'at-site' audits and suppliers deemed lower risk are often asked to complete a Self-Audit Questionnaire (SAQ) which serves to check that the basic expectations are in place (e.g., confirming that the supplier has a HACCP plan, a 'systems approach' to food safety, product release checks and traceability systems, appropriately qualified technical/food safety leadership, no track history of product issues and an appropriate level of product liability insurance).

Aside from the 'due diligence' benefits of visiting suppliers to conduct audits, such events provide a valuable opportunity for the customer to highlight to the supplier the specific quality, safety and legislative standards expected of the products being supplied. This is an opportunity to accentuate the business policies/expectations/standards to the supplier (e.g., ethical trade considerations; specific allergen control policies/requirements).

The audit can also provide a forum within which to fine tune the supply, discuss any current issues/complaint trends and check conformance/rectification actions on any problems from the past.

These points combine to help develop an increased 'right first time' approach within the supply chain, helping to ensure fewer problems and therefore greater operational stability at the final manufacturing site.

Various second-party audit 'types/occasions' include the following:

- Supplier approval audit: The customer technologist/auditor will tend to check all aspects of a business to confirm that the business is 'fit to supply'. For example, approach and control of quality, safety and legality, plus supplier capability and capacity (throughput rates, manpower, etc.). While on-site the auditor may also take the opportunity to review other audit reports and evidence of good business control (e.g., recent third party audit reports and documented enforcement authority inspections).
- Routine inspection: Once a supplier has passed the first 'approval audit' the auditor may wish to return on a routine basis (e.g., annually), on which occasions the audit policy may be to always apply the same audit approach/checklist of inspection. Alternatively, each audit occasion may be taken as an opportunity to focus on specific aspects of the business. This can be helpful in drilling down to greater levels of detail than would be possible in a time limited 'full site and all processes' audit.
- Attended production run/product launch: Such audits/site visits are focused on ensuring that the specific product/service is being progressed correctly and meets the specification required. While on-site the auditor will always be aware of and likely to assess site-operating standards more generally.
- Follow-up/Trouble shooting: In the event of a significant product or process problem (perhaps highlighted by a customer-complaint trend), the auditor may wish to visit the supplier site to review related standards, practices and procedures in order to assure themselves that the matter has been resolved.
- Unannounced Inspection: For some businesses the approach of arriving at the supplier site to conduct an audit/inspection without any notice is standard practice. Often to help facilitate the unannounced audit approach the auditor will ask the supplier to keep them updated on dates that are 'unavailable due to other commitments' over the coming months and will then avoid conducting the 'unannounced audit' on those dates (typically there will be rules for the supplier as to how many days they are allowed to block in the

auditor's diary). Some auditors and businesses will only use the unannounced audit when the supplier has caused particular concern to the extent that the auditor wishes to arrive unannounced to prevent any preparation in advance and thereby gain a clear view as to the true operating standards and quality of record keeping being practiced.

9.4 'Third-party' audits including certification schemes

A third-party audit is an audit conducted by a business/organisation which is independent from the customer/supplier relationship. Such an approach enables objectivity and impartiality in the audit process. Often this audit approach is utilised when a business is being assessed against a set standard, for example, ISO, BRC or IFS certification standards are all based on audits conducted by approved third-party audit and certification businesses.

9.4.1 Defining and monitoring 'technical compliance'

The term 'technical compliance' for a food manufacturing operation could be considered as meeting the requirements of all stakeholders engaged with matters affecting product quality, safety and legality.

Food industry businesses have many 'technical standards' with which they strive to comply. In the case of 'technical aspects' these 'standards' will typically relate to product safety, quality or legality.

Definition of the 'requirements' to be complied with can take many forms including: legislation, regulations and guidelines; customer's 'supplier codes of practice'; contracts of supply; product specifications; group or individual business operating standards including the site Food Safety and Quality Management System, certification scheme standards, etc.

The setting of and reporting on such standards can have both internal and external drivers:

- **Internal standards:** Can be set by the business or parent business group and will be a series of technical aspirations/targets set for staff, departments or the business as a whole to achieve.
- Businesses may set departmental targets for 'internal audit' scores (e.g., assessment of area hygiene standards or perhaps the extent of compliance with specific site systems when reviewed); end-product taste panel score targets; percentage compliance with microbiological specification limits' with regard to 'end-product' test results or area 'environmental' microbiological assessment.
- Typically, if a food manufacturing business owns the brand that the end product is sold under then the manufacturer will be setting the key 'technical standard' targets such as product microbiological test standards and target customer-complaint levels.

- **External standards:** These could, for example, be set by the customer and this is particularly the case when the food manufacturer is manufacturing the product under the supermarket's 'own brand', for example, 'Walmart', 'Costco', 'Tesco', Carrefour', 'Amazon' and 'Kroger'.

- Typical technical standards set by the retailers for their own brand manufacturers to achieve include: end product and environmental microbiological test criteria and targets; target number of complaints (often expressed 'per 100,000' or 'per million units sold'); percentage compliance targets for product organoleptic performance (often checked at end-product quality review panels).
- 'Legislative requirements' can also be considered as 'externally set technical standards'. For example, Regulation EC No. 853/2004 legislation is very clear on the operational and technical standards expected to be met by all food manufacturers currently operating in the European Union. As a food producer it is always advisable to have contact with the relevant regulatory body in the particular region/country of both operation and end-sale in order to clearly define the range of legislative expectations.

Business 'internal technical compliance review': In order to enable well-informed decisions and the formation of noncompliance-related action plans, it is vital that a food business maintains absolute clarity on how well it is meeting its technical requirements linked to product safety, quality and legality.

Such reviews can include the following:

- Daily site 'operations meeting' reports on technical/quality issues including:
  - department audit/performance scores and nonconformance actions required;
  - end-product quality test results (e.g., product analysis scores; taste panel results);
  - latest key customer-complaints received/investigation updates; and
  - environmental monitoring results (e.g., microbiological area standards, trade effluent consent limit performance).
- Monthly site general management meeting review of issues including:
  - Year-to-date complaints performance tracked as 'per 100,000 units sold' (or perhaps 'per million units sold', the important point here is consistency of the measurement), and compared against site targets and previous year's performance.
  - Analysis of customer complaints including focus on the major categories of concern (e.g., foreign body trends, quality complaints, microbiological problems, allergen issues).
  - Quarantined stock analysis. Such analysis provides a good insight into the problem issues being encountered within the business.
  - Routine microbiological sampling performance (both end-product '% compliance to standards' and environmental/factory area swabbing/'air plate' results).
  - Nonconformance system 'key issues' update. Some businesses will raise as a 'nonconformance report' any issue on site that adversely impacts on product safety, quality or legality, then requiring the relevant area manager to investigate the issue and report back on the corrective action taken and the measures implemented to prevent any recurrence.
  - Results of internal audits: For example, site hygiene/good manufacturing practice audits and food safety and quality management system audits.
  - Second party (customer) audit performance. Review the outcomes of any site audits that have taken place since the last management meeting. Key points raised for action and updates on progress addressing/closing any previous audit action points.
  - Third party audit performance (certification and enforcement/government agencies). Review of any such audits (e.g., ISO Certification, BRC Global Standard, Environmental Health/Trading Standards Inspections, etc.) since the last management

meeting. Consider the key points raised for action and progress updates on closing any previous audit action points.
- Supplier/raw materials control: Review of any raw material supplier issues which have or could impact on product safety, quality or legality. Trended results could include % problem issues highlighted on incoming goods checks, % site product issues attributed to 'supplier nonconformance to required standards', and supplier review/supplier audit performance.
- Technical 'customer services' update: For example, highlight new customer 'technical standards'/'code of practice' requirements, number of product specifications supplied against number requested, new product/process development challenges (e.g., new product ranges, scale-up work, shelf-life work, new process validations, etc.).

'Management review' typically involves the routine meeting of selected management team members during which key aspects of the business Food Safety and Quality Management System are reviewed to confirm that it is remaining effective. Such regular site review of factors related to technical compliance is intended to ensure full site awareness of, and focus on, the key food safety, quality and legality challenges facing the operation. This in turn is then expected to ensure a strong and ongoing site management team approach to the assurance that such challenges will be met.

'Senior' management review: Following the implementation of third party certification schemes (e.g., BRC) in the 1990s it was soon evident that a regular challenge to food safety and the continuous improvement of manufacturing operations 'technical performance' was a lack of support from the site senior management. The evident lack of support sometimes involved leaving the quality/technical team members alone to try and control food quality, safety and legality without the adequate tools and other necessary resources.

This lack of senior management commitment was often evidenced by insufficient investment in facilities and staff which would lead to a range of noncompliances against the standard required. To help address this problem the BRC Global Standard for Food developed an additional stipulated requirement for the site to demonstrate 'senior management commitment and continual improvement'.

The additional expectations (reviewed during the audit) included the necessity for the site senior management to be clearly involved in the implementation and maintenance of all practices related to the standard. In particular, the expectation placed on the senior management team to ensure that the site product safety/quality/legality systems are adequately resourced, and that 'continual improvement' is encouraged and facilitated within the organisation.

Site evidence of 'senior management commitment' would include clear and documented involvement in aspects of the operation such as site HACCP plan development/reviews, formal management reviews of business quality/technical performance, internal auditing and engagement with product review panels.

In the early years of the BRC standard, auditors and the audit results quite quickly highlighted that a manufacturing operation can employ excellent technical/quality staff, but without the backing of the business in terms of adequate resources and site operational focus on quality, safety and legality, such staff are placed in the difficult position of being ineffective and often merely 'company apologists' for the routine number of problem issues emanating from the products supplied. This is one of the reasons why the standard was developed to place greater emphasis on 'senior management commitment', as it is only when food safety, quality and legislative compliance are driven 'from the top' that a business can ensure such measures will be adequately resourced and supported at all levels within the operation.

The site 'senior management review' is required by the BRC Global Standard for Food to be conducted on a regular but not too frequent basis. Too frequent a meeting risks a lack of objectivity, and a business should expect its general management team to be closely monitoring and acting upon matters on a daily, weekly and monthly basis. The frequency of the senior management review can often be affected by

- How well the company is performing against the required technical standards.
- Senior management availability and management style. Some senior managers/directors like to be particularly 'hands on' and therefore expect an increased frequency of senior management review. In such cases a site should be cautious to avoid running the reviews any more often than quarterly in order to maintain objectivity and demonstrate to the general management team that they are trusted and expected to run technical compliance on a day to day, week to week and month to month basis.

Senior management review content requirements: The following business 'technical compliance' aspects are typically required to be formally considered/ reviewed during the senior management meeting. It is therefore prudent to ensure that the meeting agenda clearly considers each aspect in turn, often via review of 'status reports' provided by the site technical department and presented by the site quality manager/technical manager/technical director.

- Confirmation and review of the last set of senior management review meeting minutes and update on all previous action points.
- Audit results and reports. Review of the outcomes and key point findings across internal audit, second-party and third-party audits.
- Consider all/any site incidents/issues and the status of any outstanding preventative and corrective actions.
- Complaints, customer feedback and other such key performance indicators.
- Product safety system performance (e.g., HACCP plan verification reviews, CCP checks, audits, validation studies, etc.)
- Operations/production performance (e.g., review of area quality/technical issues encountered).
- Site operational and technical resources which impact on product safety, quality and legality.
- 'New information' – 'new research/science' which may impact/influence the site approach to assuring technical compliance.

This 'senior management review' should result in a series of minutes to be circulated to all relevant staff, providing clear communication and a record of the information considered and actions decided on. The minutes should also be accompanied by an action log which summarises all required actions coupled with 'who' is to do 'what', and 'by when', to address any issues affecting business technical compliance.

The benefits of such management reviews in helping to secure the adequate provision of technical compliance resources can be traced back to the requirements and recommendations of the ISO 9000 series standards for quality management which have served to encourage 'good practice', with similar approaches replicated within other food industry compliance standards such as the BRC Global Standard for Food.

9.4.2 Audit 'service provision'

The technical department of the food manufacturer is often the department best placed to organise customer technologists/auditors who wish to visit the manufacturing site in order to confirm that the process/facilities/staff are compliant with the customer's expectations. As discussed earlier such expectations/requirements are typically documented within a 'supplier manual', 'codes of practice' or other 'technical standards'. This holds true for both direct technologist visits (second-party audits) and also for third-party audits.

In both second-party and third-party audit cases the role of the technical department does not just end with organisation of the audit, they are typically expected to be the hosts while the auditor is on site, the liaison/coordinator with the rest of the business site, and also, following the audit, the collator and organiser of any corrective actions and the writer of the formal response to confirm the rectification actions taken.

It should also be noted that with the ever-increasing trend for the costs of audits to be picked up by the 'auditee' rather than the 'auditor' it is important that technical managers ensure that they have sufficient departmental budget allocation to pay for these audits and agree in advance which business party is responsible to pick up specific costs such as the audit fee, associated travel, hotel and subsistence costs.

'Product panels attendance': Customers may wish to regularly 'audit' ongoing product performance by selecting product currently within the supply chain/in-store and reviewing it against the agreed product specification.

Many food manufacturers, particularly when producing supermarket brand products, will be expected to routinely attend such product panels, often based at the customer site, to assess the end-product quality performance alongside customer technical/sales team members.

The format of such panels can vary quite widely. For example: sometimes all of the samples to be assessed are taken from the stores/supply chain at random by the customer, sometimes the supplier will be asked to bring the samples to the panel and sometimes competitor products are also assessed at the same panel.

Such panels are usually expected to be attended by a member of the business quality/technical team, and sometimes a member of the sales team and a member of the operations team. The panels can be very time consuming when also taking account of the time taken traveling to and from the panel if held at the customer site. However, the panels are a very important part of customer assurance that the business is achieving consistently correct products and that the supplier is 100% committed to the supplier/customer relationship (which can sometimes also be judged by the calibre of team members sent to attend the panel).

It can be very prudent for the supplier to send their best/most knowledgeable technical team member to the panels as sometimes there may be doubt as to whether a product is 'correct' or not. That is, does the product meet the specification required? There may be a point during a panel where if the matter is allowed to gather uncertainty there is the potential for the issue to result in product quality concern perhaps to the extent of withdrawal of the offending batches from the supply chain. Whereas if such uncertain situations are well managed and based on knowledgeable assurance that the product being viewed is correct and is within the quality tolerances allowed, such an approach can often calm and correct the customer position, potentially saving the supplier a great deal of time, trouble and cost.

9.4.3 Enforcement authority audits and interaction

Food manufacturers can be subject to many customer visits and audits over the course of the year and in addition there are likely to be inspections conducted by the local government 'enforcement' authorities. Typically, in the United Kingdom these local government bodies are the Environmental Health department and the Trading Standards department, both of which would be based within the government offices of the region in which the food business is located.

While major customer audits/visits are always likely to receive the most attention from food manufacturing suppliers who are keen to impress, it should be remembered that it is often the enforcement agencies who have the greatest capacity to impact on the food manufacturing business. This is because being government enforcement authorities they will often hold the power and responsibility to stop food manufacturing activities with immediate effect should they find that the business is operating illegally and/or posing a significant risk to the end consumer.

On some occasions auditors working on behalf of the customer may be displeased and even concerned with regard to the adequacy of the site standards in operation. Such circumstances can sometimes be influenced by commercial pressures to maintain supply and therefore often result in a very focused action plan to quickly resolve the issue rather than stop the operation completely. Whereas if an Environmental Health Officer (EHO) is sufficiently concerned as to the operating standards of food safety then a manufacturing site can be immediately closed, stock seized and products recalled. This is due to the fact that the government officer has a far more focused set of audit objectives, those being to assure public safety and ensure legal compliance.

Similarly, if a Trading Standards Officer (TSO) finds major breeches in legislative compliance with regard to, for example, product weight control or adherence to pack labeling regulations, then again, the business can be instructed to stop the manufacture and sale of these products with immediate effect.

In the same way that review of site pest control records can give the auditor an insight into the food manufacturer's control of such matters, details of liaison with the relevant authorities can provide a guide to the current site status with regard to legal compliance across a range of different key areas. Resulting from this fact, as part of their due diligence approach while assessing potential new suppliers or monitoring the performance of current suppliers, the customers of food manufacturers (e.g., retailers, food service and 'business to business' operations) can sometimes request copies of (or insight into) recent correspondence with the local enforcement authorities (usually in the form of the latest audit/visit report). Thereby providing another example of the importance of maintaining a good positive relationship. (Note: The food manufacturer would be the organisation from which a copy of the report would be requested, customers would not go direct to the enforcement authority.)

In addition to their importance from an enforcement viewpoint, authorities such as the Trading Standards and Environmental Health departments can provide a very helpful advisory service and point of contact on legislative compliance. This is especially helpful when the complexity of some laws is considered. While legislation should be unambiguous in terms of what is actually required, all too often some points come down to 'interpretation' and therefore it is usually a good idea for food manufacturers to check any uncertainties with the relevant enforcement body.

9.4.4 *Typical 'technical department' enforcement authority contact*

As the technical manager/technical department are usually required by the food business to ensure that the operation is kept aware of all legislation related to food quality, safety and legality and also expected to interpret these requirements into practical measures/systems/procedures applied to the operational practices of the site. It makes sense that the main department interacting/corresponding with the enforcement authorities relevant to product safety, quality and legality should be the technical department.

Food businesses will be routinely visited by representatives of the relevant food industry enforcement bodies in order to check and confirm that the operation is aware of and meeting all relevant legislative requirements.

Such checking will typically take the form of site audits/inspections, which can be either on an announced or an unannounced basis (often this will depend on the relationship held between the manufacturer and the relevant enforcement authority, and can also be influenced by the recent performance of the business).

To be able to competently manage such queries/discussions and provide all stakeholders with assurance that they are in safe hands with regard to their management of the issues, it is of vital importance that the technical manager takes the time to ensure their full awareness/understanding of both general food law, and any food laws which specifically relate to the particular category of product being manufactured.

With regard to food safety in the United Kingdom, while there are a range of government departments at national level with responsibility for 'food', a main government section that food businesses will deal with on a routine basis is typically their local Environmental Health department. EHOs from the department will visit the site to conduct operating standard inspections/audits at a frequency typically set on a risk assessed basis. The EHO activities are influenced and overseen by the UK FSA.

The local Environmental Health department will tend to visit the food manufacturing sites within their district on a risk assessed basis which typically results in visiting high risk sites approximately 2–4 times per year, and low risk sites around once per year.

Such local government departments are set site audit frequency targets by the FSA, and are themselves audited for compliance to the frequencies and depth of audit/follow-up conducted.

The EHO will typically visit the manufacturing site seeking to review the factory operation, staff and records which will demonstrate compliance to:

- The Food Safety Act 1990 (as amended).
- The Food Hygiene (England) Regulations 2006.
- Regulation EC No. 852/2004/Regulation EC No. 853/2004.

Aspects of the operation assessed to confirm compliance could include the following:

- Any documentation related to the site Food Safety Management System. On large manufacturing sites this could include review of HACCP documentation. With regard to smaller sites this could include review of site documentation related to more basic site checks and procedures such as those recommended in the safer food better business (SFBB) guidelines in the United Kingdom, produced by the FSA.
- Temperature control (including check records and equipment calibration);
- Customer complaints;
- Pest control;
- Training and review of staff in roles which can impact on product quality, safety or legality;
- Site hygiene (cleaning records, procedures and line sign-off/inspection reports);
- Waste control and disposal (e.g., proof of using appropriately authorised waste contractors/disposal sites, and hygienic storage/handling while still on site);

...and any other aspects of the site/operation which the government official deems to be appropriate/required to prove compliance with the necessary legislative obligations being enforced.

The local Trading Standards department is another enforcement agency that UK food businesses will typically liaise with on food quality/legality matters. Again, the Trading Standards department will seek to visit the manufacturer on a routine basis, sometimes to check the accuracy of any measuring system linked to end-product sales (e.g., weighing scales, check-weighers, pack counting systems) and sometimes to sample products and check the accuracy of product labelling, including any product claims being made such as 'low fat', 'reduced salt', 'healthy' and 'fresh'.

The local Trading Standards department will seek to routinely visit the food manufacturing sites within their district in order to assure themselves that legislative requirements are being met, typically on two general fronts:

1. Compliance with weights and measures legislation (e.g., checking the records from, and accuracy of, site scales/measurement systems from which the product is measured for sale under 'minimum weight' or 'average weight' legislation).
2. Compliance with general food standards legislation.

Having mentioned general food standards legislation, here we can see that there is the potential for overlap with the more 'food safety focused' expectations of the Environmental Health department. 'Food standards inspections' conducted by Trading Standards Officers will typically touch on the following areas/aspects of the business:

9.4.5 Product recipe/design control

- Product recipe control.
- Product manufacturing instructions.
- Specification and consistency of raw material supply (ingredients).

9.4.6 Raw materials control

- Specifications.
- Documentation (e.g., purchase notes, intake checks, certificates of conformance/certificates of analysis).
- Storage.
- Materials/articles intended for food contact (e.g., specification, appropriate materials, migration testing).

9.4.7 Production process review (and associated records/accuracy checks)

- Site fabric, facilities and equipment.
- Processes/processing methods.
- Operating instructions and training.
- Work-in-progress (WIP) control.
- Process control checks (manual and automated).

9.4.8 Quality systems

- Sampling procedures.
- Analytical procedures.
- Records.
- Acceptance/rejection/segregation systems.
- Consumer-complaint procedures.

9.4.9 Completed product

- Specifications.
- Composition and quality.
- Packaging materials.

9.4.10 Product labelling

- Labels (accuracy/correctness/legal compliance).
- Advertisements.
- Presentation.

9.4.11 Storage and distribution

- Goods in storage and distribution (e.g., clear control of products at this stage).
- Systems for segregating goods.

While these enforcement agency audit expectations can all sound like placing the food manufacturer under a high scrutiny spotlight, if managed correctly such interaction can yield great benefits to the business in terms of receipt of significant guidance and support on legislative matters, time saving when dealing with complaints from other authorities and also the business benefits from a significant boost to its demonstration of 'due diligence' due to the fact that it is openly demonstrating its compliance to requirements.

What such relationships do rely upon is having:

- A business technical team who are able to confidently articulate and clearly demonstrate the business food safety and quality management system and how it meets all relevant legislative requirements.
- Enforcement authority staff who are realistic in their expectations placed upon food manufacturing operations. For example, when dealing with large manufacturers whose products can be serving millions of consumers in the United Kingdom every week, the sheer size of the operation and volume of products manufactured means that occasionally there are going to be problems raised which require action. If in such circumstances the first reaction of the home authority is to seek to 'threaten action' rather than 'support' the manufacturer, then the relationship is likely to quickly lose the 'mutual trust'/'synergies'.

Being a good technical manager does not stop at 'just knowing the science' to underpin a range of product quality and safety decisions, clear appreciation of HACCP and prerequisite programmes, and appreciation of relevant legislation. All of these points and many more are very important, but to be a very effective technical manager all of these foundations need to then be built upon with good 'management' and 'people' skills.

In the case of dealing with the enforcement authorities (as with customers to a great extent) it is important that the technical manager demonstrates the following attributes:

- a good professional understanding of the technical and operational aspects of their business;

- enthusiasm and drive to control product safety, quality and legality;
- calm (but not casual/relaxed) when under pressure, and therefore likely to make the right decisions and act effectively in a crisis situation;
- commitment to 'customer service'; and
- tenacity (e.g., there are many ways in which a question can be asked or answered when dealing with enforcement authorities, selecting the 'right way' and not a way which triggers the undue concern of an enforcement official can be very important).

As previously discussed government enforcement agencies will wish to visit and liaise with the site across a wide range of issues, many of which can have serious impacts on the business if mishandled. Knowing the technical and operational aspects of the business thoroughly and displaying the personal attributes listed above will go a long way towards ensuring that these enforcement authorities can be confident that:

- The business will always be aware of and working to the key focal points of product safety, quality and legality.
- The business will know when to involve the relevant enforcement authority in the event of a legislative query, food safety issue or recall.

Enforcement officials have a very busy role and have to spread themselves very thinly at times [just think about how many food businesses (manufacturing, retail and food service)] there are in your region, and then imagine a job role where you are expected to oversee and interact with all of these business on matters of food safety, quality and legality. For such reasons it is often almost a relief to such officials to find a good site contact who understands the importance of compliance with all relevant legislation and who is always keen to involve the authorities at the right time.

9.4.12 Building and maintaining good relations with local authorities

Note: While many of these points may seem like obvious common sense, it is surprising how, especially when under pressure/time limitations, technical managers forget to get the basics right.

The site management should:

- Always be prompt in responding to queries raised by the enforcement body, as they are very busy and therefore usually making contact for an important reason. They will not appreciate being kept waiting or having to make contact numerous times.
- Take the time to promote the business to the enforcement officer. For example, rather than just conducting a routine site inspection where the site management waits to be asked questions, and then provides the answers. Instead seek to provide a commentary on all of the positive and proactive approaches which the business takes to control product quality, safety and legality.
- Be accommodating. For enforcement authority audits the technical manager should seek to offer the same hospitality that would be offered to a customer. If on site for inspection,

would the official appreciate a working lunch? Lunch could provide a welcome break in proceedings and give time to get to know a little more about each other's operations and experience/backgrounds.

- Empathy is very important in business relationships. A good understanding of the official's main priorities and workloads gives the technical manager a good basis on which to gain a better appreciation of their work and their expectations placed on the business. This also gives a good insight into how high up their priority list the business is.
- When new products or processes are planned for the site, the technical manager could consider involving the local authorities early in order to gain the benefit of their thoughts/observations. Local enforcement authorities appreciate (and often expect) to be kept informed of business developments that may in some way impact on product quality, safety or legality.

  Such communications may be formal or informal and could include:
  - the changing/implementation of a production process;
  - development of new products;
  - development/implementation of new product packaging; and
  - building work (either refurbishment or new building developments).
- The technical manager could look to become involved in initiatives which support the local enforcement authorities, for example, sponsorship of/involvement in awareness raising events for public food safety. For a small financial investment, the amount of publicity and goodwill that can be generated from such activities may be very significant indeed.
- Occasionally the local authorities may require access to food manufacturing sites to use for familiarisation/training of new staff or perhaps as part of a local event. Being supportive/accommodating of such initiatives can generate further goodwill.

Building goodwill with the relevant authorities will not enable a business to 'get away' with a clear braech of food law, however goodwill can sometimes facilitate an amount of leeway/'benefit of the doubt', and also helps to ensure that the business is at the forefront of the enforcement authority's thoughts when seeking to contact food manufacturers in the area/to 'tip-off' on any new sector initiatives/issues.

Reflecting on the impacts that legislative noncompliance can have on a business with regard to food safety and quality, it is vital that a good working relationship is built and maintained between the food business technical manager and the local enforcement authorities. Increasing mutual trust and respect will help to ensure an ongoing 'positive working relationship'.

9.5 Site audit preparation, performance and follow-up action

The audit process for a manufacturing site can be split into phases:

1. planning and preparation for the audit;
2. performance and processes during the audit; and
3. follow-up action to be taken after the on-site audit including the addressing of any noncompliances raised.

9.5.1 Planning and preparation for the audit

The best possible business position is to be 'audit ready, all of the time'. There should be no need for special preparations in advance of a site audit when businesses are running at their best, adhering to all site control systems and procedures, efficient and well managed with clear and continuous focus on area hygiene and product quality, safety and legality.

However, as the outcome of audits often determines whether or not the business will be able to supply (or maintain supply) to the customer, the commercial stakes are often so high that it is human nature to seek to positively influence the audit outcome via advanced planning and management of the audit day.

Technical manager considerations in advance of the audit include:

On being informed of the audit requirement...

- Ensure that there is a full knowledge of the 'scope' (requirements) of the audit. If the audit is for a certification/compliance scheme then it should be ensured that there is a copy of the audit standard available for review, and for retail/food service customers a copy of their supplier operating procedures (e.g., 'codes of practice') and audit requirements should be requested if not already held. The specific scope/requirements can then be checked against to confirm compliance in advance of the audit. This will also guard against any surprises during the audit.
- Ensure that it is clear who is conducting the audit. That is, is the auditor employed by the customer or is the auditor a third-party auditor? It should also be clarified who is paying for the costs of the audit. Third-party audits can be paid for by either the business to be supplied, or the business wishing to supply, therefore it is important that such matters are confirmed early on in the process. In addition, customers (especially the large retailers) sometimes expect the business being audited to cover their hotel and food costs while they are in the area to conduct the audit, often expecting such aspects to be booked in advance for them.
- Confirm the product or product range linked to the proposed audit. This will then help set which areas of the business are to be audited (which is especially helpful if the business has a wide range of processing options/sections) and will enable appropriate dialogue with the auditor. For example, the business can consider in advance the proposed processing methods and shelf-life trial requirements for the product range to be reviewed.
- Check whether there are any confidentiality issues involved with the proposed customer being shown specific areas of the factory site.

If there are any potential conflicts of interest (e.g., the company to be supplied is also a competitor of the businesses within some other sales categories), then the technical manager could speak to the site directors in advance to gain full clarity on the areas of the business that can be both physically reviewed and discussed. Any blocks on aspects of the business that can be reviewed during the audit should then be clearly and politely conveyed to the auditor in advance of their visit. This will help guard against any awkward moments/roadblocks being reached during the audit.

- Check whether the proposed new customer has any particular requirements that would make supply very difficult for the manufacturing operation? For example, with regard to allergen controls, some customers demand a guarantee of 'nut free' and some suppliers cannot guarantee 'nut free' due to the nature of their supply chain, handling of nuts on site for other products and/or due to shared process lines.

Clearly in such circumstances it is highly unlikely that the two businesses are going to be compatible and therefore it is probably best (and less costly) to discuss such matters via email or telephone in advance to ascertain whether there is actually any potential for a workable way forward before committing to the time and costs of an audit.

- Consider in advance, what are the particular aspects of the production site that the auditor is going to want to see? Production areas can then commence being briefed/prepared for the audit and examples of relevant production records, systems and procedures can be collated ready for audit review.

In addition, if the compliance standard that the site will be assessed against has highlighted the requirement for certain systems/procedures/checks which the operation does not currently have, then the time before the audit is an ideal time to develop and implement such systems/checks in readiness for the audit.

On this point businesses need to be careful when considering the implementation of costly systems and procedures just for one *potential* customer. If such costs can only be covered once the business has been gained, there may be a case for developing examples of the extra systems and procedures likely to be required by the customer, but the business should consider whether it would be more financially prudent to hold back from full implementation until the extra business has been secured. Of course, this all depends on what the auditor needs to see on site (and whether a record history for the checks needs to be developed), therefore advanced dialogue on such matters is important to help get the approach right.

- With regard to the production plan for the day of the audit. The business may have products that they particularly want the auditor to see, or perhaps even products that they do not want the auditor to see (such products may be confidential or perhaps may be very problematic to make/pack and therefore may not help show the business 'at its best'). For such reasons the advanced planning for the audit may also include discussion with the production planners on what are the required/preferred products to run during the audit and what products would ideally be avoided.

Some businesses also prefer not to be running their customer's products at the time that the customer is on site, as if a problem is encountered with the product online during the audit, the business may be likely to receive a far greater spotlight/scrutiny from the auditor if it is one of their products that is being packed.

(Note: When taking such 'planning steps' the technical manager should always take the time to then double check the production plan the day before and on the day of the audit, as production schedules can often change due to line availability/breakdowns and ingredient/packaging delivery delays, leading to products that were lined up for the day no longer being in the actual plan for the scheduled audit times.)

- If the business suspects that it may be weak on a particular aspect of process records/ traceability that is highly likely to be touched upon during the audit, then it may be advisable to conduct an advanced review of such records. The business will then be able to demonstrate that its own internal processes have highlighted and dealt with problem areas in advance of the audit.

If some paperwork records are currently weak it may be the case that the auditee can also have the option to switch to review and demonstrate product traceability via computerised purchasing, stock and process control systems. At least thereby showing the auditor that traceability can be maintained by the computer system in the event of any issues with the manual paperwork system. At the same time this assures the auditor that the weaknesses in the paperwork system are being addressed.

- Confirm that full records are available for the products proposed to be supplied (or which are already being supplied) to the business which the auditor will be representing. For example, microbiological results, nutritional results, shelf-life check records, product scale-up check records and sensory descriptions.
- Check who the customer account manager is for the business about to audit the site. The technical manager should seek to confirm with the account manager in advance what stage the negotiations to supply are at. Then if the project is looking unlikely to proceed due to pricing or logistics issues it may be worth delaying the audit until the foundations of the business relationship look more stable.
- If the business is already being suppled it is worth a 'health check' in advance to consider points which are likely to be raised during the audit including any recent technical issues, complaints levels and any trends in the nature of customer complaints.

Consideration of such points can enable the business to prepare in advance the points that they would like to make on any issues, and to make clear any initiatives that have been taken as a result of these matters.

When conducting such an 'ongoing business health check' review it is worth the technical manager also having a discussion with other business departments that deal with that particular customer in order to gain a full picture of the current health of the relationship. For example, the sales, development, planning and logistics departments are all likely to have routine contact with the customer and therefore can serve as a useful indicator for the current status of the supplier/customer relationship.

- Site audit preparation: The technical team should develop audit planning meetings to help raise awareness of the forthcoming audit with all related site operations departments. Such meetings can be added to the agenda of daily operations meetings, or if of particular high profile may warrant their own separate focus meetings.

At these meetings the technical manager can define a likely agenda for the duration of the audit and can select the departments required to be reviewed and represented. Such advanced planning then helps to avoid key staff being off-site/tied up on other matters and enables the best staff to be available, for example, to run process lines, review areas or discuss particular matters with the auditor. This is the business opportunity to shine, therefore it is wise to always pick the best team for the day, especially for major audits (less experienced staff may well benefit from receiving the first-hand audit experience during smaller account/less critical audits).

If there have been any previous issues with the customer, for example, process-related product complaints or issues picked up during the last site audit, such meetings are a good point at which to increase focus on ensuring that these points have been completely resolved/dealt with.

A very good point for the technical manager to make is that the audit will require the managers and area supervisors to be present within their production areas of responsibility at the time of visiting the area. The area manager will be expected to escort the auditor around the area and discuss aspects of the process (e.g., often the CCPs and any product quality checks conducted in the area).

This approach ensures that the area managers and supervisors will be fully focussed on preparing their area well for the audit, as they would be highly embarrassed in front of the auditor if anything then goes wrong.

Being spoken to directly by the area supervisors and managers should give the auditor a great deal of confidence in the business capability to consistently supply safe, quality and legal product. The auditor does not just want to hear the technical manager doing all of the talking on such points, as, of course, the technical manager is going to know the right things to say. The auditor will wish to hear from the team members who are in control of the day to day operation and wish to show their in-depth knowledge and control of such factors.

Note: If the technical manager is not confident that an area supervisor or manager is sufficiently capable of explaining to an auditor how their area and actions control product safety, quality and legality, then clearly there is a training need here to be urgently addressed. Perhaps the individual can receive coaching from the technical manager while conducting mock reviews/walk-rounds of their process area. If after plenty of coaching and experience a particular supervisor or manager is still not capable of conducting such audit reviews, this may give reason to contemplate whether the individual is in the right role for the business.

If the technical manager does not feel that the audit preparation is receiving sufficient focus, the audit planning meeting is a good time to invite along the site customer account manager, senior management or site managing director. This should send a clear message to all management and staff that the business sees the audit as very important indeed and therefore expects full commitment from all staff/areas.

- In advance of the audit the technical manager should consider the business operation weak spots. What areas are typically difficult/could show the business in a less-than-ideal light? Perhaps there is some hard to access (e.g., high level or well-guarded) cleaning that needs addressing and at this point can be planned in advance and addressed before the audit.

If certain business weaknesses cannot be addressed in advance, the technical manager should consider how such points will be explained, avoided or shown in their best light on the day.

- Out-of-date ingredients: Many food businesses have a reluctance to dispose of out-of-date ingredients, especially ingredients marked with a 'best before' rather than a short shelf-life 'use-by' date. Preferring instead to seek shelf-life extensions on such stocks set either with the supplier's blessing or on the basis of further testing/evaluation.

Some businesses will confidently explain their raw material stock shelf-life extension system to auditors and such systems are usually acceptable as long as they are well-controlled and shelf-life extensions are based on sound technical reasoning.

However, in some cases businesses have high amounts of out-of-date/to be extended stocks and are tempted to avoid the discussion topic with the auditor in the first place by placing such stocks away from likely path of auditor review.

Such avoidance tactics need to be considered carefully along the lines of. How well can such tactics be administered? What is the impression given to the auditor if they do then come across such 'hidden stocks'/issues? What is the message given to business staff who are having to help avoid such issues coming under review?

A business should be confident in its operating systems and standards, and be pleased to clearly discuss the approaches taken on all aspects of product safety, quality and legality. This includes demonstration that consideration of shelf-life extensions is conducted clearly and correctly. An auditor will find it hard to complain about well administered and controlled systems/procedures; however, they can complain about issues apparently being hidden/avoided.

- Pest control records: In advance of major audits, for extra support/assurance, request that the pest control services contracted supplier senior personnel should come on site to conduct a thorough review and at the same time check that all pest control records are up to date (e.g., electric fly killer counts and related analysis reports, site bait plans, technician visits being logged correctly and on schedule, all action points clearly addressed or 'in-progress').
- In advance of the audit, many businesses will assign a 'sweeper' whose role on the audit date will be to walk two to three rooms ahead of the audit party, conducting last-minute checks that all areas appear well prepared and taking any last-minute decisions on how to best address any unforeseen issues. For these reasons it is usually best if this person is an experienced member of the quality/technical team and is also a calm and good practical decision maker.

It is very important that the sweeper is clearly briefed on the route that the audit tour will take in order to ensure that they cover all of the required areas at the right times (including corridor routes to and from each area).

- Many factory areas prefer if possible to receive 15 min (or more) notice of when the auditor is going to arrive. This enables the area supervisors/managers to be ready and in place to meet the audit party at the entrance to their area and conduct the tour when it reaches their location.

As the technical manager will often not know exactly when the audit tour is going to start (as initial audit conversations/systems reviews may take a little more or less time than expected), some businesses will have an arrangement where when the auditor is looking ready to review the factory, a member or the hygiene or quality team is called to fetch the visitor protective clothing. At this point the hygiene or quality team member can also call ahead to the first factory areas and also let the 'sweeper' know that the tour is imminent.

- Advanced auditor details to be collated (via email or phone call) and communicated/dealt with should include: protective kit requirements (e.g., boot/shoe size, coat size), any dietary preferences or allergies to consider for lunch, and are all hotel bookings made (if the business is picking up such costs).

It is a good practice to call the auditor a few days in advance of the audit to confirm that the audit date and time is correct in all diaries, and that any arrangements such as hotel or pick-up requirements have been made.

At this point it is also useful to confirm whether the auditor is going to have any time constraints on the day? Are they going to be able to be on-site for the full day, or do they need to leave by a particular time? This will help the business team to make the appropriate plans to help review all systems/procedures/records/areas required in the time available.

- Before the audit, if not regularly present in the factory, the technical manager should start to walk the site more frequently in order to preempt any issues with regard to compliance with the audit standard and to familiarise themselves with the areas and the site tour route to be taken.

9.5.2 Late considerations (just before the audit date)

- Just before the audit date it is advisable for the business technical team to collate all of the files/records/procedures that the auditor is likely to require to review during the audit. Not only does this get all records ready in one place, it also will provide advance warning if any of the files appear to be missing, or in case advance notice is needed to request a particular file or record (e.g., perhaps the night-shift management team has the line deep cleaning positive release records and therefore requires advanced contact to organise for the files to be handed over to the daytime shift in readiness for the audit).

A checklist of 'likely to be required' files/records/procedures can be collated via review of the customer/compliance standard being audited against.

Once collated the technical team should take the time to review each of the files to double check that all procedures/systems are up to date and that all checks appear to be being conducted correctly and within required tolerance ranges (or in the event of any issues, that clear and correct rectification action has been logged).

During site audits a lot of paperwork/records are likely to be requested as the audit progresses and topics arise for discussion with the auditor seeking evidence of compliance and control. As the technical manager will typically be engaged with the auditor all day, it can help the smooth flow of information to the auditor if an experienced quality or technical team member in advance of the audit can be assigned the responsibility for 'documents and records retrieval' during the audit.

This is a key role as the person assigned needs to have a first-hand knowledge of where all site records/paperwork are held and also needs to have a clear appreciation of the level of detail expected to be presented. Then at least they will know in advance of showing to an auditor, whether or not the records will meet requirements, and if the records found are not going to be sufficient then they can also search for further records which will help to reduce the impact of the 'less-than-ideal' records initially found.

They can also give advance notice of such problem issues to the technical manager in order that the manager is at least well prepared to explain any issues found, rather than being surprised by any poor quality/insufficient records being placed in front of them at the same time as they are placed in front of the auditor.

- Other preaudit checks can include whether a visitor car parking space has been booked? Are the reception team well briefed to expect the auditor and aware of the importance of issuing the auditor with the necessary site visit paperwork (e.g., medical questionnaire, health and safety guidance and perhaps the site allergens policy if relevant to visitors).
- If the site typically has a reception 'Welcome Board', it should be considered whether it is appropriate for the company/auditor name to be displayed? (Some new potential customers may need to be kept confidential from existing customers/other visitors to the site.)

With regard to the site visitor signing-in system. If the signing-in system is in the form of a log book/list, auditors can like to take a look at who else has recently been on site (and can sometimes then ask 'how did you get on with the audit/visit last week?'). To help avoid such matters visitor log books are available which are particularly designed/formatted to keep the details of other site guests confidential/out of view. Computerised/digital sign-in/out systems can avoid such matters completely.

It is usually good practice to swiftly guide the auditor through to the audit/meeting room on arrival rather than leaving them sat in reception for too long. The auditor will be keen to get the audit started and reception areas can also be a thoroughfare for operations staff where many conversations can be held which may not be appropriate to be heard by the auditor.

Reception can be briefed in advance to guide the auditor into the meeting room and perhaps to start to prepare refreshments/teas/coffees while contacting the site technical representatives and senior management team to confirm that the auditor has arrived.

If due to unforeseen circumstances the technical representative cannot join the auditor immediately, it is advisable to send another member of the technical team to have a general conversation with the auditor to help pass the time and perhaps deal with some of the audit requirements before the main representative can get there. It does not look good to an auditor if their first experience of the technical manager is turning up late for the audit, especially if the delay is then found to be due to having to deal with site 'problem issues'.

- If the technical team are aware that the auditor is arriving in the area the night before their audit visit, it is often seen as polite (and for some customer accounts expected) to meet with the auditor the evening before.

This is a difficult matter to judge as many auditors prefer to be left alone the evening before the audit as they are typically catching up on phone calls, emails and paperwork, and some auditors will be conscious to avoid any suggestion of being ingratiated by a nice restaurant meal.

It is often good practice for the technical manager to politely make the offer while at the same time stating that it is understood that the auditor is likely to be otherwise engaged. Depending on the business, many may typically only make such an offer for major account customers, as audits can be so frequent within business operations that the technical team could soon run up a lot of time and costs being frequently engaged with preaudit meals.

- If the auditor is stopping in the area the night before the audit, the auditor may often choose to take a drive to the factory site the evening or night before the audit, both to ensure that they definitely know where they are going the next morning, and also to take the opportunity to assess the site perimeter condition and security arrangements in place.

As a result, if there are any perimeter or external door control issues that the technical team are aware of on-site, it is worth requesting area management to have a particular focus on such points in advance of the audit date and also be prepared to answer-related questions from the auditor.

- On the day of the audit it can be known for auditors to arrive intentionally very early in an attempt to either 'catch out' the site or place the site on the 'back-foot' for the audit. Therefore it is wise to ensure that all staff and management know to be 'audit ready' far earlier than the proposed start time, 'just in case'.

9.5.3 Performance and processes during the audit

On the day of the audit:

- It is often advisable for the technical manager and other key business team members involved in the audit to arrive on-site very early in order to check that the operation is running smoothly and to action any last-minute 'preaudit issues'. The technical manager/ audit lead can benefit from conducting a preaudit site walk-round on their own, speaking to area personnel along the way. This approach gives the technical lead the confidence that all areas appear well prepared and in control, and provides an opportunity to discuss first hand and deals with any potential problem issues (or to enable good explanations for issues to be ready if such problem issues are not going to be able to be resolved before the audit takes place).
- If the technical manager or any of the business team are nervous before the auditor arrives this can be positive as it shows that the team members care and appreciate what an important day this is for the business. However, it is important that this nervousness is channelled into energy to address the challenges and ensure a good audit, rather than it becoming counterproductive and affecting performance/interaction with the auditor. Audits serve to assure the stakeholders that the business is well run and under control, which would not be helped if the team members seem overly anxious/worried.

Most audits will commence with an opening meeting where the auditor will define the scope of the audit to be conducted and the preferences of the approach to be taken. It is very important that members of the senior management team attend this open meeting and indeed attend regular update meetings (and the audit closing meeting) during the course of the audit to demonstrate their senior management commitment to the project. Many audit standards will expect this rather than it just being a 'nice to have'.

- Sometimes it can be beneficial to highlight to the auditor at the start of the audit how important the business supply relationship is (or is going to be) to both/all parties.

This will help the auditor to reflect on the importance of securing ongoing supply from the site, and while this will not enable the operation to avoid any nonconformances being raised, it will help to encourage the auditor to be positive and supportive on any issues. For example, providing advice on 'best ways forward' to help the business to address matters/become compliant and thereby allow supply to begin (or supply to be maintained in the case of existing accounts).

- When the audit is a 'repeat audit', it can be beneficial at the outset of the audit to highlight all of the good work and improvements that have been progressed since the last audit from that customer/third-party audit company. This can help to place the auditor in an assured frame of mind for the audit, as effectively if the last audit secured a high audit grade, and actually there have been a lot of further improvements since then, then the auditor has an early picture in their mind of this being a 'top grade supplier' and therefore this may help influence matters during the day. Similarly, if the site was awarded a 'lower grade' last time, and a lot of work has taken place to further improve matters, the auditor will be starting to consider the site as being potentially ready for a higher grade status.
- Above all the business team should always be positive. All parties are likely to want to achieve technical compliance and consistent supply, therefore the key is to work together to achieve this.
- At the outset of the audit, it should be made clear to the auditor any site team members/managers who will be unavailable at certain times. This way if they wish the auditor can adjust their schedule to avoid any clashes. For example, if the human resources (HR) manager is only available for the next 2 hours and is then going to be off-site for the rest of the day, it would be worthwhile discussing HR-related matters in advance of the factory tour, as afterwards that individual will have left site.

This proactive approach also helps avoid auditor disappointment in finding that key staff are not available to discuss matters due to poor scheduling and communication. Such matters can swiftly lead to the auditor being concerned that the audit is not being adequately supported.

- An auditor is likely to want to see a traceability exercise conducted at some point during the audit, and often such an exercise is a major undertaking for those who have to find and piece together the related process records from each point of the production chain.

For this reason it can be beneficial to ask the auditor whether they can set the product/dates for the traceability study early in the day, so that an assigned member of the quality/technical staff can then set about finding and collating all of the records together in readiness for the 'traceability study review' later in the day. If the auditor is also on a tight schedule then this is likely to be accepted as a good idea.

Sometimes the decision as to which particular product is selected for review can be influenced by the technical manager (e.g., the auditor may ask when setting the traceability test what are the newer products being manufactured, or which products contain a good wide range of components/ingredients, or are manufactured in a particular area or packaging format). Therefore, it may be beneficial for the technical manager to take some time in advance of the audit to consider what the ideal product for a traceability exercise would be.

9.5.4 During the factory tour

The technical manager should bear in mind that auditors have a mind of their own and that often it will not be the case that the auditor will accept being led around the site to just look at the aspects of the operation that the technical manager feels are best to review and discuss. Sometimes this will be the case however often it is in an auditor's nature to be inquisitive so the technical manager should not be surprised if the auditor is soon taking the lead as to where they want to go on-site (this also applies to reviews of site paperwork too), and perhaps getting into awkward areas of the operation which are far more of a challenge to review.

A key message here is that the business should not just prepare the factory areas that are intended to be shown to the auditor, it is very wise to prepare all of the areas that the auditor may wish to visit (or may just come across during the site review). The technical manager will be very pleased that the site went to this extra trouble when the auditor decides to visit areas/aspects unexpectedly.

- If during the site review the auditor speaks to a team member and asks for their name and job title, it is likely that the auditor is going to request to review the personnel employment records of that team member, to check aspects of personnel control such as medical screening and appropriate training records for the role being conducted.

In such circumstances if the business HR or training department are likely to benefit from advanced notice to start compiling all of the required information together, then it may be helpful for the technical manager to find a timely moment during the audit tour to send a message to the relevant department giving the names of the individuals likely to have their details requested.

- Along the same lines as the point above, if during the site tour an auditor records the lot numbers or codes of materials such as ingredients, packaging, finished products, stock on hold/quarantined stock, then again it is likely that the auditor is going to ask for the process records which underpin/manage the status of such stocks on site (e.g., ingredient traceability and any related 'on-hold' records, positive release, quarantined stock, shelf-life extension records).

So again it may help smooth the flow of information to the auditor if advanced notice can be given to team members charged with the responsibility for records retrieval during the audit.

- The representative should be conscious of their body language and general mannerisms at all times during the audit. It is important that the auditor is assured that the team are being open and honest at all times. The greatest proportion of communication is 'visual'. If team members come across as nervous, defensive or quickly frustrated, then this will not assure the auditor that what they are being told/shown is correct and under control.

- Where it is easily possible, the business representatives should seek to address any audit issue/nonconformance points while the auditor is still on site so that these points can be recorded as 'closed' to assure the auditor of the site's diligent approach to such matters and possibly thereby reduce their impact on the final audit report.

For example, if during the factory tour a piece of machinery failed, leading to the auditor not being able to see a particular process that they were intending to approve, the site representatives should consider whether that machinery can be quickly repaired and the auditor taken back to the area later in the audit schedule to review and approve the now working system.

Such 'revisits' need to be handled carefully, as often once the auditor has reviewed the site for the first time some business areas may not expect the auditor to then walk back through the area again and perhaps encounter a number of process/production points (e.g., perhaps see a competitor's product being processed) that they did not encounter when the area/schedule was being more tightly managed earlier in the audit.

9.5.5 Expect the unexpected

- Experienced managers know that 'something is likely to go wrong' during the audit. Machinery breaks down, staff stop operating as they should, handwash units run out of hot water/soap/towels, etc. The auditor is likely to appreciate that such matters happen too and in these situations is going to judge the business on how it deals with the situation.

A well-controlled and swiftly addressed issue will assure the auditor that the business knows the right steps to take in order to bring matters swiftly back under control, for example:

- If a routine metal detector CCP (or other foreign material detectors) function check fails to reject the product from the line when the 'reject test' materials are put through. The following steps may well help to swiftly address the issue:

Stop the line. Quarantine all product packed since the last OK rejection check to await rechecking through a working detector. If any stock has left site (this should be unlikely as the detector checks should be frequent during production), seek to withdraw such stocks from the supply chain on the basis of foreign body control CCP failure (unless such stocks will receive a further foreign body check preconsumption, perhaps if being utilised within another food product, in which case the issue should be corresponded to the customer technical team to request special dispensation and additional focus on these checks when they are processing the product). Call engineering to address the unit fault and do not pack any more product down that line until the issue is resolved.

- If there is no soap in an area handwash dispenser when the auditor enters the area and seeks to wash their hands:

Stop the audit tour. Ask the area supervisor to fetch some soap and wait. (If the tour continued, missing a required handwash, this would send a poor standard signal to the auditor). Ask the area manager for assurance that the unit is typically frequently used and topped up to dispel any auditor concerns that this handwash station is not used regularly within the area.

- A staff member is seen not adhering to site rules and regulations (perhaps around correct protective equipment, operating or hygiene practices):

  The technical manager could either call the individual over and deal with the matter themselves clearly, politely and professionally, or request that the area supervisor/manager swiftly deal with the matter. Either way a clear signal is sent to the auditor that site rules are well maintained on site.
  - A technical manager should always be calm. There is usually a good explanation for everything, and a clear calm mind will help in thinking of actions/solutions quickly and rationally rather than being upset by a difficult audit situation.
  - The pressure of an audit situation can make usually rational/calm staff and managers suddenly say things that can mislead or concern an auditor.

This is especially so if something goes wrong or if the auditor seems particularly 'pressurising' in their drive for information. Therefore, even when the technical manager has handed over the auditor in a production area to the production staff member/supervisor/manager, it is very important that the technical manager still keeps listening to the conversation, ready to help guide 'back on track' if anything incorrect or misleading is said.

9.5.6 *'You know your site better than anyone'*

This is both a strength and an added responsibility.

- When reviewing the site areas, systems, procedures and records, it is likely that the technical manager will spot far more faults/issues than the auditor will be able to.

The technical manager should consider the auditor's position. They have arrived at a manufacturing site which they possibly have never seen before and they have a relatively small window of time within which to gain sufficient understanding of the processes and practices to then form a judgement as to whether the site is controlled and compliant to the required standards.

There is unlikely to be a large amount of time to gain in-depth information on how each specific procedure should be followed. Whereas the technical manager knows exactly how the site should look and exactly how all systems and procedures should work and be recorded.

This therefore places an added responsibility on the technical manager to keep a mental or written record of all of the 'issue points' that they spot alongside recording the points that the auditor raises. To aid continuous improvement such additional points should then also be addressed alongside the 'official audit points'. The technical manager is also best placed to decide whether such additional points should be shared with the auditor.

To help ensure clarity the technical manager may, on completion of the audit, decide to circulate two separate reports: the 'customer audit report' and the 'internal audit review report'. This will help ensure that site management and staff do not become confused as to which issues are under customer focus and which issues are to be internally dealt with.

Circumstances where the auditor has been impressed during their visit but internally the business knows it could have done far better can lead to a technical manager feeling deflated at the end of a successful site audit. Even though the site has passed with a good grade, the technical manager knows how many 'problem issues' were actually present and by good management or good luck were avoided from becoming major issues in the eyes of the auditor. In such circumstances the technical manager may choose to console themselves with the fact that they can now work hard to address these 'internal issues' alongside the 'auditor issues' raised during the day.

9.5.7 The audit summing up/closing meeting

- If the audit has not gone well, with a number of issues requiring resolution which will necessitate senior management commitment/financial resources, it is often wise (and may actually be expected under the particular audit standard) to invite site senior management/directors into the closing meeting. Having these senior managers present ensures that they can hear of the problem issues encountered first hand from the auditor, and have an opportunity to discuss acceptable measures to address these points.

Such an approach takes away the opportunity for senior management team members to ignore audit rectification action requirements after an audit in an attempt to avoid expenditure/change.

- The technical manager should make sure during the closing meeting that they are very clear on every single nonconformance point raised and what the auditor's expectations are with regard to how each issue should be addressed, and agree the timescales for such actions to be completed.

If unsure as to the required timescales for an action to take place (e.g., perhaps the point needs significant expenditure or operational downtime to address) then the technical manager should avoid committing to timescales that they are unsure can be met. Instead it may be appropriate to request guidance from the auditor on how swiftly they would expect to see the issue addressed. The technical manager can then state that they will respond back to the auditor with timescale feedback once they have had chance to discuss the matter with the appropriate business team members.

Once the auditor has left:

- If there have been noncompliance matters raised, it may be important for the sales account manager (of the customers affected) and senior management team to be briefed on the issues as soon as possible. Actions may require to be taken with the customer in order to provide assurance and secure time to address particular issues.
- The site technical management should look to swiftly correspond the audit result and any noncompliances to be addressed to the relevant team members within the organisation.
- If the audit result is a very good one, the technical manager should ensure that themselves and their teams take their fair share of the praise (it may not be often in the business that the technical/quality team are seen as the bearers of good news). Also, it should be

ensured that particularly well-performing departments/teams receive appreciation and praise, as this raises morale, confidence, and a sense of pride which will then underpin further audit successes in the future.

- The technical manager should make clear any additional points which have been picked up but which were not logged by the auditor. It is important to ensure that these are addressed just as rigorously as the auditor points as such issues are weaknesses in the current food safety/quality/legality approach and could become major problem issues if they go unaddressed. This approach will also send a strong message to all team members that the site is looking to run correctly and 'fully compliant' all of the time, and is not just seeking to 'get by' from audit to audit.
- It is often helpful for the technical team to set up a postaudit session where all area management/supervisors can be briefed in detail on their area's performance and on any improvements that can be made for future audits.
- The technical manager should highlight and speak individually to any managers or staff who were felt to have performed particularly well during the audit. Team members will respond well to praise, building further assurance and confidence for the future.
- Similarly, the technical manager should look to provide constructive feedback and coaching to any managers or staff who clearly struggled during the audit.
- Each audit noncompliance point should be assigned an owner/champion to address the issue. It is important that these 'issue champions' are aware of and agree to the 'complete by' date and to track progress up to the completion date to ensure that all issues will be closed within the auditor timescales required.
- The technical manager will be expected to formally respond to the auditor report describing the actions taken to address all nonconformances within the timescales agreed. Ideally the business should look to complete feedback reports early in order to move the customer to a point where they are comfortable that all key points have been addressed.
- If the auditor decides that a follow-up visit is necessary to 'close' some of the audit points (i.e., confirm that the required action has been taken), then the technical manager should ensure that such return visits are as tightly managed as the initial audit. Even if the auditor says that they just need to drop by to see one area it is appropriate to still prepare the whole site, as if the auditor is going to go to the trouble to come back to the site, it is likely that they will also take the time to complete another site review.

Further reading

Audit standards and associated support information for BRC Global Standard and IFS (International Featured Standard) are freely available via the relevant websites.

www.brc.org.uk.

www.ifs-certification.com.

Alli, I., 2003. Food Quality Assurance, Principles and Practices. CRC Press.

Yianaas, F., 2009. Food Safety Culture—Creating a Behavior-based Food Safety Management System. Springer.

Yianaas, F., 2015. Food Safety = Behavior. SpringerGlobal Food Safety Initiative (GFSI). Available from: www.mygfsi.com.

ISO, the International Organization for Standardization develop and publish International Standards including the ISO 9000 series Quality Management standard, ISO 14000 series Environmental Management standard and ISO 22000 series Food Safety Management standard.

http://www.iso.org/iso/home.html.

Safer Food Better Business (SFBB) guidelines are produced by the UK Food Standards Agency (FSA).

https://www.food.gov.uk/business-industry/sfbb.

Kill, R., 2012. The BRC Global Standard for Food Safety: A Guide to a Successful Audit, second ed. Wiley-Blackwell.

Nonconformity, recall and crisis management

10

Mark Swainson
National Centre for Food Manufacturing, University of Lincoln, Holbeach, Lincolnshire, United Kingdom

10.1 Introduction

In the food sector, consumers rarely purchase products on the basis of food safety or 'compliance to legislation' as both factors are expected and assumed to be provided as 'standard'. Therefore, it is usually 'product quality' that is the key point on which a product will be judged by the consumer. 'Price' could also be considered as influencing the degree of 'quality expectation'.

Each product category will have typical 'quality variables'. For example, such variables could include: the consistency of ingredients; focus and ability of employees; the performance of process machinery; and the control of the onward supply chain. Many of these factors can vary over time leading to variability in product quality/consistency between production runs or batches. It is therefore highly likely that most food production businesses will need to regularly take decisions on the acceptability of end-product quality.

10.2 Quality management of nonconformance

There are many considerations related to investigation and decision making on product quality issues, starting with the importance of defining 'What is a quality fault'?

Defining a quality fault: The ISO 9000 family of quality management systems define 'quality' as the 'degree to which a set of inherent characteristics fulfils requirement'. The term 'quality' is also often linked to the 'grade', 'standard', 'class' or 'degree of excellence' of a product.

In food manufacturing the product quality aspects often being measured include:

'Organoleptic quality' attributes of:

- appearance (e.g., colour, shape and size);
- aroma;
- taste/flavour; and
- texture/viscosity.

Swainson's Handbook of Technical and Quality Management for the Food Manufacturing Sector.
DOI: https://doi.org/10.1016/B978-1-78242-275-4.00010-1

'Internal'/'intrinsic' attributes of:

- microbiological performance;
- food chemistry – including nutritional properties; and
- physical standards (e.g., defining the amount of extraneous matter acceptable).

And other attributes perceived to define product quality which include:

- provenance/origin of ingredients; and
- packaging and label function/appearance.

Therefore a 'quality fault' could be considered to be a failure to fulfil any particularly required/expected 'quality attribute'. Typically, in 'business-to-business' transactions the required quality attributes will be defined and agreed between the seller and the purchaser. Often the final consumer will have also influenced these quality requirements either directly or indirectly.

With regard to many foods the expected 'quality' is covered in food legislation/regulations, sometimes on a broad basis of legislation expecting foodstuffs to be 'of the nature and substance intended' and in other cases defining specific prescriptive standards expected in order to both help protect the consumer and provide a consistent foundation on which businesses can trade locally, nationally and internationally.

The Codex Alimentarius Commission (CAC) is a joint intergovernmental body of the Food and Agriculture Organization (FAO) of the United Nations and World Health Organization (WHO). CAC has worked since 1963 on the provision of harmonised international food standards which seek to both protect the health of consumers and ensure fair trade practices. The CAC provides a collection of such food standards, guidelines and codes of practice recognised by the World Trade Organization (WTO) as 'benchmark' standards for product safety and quality.

These science-based standards are adopted through global consensus forming the basis for international food trade. http://www.fao.org/fao-who-codexalimentarius/codex-texts/list-standards/en/.

10.2.1 Sources of information (for 'quality issue' decision making)

'Accurate information' is vital when considering the appropriate actions to take with regard to a product quality issue. For the quality manager there are a wide range of information sources which can be utilised when seeking to make a product quality decision. These include the following:

- *Product descriptors/specifications:* Businesses can produce a wide variety of products for many different customers. What may be a quality issue to one customer may be perfectly acceptable to another customer (it is worth noting here that the product marketing point/price will also sometimes have an influence). The key documented quality standards to refer to when seeking to assess the 'quality correctness' of the product are the product specific sensory description and the end product specification, both of which should have been initially agreed with the customer as defining the ongoing requirements of the product (including any acceptable tolerances/leeway upon specific quality assessment points).

- *Product/process records* will often provide an insight into the circumstances that have led to the occurrence of the quality issue, for example, raw material checks, time/temperature logs, sensory assessment records, food chemistry/microbiological check results, etc. It can also be helpful to review the 'quality problem' batch records compared against batch records drawn from products that have been processed without any problem, as any differences may provide an insight into the root cause of the issue.
- *Factory process staff and management:* Can provide information and particular detail on what appears to have gone wrong and the extent of the quality fault/issue. It is therefore vital that the business has developed a culture of trust where employees can be sufficiently confident to discuss any errors that they or their team may have made without fear of any repercussions (otherwise the quality manager is likely to receive little or no support in trying to get to the root cause of a particular issue).
- *Sales and development staff:* Can provide an insight into the key product quality aspects being sought by the customer (this can be especially helpful if there is any ambiguity within the product sensory description or specification). They may also have useful knowledge of when that particular product is going to next be required for review by the customer (e.g., forthcoming customer range reviews/store checks) as part of the end-product quality decision may be influenced by how likely it is for the issue to show up on the customer's supplier quality checks.
- *The customer*: While care has to be taken not to be of concern to the customer, if done correctly and in the right circumstances customers can be consulted on quality issues. Often customers would rather have 'slightly less than ideal quality' available on their shelves compared to no product at all.
- *Customer complaints*: Can provide an insight into whether the issue encountered has also been noticed in the past (or is being particularly noticed now) by the consumer (or perhaps by the retailer in the form of product sensory/benchmarking panel feedback).
- *The law*: The quality manager will need to understand whether there is any potential to be in breach of the law on a particular quality point. Such points could range from a product not being deemed to be of the nature and substance intended (e.g., the product does not meet the customer specification) through to weights and measures faults in the case of a packing quantity/consistency issue).
- *Environmental Health and Trading Standards*: Local enforcement officials can be called upon to provide advice and guidance on specific quality points. As with customer interaction, such activities have to be handled carefully so as not to raise any undue concerns.
- *Trade associations and membership bodies* employ food category and process experts whose roles are often partly related to answering members questions on food quality issues. Expert/experienced advice can often be invaluable when seeking to resolve a quality fault/issue.
- *Personal network:* Over a career in the food industry the quality manager is likely to meet many people with many skillsets. Every so often if the quality manager has maintained their contacts network, they may be able to ask a favour and call upon a specific skillset/resources in order to help deal with a particular quality issue.
- The *business senior management* team/directors/owners: Often 'quality decisions' are not made on a clear cut 'yes/no', 'right/wrong' basis. There is typically an element of cost or business risk in most quality decisions and therefore it is important that the quality manager has a full appreciation of both the financial position of the business and the extent to which the business is willing to accept a greater element of risk. In circumstances of doubt

the quality manager is always well advised to gain the backing of the site senior management/directors before taking the final quality decision on the actions required to resolve the matter.

With such a wide range of optional sources of information available to aid the decision-making process on product 'quality issues' the challenge is often that a thorough review will take a significant amount of time. When a quality issue occurs the matter often requires addressing swiftly and therefore to facilitate a rapid response it is advisable for the technical/quality manager to prepare in advance, taking the time on a routine basis to ensure that their 'network of information sources' is clearly understood, well mapped and maintained, for example:

- The business should hold and maintain an up to date and formally agreed/approved customer specification for every product produced.
- Raw material (including ingredients and packaging) specifications should be up to date and held on site for all previous and current suppliers.
- Routinely check that the business product traceability system (and associated product recall system) is clear, quick and easy to use.
- Confirm that product process records, from raw material specification and intake checks through the production process and up to the point of end-product despatch and delivery, must be comprehensive and able to be accessed swiftly in the event of an issue.
- Ensure that the technical/quality team are receiving copies of all customer complaints and product panel quality feedback (both internal and customer panels).
- Ensure familiarity on how to reference/access sources of information including online legislation access.
- Building and maintaining very good, trusted relationships with key stakeholders including customer quality/technical representatives and local enforcement officials.
- Technical/quality staff should ensure that they gain and maintain the trust and confidence of the business personnel and managers at all levels.

These and other ongoing 'support network development activities' will often contribute toward quicker, smoother management of issues and better decisions being made.

10.2.2 Quality management of nonconformance: decision processes

When faced with a quality issue during the manufacturing/processing operations the quality manager is usually tasked with ensuring that the following is achieved:

- *Full clarity*: Is required as to 'exactly what has occurred' and the extent of the quality fault/issue (including checks to confirm that the highlighted issue is actually a quality fault).
- *Form an effective* 'correct' *decision* on what action should be taken to address the situation. For product which has just been manufactured on site this could include a decision to 'dispose', 'rework' or 'release the product for sale'.
- *Control the product stock affected* while investigations are underway and decisions are being taken. Typically, such control is achieved via a site quarantined stock or 'product-on-hold' system.

To help form a decision as to whether a product with a 'quality issue' is still fit for sale the quality manager not only needs to understand the full nature and extent of the issue but also requires a good appreciation of the 'customer expectations' and current 'sales environment/climate'. For example, is the product a routinely sampled/assessed flagship product within a product range? Is the product already under particular customer focus due to any previous issues/complaints?

- *Conduct follow-up actions*, often including insight, information and action support from other managers in the business where their operational areas are linked to (or perhaps caused) the issue. The follow-up actions taken should also involve the implementation of additional procedures/systems designed to help prevent a recurrence.
- *Complete interim (if required) and final reports* detailing the issue encountered, the actions progressed to resolve that specific incident and also the actions/systems implemented to help avoid any recurrence. Such reports are often required for the assurance of the customer affected by the issue, and also circulated internally for the benefit of communication across the business and further management/senior management review.

10.2.3 Quality management of nonconformance: corrective and preventive actions

The following definitions are provided by the Quality Management Standard ISO 9001:

> 'Correction' – action to eliminate a detected nonconformity.
> 'Corrective action' – action to eliminate *the cause* of a *detected* nonconformity or other undesirable situation.
> 'Preventive action' – action to eliminate *the cause* of a *potential* nonconformity or other undesirable situation.

These terms cover key practices that the quality/technical team are likely to be regularly involved with. 'Nonconformities' could be customer complaints or on-site problems encountered with the product not meeting the specified requirements (e.g., organoleptic, microbiological or chemical/nutritional standards).

Steps to conduct a 'correction' could, for example, involve taking particular additional steps in the manufacturing process to bring a nonconforming product back into line with the specified requirements. Or if this is not possible then making a decision to regrade or rework the product.

While 'correction' deals with the immediate product issue, 'corrective action' then seeks to address the root cause (or causes) of the issue in order to help avoid/prevent a recurrence. Common examples of corrective actions include responding to customer complaints or internal process/product issues by implementing additional check procedures, training and retraining staff on how to avoid or address the issue in future, and adjustments/repairs to processes or equipment that may have contributed to the issue.

'Preventive' actions seek to move the operation into being 'proactive' rather than 'reactive' on quality issues. Such actions are taken to preempt and therefore help avoid an issue that the business can foresee could happen if left unaddressed.

Due to their importance to the quality management process, many compliance standards will expect producers to both establish and maintain procedures for implementing and maintaining corrective and preventive actions.

Documented procedures are good practice and typically expected to cover corrective action activities, including the recording of the results/outcomes of the corrective actions progressed.

Operational/compliance standards, often linking to 'senior management commitment', expect that the site senior management team will ensure that the 'corrective action' approach is utilised as a system of 'continuous improvement'. The business is typically expected to be able to demonstrate that it is clear as to the range of information sources available to define its nonconformities (e.g., customer complaints, on-site product quality/safety checks, laboratory analysis results, etc.), and have a responsive and routine approach to the review of this information in order to form its corrective actions (designed to eliminate the causes of the nonconformities and thereby prevent their recurrence).

In technical management it is better to be proactive rather than reactive wherever possible. The challenge for the proactive technical manager is often in ensuring that the right proactive measures are implemented (without overburdening the existing control system) in a manner which is both resource efficient and cost-effective.

'Corrective actions' on a product quality issue could be divided into two parts:

1. dealing with the product issue in hand; and
2. working to ensure that a similar issue cannot recur in the future.

Dealing with the product issue in hand as the quality manager may well involve progressing and managing the investigation, arranging for further product testing/assessment and also steering the decision-making process. Overseeing the quarantine control of the affected stock, including administration of any subsequent release from quarantine (if the stock is later deemed suitable for despatch) or perhaps administering actions to either rework or dispose of the product. Plus, if required the role will involve writing update reports and summary reports for submission internally and also perhaps for issue to the customer (including customer complaint responses).

It is very important for business efficiency and customer assurance that specific quality issues are not allowed to recur. Working to ensure that a similar quality issue is unlikely to be encountered again in the future can often be supported by the use of a nonconformance system which captures every issue/problem as it occurs and typically tasks the relevant area manager with the expectation that they need to formally review the matter in relation to their process area, equipment and staff, define the root cause/s and implement steps to help provide increased assurance that the matter will not recur.

Such corrective action may include briefing and retraining personnel, adjustments to manufacturing procedures/methods, implementation of new checks and perhaps investment in new machinery/test equipment. Fig. 10.1 shows an example of a nonconformance form which can help to facilitate and document this process of issue review, action and follow-up.

ABCD FOODS Ltd: Area Non-Conformance Investigation and Action Record.

Please investigate the non-conformance issue highlighted and provide feedback to the Quality Dept. on the findings and further actions taken to address the matter.

Non-conformance No.	000123
Date of issue:	21st June 2018
Issued to:	A.Rhea - Manager
Position:	Packing Department Manager

Non-conformance issue	Specific Non-conformance.	Timescale & responsibility	Corrective action	Manager signature	Quality dept. follow-up.
Customer complaints received of incorrect pack counts present within cases upon delivery to depot. Customer Regional Depot. Delivery 8th June 2018. 18 cases of Cheese 30 x 200g packs delivered. 4 cases incorrect pack counts. 25 23 22 and 26 packs in each of the 4 cases. 4 Cases rejected. On hold at customer awaiting pick up or disposal.	Cheese case count specification requires exactly 30 packs in each case.	A.Rhea Manager to complete investigation/ follow up actions and feedback by 26th June.	***Manager, please record here the investigation findings and further actions taken.*** ***Where appropriate please include timescales and specific names of staff retrained, and whether any further action was taken etc.***	***Area Manager to sign and date here that the matter is complete/ addressed.***	***Quality Department check to be recorded here to follow up and confirm that all corrective actions noted have been checked to confirm completion.*** ***Record full details of findings, follow-up auditor name, date, confirmatory checks conducted.*** ***Record any recommendations for subsequent follow-up or monitoring systems adjustment.***

Figure 10.1 Nonconformance, area management investigation & action record (example).

10.2.4 Quarantined stock control: systems and procedures

The food manufacturing sector is a very dynamic operating environment: high expectations of quality assurance and 'right first time', all often coupled with 'JIT' (Just In Time) production planning. Commercial pressures demand that most focus tends to be placed on a continuous and fast-moving production drive from the raw materials arriving on-site right through the process and on to the end customer delivery.

Within all such operations, for reasons that will be discussed in this section, there will be times when a business does not want the stock to go anywhere at all, and therefore the business requires a 'quarantined stock' control system.

'Quarantine' definitions often refer to a period of time during which material suspected of carrying imperfections/issues is detained at a location under enforced isolation to prevent onward movement. In the case of substandard food material, the quarantined stock system should be designed to prevent any further onward movement both onsite or into the food supply chain.

Reasons for quarantine: Products can be placed on quarantine to prevent their movement/despatch for a number of reasons including:

- There is a quality fault with the product.
- There is a safety issue (e.g., physical, chemical, microbiological or allergen risk).
- There is a legality issue (e.g., perhaps the packaging or pack weights are incorrect against those declared).
- The product recipe/packaging format has changed and therefore this version is no longer saleable but could be mistakenly picked if still listed on 'live stock'.
- The product is a trial to be used for sampling/reference only, not for sale.
- Product is on 'positive release', that is, the stock must not be despatched until confirmed to be correct/acceptable/'as required' on the basis of test/check results (e.g., microbiological testing, organoleptic testing/taste panel).

The 'reasons for quarantine' list can become very long as basically any reason at all for not wanting the product to go anywhere may result in the product being placed 'on quarantine' (sometimes also described as 'on hold' within the business system).

It should be noted that within businesses the product to be placed on quarantine will not necessarily always be in the finished product format. That is, a business may want to place a raw material/ingredient on hold, or perhaps 'work in progress' (e.g., mixed/cooked ingredients which have not yet been packaged). Therefore, businesses need to consider their particular processes and how such intermediary phases are managed.

(Note: Ingredients/raw materials are often typically administered by an 'on hold' system which is maintained separately from the end-product 'quarantined stock' system as usually such stocks will be in very different locations both physically and also from a computerised stock control viewpoint.)

10.2.5 Key points of the quarantine system/process

Placing material on quarantine: Within food businesses, to stop material/product from going anywhere it typically needs placing on quarantine in two separate places:

1. physical quarantine; and
2. place on quarantine/hold on the (often computerised) stock control system.

Physical quarantine will usually involve the following steps:

- Collation of the affected stock.
- Physical segregation of the stock involved (typically the stock is expected to be in a marked and dedicated quarantined stock area to add further control).

- Physically marking the stock with clear notices stating that the product is 'on quarantine'. Such notices should be affixed to the affected stock to prevent their movement and should be clearly visible from all approach angles. Ideally the notices will be coloured to attract additional attention.
- The quarantine notices should state the reason (or a 'reason code' to help ensure confidentiality if required) for the quarantine and/or should provide a quarantined stock 'unique identifier'/'reference number' which will lead the reviewer to the appropriate information/ paperwork on the product.

Stock control system quarantine will typically involve:

- Highlighting the quarantined stock status on the (often computerised) stock control system.
- Attaching information to the stock listing to provide the reviewer with instant information as to why the product is on quarantine and describe the further action being taken.
- On some computerised stock control systems, when the product is assigned to 'quarantine' status, the product disappears from the main stock holding screens and is only visible on a dedicated quarantined stock section/screen to prevent accidental use/selection. Other systems simply lock the quarantined product listing so that it cannot be subject to any movement/picking instructions until released by an authorised member of staff.

The key point is that each business should reflect on the layout, pace and staff working in their operation and apply the extent of quarantine stock markings/notifications sufficient to achieve the appropriate level of control/assurance.

Quarantined stock records: The business 'quarantine stock control system' should involve clear documented records for each item/product quarantined. These records should be continuously available to all relevant personnel (on the computer stock control system or via a written hard-copy file) and should clearly state the reason for the product being on quarantine, the further actions being progressed and the status of these actions (including times/dates and the key staff/department dealing with the matter).

All product actions should be noted on these records both for easy review by the business team members and also as an auditable due diligence record that all such stock issues are being thoroughly controlled with clearly recorded results/decisions and actions.

10.2.6 Control and decision making upon quarantined stocks

Due to the wide range of reasons why a product may be given 'quarantined stock' status there is usually no 'one-size-fits-all' approach to decision making on such materials. However, the following points are often relevant:

- Quarantined stocks are typically administered by the quality/technical department to ensure that the appropriate decision-making skill set is in control. This also provides customers, enforcement agencies and other stakeholders with added assurance of impartiality.
- To help ensure full and appropriate decision-making control, while anyone within the business should be able to place a product on quarantine, only the quality or technical manager/team should be authorised to then remove stock from 'quarantine' status.
- A decision on the appropriate actions to take with regard to the quarantined stock will typically depend on the outcome of an investigation/review/further tests. Sometimes the

quality/technical team will be comfortable with then taking the 'further action decision' independently and other times they may need the input of business team members such as the commercial sales/senior management team/directors (e.g., circumstances where the issue is deemed a 'minor quality issue', and yet if released may have the potential for some negative impact on the business).

- When a product has been quarantined due to a potential fault, decisions on quarantined stocks will often result in the product being either released for sale, reprocessed/reworked to correct the fault (or reduce the fault to an acceptable level), regraded, or result in the product being disposed of.

10.2.7 Typical pitfalls

Aspects that tend to 'go wrong' with regard to quarantined stock systems include:

- Not all of the product to be placed on quarantine has been captured, clearly marked, and placed in a segregated area (leading to accidental use/despatch of the product). When placing stock on quarantine staff need to be particularly vigilant to ensure that all product affected is rounded up, fully accounted for and clearly placed under control.
- Staff still pick from quarantined stock. For example, due to being unaware that the stock is on quarantine (maybe the notices on the stock have fallen loose, or a person/fault on the computer stock control system has accidentally released the product for use), perhaps due to a lack of training, rushing or due to a mistaken belief that the product is now no longer on quarantine.
- Stock placed on quarantine exceeds its use by/best before date before a decision is reached on whether it is saleable/reworkable/regradeable. For example, perhaps due to other quality/technical management priorities leading to a lack of quarantined stock focus, or due to the fact that the testing/assessment required to form a decision takes a long time. Therefore, at the point of placing materials on quarantine it is helpful to also assign a 'decision by date' to help the quality/technical team prioritise their work in progressing quarantined stock decisions.

10.2.8 Reporting quarantine stocks and use of quarantine data

There are a number of stakeholders within the business who need to be aware and kept up to date on quarantined stock movements:

Despatch/customer service: Need to be aware of what stock holding they have available for sale to the customer (and the likelihood of any quarantined stocks being released for sale).

Production planning and purchasing departments: Need to consider replanning stock manufacture to replace quarantined product if these quarantined stocks are not going to be released in time (if at all) to meet the customer orders.

Production/operations management need to be aware of the 'losses'/'costs' caused by quarantined stocks in order to then take further action within the business if necessary. For example, conduct root cause analysis and apply corrective actions; place additional focus on regular/repeated product faults and seek to reduce the size/volume of 'production trial' batches (which will subsequently always be quarantined) to reduce the ingredient, packaging and processing time costs of such trials.

End customer: Some customers (in their monthly supplier reports or during audits) will request that the supplier provides an insight into the amounts of quarantined stock being incurred/processed, as the customer can consider quarantined stocks as an indicator of the efficiency and focus in the business (i.e., if a lot of product is being placed on quarantine it can suggest a lack of sufficient control being applied on site).

Trend analysis: The reporting of quarantined stocks can provide a daily/weekly/monthly figure for the total amount of material on quarantine. Also to give a clearer reflection to the business of where the key quarantined stock issues/challenges are, a business often needs further information/analysis which may include:

- Definition of the main reasons why volumes are being placed on quarantine: Such categories can include: 'trial', 'positive release', 'process fault', 'ingredient fault', 'packaging fault', etc. Such additional trending/information over time will help to drive further corrective action on the issues that are causing the most losses within the business.
- Information on percentage quarantined stock eventually released, percentage stock disposed of, percentage stock reworked/regraded.
- Information on the turnaround timescales for decision making and further actions taken with regard to quarantined stocks.

10.2.9 Quarantine system summary

When addressing material/product faults the quality manager will focus on overseeing/conducting the 'issue investigation', arranging for any further product testing/assessment required and steering the decision-making process.

While progressing such matters to their conclusion it is vitally important that the product affected remains firmly under quarantine control to avoid accidental use/despatch of substandard (and perhaps in some cases potentially harmful) product. Therefore, controlling the quarantine stock control system, including administration of any subsequent release from quarantine (if the stock is later deemed suitable for use) or perhaps administering actions to either rework or dispose of the product, provides the quality manager with a useful tool via which to maintain control. The quarantined stock system also will provide the quality manager with highly useful information/data which can be used to direct and prioritise continuous improvement initiatives within the business.

10.3 Complaints management

Ideally products for consumption should be manufactured to such a consistently high standard that there would never be any requirement for complaint. In reality all foods from a wide range of competent manufacturers have the potential to cause routine product complaints, leading to the necessity for complaint investigation, root cause analysis, completion of rectification actions and formation of a formal customer complaint response.

10.3.1 Cause of complaints

There are many aspects of consumer 'product expectation' that can result in a complaint, including:

Harmful complaints: The first of these expectations is the reasonable expectation that the food is not going to contain anything harmful. Review of the standard 'food hazard groups' highlights a range of potential complaint issues including:
Physical – For example,'foreign bodies' including pieces of packaging (perhaps from the ingredients or the end-product packaging), metal, stones, glass, etc., could be present due to a fault occurring at some point in the raw material supply chain or in-house process.
Chemical – for example, pesticide or veterinary residue presence on the raw materials at point of use; cleaning chemicals not sufficiently rinsed away from product contact surfaces prior to commencing the processing of product, leading to a chemical taste taint remaining.
Allergens – for example, allergenic ingredients accidentally added to the batch as part of a recipe/process fault, leading to customer reaction to the allergenic material.
Microbiological – for example, customer feels ill after eating the product due to pathogen presence caused by insufficient processing or in-process/postprocess contamination. Another microbiological food safety hazard could be that the product has been incorrectly use-by labelled with far more shelf life than is appropriate (the first evidence of this may be off-flavours or signs of deterioration/mould growth).

'*Personal injury*': Can result, not just from the hazard categories above, but also via the handling and eating of the end product, for example:

- packaging becoming flexible/unstable during consumer heating causing spillage and burns;
- choking on a bone/gristle/fruit stone present in the product; and
- packaging so hard to open or so brittle/sharp that injury is caused during opening.

Food manufacturers often try to protect themselves from such complaints via on-pack warnings. For example, 'warning, packaging becomes flexible when hot, handle carefully at all times'; 'while all care has been taken, some bones may remain'.

Dissatisfaction complaints: Can arise when the consumer feels let down by the product. Such circumstances can arise for many reasons including

- wrong or unexpectedly low quantity of product in the packaging;
- product does not look like the picture shown on the packaging/in advertising; and
- customer preference… the consumer simply does not like the product.

With regard to dissatisfaction complaints, it is likely that the more a consumer has paid for a product, the higher the chance that the consumer is going to complain if there are any points on which the consumer feels let down. With higher prices come higher expectations. There are also some customers who enjoy complaining and are unlikely to ever be satisfied. A challenge to an organisation/individual working in customer services and quality assurance is often to avoid becoming cynical/blasé when answering complaints.

10.3.2 Compounding complaints

There are circumstances surrounding product issues or complaints that can lead to the consumer becoming even more irate and frustrated. Examples include the following:

- The nature of the product fault being a different recipe/product than the label suggests (perhaps due to a packing or labelling fault) leading to the product containing ingredient material deemed emotive/offensive to some groups. For example, meat found in what should be a vegetarian product; finding alcohol is unexpectedly contained within the recipe (e.g., perhaps a mix up when packing 'gravy' accidentally using a recipe for 'red wine gravy').
- Issue has caused physical injury or distress. For example, presence of a foreign body causing either a broken tooth or perhaps the distress of choking.
- Issue has caused an event to be spoilt. For example, perhaps the inedible/failed product was intended to be the centrepiece of a special occasion/party.
- Issue has caused damage to property. For example, a packaging failure fault leading to spillage in the car or on a costly carpet.
- Delay in responding to the complaint, or perhaps receiving no response at all.
- Answer to the complaint appears dismissive. (Team members responsible for answering customer complaints are well advised to bear in mind that while their business operation may make millions of meals successfully every week/month, each single complaint received is extremely important to that one person affected.)

10.3.3 Complaint management/investigation

Within a business the process of managing customer complaints is often the responsibility of the quality/technical department or a dedicated customer services team. Key points with regard to these processes include:

- For clarity and consistency, the business should hold and maintain policies and procedures related to the handling and management of complaints.
- Maintaining clear records and ensuring sufficient staff resources to enable swift investigation and response are vital factors when seeking to efficiently address product complaints.
- The outset of the business complaints handling system should involve the recording/logging of the complaint, ascertaining the full details of the issue including product specific name, lot/batch coding, complaint detail/circumstances, purchase location, etc.
- When commencing the investigation of a complaint it is helpful if an experienced team member can give due consideration to the appropriate approach and depth of the investigation, including setting the areas of the operation most likely to require review.
- It is always important that the complainant is kept informed with regards to the progress of the investigation.
- Trend analysis of the complaints received and of the subsequent root causes found is very important to help highlight whether there are any routine/regular issues which require addressing.
- It is important to take swift and thorough corrective action on the findings of the complaint investigation. This will help reduce the possibility of any recurrence. It is also necessary to swiftly inform the complainant as to the results of the investigation in writing.

- Types of business documentation which are helpful to support a business 'due diligence defence' in the event of complaints being received can include product quality assurance and HACCP (Hazard Analysis & Critical Control Point) check records, plus similar records from raw material suppliers. Traceability records for all raw materials, in house processes and end-product despatch/onward journey in the supply chain will also be vitally important to both the investigation and the demonstration of due diligence/control.
- In the event of a food safety issue it is particularly important that the relevant food safety procedures of the business are reviewed by the appropriate management team members to either confirm that these procedures are still sound or enable these checks to be reviewed and adjusted for correction/added assurance.
- If it is suspected that the root cause of the complaint may involve malicious contamination then it is important that the investigating staff member realises the importance of referring such a matter to the relevant authorities for further advice. These relevant authorities could include the local government food law enforcement officers and the police.

10.3.4 Root cause analysis

The 'initially thought' answer/reason for a complaint may not necessarily be the end of the matter. Root cause analysis is a term which suggests the need to investigate further in order to find the actual underlying (the root beneath the surface) causes of the complaint/issue, as only then can effective measures be taken by a business to help prevent a recurrence.

The core considerations from a root cause analysis approach are to ask/review
What is the specific issue/problem?
Why did it happen?

What actions can be taken to prevent the issue from happening again?

For example, a complaint of 'wrong food product present in the retail packaging' could easily be blamed on the production line operator who in preparation for packing the product placed the incorrect packaging in the packaging machine. However,

- the stock of product specific retail packaging may have been incorrectly labelled by the supplier; or
- the operator may have been following a production schedule which suggested that a different product was to be packaged first.

By digging deeper and not stopping at the first explanation, businesses and their control systems will become far stronger. Taking the opportunities to truly 'learn from complaints' is a vital factor within complaint handling systems.

Root cause analysis utilises 'cause mapping' methods and involves the following three steps to help address problem issues:

1. 'Define the issue': This can be achieved by stating how the issue has impacted on the business objective/goal (e.g., customer complaint has highlighted a failure to produce a foreign body free food). It is important to ensure absolute clarity on the specific issue to be addressed as business team members can sometimes have different perspectives/definitions of the problem.

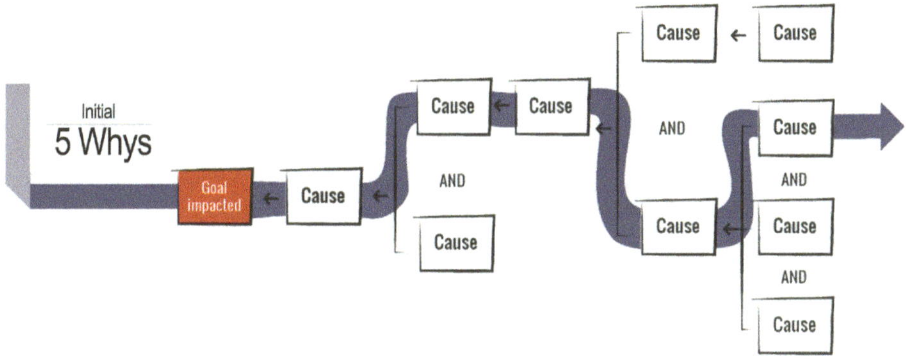

Figure 10.2 5 Whys on a cause map.
Source: Think Reliability Ltd. https://www.thinkreliability.com/cause-mapping/cause-mapping-method-explained/.

2. 'Break the problem down': The approach here is helped by asking 'Why' questions and capturing the answers/results in a visual map. Creating this 'cause map' helps to reveal and define all of the causes involved which resulted in the issue/problem. A 'cause' can be any factor which produced the effect (i.e., contributed to the issue/problem encountered).

The approach of asking the '5 Whys' (Fig. 10.2) is often applied here, the theory being that when investigating 'Why' something happened, it is only when the investigator has asked the 'but why?' question 5 times that they will truly find the root cause (or causes) of the issue and pave the way toward a clear and effective solution. For example, '1st Why' – Why is the manufactured chilli sauce too hot? '2nd Why' – Why was there too much chill powder in the recipe? '3rd Why' – Why was too much chilli powder prepared in the preparation room before cooking? '4th Why' – Why were the weighing scales inconsistent/inaccurate? By the '5th Why' it is found that the weighing scales had not been maintained and serviced in accordance with the site preventative maintenance schedule and supplier recommendations.

3. 'Formulate the best actions/solutions to help prevent (or reduce the potential for) a recurrence of the issue/problem. Such actions will be dependent on the particular problem and may often involve adjusting the working practices of the operation, or liaising further with other stakeholders in the supply chain.

10.3.5 Complaint response

Many complaints are still responded to by letter in the food sector, though emails are likely to continue their increase in popularity over the coming years. Ensuring that complaints are answered in writing is preferable in most cases (rather than responding verbally) as there is then a clear paper-trail/traceability linked to the investigation and clarity on what information has been corresponded to the complainant.

In general, well-written customer complaint response letters need to achieve the following:

- reply within an appropriate and fair timescale;
- convey to the complainant that the business has taken their specific complaint seriously;
- highlight that the business is sorry that they have felt the need to complain (whether the business decides to apologise for the actual issue and thereby potentially admit liability is another issue which can only be decided upon on a case by case basis);
- provide information as to the results/outcome of any investigation into the matter; and
- thank the complainant for taking the time to highlight the complaint which in turn will help the business to improve its processes.

The business may also choose on a case-by-case basis to compensate the customer for the complaint, perhaps via vouchers or perhaps via a cheque. This should be at the discretion of the technical/quality manager and will often be quantified by factors such as the nature of the complaint, how liable the business was for the complaint and the cost/inconvenience to the end customer.

'*Without prejudice*': Dependent on the country/law applicable on customer complaint responses the term 'without prejudice' is sometimes stated at the top of the response letter. This is a legal term via which the business is highlighting that it is trying to resolve the issue/dispute with the customer, but that if these attempts to resolve the issue to the satisfaction of the customer ultimately fail, the business does not expect the correspondence entered into (and clearly marked 'without prejudice') to be ultimately used as evidence in a court of law. Effectively the business is highlighting that any concessions or apologies made in the correspondence are being made in order to assist in bringing the matter to a conclusion which satisfies both parties, rather than conceding these points in legal terms.

It may be an obvious point, but businesses should always be cautious as to the information placed in complaint response letters, even if marking as 'without prejudice' as this does not automatically guarantee exclusion from a court case, in the event of the matter escalating that far. As ever on legal matters the advice of an expert should always be sought.

10.3.6 Involvement of enforcement bodies/local authorities

While there is a great deal of food law that can be directly related to the potential for food product-related complaints (product must not be injurious to health, must be fit for human consumption, must be of the nature, substance or quality intended, etc.), unless a complaint is very serious, enforcement bodies/local authorities are unlikely to become involved in a food complaint unless the matter has been directly referred to them by the complainant.

Direct referrals by the complainant tend to arise when:

- The complainant is unsure who to contact with regard to the complaint. (To help avoid such uncertainly a manufacturer should always mark on the packaging how to make contact 'in the event of not being entirely satisfied' by the product.)

- The complainant has become frustrated at awaiting a satisfactory (or any) response following making their complaint to the business.
- The complainant feels that the issue is sufficiently serious to involve the authorities.

Enforcement bodies/local authorities will usually have a responsibility to investigate any food complaints passed on to them by consumers.

Typically an enforcement official will investigate food safety/quality issues, with the key stages commencing with a receipt of complaint details from the complainant, inspection of the food complaint/gathering information via inspection and testing of the product and interviewing the complainant to ascertain further details if necessary (times, dates, proof of purchase, original packaging and a detailed written statement of the issue encountered, are all useful at this stage to support further investigations).

The enforcement official will usually have to quickly give consideration to the likely extent of the problem issue, sometimes before they have had chance to receive any further analysis results, and perhaps even before they have had chance to speak to the manufacturer. They will consider a range of points including whether the problem is likely to be widespread or just isolated to the issue reported so far? The officer will relate to relevant 'food law' and 'codes of practice' which will provide them with guidance on the appropriate steps to follow in considering such matters. If the complaint also appears to involve illness/food poisoning then the officer will also be likely to open a formal food poisoning investigation.

The officer will usually wish to speak with and (if sufficiently concerned) visit the food handling operation in order to review the approach and controls in place which are intended to help avoid such issues from occurring. They will then be in an informed position from which to take any further enforcement actions with the business or make recommendations to help avoid any recurrence.

10.3.7 The 'due diligence defence'

The 'due diligence defence' often cited in food law when recognised as a legally admissible defence in the event of prosecution for a food safety-related offence, in that the business will seek to demonstrate that it took 'all steps reasonably practicable' to prevent the issue from occurring.

A clearly considered and well implemented/monitored business Food Safety and Quality Management System should help provide a business with a robust due diligence defence, especially records related to the business HACCP system and prerequisite programmes including product/process control check records and operator training.

10.3.8 Management review

As part of the monitoring and control of customer complaints the site quality/technical management should be frequently reviewing the range of customer complaints received in order to then form appropriate action plans to deal with the issues being encountered.

In addition, routine management team review of the complaints received helps to increase site awareness/'buy-in' to the improvement areas being suggested, and finally 'senior management review' is also required upon complaints on a routine basis in order to ensure that complaint trends are being appropriately dealt with utilising the required resources.

To achieve a clear review of general complaint trends and also to enable further analysis of the key causes of the complaints received, clear graphing/display techniques are required. Businesses usually find it beneficial to generate a general trend graph of 'complaints per million units' sold (CPMU) over time (or complaints per thousand units sold for smaller businesses), and then tend to have specific charts focusing on particular complaint types where additional trends may be noticed.

10.3.9 Pareto analysis

'Pareto analysis' of graphed complaints data can provide a clear view as to which issues tend to cause the most complaints and therefore should be the main areas on which to focus resources for maximum benefit to the organisation. This approach is known as the 80:20 rule, the theory being that 20% of the business complaint categories are likely to be causing approximately 80% of the customer complaints. Therefore, in the case demonstrated in Fig. 10.3, out of the nine separate 'cause categories' of customer complaints, the pizza delivery business will be well advised to focus resources on particularly addressing two of these categories – 'speed of delivery', and 'temperature of product on delivery' (both of which may be interrelated).

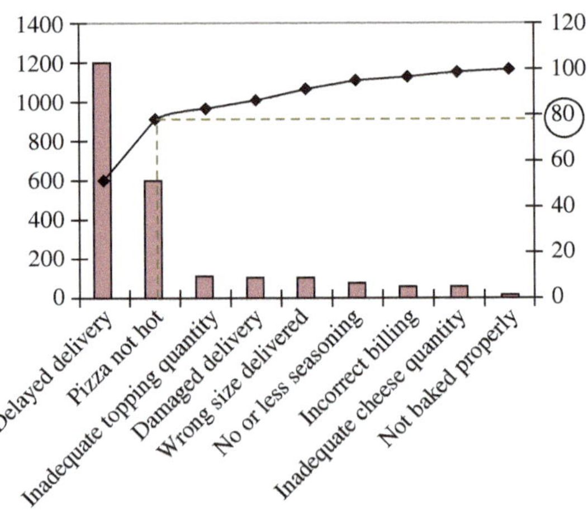

Figure 10.3 Pizza delivery service bar and line graph facilitating Pareto analysis of customer complaints. The bar represents the count of each complaint. The line illustrates the cumulative complaint count percentages.
Source: Discover6Sigma.org.

In summary: The effective management of customer complaints can be a great strength and add significant value to the ongoing business, as such processes can support and assure customers, highlight issues and ultimately lead to business improvements.

10.4 Incident management and product recall

Introduction: Sometimes in food handling operations by the time a nonconformance problem/issue has come to light the product is already in the onward food supply chain. These incidents therefore require quick and careful handling in order to bring them under control and will often involve processes of 'product recall'. 'Crisis management' is also considered in this chapter and is typically associated with the control of a wider range of business incidents and issues.

10.4.1 Product recall

The process of 'product recall' involves the removal of the relevant food product (or products) from the supply chain, including the storage and distribution phase, points of sale and use/consumption). Recalls are usually instigated due to a product or process fault which has led to the product having an unacceptable/incorrect quality or has led to the product being potentially harmful.

The 'fault' may often have come to light via the following routes:

- The food manufacturer: Perhaps due to product analysis test results received or a process control problem issue coming to light.
- The supplier base: Supplier of materials to the food manufacturer highlighting an issue with the ingredients/packaging which would render the end products harmful/unacceptable (e.g., microbiological, physical, chemical or allergen contamination).
- The end-product supply chain: The retailers, food service outlets or further processors highlight a product issue/fault while assessing/handling the product.
- The end consumer: Highlighting a product fault/issue, perhaps as a result of injury or illness caused by the product.
- Enforcement authorities: May have been informed of an issue via consumer complainants or perhaps as a result of food safety monitoring on a local/national/international basis.

Most food manufacturing businesses are involved in the sourcing of raw materials and the production, processing and distribution of finished goods/products. The product safety threats to such an operation are therefore most likely to arise from the major food hazard categories of physical, chemical, microbiological or allergen contamination.

10.4.2 Legislative considerations

In the European Union as of the 1st of January 2005 EU food businesses are legally expected to withdraw from the market any food that is unfit or injurious to health, and are also legally required to notify the 'competent authority' of the issue.

Before 2005 product recalls in the EU could be routine occurrences, however the 'competent authority' (in the United Kingdom typically a combination of the Food Standards Agency and the local Environmental Health or sometimes Trading Standards department) would often be unaware of the issue. The introduction in 2005 of Commission Regulation on the General Principles of Food Law (178/2002) meant that major issues related to the supply of food products should always be highlighted to the authorities and therefore increase the potential for focus to be placed on the manufacturer/issues in question.

Being directly notified of such instances helps the enforcement authorities to maintain focus on manufacturers in their area (providing an option to further review those operations which may be having problems), and is also a benefit at national level as such information can be collated in order to monitor for any national trends/issues and also to inform future food policy.

10.4.3 Responding to a food incident

If a food handling business has reason to believe that a product is unsafe/unfit for human consumption then all steps should be taken to prevent the product from being consumed and in many countries the business will be legally required to inform their enforcement authorities of the issue.

In circumstances where a product does not meet the requirements of 'food standards' requirements (e.g., labelling or quality faults), but where there is not a related food safety issue, the business may still decide to ensure that products are withdrawn from the supply chain, rather than risk prosecution.

Product recalls can often be categorised on two distinct levels:

'Trade withdrawal' involves the recovery of the product from the supply chain, up to and including the point of sale. For example, distribution centres, hauliers, retail shelves and restaurant/canteen stores. In these circumstances the withdrawal stops at this point and customers are not asked to return or destroy the product.

Consumer 'public recall' involves removing all of the affected product from the market. In the circumstances where the food has reached the customer/consumer, they must be advised to either return or destroy the product. Reaching the customer can be a major challenge and will often be via numerous communication channels including public notices, news features and social media. The recall advice given to the customer should always include the reason for the recall in order to convey the seriousness of the matter.

10.4.4 The recall process

While all food businesses will have differences in the products manufactured, supply chains and end consumers, and every recall scenario/issue will tend to bring its own specific challenges, the recall process will usually involve the following broad themes:

- The receipt of information which suggests that a product/batch may have such a serious fault/issue that it may therefore require recalling. Reporting measures should be in place to ensure that such information finds its way swiftly to the site quality/technical management, as they are typically expected to be best placed and qualified to lead the business response on such matters.

- A rapid investigation will need to take place to ensure that the initial information received is correct, and to ascertain the extent of the issue. For example, such measures may include retesting the product, inspecting the stock in question (both on site and in the supply chain), and possible further discussion with the complainant.
- Decision to be taken as to whether or not to recall. If the decision is to recall then there also needs to be a decision upon whether the recall will be a 'trade withdrawal' or a 'public recall'.
- The business will need to ensure that it meets any legislative expectation of informing the relevant 'competent authorities'.
- The relevant points in the supply chain will require contact with clear information as to the product, the issue, the batch codes, date codes, supply dates and quantities affected, in order to effectively progress the recall.
- A clear record is required to be kept of the whole recall process in order to help account for all stocks affected and in order to demonstrate 'due diligence' of the approach taken.
- Rectification actions in order to prevent a recurrence will be required to be taken swiftly and reported upon to all stakeholders (e.g., customers and the relevant enforcement authorities).

10.4.5 The business 'product recall plan/procedure'

The business product recall procedure is a document (ideally held in both printed and electronic format for speed of access by key personnel both on and off site) which details to the user exactly what a recall situation is, the thought processes to be followed in decision making and resolving the issue, logical step-by-step guidance on how to go about progressing the recall if required, and a full list of up-to-date business contact names and numbers (both 'in-hours' and 'out-of-hours') to be contacted in the event of the recall affecting them (typically affecting their supply chains/outlets).

The business recall plan should be developed and written by a multidisciplinary team drawn together within the business to cover aspects of:

- Technical/quality – Experts on the likely food safety/quality issues that could result in a recall, plus have appreciation of customer and certification scheme audit expectations of the site recall plan/procedure.
- Customer services/account management – Appreciation of the product supply chain and contact points.
- Senior management/managing directors – Can approve the proposed plan, serving to assure the technical/quality management that the steps to be taken in such circumstances are formally approved.

Traceability is a vital factor in all food business organisations, and particularly in the context of a food incident where it is vital for the effectiveness of the investigation and subsequent recall that the business traceability system yields quick, accurate information upon all end product, raw material and in-house process movements associated with the foodstuffs affected.

Alongside clear visibility/traceability of the supply chain, 'internal traceability' at the food handling operation is also vital to enable links to be quickly made between the despatch and sale of end products and the sources of the materials used to produce them. Such internal systems will facilitate increased consumer protection via faster and better targeted recalls/withdrawals.

To ensure that recall responses remain up to date and effective food handling operations should routinely consider and regularly review their:

- reliability of site traceability records (both internal and external);
- accuracy, speed and extent of traceability and recall procedures;
- due diligence steps taken to confirm the reliability of the traceability systems operated by suppliers to the business;
- capacity to narrow down issues/problems to the affected food product batch or batches (e.g., product batch/time coding and links to check/test records).

10.5 Crisis management

Crisis management procedures should be in place in food handling operations to help ensure a particularly well-focused management team response to situations of potential major business impact/disruption. Such situations can cover incidents related to the products being sold and also incidents related to any other major business impact such as fires, floods, terrorist/malicious threats, media expose's, etc.

In crisis management plans businesses can sometimes choose to categorise the nature of the issues being dealt with and link these categories to which team members should therefore be involved, as listed below.

10.5.1 Crisis management issue – level 1

- Incorrect product pricing/labelling
- 'Minor' product contamination issues (unlikely to cause serious injury)
- Minor issues involving enforcement agencies

It is accepted that these could be viewed as 'routine issues', which will as part of standard operations be handled by the quality or technical manager, but which may sometimes require the focus of a crisis management team in support.

10.5.2 Crisis management issue – level 2

Escalation of level 1 issues:

- health scares (e.g., related to food pathogen presence);
- product recall;
- serious quality issues causing significant customer concern/disruption;
- serious contamination issues; and
- serious enforcement authority issues.

Typically in such situations the technical/quality manager alongside the key business directors will form the crisis management team.

10.5.3 Crisis management issue – level 3

- Media involvement
- Serious injury/illness of customer/site employee
- Extortion
- Deliberate contamination
- Site utility failure (e.g., electricity, water, gas)

The site directors will progress the crisis management of such issues, drawing together a crisis team which they feel best comprise of the skill sets required to deal with the particular issues faced.

10.5.4 Crisis management issue – level 4

- Fire
- Explosion – including bomb threats
- Loss of life
- Terrorist threat
- Major environmental issue

Such incidents will typically require a full crisis management team procedure to be utilised.

The business crisis management document should detail all necessary actions and contact details. The document should be held in both printed and electronic format for speed of access by key personnel both on and off site.

It should be noted that as a 'business crisis' is not necessarily going to be food product related, it is not always the business technical/quality function that is expected to administer and maintain the crisis management plans for the business, often the role will be held by the human resources department.

With regard to incidents such as those listed in the level 4 examples, full crisis management plans will tend to assign a range of duties across the management team in order that in the event of a major incident, each management team member will know what their assigned role is and can swiftly commence action.

Roles assigned within the crisis management team/plan can include the following:

- Incident coordinator: Will take the lead role within the incident/crisis team seeking to monitor and coordinate the actions of the team.
- External relations team: Deal with the media upon any issues, arranging business statements. Ensure that the team can operate with appropriate communications resources (at the very least laptops and mobile phones, and ideally organise use of an appropriate location).
- Personnel team: To account for and deal with all staff matters, communicating personnel policy to all staff, and handling personnel issues.
- Salvage team: Required to attend the incident (if applicable) and liaise with the emergency services, assessing the damage and residual capability and recovery of salvageable equipment and stock.
- IT (information technology) team: IT team will agree information and hardware recovery for the business, will seek to reinstate key systems and intercompany communications.
- Planning team: Are required to assess business capacity/capability, review salvageable materials and stocks and liaise with business suppliers and contractors.

With regard to major damage/disruption to the business site the crisis management plans should contain details of disaster recovery measures and processes pertinent to a range of possible situations.

The business insurers will often want to understand the extent of business planning to address potential major crisis situations and can sometimes offer the support of guidance documentation or relevant staff to help develop such systems. This is in their interests as well as the business interests, as a well-managed crisis situation will typically limit the losses incurred by the business, and ultimately also incurred by the insurer.

In summary: Businesses cannot possibly foresee every potential recall or crisis scenario, but what they can do is set in place thorough and robust systems and procedures for the management of such instances. Advanced planning and briefing of key management team members can easily make the difference between a 'business disaster' and well a 'well managed, controlled incident'.

Further reading

British Retail Consortium (BRC) – https://www.brcglobalstandards.com/

BRC Global Standard Food Safety Issue 8 http://www.brcbookshop.com

Chilled Foods Association. http://www.chilledfood.org

Codex Alimentarius Site Covering HACCP. http://www.fao.org/docrep/005/Y1579E/y1579e03.htm

Codex Alimentarius Commission Food Standards: http://www.fao.org/fao-who-codexalimentarius/codex-texts/list-standards/en/

Food Hygiene (Basic Texts) Fourth Edition – Codex Alimentarius. http://www.fao.org/docrep/012/a1552e/a1552e00.htm

Food industry quality control. Mark Clute, 2009. CRC Press.

Food Quality Assurance, Principles and Practices. Inteaz Alli, 2004. CRC Press.

Hygiene for Management. R.A.Sprenger. 19th Edition, 2017. Highfield Publishing.

International Featured Standards (IFS) – www.ifs-certification.com

International Organization for Standardisation.

ISO 9000: 2015 Quality Management Systems: https://www.iso.org/standard/45481.html

ISO 22000 Food Safety Management: https://www.iso.org/iso-22000-food-safety-management.html

Sensory analysis for food and beverage quality control: A practical guide. Edited by D Kilcast, Consultant, Food and Beverage Sensory Quality, UK. 2010. Woodhead Food Series No. 191.

The BRC global standard for food safety: a guide to a successful audit/Ron Kill. 2008. Chichester: John Wiley & Sons.

Managing the technical department

11

Mark Swainson
National Centre for Food Manufacturing, University of Lincoln, Holbeach, Lincolnshire, United Kingdom

11.1 Introduction

The technical and quality functions of a food business are expected to provide 'quality, safety and legality' guidance and assurance across all areas of the operation. With this breadth of responsibility it is easy to overlook the fact that these are also business departments which themselves require effective manage'ment and coordination. Business technical/quality management need to be equally effective operating externally with all relevant stakeholders and internally within their departments, always ensuring the focus, commitment and performance of their own team members and managing any externally sourced services within their technical/quality remit.

11.2 Team management in the technical department

The role of managing the technical / quality function is very demanding, as the holder must be competent and effective across a very wide remit. From a detailed knowledge of the food science and technology relevant to their operation through to strong decision-making ability and interpersonal skills, this key leader is expected to be the figurehead for all business activities related to food quality, safety and legality.

In addition to providing technical direction and decision-making, a crucial function of this role is to manage the technical and quality-related personnel, and also to promote the furtherance of business goals in terms of food safety/quality/legality.

Depending on the size of the business, the market to which it caters and the quality/technical ethos applied on site, technical teams or team members will typically fall into the following broad categories:

- raw materials technologists;
- process technologists;
- site factory/process-related quality assurance or quality control staff;
- end product customer services and specification-writing staff and
- hygiene department staff (if applicable – in some businesses, the hygiene department is not controlled by the site technical function).

Swainson's Handbook of Technical and Quality Management for the Food Manufacturing Sector.
DOI: https://doi.org/10.1016/B978-1-78242-275-4.00011-3

Recruiting and developing employees with the right skills to fulfil such a diverse range of roles is a very important part of the technical manager's remit.

11.2.1 Building a technical team

A major factor in the success of most businesses is the competence and diligence of their staff members. Within food sector technical departments, the skills and experience required for each role are often specialist in nature. For example, raw materials technologists should have an in-depth technical knowledge of the range of raw materials (both ingredients and packaging) being used by the manufacturing business. They will also need good attention to detail and an appreciation of both the business manufacturing process and end customer expectations, in order to reflect these factors within specifications provided by the supply base. Good interpersonal skills are also required to liaise with suppliers, often in time-pressured situations such as requesting swift feedback on issues, or requesting raw material information in order to check that it will satisfy both business and customer requirements

By contrast, quality assurance and quality control staff will require an appreciation of food safety, quality and legality expectations, plus an understanding of how these must be met by the technical/quality systems of the food manufacturing business. An appreciation of food science and technology, internal auditing and good interpersonal skills are often important within these roles, as is the ability to take decisions and work quickly and effectively under pressure.

Staff who handle end product scale-up/launch (process technologists), customer services and specification writing require a thorough understanding of customer codes of practice and other expectations. Awareness of relevant legislative standards and the site food manufacturing operation controls will enable a thorough job to be achieved when completing production trials, customer specifications or information requests. In addition, due to the customer facing nature of these roles, very good communication and prioritisation skills are an absolute necessity.

Hygiene department staff require an in-depth knowledge of the food manufacturing site processes and machinery, together with robust awareness of the importance of cleaning and avoidance of cross-contamination. A thorough understanding of the appropriate application of cleaning chemicals, health and safety awareness, and personal protective equipment (PPE) requirements are also crucial.

11.2.2 Team recruitment considerations

It is vital that the technical manager is able to clearly understand and define the particular skill sets required for each role within the team, together with the ideal experience that should be held by the person undertaking each role. Creating a clear and considered job description will prove invaluable in both the recruitment and the future management of the individual. When seeking to build or add to a technical team, essential and desirable skill sets can be provided to a recruitment agency or written into a job advertisement, either of which should then provide a range of potentially suitable candidates. The technical manager will then be required to use interviewing skills to discuss the proposed job with each prospective candidate and assess whether they have the necessary technical knowledge, skills and ability to be effective and successful within the role.

The technical manager will also need to evaluate 'soft skills' such as the right personality to fit into the team and the ability to function well under various pressures including time limitations and customer demands. Setting competency tests or question and answer exercises can help to ensure a consistent assessment of knowledge, ability and other attributes across a range of candidates.

Personality assessment measures and interview techniques can be applied to help judge and select the type of character most suited to a particular role. They can also help to analyse whether an individual is likely to fit well within an existing team. Human resources personnel can advise on available techniques, such as psychometric and aptitude testing. Facilitators such as Belbin Associates (www.belbin.com) and Thomas International Ltd (www.thomasinternational.net) can also be consulted to help with this process.

11.2.2.1 Other recruitment considerations

Tenacity: A technical manager must have the ability to justify an increase in departmental staff or staff salary costs. Most business departments will say that ideally they would benefit from additional staff members. As technical and quality personnel are often viewed as 'indirect' or 'overhead' costs to the business, the technical manager must be able to build and articulate a particularly strong business case to justify additional staff members and secure approval to proceed with recruitment.

Recruitment costs: The process of recruiting staff can cost a significant amount of money. The technical manager will need to calculate and secure approval for recruitment costs, such as placing vacancy advertisements in local press and relevant trade journals. For particularly specialist roles which are therefore difficult to source, businesses could also utilise recruitment agencies to seek appropriate candidates. Such services are usually charged for upon securing the required individual.

Working within fixed budgets: It is an ideal position to be able to 'buy in' the exact skill set and experience required for a particular role. The new staff member can then deliver effectively against the key performance indicators (KPIs) of the role almost from day one. However, it is often the case that, due to business circumstances or departmental wage structures, there is not a sufficient budget available to be able to hire the individual the manager would ideally wish to place in the role. In such cases, it is essential that the technical manager is able to compromise. This might involve hiring a less experienced team member on the understanding that they will need to be developed over time to acquire the ideal skills and ability required.

Such an approach requires the technical manager to accept and progress an increased mentoring and training role while developing the staff member. It also requires the manager to accept that in the short term they (or other members of the team) will need to personally carry more responsibility and perhaps fill some 'knowledge gaps' in support of the newly hired individual.

Effective induction into the business: Introduction to employment is a vital time for a new technical team member. From their very first day a new recruit will benefit from feeling valued and supported. Therefore a well-planned induction programme should be designed to provide a clear briefing on all organisational policies and procedures, plus a thorough introduction to all key business team members, areas and departments. During the induction process – and possibly on an ongoing basis – it is often beneficial to appoint a 'buddy' or 'mentor' to the new employee in order that they have a point of immediate contact and support during this early, and potentially uncertain, phase.

11.2.3 Managing the technical team

A business may have world-class technical systems and procedures, however, if these systems are not maintained by the correct level of committed technical staff, they will fail to assure product safety, quality and legality. When managing a technical team it is therefore vital to safeguard staff quality and commitment. No matter what role a staff member holds, it is critical that the technical manager is able to make every individual feel valued, developed and appreciated. This is the best way to generate focus and commitment.

Technical managers can go some way toward achieving this by taking a genuine interest in the work of each team member, and routinely highlighting the importance of this work to both the individual in question and others within the organisation. Celebrating and highlighting the successes and good work of team members is a significant factor in individual and team development. This approach can also be used to raise the profile of the technical department across the wider business.

The opportunity to learn or practice new skills can help staff to feel that attention is being paid to their development. This is especially the case when development has been requested to help underpin an individual's current role or advancement within the business. Such training opportunities can often be provided internally, utilising existing skill sets within the business or team. However, where the training requirement is new, or staff time is precious, the objective can be achieved using external sources, for example, by sending team members to off-site training courses.

The technical manager provides the interface between their staff and the higher tiers of business management. Provision of business information to the technical team will be seen as an important communication factor by team members. Regular team briefings are an ideal tool for the technical manager to use to bring the team together and ensure that everybody is clear on current business performance, challenges and future plans. Such meetings can also be used to ensure that members of the team understand and appreciate the work and priorities of colleagues, sometimes with the benefit of being able to help support each other in times of peak workload.

It is vital that all team members feel supported. When workloads are particularly high, staff may value direct help or prioritisation of tasks. In the event of an individual encountering negative feedback on the quality of their work, a good manager will be able to judge whether the criticism is justified and whether the team member requires assurance or perhaps further development/skills training to help address the issue.

Other circumstances in which a team member might need support include times when the team member would benefit from flexibility in their work pattern, perhaps working particular hours for a few weeks while caring for dependents at home or dealing with a family issue. High levels of managerial support will typically be repaid in terms of staff loyalty and commitment. This can be a powerful basis for achieving and maintaining a strong team ethos within a department.

11.2.3.1 Technical staff appraisals and reviews

Work recognition and feedback should be provided on an ongoing basis. However, taking the time to conduct a formal appraisal process can provide a very useful routine focal point at which to summarise, take stock of current performance and agree future targets and any associated development goals or training needs.

Key themes when formally reviewing a team member's performance include:

- Encouraging the individual to reflect and feedback on their own views regarding their performance within the role over the agreed time period.
- Reflecting on performance against clear goals/job description points.
- Considering how to best align personal performance with current and predicted business requirements.
- Setting clear action plans for any improvement points required (including timescales for further follow-up/review).
- Planning further personal development/training requirements to underpin current or future roles within the department/business.

Development initiatives and goals set should always be 'SMART', that is, specific, measurable, achievable, realistic and timely. Such goals or initiatives should always be recorded to avoid any ambiguity and followed up at regular intervals to maintain momentum and focus.

11.2.3.2 Managing substandard performance

Having taken all of the right steps in the team planning, recruitment and staff development phases, there will be points in a technical manager's career when a team member's performance is unacceptable and requires action to address the situation. There is an endless list of possible reasons why a staff member's performance may give cause for concern. The issue could be absenteeism, quality of work, misuse of company resources or the staff member's attitude toward other members of staff. It is at this point that managers who have allowed themselves to become too friendly with their team may regret not maintaining a polite and professional distance, as a firm discussion on the nature of the performance issue is now required. The technical manager will also need to make clear to the individual the departmental or business expectations to resolve the issue swiftly.

The manager should always reflect on their self first in this situation considering whether they have done everything that they could have and should have to help ensure that the team member's performance was effective and up to the required standard? If the answer is 'yes', then the manager should feel comfortable in placing the onus upon the individual to improve their performance. However, if on reflection it is evident that the individual has perhaps been placed in a position of responsibility too soon, has been overloaded or has not received sufficient training or development to be able to consistently fulfil the tasks being expected of him or her, then the manager must address all potential root causes before judging the adequacy and ability of the team member in question.

Underperformance can be challenging to manage, as due to the nature of the business the process will often play out under the scrutiny of other departments and managers. The technical manager therefore needs to be calm and controlled while acting swiftly to address the issue. Support systems should be put in place to ensure that the quality of the individual's work is monitored and supported by other team members while they are being brought up to a position of self-sufficiency.

Unfortunately, businesses/departments can rarely afford to carry underperforming staff. In fact such an approach would usually be detrimental to other team members. If an individual is still not meeting the job role requirements after skills, training, ability and workload have all been allowed for, the technical manager may be expected to manage the individual out of the role, possibly out of the department and maybe even out of the business. This can be approached via formal capability/disciplinary processes or via a confidential discussion outlining how and why the role is not working out, to the extent that the position has become untenable.

Assuming that there is no negligence on the part of the individual, and the manager is convinced they are genuinely a committed worker who just is not right for their current technical department role, a good way forward could be to try to find the individual a more appropriate position within the department or business. Such a move should always be conducted with the addition of a new probationary period to ensure that the individual will be closely managed and reviewed to confirm their suitability to their new role before being permanently placed.

11.2.3.3 'People skills' and pressure

Technical team members will often be in time-pressured roles. Because these roles usually impact on the quality, safety and legality of the product, and can often be customer facing, the potential for stress to impact team members is very high indeed. The technical manager should therefore always bear in mind that pressure is likely to affect team members in different ways. A high workload coupled with tight timescales and unexpected problems is likely to pose a significant test of character for some individuals. An important aspect of the technical manager's role is to be a strong, focused and calm performer under pressure. However it is vitally important that the manager does not assume that all of their team are equally capable of carrying out their roles consistently when operating in very challenging conditions.

Colleagues who work closely with each other will gain an insight into how their fellow team members react in very demanding circumstances. If there is any doubt as to the ability of an individual to respond well to pressure, the technical manager may consider providing coaching, training or tips on handling difficult situations. Such development could be provided by more experienced members of the technical team, or perhaps using books on pressure and stress management. External 'management coaching' consultancies are also available to support. Ultimately, those who cannot cope and make mistakes when under pressure should not be placed in roles where they will regularly encounter such challenges, both for the good of their health and for the benefit and protection of the business. It probably comes as no surprise that when choosing candidates for progression in a technical team, individuals are often selected not only on the basis of their skill sets and knowledge, but also with due consideration of their ability to respond effectively under pressured and challenging circumstances.

11.2.4 Summary of considerations for technical team development

A food manufacturing business can only be successful if it is staffed by competent, hardworking team members. Building and maintaining an effective and efficient quality or technical team requires strong technical management skills, and much of the 'magic formula' for doing so can be defined by the following ingredients:

- Take the time to define job roles and related skills, qualities and experience, and follow this template to appoint the appropriate people.
- Develop each team member to optimise their performance within their current role.
- Envision and prepare team members for their development/future roles within the business.
- Monitor all staff performance closely against agreed KPIs and swiftly address any potential issues/shortfalls.
- Always be supportive and demonstrate good leadership skills.
- Lead by example.

This is all very easy to say but can often prove difficult to achieve consistently in practice, especially due to the many other calls upon a technical manager's time and availability. Experience, as well as building sufficient technical resources over time, will certainly help in this regard.

11.3 Technical budget management

As food manufacturing operations vary significantly from business to business in terms of reporting structures, range of departments and accounting practices, there follows a general guide to the setting and management of financial budgets related to the technical function. This section should be read with flexibility, bearing in mind that individual organisations may differ in their approaches.

11.3.1 The technical function budget

Depending on the business structure and accounting/budgetary preferences, the site technical function budget is likely to comprise some or all of the following sections:

- staff-related costs,
- Transport and travel costs,
- Information technology (IT)/mobile phone costs,
- Departmental administration costs,
- Memberships and subscriptions,
- Laboratory/analytical costs,
- Technical equipment budgets,
- Consumables,
- Customer or third-party technical audit costs and
- Potentially also other associated departmental costs such as a hygiene department.

Reviewing these budgetary areas in more detail, the following are examples of the wide range of considerations required when managing the costs of such technical function responsibilities.

11.3.1.1 Technical staff costs

Technical staff wages:

Payroll accounting terminology can sometimes divide technical staff into two categories: 'direct costs' and 'indirect costs'. 'Directs' tend to be team members who are directly linked to the production operation. It is often easier to justify increases in 'direct' staff numbers when the business expands, as these roles are usually linked directly to operation size and output volumes. Examples of 'direct cost' staff include factory production line quality assurance operatives.

'Indirects' are employees who do not directly produce products, but who make their production possible or more efficient. These tend to be office-based staff, whose numbers are not linked directly to the size of the manufacturing process. Examples of 'indirect' staff in the technical function could include raw materials technologists and customer service technologists.

Company budgets are often tracked and presented on a monthly or financial period basis. It is good practice for the technical manager to monitor their departmental wage bill and track actual performance against predicted monthly budgets, routinely reporting on any under or overspend. If the business intends to pay a bonus to staff at particular points during the year, it is good accounting practice to accrue (build up over time) funds for the bonus to be paid at the appropriate time. Sometimes businesses will prefer to accrue bonus payments in central accounting pools. However, if bonuses are set at departmental level and awarded on the basis of departmental performance against KPIs, then it may be more appropriate for the technical manager to accrue funds for bonus payments to staff within their own budget pool.

While some managers may expect pensions and other wage-related costs to be combined and included within the overall staff wage costs, budget holders are well advised to check in advance of finalising budgets. Business accounting teams often prefer to separate these costs from the main salary figures. Typically, it all often depends on how closely the departmental budgets are integrated with the master budget files held by the accounts department.

Recruitment costs:
Technical staff can cost a significant amount of money to recruit, due to the skill sets, qualifications and experience required. Trade press and local press advertisements can be expensive, and with the cost of recruitment agencies sometimes being charged at 15%–20% of the first-year salary, any recruitment proposals for the year ahead should be underpinned by budget planning for the true costs of filling these roles. Agency labour or consultancy costs may also have to be factored into the technical budget if it can be foreseen that such services will be required, for example, at particularly busy times of year, to provide holiday cover or to write urgent customer launch specifications linked to new business initiatives.

11.3.1.2 Transport/travel and subsistence costs

Technical functions within food manufacturing businesses often have significant potential to incur travel costs. Therefore, the following costs will need to be factored into the technical function budget and split into the predicted period expenditure totals.

- *Company car lease/hire* provision to managers can be a part of the role salary package. Fuel costs can also sometimes be integrated within this. Some businesses prefer to have such costs aggregated within technical staff costs rather than reported within a separate transport and travel section.
- *Cost of occasional hire cars, trains, air fares, parking, tolls, etc.* This is particularly relevant for audits or visits to suppliers, training conferences and customers.
- *Hotel and subsistence costs* also require factoring into such activity.

Staff events/welfare:
Depending on the business ethos, there is sometimes the opportunity to also accrue departmental budgets to cover staff events such as team-building sessions, or to pay for an end of year party or meal. Such benefits can be contentious within businesses if one department has allowed for funding and others have not. It is therefore usually good practice for site directors to define the company position during the budgetary process, and to ensure that this position is consistently adhered to by all departments.

11.3.1.3 Information technology and mobile phone costs

With 24/7 Internet access and mobile contact a standard part of 21st century technical department life, the technology to enable such 'connected' working practices can add a significant cost to the business. Depending on the business position, information and communication technology (ICT) costs may be allocated to a central IT department. However, these cost centres are often devolved to the departments that tend to use them the most. Businesses may also provide key staff with home working facilities such as a home phone line, broadband and a desktop computer (though laptops are often preferred for portability). All such costs should be considered when setting the technical budget for the year ahead.

11.3.1.4 Departmental administration costs

It can be surprising in this digital age just how much a technical function can incur in printing and paperwork costs. Raw materials and end product specifications often have to be printed and placed on file or posted to customers and suppliers. Lengthy supplier self-audit questionnaires may be completed and submitted digitally but will often still require printing out for checking and signing off by the relevant technical team member. The costs of printing, paperwork and postage therefore need to be factored into departmental budgets.

11.3.1.5 Memberships and subscriptions

In support of their food safety, quality and legality remit, technical departments will often maintain memberships of trade associations and research/consultancy organisations. Subscriptions may also be set up for routine services and updates such as food law and research news bulletins. In addition, it is sometimes the case that a business will pay for an employee's personal subscriptions to sector-related organisations, especially where such engagement confers advantages to the business as well as to the individual.

11.3.1.6 Laboratory and analytical costs

For many food manufacturers, analytical testing is a significant cost of new product development (NPD) and ongoing production monitoring. These services may be provided by contract laboratories or in-house testing facilities. Typically, services are charged on a 'per test basis'. Therefore, the extent of the costs incurred is usually related to the range of tests required (sometimes termed a 'test suite') and the frequency of the testing, both of which are often customer-dependent factors.

Technical managers will need to consider the following.

Microbiological testing budget:
Product microbiological testing can include:

- NPD scale-up work, ongoing production and environmental monitoring.
- process validation work, for example, microbiological testing of trial products to confirm that process machinery is working correctly.

Such testing may require a separate budget to be set at the point of the new machinery or new process initiative being considered. This will provide full cost visibility and also avoid impacting on the accuracy of the NPD and production testing budgets.

Chemical testing budget:
Chemical testing could include:

- nutritional tests for pack declarations and specifications, for example, energy, salt, carbohydrate, fibre, fat and protein;
- vitamin testing, for example, where specific product claims are being made;
- monitoring of product chemical levels, for example, maximum residue levels (MRLs) for pesticides on produce or veterinary medicine residues in meats and
- meat speciation testing.

For increased visibility and control of costs businesses will often split chemical and microbiological testing into a number of subcategories. For example:

- Prelaunch product development/scale-up testing, for example, in-house trials, customer assurance/due diligence and packaging information-related.
- Postlaunch factory product monitoring, for example, batch microbiological testing and routine product nutritional analysis/monitoring.
- Environmental monitoring, for example, the use of swabbing and air plate techniques to monitor the microbiological integrity of the factory processing environments on a routine basis.
- Raw materials monitoring, for example, routine testing/surveillance of raw materials (ingredients and sometimes packaging) to ensure that microbiological and chemical criteria are being maintained to specification.
- Postlaunch product surveillance costs. Major retailers and food service groups can instruct their product manufacturers to meet the cost of their products being routinely independently tested by particular laboratories (usually nominated by the customer). For example, this may involve a 'contract laboratory' being paid to sample the company's products in-store or in-depot. The laboratory will conduct a full microbiological and chemical analysis, report the results against the customer specification to both the manufacturer and the customer, and then charge the costs to the manufacturer.

Clearly maintaining separate cost categorisations is critical to ensure that budgets relating to all aspects of product testing can be monitored. Difficulties in setting such budgets accurately can easily occur. For example:

- A company that has a strong focus on NPD will often require a great deal of underpinning microbiological shelf life work, which cannot be easily predicted at the start of the financial year. For example, the business may not be dealing with a particular customer at the start of the year, but then take on a technically demanding customer in-year. Chemical and microbiological testing budgets can soon feel the strain of these additional requirements.
- Where businesses implement new processes or machinery in-year, which require extensive validation review via product testing.
- Process or product compliance issues can lead to a significant number of extra raw materials and finished product sampling / assessments being undertaken, in order to identify and address route causes and provide assurance that ongoing control has been achieved.
- Changes to customer policy with regard to the requirement for third-party product surveillance testing – and the frequency and extent of such testing – can quickly impact on budget accuracy and costs.

11.3.1.7 Technical equipment budgets

Technical equipment budgets can be broken down into purchase costs and ongoing maintenance and repair costs. Equipment falling within technical department budgets might include, for example, temperature probes; thermal process validation equipment; swabbing and hygiene assessment equipment; texture and viscosity measures; colour charts; pH meters and tools such as microscopes and digital cameras.

11.3.1.8 Testing consumables

Testing consumables also need to be considered as an ongoing expenditure. pH meter buffer chemicals, microscope slides and rapid hygiene ATP swabs are just some examples of materials that could be planned for within technical expenditure budgets.

11.3.1.9 Customer/third-party technical audit costs

The cost of third-party audits is a substantial component in many technical budgets, especially when auditor travel and accommodation have been factored in. Related 'unknowns' at the time of setting the annual technical budget include additional customer expectations. This is especially true in the case of new customers, who may require the manufacturer to cover their audit approval and ongoing monitoring costs, or perhaps require new standards to be applied to the site, such as organic, halal or kosher approval.

Audits of the manufacturing operation can also incur preparation and entertaining costs such as lunch and evening meals for the auditor. All of this can certainly add up if a business is routinely reviewed by a wide range of customers, third-party auditors and enforcement officials.

11.3.1.10 Pest control contract

When assessing the cost of pest control, the technical manager should always take time to clearly consider and define the range of pests and extent of service/review visits covered by the site pest control service contract. Additional factors such as whether the pest control budget includes electric fly killer (EFK) hire can make a big difference to the overall cost. Some businesses lease their EFKs from the pest control company whereas others purchase EFKs outright and then pay the pest control contractor a fee to service the units.

Additional visits or treatments due to unforeseen pest activity can quickly cause overspending within this cost centre during the course of a financial year.

The subject of pest control is further covered later in this chapter and also in Chapter 7, Operations and process control.

11.3.1.11 Training budget allocation

Training plays a key role in both the development and retention of technical staff. A good technical manager will have a professional development plan for each of their staff members. Often this will involve the staff member being given access to a wider range of experience and perhaps mentoring by more senior or experienced colleagues. However, in some cases it may be necessary to send staff on technical courses to update their knowledge or learn new skills. The provision of training courses is a key investment, which the technical manager can use to ensure that their staff feel supported and appreciated.

Many businesses have centralised training budgets, typically held by either the human resources (HR) department or the site training department. However, due to their specialist nature, training courses for technical staff often cost far more on average than training courses for production related staff. This can cause problems, as the centralised (nontechnical) budget holders may not allocate sufficient training funds for the technical department due to a lack of appreciation of the 'going rate' for specialist training. To avoid such scenarios, some technical managers prefer to hold a specific budget allocation within the technical function budget for their staff training. By adopting this practice, the technical manager can ensure that sufficient amounts are justified and allocated each year. All promises made to staff (perhaps during annual/routine reviews) for further development can then be delivered on, rather than having to explain that the central training budget has run out which can have a demotivating effect.

11.3.1.12 Budgeting for the costs of customer complaints and recalls

The business costs of customer complaints and recalls usually tend to be absorbed within commercial/marketing budgets as part of the ongoing costs and risks associated with supply. When supplying major retailers, such costs can be very high indeed if a company is manufacturing and supplying for a retailer's own brand. Wherever such costs are placed, given the major cash flow impacts of a high volume of complaints or a recall fine, it is prudent for businesses to accrue funds and particularly to have a plan in place should the worst happen, and for technical managers to be aware of this.

Business insurance can also play a major role in easing cash flow during recall circumstances. However businesses must be careful to ensure that they have an appropriate level of insurance, which covers all recall costs, as some policies only cover the basic recall fine costs without consideration of the overall impact on the business, including costs to remake/resupply.

Customer complaints can incur the cost of reimbursing or compensating customers and can also include the cost of investigating complaints. For example, in some cases, additional product or process analysis may be required.

11.3.2 Associated departments: the hygiene department budget

Depending on where the responsibility for hygiene is placed within an organisation, the hygiene department budget could sit within either the technical or operations budget. It could also be a completely separate cost centre. Annual/ongoing hygiene department cost considerations may include:

- *Direct and indirect staff costs.* Sometimes hygiene departments will also differentiate between factory cleaning staff and office cleaning staff.
- *Cleaning chemical supply and related services.* Often a major proportion of the hygiene department budget, the costs of cleaning chemical supply, are set within annually reviewed contracts with major hygiene service suppliers. Sometimes food production businesses will commit to longer term contracts in return for an agreement that the cleaning chemical company also provide supporting equipment free of charge. This includes bulk

tanks/bulk handling equipment (bunded areas, pumps, pipework, etc.), which forms part of the contract negotiations.

- *Consumables*, for example, cleaning PPE (Personal Protective Equipment).
- *Contract cleaning services*. Separate hygiene businesses may be utilised for additional deep cleaning or specialist jobs, for example, "out of hours deep cleaning", high access cherry picker/scaffolding jobs, lift pits or confined working conditions.

11.3.3 The budget setting process

The typical business financial year runs either from April to April, or January to January. A number of months before the financial year is due to start the business accounts team or finance director will ask all department heads to submit their budget proposals for the year ahead. Accountants need this projection data, not only so that they can review and attempt to reduce costs where possible, but also to facilitate forward planning of business cash flow/expenditure requirements. Directors can then model and arrange any further finance that may be required from investors or parent group companies within the coming year.

The high level of scrutiny to which budgets will be subjected, means that in all likelihood the first technical budget submitted will be challenged on both 'accuracy' and 'necessity' grounds. Therefore, the technical manager should ensure that all figures are based on a series of clearly documented facts or assumptions. Such a document is called a 'budget narrative' and can be submitted alongside the proposed budget facts and figures to provide additional justification/assurance.

There are manifold potential budget cost drivers for the technical function. Some examples are given in Table 11.1.

Table 11.1 Key technical function budget drivers

Business factor	Example aspects of technical budget impacted include
Predicted sales volumes for the coming year	Microbiological and nutritional testing budgets Direct factory quality staff numbers Cleaning chemical utilisation and hygiene services
Number of new products to be launched	Microbiological and nutritional testing budgets related to NPD Amount of process technologist time required
Increasing/decreasing hours of business operation	Direct factory related staff costs
Increasing/decreasing raw material supplier base	Number of staff, for example, raw materials technologists
Increasing/decreasing customer range	Number of staff, for example, specification/customer service technologist resources

11.3.3.1 Budget progress reporting and reforecasting

Although financial focus varies from business to business, it is always good practice for the technical manager to conduct monthly or quarterly reviews of their department budget, to check how it is tracking against predicted expenditure. This is beneficial as it ensures that any overexpenditure is highlighted early, which means that root causes can be identified and addressed. In practice, this will mean either reduced expenditure, or where outlay is found to be unavoidable, accountants may simply accept the overspend.

An example of a 'good overspend' could be that the business has sold far more product than anticipated. Therefore, the technical microbiological testing cost is higher than expected to cover the additional volume. This is typically a good problem as the extra sales should easily cover the extra testing expenditure. An example of a 'bad overspend' could be where the budget is found to be incorrectly set and does not allow for the true costs encountered due to a lack of sufficient budget preparation or accuracy faults (Fig. 11.1).

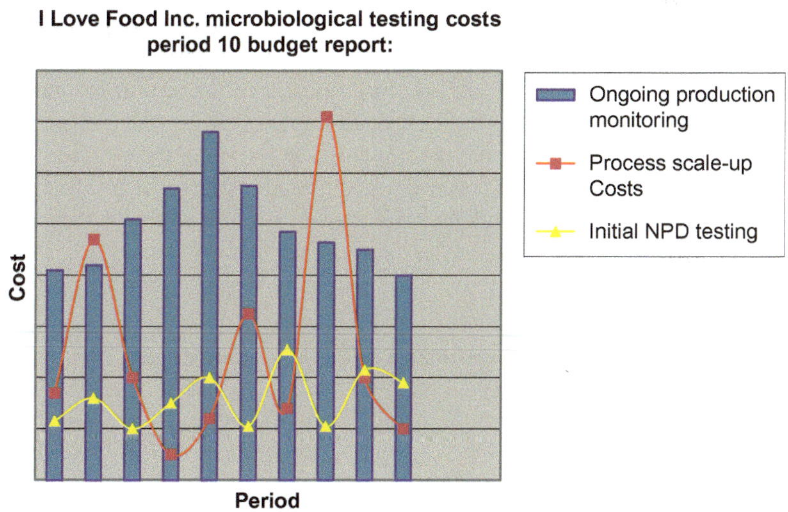

Figure 11.1 Example of a basic Excel spreadsheet generated graphical report which could be used to report business microbiological expenditure by financial period.

11.3.3.2 Higher sales volumes/new customers: budget impact analysis and forecasting

When a food manufacturing business seeks to secure a new major customer there are likely to be a range of budget impacts that could not be foreseen at the start of the financial year when budgets were originally set. A new customer will bring increased business costs in terms of ingredients and packaging, production time and transport. These increases should all be easily accounted for in the purchasing, production and logistics cost centres, provided that the business and its suppliers have the capacity to produce the increased volume or range of products. However, with regard to the technical function, assessing and planning the budgetary impact of new business may not always be so straightforward. For example, different customers will have varying levels of expectation with regard to the following:

- Site audit/accreditation standard expectations. New standards may need to be achieved which may well require further site investment and audit costs.
- New launch costs associated with NPD/scale-up testing.
- Product testing criteria (e.g., range of tests, frequencies, third-party surveillance).

It is important that the technical manager identifies and calculates the true costs of supply within their function when assessing the impact of taking on a new customer.

11.3.4 Summary considerations related to technical budget management

Businesses whose finance or managing directors gain a reputation for setting unrealistically low departmental budgets by slicing a significant amount off every budget proposal submitted should not be surprised when department managers become aware of their 'standard reduction' approach and automatically inflate their budget requirement predictions to allow for the 'imminent axe'. The phrase 'garbage in, garbage out' is very applicable to the budget setting process.

Taking the time to gather robust forecast sales and volume figures from relevant departments and ensuring that all key supply contract costs are firmly 'set in stone' will help ensure that an accurate technical budget is set and tracks well against predicted spend throughout the year.

11.4 Provision of customer 'technical services'

11.4.1 Introduction

There are a vast range of interactions that technical/quality departments and their managers can have with external customers.

Technical focused 'customer services' can include:

- product specifications;
- response to complaints, for example, investigation reports and responses;
- 'issue management', for example, incidents, recalls and crisis management;
- audit service provision, for example, arrangement of customer audits or required third-party audits;
- NPD and launched product review sensory panel attendance;
- creating and checking product marketing claims, for example, 'low fat', 'high fibre';
- packaging design initiatives and innovation;
- provision of customer technical KPI reports, that is, completed monthly/quarterly customer technical reports covering performance against preset customer KPIs such as microbiological testing data, complaints trending, development activity, etc.;

- certificates of conformance (CoC) and certificates of analysis (CoA);
- rapid response to emerging issues requiring information/feedback; and
- keeping stakeholders generally informed, for example, customers and enforcement authorities.

11.4.1.1 Provision of specifications

Specifications are a fundamental aspect of a business relationship with the customer. This is especially true when the food manufacturer is producing a bespoke product on behalf of the customer or supplying another food processing operation with an ingredient. In such circumstances, the specification is effectively the blueprint for what must be supplied. Consequently it forms a legally binding agreement, and a point of reference in the event of an issue or dispute arising.

Specifications can influence or assure a number of different business/product areas, including:

- Product 'pack copy brief' and artwork for retail packaging. The specification is typically used to inform the pack copy/artwork team of the key details to be printed (e.g., product name, product description, weight statement, ingredient declaration, allergen statement, pack warnings, nutritional data and product claims).
- Product safety: The specification can contain statements on product safety assurance (e.g., critical control points (CCPs)/application of hazard analysis and critical control point (HACCP)), product warnings (e.g., bone warnings, allergen warnings), microbiological criteria and limits, and legislation adherence (e.g., pesticide and veterinary MRLs).
- Product quality: for example, organoleptic description. This often includes a definition of the expected appearance, aroma, taste, texture, colour, viscosity, etc. This explanation should also contain tolerances in order to ensure that the quality specification is consistently achievable, allowing for acceptable variations in raw materials and processes.

As the product specification is such a critical and binding document its accuracy is vital. However, as the specification drives product customer approval and related artwork, it is typically compiled and approved to strict deadlines. Consequently, this necessity for accuracy coupled with strict time pressures can be a very challenging combination, potentially leading to problems such as:

- collation and inclusion of incorrect allergen data, leading to missing (legally required) allergen warnings from the packaging and
- incorrect ingredient declarations or weight statements, rendering the product illegal.

Not only do these examples place the consumer at risk if they go unnoticed, but once the error has been picked up it will cost the business dearly in terms of product recall and either disposing of incorrect stock/packaging or correcting (perhaps by 'overstickering') the incorrect packs.

Given the high costs associated with specification inaccuracies and the speed at which specifications are expected to be completed, many food businesses seek to automate the process via the use of specialised specification systems. These link raw material specification data and preset process/product information, to form end product specifications which can be supplied in a customer's preferred specification standard format.

Specifications Achilles' heel: No matter how good the software and no matter how competent the specifications technologist, the key to compiling a fast and accurate end product specification is to ensure that accurate raw materials information is collated and held on site or is swiftly available from competent suppliers. Key ingredient information used within an end product specification includes:

- detailed ingredient declarations;
- allergen status data;
- 'suitable for' data (e.g., vegetarians/vegans/halal/kosher) and
- product safety information (e.g., food safety systems and tests applied).

End product specifications cannot be completed without the necessary details. This can lead to problematic delays or – worse still – time-pressured staff being tempted to make a 'best guess' on the required data, convincing themselves that they will update the specification when the data finally arrives from the supplier. This can cause major problems with inaccuracies, often leading to costly packaging reprints, product safety or legality flaws and consequent product recalls.

11.4.1.2 Response to complaints

The technical/quality department will usually be tasked with the receipt, investigation and response to customer complaints. Most food safety and quality management systems will require reference to the customer complaint handling methodology. The process of interaction with the customer can work in a number of ways including:

- *Retailers (supermarkets)*: Retailers will often screen customer complaints made directly to them, then forwarding on to the manufacturer:
  - A general list of all complaints.
  - Individual complaints, which they feel warrant a specific investigation and response.
  - Serious complaints where the retail technologist will contact the manufacturer in order to personally oversee or remain closely in touch with the investigation and formation of a response. Depending on contractual arrangements, the manufacturer can be charged a complaint handling fee by the supermarket.
- *Food service (restaurants and canteens)*: Such businesses usually follow the same approach as retailers. The customer technologist involvement may vary on a company by company basis, due to the level of technical resource available.
- *Direct consumer dialogue*: Where a food manufacturer is selling its own brand there will not be a buffer or filter between the consumer and the manufacturer, which means that all complaints are sent directly to the food manufacturing site. This can place a great administration burden on quality/technical departments, especially in smaller businesses that may have only just ventured into the branded market. As a business brand grows, the site directors will typically look to invest in dedicated customer service staff, who will either report to the quality/technical department or to the sales department.

At times, government food standards agencies may become involved in the response to a complaint. For example, if the consumer feels particularly concerned with regard to the nature of the product fault then they are likely to raise the matter with the authorities. Perhaps the consumer has felt ill after eating the product or has found a particularly concerning foreign body, or perhaps the packaging amount is felt to be excessive or the pack is believed to be short on content/lightweight. Such complaints when made to the relevant authorities can be very time-consuming due to the in-depth nature of the enforcement agency investigation required. This often involves a site visit or inspection to discuss the issue and perhaps to review the areas of the business where the alleged fault may have occurred.

While complaints are generated out of disappointment or concern and are therefore a negative aspect of running a food business, they also provide a valuable source of information as to the performance of the business quality/safety/legality controls. An often-overlooked aspect of complaints is that ultimately they can be used to help improve the business. It is good practice to conduct routine and detailed reviews of the business complaints using statistical (e.g., Pareto analysis) and trend analysis, in order to gain a good insight into the overall performance of the business and highlight particular priority areas for improvement.

Chapter 10, Nonconformance, recall and crisis management, provides more detail on this subject, including causes of complaints, complaint management and investigation, root cause analysis, complaint response, involvement of enforcement bodies/local authorities, the 'due diligence defence', management review and Pareto analysis.

11.4.1.3 'Issue management': incidents, recalls and crisis management

Although the circumstances would rather be avoided, being effective, controlled and professional during major business incidents and other issues is an important service required of the quality/technical department.

The technical department should always seek to handle these situations as smoothly as possible, coordinating site review and rectification actions while communicating to the customer and other stakeholders that the matter is in hand and being dealt with in a manner that will identify the root cause, swiftly implement robust measures to regain control and help prevent any recurrence.

Product withdrawals or recalls can be necessary for a wide range of reasons including microbiological, chemical or physical contamination, quality faults, packaging faults or failure in some other way to meet legislative requirements (e.g., a product labelling issue). Incident or crisis management can be necessary due to direct issues with specific products, and can also be required for a wider range of problems such as failure of key processing equipment, loss of utilities (e.g., electricity, water, gas), factory area damage or health and safety issues impacting on production continuity, media issues, health scares and malicious contamination (or threats of such).

Due to the numerous potential causes of an incident, recall or crisis, a business is well advised to have documented, implemented and tested clear plans in readiness for dealing with such issues. The technical team will usually be expected to play a key role in this advanced preparation, including:

- Trade withdrawal, recall and crisis management plans prepared and in place, fit for operation 24 h per day, every day of the year, including holidays.
- Maintenance of clear, quick and reliable product traceability systems which should be effective at tracing products both forwards and backwards through the supply chain.

- Up-to-date and swiftly accessible lists of supply chain key contacts (those points in the chain to be contacted in the event of an issue to recover the product). This includes out of hours contact numbers for both internal key team members and external contacts.
- Withdrawal and recall procedures must be checked regularly by conducting a test withdrawal or recall. Such testing should be conducted at least annually. Adding this requirement to the business 'systems audit' schedule can help in ensuring this beneficial approach.
- All procedures associated with the management of incidents and issues should be referred to in the business Food Safety and Quality Management System manual.

Chapter 10, Nonconformance, recall and crisis management, provides an in-depth review of 'incident management and product recall', including the associated topics of product recall and trade withdrawal, legislative considerations, responding to a food incident and the recall process, the business 'product recall plan/procedure' and crisis management.

11.4.1.4 Audit service provision

The food business technical department is often expected to organise and host customer technologists/auditors who wish to visit the site to confirm that the operation is compliant with their expectations. This is the case for both direct technologist visits (second-party audits) and for third-party audits. With regard to both second- and third-party audits, the role of the technical department does not end with the organisation of the audit. The department will typically be expected to be the hosts for the audit day/s, the liaison/coordinator with the rest of the business site and also the collator and communicator of all corrective actions and the writer of the formal response to the auditing party on the rectification actions taken.

Chapter 9, Audits, reviews audit processes and practices in more detail including 'first party', internal audit practices and systems; 'second party', supplier and customer audits; 'third party', audits including certification schemes; defining and monitoring 'technical compliance'; audit 'service provision'; enforcement authority audits and interaction; site audit preparation, performance and follow-up action.

The technical department is usually also responsible for coordinating the business internal audit requirements which are particularly important to ensure that the product quality, safety and legality-related systems and procedures operated by the business are operating effectively, and in some cases identifying areas for further improvement.

These internal audit requirements include regular hygiene and housekeeping audits and 'systems audits' covering a variety of areas including (typically covered over a 12-month period): senior management commitment and continual improvement; quality policy; quality manual; organisational structure, responsibilities and management authority; contract review and customer focus; internal audit; documentation control; specifications; record completion and maintenance; complaint handling; external standards; internal site standards; utilities; pest control; foreign body detection; product packaging; control of nonconforming product; control of operations; calibration and control of measuring and monitoring devices; training; access and movement of personnel; medical screening; protective clothing.

11.4.1.5 NPD and launched product review: sensory panel attendance

Often food manufacturers will be expected to routinely attend product review panels based at the customer site to monitor/assess the ongoing product quality performance.

These panels are more common when the business is producing for the customer's brand (as the customer therefore has an added requirement for product monitoring and demonstration of due diligence). Such panels are usually attended by members of the customer technical, development, production and sales teams. The business is always best advised to ensure that staff in attendance are knowledgeable, confident and diplomatic.

Chapter 9, Audits, considers 'product panel attendance' in more detail.

11.4.1.6 Creation and checking of product claims

A common occurrence within many food manufacturing businesses are routine requests from the sales or development departments to make specific product-related claims, in the hope that such claims could lead to increased product sales. Claims could include low fat, low salt, high in omega-3 fatty acids, high in fibre, etc. It is usually a legal requirement that product claims need to be underpinned by clearly related facts, therefore the technical department is often asked to define and substantiate that the desired claims can actually be made. If the desired product claims cannot be made then the technical department may sometimes also be asked to advise on the product reformulation required to be able to achieve the desired claims.

Such tasks typically require:

- clarity on the customer specification and legal framework for the claim to be made (e.g., what percentage of fat within the product would allow the product to be termed 'low fat'?);
- understanding the tests and/or calculations required to enable the claim to be made; and
- the resources to underpin the claim with analytical work where necessary for further confirmation and to demonstrate due diligence (e.g., taking appropriate batch samples and sending to an accredited analytical laboratory for testing).

Sometimes the claim terms proposed are not clearly defined by guidelines/legislation. Examples of potentially ambiguous terms include 'fresh', 'wholesome' and 'traditional'.

These types of claims often require further research to clarify what the term typically is understood to mean within the context of the product being manufactured. Often with regard to product claim terms that could be ambiguous it is worthwhile protecting the business position by requesting the approval of the relevant enforcement authority, as government officials will be tasked with ensuring that sales terms are not misleading to the consumer.

Some food manufacturing businesses take the view that they would rather progress with making such claims and wait for the relevant enforcement authorities to catch up with them if there's a problem. However, the financial implications of suddenly being told that the product claim cannot be made can be very significant indeed, especially where product stock levels and preprinted packaging are issues. Therefore, it is prudent to always confirm all product claims in advance of them being made.

It is worth noting that retail/food service operations will often be interested in encouraging the food manufacturer to make product claims which they feel will be beneficial to product sales. It is important in these circumstances that the manufacturer does not succumb to pressure to make product claims that it does not feel can be 100% substantiated. The major retailers can sometimes employ ex-enforcement authority officers to advise them on the legality of product claims, accuracy of product photography and in-store sales notices; therefore a food supplier will be well advised to request that they can also utilise such resources when they are available.

Photography and design for retail product packaging is not usually the remit of the technical department, however, the implications of a misleading product picture being printed on the product packaging will find its way back to the technical department via customer complaints and interest from the relevant enforcement authorities. As a result, technical managers are well advised to take a strong interest in the business product sales photography approach and to appropriately guide proceedings if they can see that the sales department is taking 'artistic licence' a step too far.

11.4.1.7 Packaging design initiatives and innovation

Product packaging is regularly reviewed within most food manufacturing operations for many reasons including:

- cost reduction (e.g., change of supplier, or weight reduction of materials used);
- 'lightweighting' (e.g., reduction in the amount of material used per pack, often as both a cost saving and environmental initiative);
- increased product sales/quality (e.g., highlight via improved pack visual appearance); and
- increased product shelf life (e.g., via optimised packaging barrier properties).

Making changes to product packaging can have wide ranging implications. Lightweighting initiatives to save material costs (for either commercial or environmental reasons) can lead to a weaker pack and therefore increased packaging failure during product processing and/or in transit, resulting in significant costs being incurred rather than the desired cost and environmental savings. Other adjustments such as varying the pack barrier properties while seeking increased shelf life can also lead to unexpected impacts on product quality and safety.

For these reasons it is usually the technical department that is given the responsibility of progressing/overseeing the packaging related change initiatives as they are seen as the best qualified to predict or understand product safety, quality and legality implications. However, a good technical manager does not necessarily make a good packaging technologist and therefore it is important for technical management to know their limitations in terms of skill sets and packaging knowledge, and realise the importance when necessary of utilising support from the packaging supplier base, independent packaging organisations and packaging consultants.

11.4.1.8 Provision of technical KPI reports

It is good practice (and therefore a requirement of many compliance and certification standards) for a business to conduct routine 'senior management review' of the business technical and quality performance. Such reviews are usually facilitated by monthly or quarterly reports compiling business information and updates related to product quality and safety performance.

Chapter 9, Audits, covers 'business' 'internal technical compliance review' including 'senior management review'.

Many retail and food service businesses will expect their food suppliers to complete and submit technical KPI reports, typically on a monthly basis.

As the content of these reports is mainly focused on the monitoring of product quality, safety and legality, the responsibility for the collation and provision of such reports is usually held by the technical department (which will already be handling most of this detail in the progression of their work and in preparation for internal senior management reviews).

If a food manufacturer is producing foods which are to be sold under their own brand then they will typically concentrate on ensuring that their own internal technical compliance standards are maintained and regularly reviewed. Such businesses will only expect to answer to the end consumer and the regulatory authorities in the event of a product issue. The retailer will not usually get involved in these circumstances as the supermarket name is not associated with the product, therefore all of the focus and responsibility will fall upon the manufacturing company/brand owner.

However, any food manufacturer producing foods for the major supermarkets and food service groups, where these foods are sold under the name of the particular supermarket/food service outlet, is highly likely to be expected to routinely report on a range of 'technical compliance' KPIs. This is in order to regularly assure the brand holder that their supplier is in control of product safety, quality and legality. Such suppliers can also expect to be visited and audited on a routine basis to further assure the brand owner of site technical compliance.

Every customer is different in terms of the level of technical compliance assurance they require. A typical model is for a major retailer to expect key suppliers to complete and submit a technical compliance report on a monthly or quarterly basis, providing business details on a range of technical compliance KPIs defined by the customer, and typically to be presented within a standard report format also designed by the customer.

The technical KPI content of such internal and external reports to customers typically includes review of the following technical compliance aspects:

Audit controls:

- Site certification audits undertaken (e.g., British Retail Consortium (BRC), International Organisation for Standardisation (ISO), International Featured Standards (IFS), organic, halal, kosher, ethical trade, etc.).
- Scores achieved at the most recent customer/s technical audit (including progress on corrective actions).
- Supplier audits progressed, results and site adherence to the supplier audit schedule.

- Site internal audit results (both 'systems' audits and 'area compliance' audits).
- Outcomes of any audits progressed by food-related enforcement authorities.

Product monitoring:
- Business end product taste panel results (e.g., percentage judged to be 'in compliance' with specification).
- Performance at routine customer quality reviews (e.g., number of issues raised).
- Chemical/analytical testing undertaken and percentage compliance (e.g., nutritional health claim checks, speciation checks and pesticide residue checks).
- Microbiological reports (e.g., end product compliance to specification and factory area environmental swabbing results).
- Store visits undertaken to check product appearance/conformance in-store.
- Competitor assessments undertaken and the results collected.

Customer complaints and product withdrawals:
- Customer complaint levels and analysis (e.g., often presented as complaints per million units sold to facilitate trend analysis).
- Detail on any major complaints of particular concern.
- Any product recalls or trade withdrawals required.

System reviews:
- HACCP reviews and updates.
- Traceability checks and mock recalls.
- Specification reviews undertaken.

Submitted reports will also be expected to provide full detail on any areas where compliance is in doubt or outside of the customer-specified tolerances. This should include a review of the actions taken to address the issue and to avoid any recurrence. For example, if a customer-compliant trend of a 'flexible plastic foreign body' has been highlighted, the business will be expected to state what investigation has been conducted internally into the presence of that foreign material, the conclusions reached and any further actions taken to prevent similar issues in the future. For example, if the foreign material is traced to ingredient packaging ingress, the business may report that it has improved the grade of the supplier ingredient packaging and has also retrained the preparation staff on careful handling of ingredient packaging during the preparation and weighing processes.

11.4.1.9 Certificates of conformance/certificates of analysis

Customers may require the provision of CoC and/or CoA to accompany (or be sent in advance of) deliveries, to assure them that the product supplied meets their requirements. The technical department will often be the department checking that the product conforms with the required CoAs/CoCs and will be expected to ensure that the tests are conducted thoroughly and in time for the delivery schedule required.

Chapter 4, Raw materials and packaging supplier control, provides further detail on CoC, CoA and their use from a 'customer assurance/supplier' perspective.

11.4.1.10 Rapid response to emerging issues

Food manufacturer technical departments can sometimes be expected to rapidly respond and act on emerging issues within the food industry. Such issues may come to light via the news, customer or supplier relationships or via the relevant enforcement authorities.

When supplying the major retailers and food service groups, businesses will often become aware of emerging food scares and potential problem issues in advance of the locally based enforcement officials. This is due to the often high-level resources which large retailers and other food groups can draw upon and the fact that their representatives are present on advisory boards and other panels, regularly interfacing with government.

In such cases typically at the outset of a 'food sector emerging issue', the food manufacturer's technical department will receive an e-mail or telephone call from the customer technologist requesting that they check the food 'business position/exposure' on the particular issue and feedback swiftly within a set timescale. Following receipt of the supplier's feedback, the customer can then form their own 'position' on the matter, continue to monitor the situation and act at the appropriate time.

Some examples of 'emerging issues' that have given rise to supplier–customer dialogue in the past include melamine contamination of milk products (China); Sudan 1 dye contamination of Worcester sauce containing products (United Kingdom); food fraud – horsemeat labelled as beef (United Kingdom), fipronil contamination of eggs (EU). Such issues typically trigger a large amount of correspondence between the customer, supplier and supply base, and sometimes lead to products being removed from sale.

To be able to deal with such circumstances rapidly a technical department will benefit from:

- robust and rapidly interrogatable traceability systems;
- well-developed raw material supplier contacts to ensure swift progress and feedback;
- up-to-date raw material and finished product specifications immediately accessible on site;
- sufficient manpower resources to enable issues to be quickly followed up and reported upon; and
- a calm and controlled approach, thinking ahead and communicating internally on the emerging issue and how the issue may affect current raw material and end product stocks, including those stocks currently in the supply chain.

11.4.1.11 Keeping stakeholders informed

Customers and local enforcement authorities will usually expect to be kept informed on business developments that may impact on product quality, safety or legality. It is therefore good practice for the technical department of a food manufacturer to manage the communication of business developments to customer technologists and the relevant enforcement authorities. Such communication may be formal or informal and could include:

- the changing or implementation of a production process;
- moving into new product categories;
- development or implementation of new product packaging;
- building work (either refurbishment or new building developments); and
- disruption to the operation (e.g., utility supplies, labour availability).

Sometimes the customer or enforcement agency will need to be assured or convinced that the matters raised will be controlled and will not adversely affect product quality, safety or legality. The technical manager should therefore take the time to form a strong and considered case for the changes before addressing the matter with the relevant stakeholders. Assurances may include the fact that any product or process changes will be trialled and validated before going live, or that all products will only be supplied on a 'positive release' basis or having cleared and passed all tests/quality checks.

Customer technologists and local authorities can prove to be very helpful allies on the development of new products or processes, offering advice and guidance based on their experience and also potentially tapping into the resources of their own operation.

Feeling involved from the outset usually reassures stakeholders that the business is in control and thinking ahead, as it can be a very frustrating situation for a customer technologist or enforcement official to visit a site or receive a complaint and only then realise that significant changes are taking place (or have taken place) on site that may have adverse effects.

11.4.2 Summary of considerations related to customer 'technical services'

When providing technical services to the customer, the technical department has to meet a very varied set of requirements. There is great importance to be placed on the technical department staff being good communicators and taking the time to build a rapport and trust with the customer and enforcement authorities. Such factors are invaluable as often during communication suppliers will benefit from a little leeway and patience, which is more likely to be given if the stakeholders feel that they have been well looked after over time and are dealing with consistent and competent team members.

11.5 Engagement with outsourced services

Alongside regular liaison with raw material suppliers, internal functions, customers, audit/compliance bodies, enforcement officials and other stakeholders, the technical department of a food handling operation will also have engagement with and sometimes oversight of a range of services which have been outsourced by the business.

Technical and quality department links to 'outsourced services' can include:

- pest control,
- laundry,
- hygiene-related services,
- storage facilities,

- product transport and distribution,
- catering,
- waste,
- selection and management of analytical services, and
- calibration for key equipment/processes.

11.5.1 Pest control

Pests such as insects, rodents and birds can pose a risk to the food safety, quality and ultimately legality of the food manufacturing process. It is therefore vital that an appropriately structured and administered pest control system is operational on food manufacturing sites as part of the food safety prerequisite programme.

Many businesses engage pest control contractors to provide these specialist services and for added assurance customer codes of practice can sometimes state an expectation for the contractor to be a member of a relevant trade association. Reputable pest control businesses will typically have appropriate staff training programmes, be adequately resourced, maintain clear and detailed records of the status of pest control activities on site and have sufficient liability insurance.

Key pest control contractor staff who will regularly engage with the technical department include service technicians (who tend to conduct routine visits, monitoring and maintenance of the site pest control measures) and more specialist roles such as field biologists.

Typical pest control equipment and services include EFKs and related catch tray analysis; proofing (materials and fittings for the prevention of pest access); toxic perimeter baits for rodents; rodent/mouse traps; moth/pheromone traps; crawling insect strips.

The 'service schedule' applied in terms of pests covered and the service/visit schedules will vary depending on business compliance requirement and cost. The pest control contractor is providing a specialist service on site and therefore will be expected to ensure that the business is operating within the required legal frameworks and also adhering to specific customer requirements.

For example:

- All pesticides/fumigation materials used must be on any required 'approved lists'.
- Methods of capture/kill must be recognised as acceptable/good practice.
- If the site is manufacturing organic products there are likely to be extra stipulations with regard to the necessity and nature of treatments and practices proposed to be used.
- Some 'pests' may have special protection under law and therefore cannot be simply killed and removed from site, for example, some wild birds (e.g., flying into a warehouse) may be classed as a 'protected species' and therefore should not be harmed, instead seeking to usher the bird out of the building over time. An approved pest control contractor will know the appropriate way to deal with all such matters, including formal application to the relevant authorities for special licences to remove protected species pests in certain circumstances.

While there is the potential for food manufacturers to maintain their own pest control programmes without the use of a contractor, in reality the specialist nature and level of expectation placed upon such activities often results in the fact that most businesses are best advised to appoint the services of a specialist pest control business.

Chapter 7, Operations and process control, further considers these matters in 'pest control: measures to prevent pest-related contamination'.

11.5.2 Laundry

Protective clothing is of key importance in food handling operations to protect both the food and the employee, therefore technical departments will usually be expected to advise on the standards required and monitor the ongoing adherence to these standards. Such materials should either be 'single use' (disposable) or laundered via a facility which has been audited and verified to be effective in its cleaning processes. Many businesses use contract cleaning companies and others set up and utilise their own in-house laundry services.

Chapter 8, Personnel control, considers these matters in 'personal protective clothing and equipment'. The provision of appropriate facilities and factory clothing materials/PPE is also covered in further detail in Chapter 5, Site standards.

11.5.3 Hygiene-related services

Maintaining good hygiene standards is fundamental to food safety assurance and legal compliance in food handling operations. The technical department will therefore be required to maintain very close engagement on any aspects of hygiene control placed with suppliers to the business.

Outsourced services related to 'hygiene' can include:

- Cleaning chemical supply and related 'hygiene services' (including provision of cleaning instructions, support on cleaning validation and troubleshooting).
- Consumables (e.g., supply of cleaning materials and PPE).
- Contract cleaning services. Separate hygiene businesses may be utilised for additional deep cleaning or specialist jobs, for example, high access cherry picker/scaffolding tasks, lift pits or confined working conditions.

Matters related to hygiene and its control are considered across many chapters and particularly in the guest chapter later in this book covering 'hygiene management'.

11.5.4 Storage facilities

Food manufacturing operations, usually as a result of short- or long-term growth, often need to secure additional storage capacity for raw materials, intermediate 'work in progress' or finished products. The technical team can support on the challenges involved in such scenarios and are key to setting the required standards and ongoing monitoring of such facilities.

Chapter 7, Operations and process control, provides technical consideration of 'storage facilities' including construction materials, area design, setting the operating parameters, records, rental of 'off-site' storage and building or purchasing an additional storage location.

11.5.5 Product transport and distribution

Having produced sufficient stocks to fulfil customer orders, the product requires transferring in safe hygienic conditions to the customer. For many food manufacturers the 'customer location' is often a distribution depot for a retail or food service chain (where the bulk delivered product will then be broken down into smaller orders to fulfil the requirements of individual stores/restaurants, to which the product will then be transported alongside goods from other suppliers).

As this product supply phase has the potential to greatly affect product quality, safety and legality, it is important that the technical department is involved with standard setting, audit/approval and ongoing monitoring for these operations (whether owned by the business or supplied by a contract distribution company). In the European Union the requirements of the transport system selected by the food manufacturer are well summarised by Regulation (EC) No. 852/2004 on the Hygiene of Foodstuffs. 'Chapter IV (Transport): Conveyances and/or containers used for transporting foodstuffs are to be kept clean and maintained in good repair and condition to protect foodstuffs from contamination, and are, where necessary, to be designed and constructed to permit adequate cleaning and/or disinfection'.

Chapter 7, Operations and process control', considers 'product transport and distribution' in more detail including transport and distribution approach and standards, temperature control, transit tests, Modified Atmosphere and Controlled Atmosphere systems, packaging 'tamper evidence' and environmental considerations.

11.5.6 Catering

If a food handling operation provides catering resources to the site staff then such food production processes should adhere to the same hygiene and product safety control standards as the rest of the operation.

Technical and quality teams will therefore often be involved with such catering services, helping to set and maintain the required standards including monitoring/inclusion on product testing and audit schedules.

Chapter 5, Site standards, considers 'staff catering and rest areas' in more detail.

11.5.7 Waste

Waste can take many forms in a food business including waste ingredients, end product and associated packaging. Waste food can also be in the form of trade effluent created when cleaning the food processing areas.

With links to site hygiene and pest control the appropriate management of waste in a food handling operation is required as part of 'good manufacturing practice' and to ensure adherence to customer and legislative standards. Such matters therefore require the regular input and engagement of the technical/quality department.

Chapter 5, Site standards, considers 'waste management' in detail and Chapter 7, Operations and process control, provides added details from a pest control and general food safety assurance perspective. Chapter 1, Food sector challenges and the role of technical and quality management, also reviews the environmental impacts of food processes.

11.5.8 Selection and management of analytical services

In the food sector 'analytical services' are considered to include the use of microbiological and chemical laboratories. These product testing facilities are typically used to assess and assure product and process safety, quality and legality and therefore are vital operations within the quality/technical department remit.

Examples include end product analysis, raw material assessment and process/environmental assessment. For large businesses the economies of scale may work in favour of building and staffing their own laboratory services which will meet the testing requirements of the operation and of customer codes of practice. However, with regard to microbiological and chemical/nutritional analysis the costs of setting up, staffing, gaining appropriate accreditation status and then running such services can be very high, leading to most small to medium-sized businesses deciding to instead use a contract laboratory in order to benefit from the economies of scale and reduced complexity offered by such an arrangement. Larger food groups will often take a similar approach by building and staffing one laboratory which then services numerous sites within the group. Such 'central laboratories' tend to be in a separate building to the main manufacturing operations in order to address specific concerns around the control of pathogen testing/protection of the food manufacturing operation.

During the consideration of a contract laboratory, before conducting the review/approval audit phase the food manufacturer should check any customer or compliance standard expectations with regard to the type of accreditation required to be held. They should also check whether all of the required tests will be covered by the laboratory accreditation, and where certain tests are excluded from the accreditation held, consideration should be given as to whether this will be acceptable or whether such tests should be conducted by an appropriately accredited subcontractor.

When setting up products for ongoing microbiological assessment it is important that the business sets systems which ensure that the required product test frequency (how often the tests are to be carried out on that product) is maintained. Systems should also set the range of microbiological test organisms and their required test result 'target' levels and 'action' levels.

Actions required on the occurrence of nonconforming products should be defined/referred to an appropriately qualified site technical/quality manager for decision-making.

When setting product nutritional testing with a laboratory the same general principles apply as with microbiological testing. For example, laboratory accreditation requirements and the range of tests and test frequencies need to be defined. The most common tests for a product nutritional assessment include energy, protein, fat and carbohydrate, with additional tests providing product content detail for sugar, saturated fat, fibre and sodium.

Chapter 6 reviews microbiological and chemical testing with regard to 'product control and HACCP considerations' including validation and shelf life assessment.

11.5.9 Calibration for key equipment/processes

Food manufacturing processes usually require accurate measurements to be taken at various points in the production process, often in order to monitor/control/assure product quality, safety or legality. Temperature gauges/probes, weighing equipment, pH meters, volume meters, timers, pressure gauges, conductivity meters, etc. are all accurate devices that can over time either break down or drift in their accuracy. Such circumstances would then threaten product quality, safety or legality and therefore need to be guarded against via the use of 'calibration' related systems and procedures.

'Calibration' can be defined as the act of checking or adjusting, via comparison with a 'standard' (e.g., a measurement instrument or system with a known accuracy), the accuracy of a measuring device/instrument.

The general expectation within a food manufacturing business is that any measurement activity within the operation that can have an impact upon product quality, safety or legality must be subject to calibration control.

Calibration methods should check the accuracy of measurement against a set/appropriate standard. Such 'standards' will vary dependent upon the nature of the equipment/measurement. For temperature probes, the accuracy check is typically expected to be traceable back to a recognised 'national standard', usually achieved by having a formally 'calibrated' temperature probe held on site, against which all of the business probes are checked for accuracy on a routine basis.

A similar approach can also be taken for the calibration checking of test weights, where a formally 'calibration checked' set of weights can be held securely on site and used on a routine basis to check all factory 'in-service' weights against to confirm that they are still accurate (e.g., they haven't been subject to damage that could have altered their weight).

Calibration checks should be conducted at a set/appropriate frequency and the results of these checks plus any further corrective actions should be recorded. (For example, temperature probes used to monitor factory CCPs often are expected to receive a daily calibration check due to their importance to food safety assurance).

As there is the potential for a calibration check to highlight that an item of measuring equipment is no longer sufficiently accurate it is very important that a business ensures that it has 'back-up' equipment/systems available in order to cover the required operation control points while the inaccurate piece of equipment is repaired/recalibrated. Clear business follow-up procedures are also required in order to address any stocks that may have been affected by the calibration fault.

Calibration underpins the accuracy and reliability of a wide range of operational checks/monitoring procedures. Given the importance of 'accurate measurement' to

Quality Management Systems technical standards such as ISO 9000, ISO 22000, BRC Global Standard for Food and FSSC 22000 all contain expectations/requirements upon the subject of the operational control of 'calibration'.

11.5.9.1 Summaryof considerations related to outsourcing services

Outsourcing of services which impact on product quality, safety and legality can be of great benefit to the technical department and wider business providing that these services are properly engaged and well managed. It is therefore certainly worth the time of the technical manager and their team to commit to engagement with and control of these key outsourced resources.

11.6 Sources of technical information and support

The technical/quality manager within a food manufacturing business is often looked to as being the 'oracle of all knowledge' with regard to all matters related to quality, safety and legality of products. However, the 'technical' subject area is so vast, with global food markets, variations in international law, rapidly developing national markets, routinely updated compliance standards, new research findings and ever-changing customer expectations, today's technical manager is unlikely to be able to be self-sufficient, with instant personal knowledge and recall of all of the information required.

It is therefore a useful skill for those in roles related to technical management to be able to form networks of both information provision and personal/business support. The following range of 'information and support' advice should prove useful in this regard.

11.6.1 Sources of technical information

Clarity of 'technical requirements' for a food manufacturing operation will enable the technical manager to form a detailed plan/'blueprint' for how the business should deliver on objectives related to quality, safety and legality.

The following are example sources of information and advice which can prove helpful when defining, implementing and reviewing business technical and quality requirements:

Customer standards – Almost all food retailers and many food service operations/food businesses have compiled their own supplier manual/codes of practice/documents highlighting all aspects required of their suppliers. Many of the retailer supplier manuals are held online and require password access to be set up for each supplier to ensure that authorisation levels are adhered to. Other manuals are forwarded to suppliers in the form of downloadable files, discs or printed copy.

Compliance/accreditation standards – In the food sector there are many standards available for businesses to align themselves with, either of their own volition or upon the insistence of customers who expect a particular standard to be achieved and maintained as a prerequisite of supply. Examples of such standards include:

SFBB – Safer Food Better Business, http://www.food.gov.uk/business-industry/caterers/sfbb
SALSA – Safe and Local Supplier Approval, http://www.salsafood.co.uk
BRC – https://www.brcglobalstandards.com/brc-global-standards/food-safety/
IFS – http://www.ifs-certification.com/index.php/en/
ISO 9001 – http://www.iso.org/iso/iso_9000
ISO 22000 – http://www.iso.org/iso/home/standards/management-standards/iso22000.htm
Organic – https://www.soilassociation.org/organic-standards/
Kosher – http://www.koshercertification.org.uk/whatdoe.html
Halal – http://halalfoodauthority.com/
Vegan – https://www.vegansociety.com/your-business/vegan-trademark-standards

Good manufacturing practice and HACCP guidance – Codex Alimentarius Commission is internationally recognised having worked for many years to define standards of GMP and HACCP alongside provision of many other technical guidelines. Codex guidance is available online via: http://www.codexalimentarius.net

Food Industry Trade Associations, for example, American Frozen Food Institute (AFFI), http://www.affi.org*; UK Food & Drink Federation (FDF),* https://www.fdf.org.uk*; Chilled Foods Association (CFA),* http://www.chilledfood.org*; British Frozen Foods Federation* (BFFF), http://bfff.co.uk*;* Australian Food and Grocery Council (AFGC), http://www.afgc.org.au/.

Specialist networks/societies – There are a range of specialist networks/societies based on the internet whose skills/membership may be of particular relevance to specific business requirements, for example, The International Food Safety and Quality Network, http://www.ifsqn.com provides a forum for food sector professionals; The Anaphylaxis Campaign promotes awareness and the need for allergen control. More details on allergens and their control can be found at the anaphylaxis campaign website, www.anaphylaxis.org.uk.

Books/literature/podcast reference materials – Food sector-related books are plentiful providing very good sources for both summarising a vast amount of food industry information and provision of many specialist guides.

Google scholar – Aside from standard web search engines Google Scholar provides a more 'academic/scientific focused' search, including journal papers on the most up-to-date topics in Food Science & Technology, http://scholar.google.co.uk.

Analytical and technical services – Contract testing laboratories utilised by the business are likely to have specialist scientists who can be consulted on related issues (e.g., product microbiology/chemical/nutritional standards and environmental microbiology standards).

Consultants – Can be engaged to provide extra technical support or analytical services, as can specialist institutes/research associations, for example, packaging advice/support could be sought from the 'Packaging Society' (based within the The Institute of Materials, Minerals and Mining - IoM3): http://www.iom3.org/packaging-society. Specialist consultancy services can also include market and media communications advice provided by specialist PR consultants, and crisis management advice provided by specialists in the field.

Legislation advice – Local food standards enforcement authorities should not 'just enforce', they are also likely to be there as a service to provide advice and guidance. Therefore, if the manager has a particular legislation query with regard to food safety, quality or legality, it is highly likely that the designated enforcement official may well be able to help. The enforcement authorities are also likely to have online guides available covering legislative expectations. In the United Kingdom the Food Standards Agency website is an excellent source of up-to-date food legislation: http://www.food.gov.uk.

In the event of the food business requiring legal representation (e.g., perhaps in the event of being sued over a customer complaint or challenged in the courts by specific authorities) the business should ensure that the lawyers selected to advise and represent the business have sufficient 'food specialist' knowledge.

It is sometimes also helpful in such circumstances to appoint the services of an 'expert witness' – this is an individual who has specialist knowledge (usually through skill, education, training, or experience) which will aid the 'trier of fact' (e.g., a jury, a judge, an arbitrator) in reaching a proper decision.

Knowledge in the supply chain – Suppliers to a food business will be specialists in their particular ingredient or packaging field. This is a powerful source of information to be drawn upon when specific information or insights are required. In addition, the process machinery suppliers to the business are likely to be supplying similar processing operations elsewhere and can sometimes therefore be a source of helpful information/contact advice. Similarly, the end customer is likely to be employing technologists who will have a wide range of experience on raw material, in-house processing and supply challenges.

Technical and quality managers are well advised to build their own personal list over time of key contacts who can be called upon when required. A manager who has developed a very good range of contacts over the years will benefit greatly from this extended support network when needing to move swiftly and effectively on a particular issue.

11.7 Personal/business support

The role of the technical/quality manager is usually very busy and demanding, yet often can seem like quite an isolated position. The following approaches can help the individual to build a level of support while conducting this key role for their business:

- An excellent resource to the quality/technical manager is the business factory staff. They will tend to know what's right and what's wrong both with the products and with the general business approach and usually just need encouragement to share their insights. When in the process areas for a walk round/internal audit/trial the technical manager could consider taking time to ask how things are going and whether there are any current issues that the area needs help or support upon. Any particular recent complaints could be discussed to gain the benefit of the staff opinions. The business canteen is usually a good place within which to catch up with team members and ask their opinions outside of the busy work environment.
- Encourage/orchestrate a 'team approach' among the technical team or the business management team toward issues/problem-solving. Accumulated brainpower can lead to better solutions.
- Always look for 'synergies'. The manager should consider who are the natural site departmental allies upon business technical/quality-related issues?
- Be consistently supportive and fair with fellow team members and the business management team. Criticism should always be constructive and ideally given behind closed doors (unless there's a particular person or aspect of general business culture which specifically needs shaking up).
- Work hard to ensure the trust and support of line management and directors. The technical/quality manager should always take the time to explain why particular technical decisions and actions are 'well considered' and 'thought through'. From the outside 'technical' teams can sometimes appear to be a 'sales prevention force'. It is important to try to turn this image around to be one of a 'brand protection and business support force'. Ideally the technical function should provide a strong backbone to all of the business activities, helping assure and guide upon product safety, quality and legality matters, while also supporting on wider business initiatives including operational efficiency improvements.
  - If the food business is associated with issues which could be particularly contentious, for example, its position on use of genetically modified materials, irradiation, allergens, etc. It is worth the technical management taking the time to consider the 'business position'/'statements' on such points with the business directors. This will help the business to take a well-considered position and the technical representatives will feel added assurance and confidence that they are fully in tune with the 'company line' when discussing such matters with customers/auditors/enforcement officials.
- Technical team members may find it helpful to consider 'who used to do the job?'. If they're still in the organisation what can be learnt from their experience? If they left, why did they leave and what can be learnt from that? If they left and it was on good terms it may be helpful to have a telephone discussion at some point to benefit from their experience on matters. For example, many business quality and safety systems are rooted in decisions taken 'years ago' often before the current management were in place, therefore it is sometimes very useful to review the rationale of such business decisions to provide insight into whether it still is the most appropriate approach for the future.
- Interest groups/panels/trade associations. Both regionally and nationally managers can meet routinely to discuss issues of mutual interest. For example, shared technical interest points could include best practice, emerging issues, new legislation, training and information initiatives.
- Technical management may consider joining a society linked to their specific sector/interest areas/profession. For example, the Institute of Food Science and Technologists provides a very good range of services and benefits to its members (conferences, training, news bulletins, advice): http://www.ifst.org

11.8 Summary thoughts on the importance of technical information and support

An effective ability to cope with the whole range of quality, safety and legality issues which could potentially arise at a food operation stems from a combination of personal knowledge, drive, competence to swiftly network and compile key information and finally the ability to take decisions and coordinate/complete the actions required.

It is important for technical and quality managers to take the time to reflect upon their skills (and skills gaps) in order to then undertake personal development initiatives. Also of importance is to build personal networks as an invaluable way of helping to ensure that good support is always on hand when required, and to benefit career development within the food sector.

Further reading

Codex Alimentarius Commission – http://www.codexalimentarius.org

Food Quality Assurance, Principles and Practices. Inteaz Alli. CRC Press (2003).

Food Safety Culture – Creating a Behavior-Based Food Safety Management System. Yiannas, F., 2009. Springer.

Food Safety Behavior. Yiannas, F., 2015. Springer.

Global Food Safety Initiative (GFSI) – www.mygfsi.com

ISO 9000 Quality Management System. International Organisation for Standardisation – www.iso.org

ISO 22000 Food Safety Management System. International Organisation for Standardisation – www.iso.org

Psychometric and aptitude testing. Belbin Associates, www.belbin.com; Thomas International Ltd, www.thomasinternational.net

Publicly Available Specification PAS 96:2017 'Guide to protecting and defending food and drink from deliberate attack'. https://www.food.gov.uk/sites/default/files/media/document/pas962017.pdf

The BRC Global Standard for Food Safety: A Guide to a Successful Audit, second ed. Kill, R., 2012. Wiley-Blackwell.

The Seven Habits of Highly Effective People. Covey, S. https://www.stephencovey.com/7habits/7habits.php

Part II

Specialist Topics

Hygiene management

12

Belinda Stuart-Moonlight
Moonlight Environmental Ltd – Scientific & Legal Consultancy, Folkestone, Kent, United Kingdom

12.1 Introduction

In this chapter I'm going to share some of my experience from observing hygiene management as well as trying to implement or improve it in food manufacturing operations. I shall also contextualise relevant aspects of hygiene by reviewing the factors affecting bacterial attachment and survival on surfaces.

As an enforcement officer (UK Environmental Health Officer, EHO) of food businesses and a certification auditor, I have assessed many food businesses, both good and bad. As an expert witness, I have encountered numerous cases where the management of hygiene has been on trial, usually due to food-borne illness but sometimes from other forms of contamination. In giving consultancy, I have been in the privileged position of seeing from the inside how businesses manage hygiene – from the boardroom down. This has given me an insight into the priorities, tensions and decision-making that contribute to the success, or otherwise, of hygiene management.

This chapter will outline some of my encounters and I will use these, as well as some theory, to illustrate a variety of aspects of hygiene management in practice. I hope this will give readers food for thought to assist in avoiding the expensive mistakes that are shown in these examples. Of course, there are plenty of texts that detail and list technical aspects of hygiene requirements. However, I hope to provide the reader here with a more kinaesthetic approach to hygiene management: what 'good' looks like, what 'poor' looks like; why it is so much better to consider hygiene management at the beginning of a project rather than during or after. Further, what one can deduce from certain indicators about hygiene management generally.

Hygiene management is inextricably linked to culture. Where a business is good at one aspect of food safety they are usually good at others. For example, good practice in cleaning and disinfection links to good trend analysis for contamination complaints. A culture where senior management buys into the understanding that poor hygiene management ultimately closes a business is likely to lead to success. This is because allocating the correct resources, judging performance, and providing timely interventions appropriately enable high standards to be maintained consistently.

In considering hygiene management, one needs to account for the fact that food companies grow and evolve over time. It is not until quite far down the line that one can sit down with a pot of money and design the best factory. Financial restrictions often mean that facilities expand on one site into whatever buildings or lands are around the business, even if this is not optimal. Hygiene management here can

Swainson's Handbook of Technical and Quality Management for the Food Manufacturing Sector.
DOI: https://doi.org/10.1016/B978-1-78242-275-4.00012-5

be successful within such situations but plenty of pragmatism and forethought about the 'what if' scenarios (and willingness to accommodate additional hygiene measures where necessary) is needed. Hygiene management evolves as a business grows, but a lack of interest and investment in good practice will quickly cause a food business to fail.

12.1.1 Definition of hygiene management

The World Health Organization (WHO) define 'hygiene' as 'conditions and practices that help to maintain health and prevent the spread of diseases'. The Concise Oxford Dictionary (Allen, 1990) defines management as 'the process or instance of managing or being managed': 'managing' is having executive control or authority. Different people will have different views on what the definition covers, some use the terms 'food hygiene' and 'food safety' synonymously, particularly in the enforcement and catering sectors, whereas in manufacturing, 'hygiene' is often reserved for areas of safety control related to cleaning and disinfection.

In this chapter I view hygiene management as managing cleaning and disinfection, including the management of wider elements to enable the process to be effective. This refers to the siting, design and layout of the food business to enable microbial contaminants to be controlled. Further, the process of cleaning allergen soils from surfaces in contact with foods defined not to contain those allergens.

12.2 Facility design for successful hygiene management – location, layout, services and equipment

The location, layout, services and equipment are key in successful hygiene management. Specifying them correctly at the design stage of a new build or a refurbishment is good insurance against product contamination as well as wider business and personnel risk. Retrofitting often provides a less successful result for these physical elements. I frequently get invited to factories to correct practices where the layout is wrong. Amelioration in this way is never as good as a design that is well thought out initially.

12.3 Location

Location or siting is considered a point of conformity both in legislation (EU Commission, 2004: Chapter 1) and in industry certification standards (BRC – British Retail Consortium, SALSA – UK 'Safe and Local Supplier Approval', etc.). Below are two example businesses where location impacted on hygiene management due to elevated risks associated with environmental features of the specific food factory locations.

12.3.1 Meat factory

The business, based in a picturesque farm, raised its own elite breed animals (beef, pork, lamb) and sold its meat and meat products through high-end shops, including restaurants and delicatessen outlets. The meat factory was on the farm. The farmhouse had been extended to convert outhouses into a butchery and cold store. On the opposite side of the farmyard was the meat products factory, a converted barn where high-care, ready-to-eat products such as pies, hams and sausage rolls were made.

The farmyard was flanked by both the raw and ready-to-eat building elements as well as other farm outhouses and barns. These contained animals, farm equipment, staff welfare facilities (in the form of bales of straw) and a tray wash booth. Trays were used for product transport: unlined for raw meat or lined with greaseproof paper for loose, unwrapped ready-to-eat product. The farmyard itself was a traffic hub where all farm vehicles mixed with staff cars as well as the meat business lorries. It was also the area where products were put into vehicles for dispatch.

The yard was an uncovered mudbath with doubtless significant levels of *Listeria* and pathogenic *Enterbacteriaceae* including *Escherichia coli* from the environment and proximal manure. As staff moved product around, they traipsed the contamination primarily on their boots but also from splashes on their aprons into the raw side butchery and raw meat chillers. Personnel (in protective clothing) moved across the yard for various reasons, such as collecting washed product trays, putting product in kilns for smoking and taking butchered raw meat to the high-care side to be made into ready-to-eat finished products. In this high-care side, there was a strict change protocol that worked very well in keeping out personnel-related contamination. A full clothing change and swing over a bench (to put on captive footwear) had been built into the design. However, the control of contamination of raw meat had not received as much attention.

Despite the tray washing taking place in a dedicated booth with cleaning and disinfecting chemicals and hot water, the process still took place in the farmyard. Mud splash from vehicle wheels, staff footwear, etc. therefore made disinfection attempts pretty futile.

12.3.1.1 Meat factory risks due to the location of workrooms in proximity to a working farmyard includes:

1. Pathogens from the farmyard ultimately *contaminating* raw meat, raw meat preparations and cooked meat *products*.
2. Spoilage microorganisms from the farmyard getting into and onto raw meat, raw meat preparations and cooked meat products *reducing shelf life*.

12.3.1.2 Hygiene management solutions

1. Short term – provision of different, clean-assured transport trays for ready-to-eat products.
2. Short term – captive footwear for raw side operation.

3. Short term – personnel movement control.
4. Short term – relocation of tray washing so postclean recontamination cannot occur.
5. Short term – ensure no raw meat is destined for rare (undercooked) products (because of the assumption of contamination).
6. Short term – cover all ready-to-eat products during van loading.
7. Long term – relocate meat factory away from the farmyard or move the farm activities elsewhere. Resurface the yard.

12.3.2 Fish-based Deli Factory

This business was located on an industrial estate in London, United Kingdom. The unit occupied by the business was c. 50 m from a river, behind which was a narrow strip of land and a reservoir.

Whenever there was rain, runoff would sweep from the vegetation and land near the river towards the factory unit. The factory was at a level in relation to the river that made it prone to flooding and water ponding.

Listeria was frequently isolated both in environmental site swabs and in the finished product. A good deal of raw herring was processed through the factory but as well as *Listeria* being associated with this raw fish in particular, the runoff collecting within and then spreading throughout the factory would have been another major source.

12.3.2.1 Fish factory risks due to the location of the factory in the path of runoff from a riverbank includes:

1. *Listeria-contaminating* ready-to-eat products directly from splash as well as indirectly through vectors via a contaminated environment. NB: Once *Listeria* is resident in a food production environment it is very difficult to eliminate.

12.3.2.2 Hygiene management solutions includes:

1. Short term – bunds to stop runoff penetrating the factory environment.
2. Short term – focus on cleaning and disinfection, and championing drying in the factory environment.
3. Short term – captive footwear.
4. Short term – personnel movement control.
5. Long term – relocate factory to a site less *Listeria* prone.

In both these examples, the environment in which the facility was located impacted on food safety and shelf life because it acted as a reservoir of pathogenic and spoilage microorganisms. Additional hygiene measures were needed to keep the product safe. In both cases, the shorter-term mitigation measures worked to keep the food safe and in the longer term, both food operations successfully relocated.

12.4 Layout and services

Designing a factory is best done with a team of professionals including architects, electricians, equipment suppliers, food safety consultants, flooring specialists, construction company representatives, heating, cooling and ventilation suppliers and a separate refrigeration supplier, if chilling is a significant part of the project. The importance of investment in design cannot be emphasised enough because the ultimate costs of getting this wrong can be so significant. In the same way that the team approach to devising a hazard analysis and critical control point (HACCP) plan is so useful in amassing a variety of perspectives on food safety, the same principle applies when designing a food factory. Each of the professionals brings an expertise in areas that others may not have considered.

A simple example is the refurbishment of a light industrial unit for use as a bakery. Without the electrician examining the incoming main and so determining the maximum power output it can deliver, a refurbishment may go ahead that is blind to the power-hungry requirements of rack and deck ovens. It may be that a young business has invested heavily in equipment and a newly refurbished site but cannot deliver anticipated volumes of finished product. Significant delays – usually months – result from getting a supply upgrade, by which time the creditors are calling in the debt, and the project fails. Although this example is not specifically one of hygiene management, it highlights the wider impact of a lack of a systematic approach at the design stage.

When I am designing a food factory I use some very fundamental principles for layout:

1. Have a linear product flow, from raw material intake to dispatch.
2. Separate the raw and ready-to-eat parts of the operation, ideally by a partition, where raw comes in from one side and product comes out of the other. This may be through double-doored appliances in a wall, for example, a kiln for smoked fish or an oven for pies.
3. Ensure separate change areas for staff working with raw product and those in the ready-to-eat sides. Alternatively, if there is just one change, an additional change on entry to the ready-to-eat side should be accommodated along with additional protective clothing provision and change protocols.
4. Account for waste to ensure rubbish doesn't build in food rooms and staff don't contaminate themselves when handling it. Ensure bulk storage design accounts for reducing harbourage for pests.
5. Account for wash up. Potential for cross-contamination should be considered along with accommodation of dirty and clean equipment.
6. Account for wastewater flow. Ensure that clean flows to dirty and that this cannot be reversed – allowing wastewater from the raw area to flow into the ready-to-eat areas.

Services include electric, gas, water, drainage, lighting, ventilation, hand and equipment washing facilities and waste collection. It is helpful to list each of these and how they will be accommodated in practice in a new unit or refurbishment plan. Additionally, I like to use a base layout and overlay additional layer plans each for (1) equipment, hand and wash up locations, (2) product flow, (3) waste flow, (4) water flow and (5) staff movement.

One area of design failure crops up very regularly in food factories – badly designed wash up facilities. Here is one example where poorly designed wash up facilities impacted on hygiene management, and the elevated risks contributed to a significant outbreak of *Salmonella* food poisoning.

12.4.1 The Peckham Poisoner

'The Peckham Poisoner' was one of my first ever consultancy clients. As an innocent private sector consultant, I was pleased to get the contract. As weeks passed, however, I began to realise that the owner had not been completely open with me. It transpired that the business had poisoned around 50 people with *Salmonella*-contaminated sandwich fillings over a period of several weeks, just prior to my appointment. The men in suits in the office were the receivers desperately trying to keep the business afloat.

The outbreak occurred as a result of a number of contributory factors:

1. The company was cooking poultry meat in large volume pouches (c. 5–7 kg). The HACCP plan had been developed on the basis of critical limits associated with a volume of 2.5 kg in each pouch. The product was therefore undercooked and bacterial survival was assumed to have occurred.
2. The pouches were filled with raw poultry meat and seasonings. These were tumbled for several hours to tenderise the meat, they were then cooked. Once cool, the content was removed from the pouch and sliced.
3. *Salmonella* from the undercooked meat got into nooks and crannies within the dedicated ready-to-eat product meat slicer.
4. The wash up sinks were not designed to enable full submersion of the large elements of the dismantled slicer. *The bowls were too small.*
5. The staff were not properly trained in cleaning and disinfection of the slicer. They did not know how to take it apart or how to clean it effectively. The slicer was often not dismantled but sprayed and wiped in situ.
6. Biofilms (see Section 12.8) grew on the inaccessible parts of the slicer, and wiping during and at the end of the day removed the majority of pathogens. However, overnight residual levels grew.
7. The first few slices of meat that were passed through the slicer first thing in the morning were heavily contaminated with levels sufficient to act as infections doses.
8. The outbreak went on for several weeks until epidemiological data linked the unusual strain to the common factor of sandwiches sold from certain outlets. Various fillings such as minted lamb as well as coronation chicken were contaminated, confusing the initial picture.

Here, hygiene management failures include the lack of account for cleaning and disinfection of equipment. The case highlights that where businesses fail in one area of hygiene management, they often fail in others. Root cause analysis is being used increasingly to see the immediate as well as underlying causes. Additionally, it can be useful to look at the issues in terms of human failures (e.g., failed to disinfect adequately), physical failures (e.g., not enough space in sinks to fit equipment to wash and disinfect adequately) and organisational failures (e.g., negligence in choosing to ignore the HACCP critical limits).

12.5 How surfaces become contaminated

In order to manage hygiene effectively, it is useful to have a background understanding of how microorganisms interact with surfaces and the factors that influence resilience within the food factory environment.

Attachment of bacteria to structural and food contact surfaces is of concern because of the potential for desorption and subsequent food contamination. The probability of this happening can be reduced by the selection of surfaces with appropriate performance criteria, including reduced attachment potential, ease of decontamination and durability. For different sectors of the food industry, the importance of individual criteria may vary according to the types of microorganisms likely to be encountered, legal constraints, and practical considerations such as the need for hosing down, or the use of cleaning 'in place' systems.

Microbial loads can be controlled by sanitation, although the effectiveness of the process will be governed by the design and management of the systems as described earlier. Where cleaning is inadequate, biofilms may develop and bacteria within surface films may be more protected from chemical cleaning treatments (Costerton, 1999; Mosteller and Bishop, 1993).

12.6 Microbial contamination of surfaces: organism attachment

Studies of bacterial attachment to substrata have revealed that the process is complex, being affected by properties of the bacterial cell surface, substratum surface and the environment in which the interaction occurs (Fletcher, 1991). Despite extensive research, it is still not possible to predict the extent to which microorganisms will attach to surfaces and the precise role of particular factors in the attachment process. Two general approaches have been used in studies of attachment (Fletcher, 1987; Hood and Zottola, 1995). In the first, microbial populations are analysed. The second approach is physicochemical and considers the bacterium as an inert colloidal particle. Attachment is considered in terms of electrokinetic theory and related to net surface charge, the concentration of electrolytes in solution, and surface energy and related parameters such as hydrophobicity. However, the interpretation of all studies is limited by the diversity of microbial surface structures, the influence of microbial metabolism and modification of the substratum due to macromolecule adsorption or dissociation (Fletcher, 1987).

12.6.1 Stages in the attachment process

Various authors have described the stages involved in microbial attachment to surfaces (Busscher and Weerkamp, 1987; Fletcher, 1987; Marshall et al., 1971; van Loosdrecht et al., 1990; Zottola and Sasahara, 1994). Cells arrive close to surfaces as a result of diffusive transport, convective transport or by active motility (van

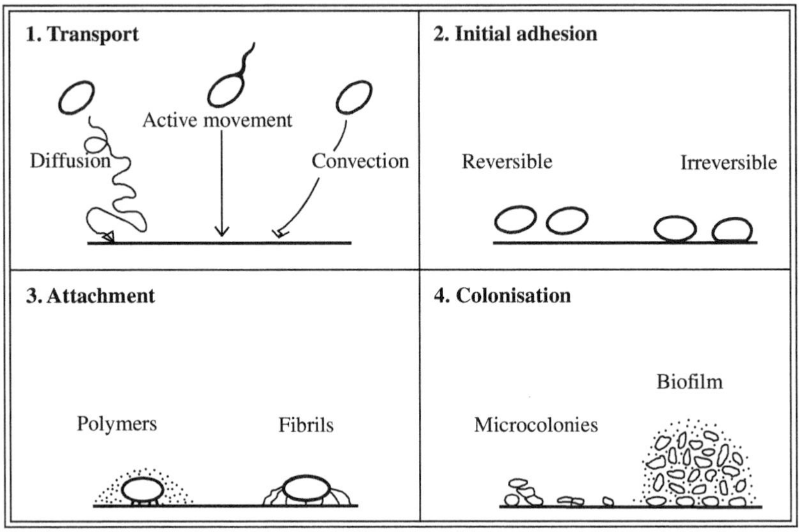

Figure 12.1 Schematic representation of the sequence of steps in the attachment of microorganisms to surfaces.
Source: Adapted from van Loosdrecht, M.C., Lyklema, J., Norde, W., & Zehnder, A.J., 1990. Influence of interfaces on microbial activity. Microbiol. Rev. 54 (1), 75–87.

Loosdrecht et al., 1990). If attachment is viewed as a thermodynamic process, minimisation of surface free energy is the driving force. The initial step involves the approach of the bacterium to the surface. At distances >50 nm, long-range forces such as electrostatic and van der Waals forces operate with the association between bacterium and substratum being reversible. Between approximately 50 and 15 nm, electrostatic attraction or repulsion forces (e.g., due to hydrophobic interactions) become stronger. Provided attractive forces are sufficiently strong to overcome localised repulsive forces, binding becomes increasingly irreversible, and cells are referred to as being adsorbed. At distances <15 nm, specific attachment occurs with covalent interactions anchoring the bacterium to the substratum (Bower et al., 1996; Hood and Zottola, 1995).The various stages of the attachment process are shown in Fig. 12.1.

12.6.2 Factors influencing attachment

12.6.2.1 Time

Attachment times are influenced by various factors and differ with the organism. The number of bacterial cells attaching to surfaces generally increases with time and this is often associated with cell surface modification and the production of extracellular adhesive materials by microorganisms already attached (Fletcher, 1987). The time taken for the first cells to attach to a surface appears to vary with substratum and environmental conditions and may be as short as 2.5 min or less

(Piette and Idziak, 1992). Mafu et al. (1990) report initial attachment of *Listeria monocytogenes* being influenced by temperature: 20 and 60 min respectively at 20°C and 5°C on stainless steel, glass, polypropylene and rubber. A number of other researchers refer to 15–60 min as typical for initial bacterial attachment for instance (Gelinas and Goulet, 1983; Piette and Idziak, 1992).

Significantly, although attachment may be time-related and associated with physiological activity, it is not always physiologically driven. Piette and Idziak (1992) report that attachment of Pseudomonas *fluorescens* to meat surfaces was not suppressed when cells were killed.

12.6.2.2 Substratum surface

The substratum surface can affect the rate and number of cells attaching and the production of extracellular material. Inherent properties of the surface such as topography, surface charge, hydrophobicity, chemical constituents and adsorption of macromolecules have all been shown to affect attachment. Various studies have investigated attachment of mixed cultures of representative genera on different surfaces, such as Suárez et al. (1992). Glass and brass are noted as surfaces where lower levels of bacterial attachment occur, such as Whitekettle (1991). Stainless steel and wood were among the surfaces where higher levels of attachment occurred such as Wolfaardt and Cloete (1992). Surface treatment with antimicrobials such as nisin has also been demonstrated to reduce colonisation (Bower et al., 1995).

12.6.2.3 Substratum surface conditioning

When a surface comes into contact with certain substances such as milk or other proteinaceous fluids, macromolecules can be adsorbed by the surface and alter its properties either facilitating or inhibiting microbial attachment. Once adsorbed, macromolecules may be displaced with other dissolved components that have a higher surface affinity (Marshall, 1985). Macromolecules can therefore be considered to be dominant or subdominant. Their role in the food industry is particularly important because equipment surfaces are quickly covered with foods, and surface modification can often occur soon after a new surface has been introduced into a food area.

12.6.2.4 Cell surface

Cell surfaces vary between bacterial species and can change with time and environmental conditions. The chemical constituents of the cell envelope will affect hydrophobicity and cell surface charge, and these factors appear important in influencing attachment. Surface functions such as the production of extracellular materials (Fletcher, 1987) and appendages such as flagellae, fimbriae or the hair-like structures of spores may also facilitate attachment (Bower et al., 1996; Gehin et al., 1996).

12.6.2.5 Cell background and environmental conditioning

Environmental conditions during cell growth (Hood and Zottola, 1997), culture age (Gilbert and Watson, 1971), and cell preparation procedures (Chakrabarti and Banerjee, 1991) affect physicochemical properties of the cell surface and may thus influence attachment. The presence of metabolisable substrates also influences attachment along with temperature. The rate of flow of bacterial suspension over surfaces is a further factor (Wolfaardt and Cloete, 1992) reporting greatest attachment occurs at zero velocity.

12.7 Microbial biofilms

Biofilms are composed of bacteria that colonise surfaces, their extracellular material and any other substances trapped in the resulting matrix (Hood and Zottola, 1995). In describing biofilms, Costerton et al. (1994) likened the structure to a primitive circulatory system with 'metabolic specialisation beginning to approach that seen in primitive multicellular eukaryotes'.

Microbial biofilms are formed at interfaces where cells interact with inert material. Examples of sites include pipework associated with liquid distribution systems, drainage channels and difficult to access parts of complex machinery.

Although it is acknowledged that simple laboratory microcosm studies may have shortcomings as models for complex natural systems (Warren et al., 1992), it is clear that bacteria in biofilms react in a different way to suspended cells in batch culture (Costerton et al., 1994).

The point at which surface-attached cells become 'biofilms' is gradual and variable and has not been clearly distinguished in the literature. Zottola and Sasahara (1994) suggest that biofilm development commences once cells are irreversibly bound to a substratum. However, according to Hood and Zottola (1995), a minimum of 48 h is necessary for complete biofilm development. Maturity to a point where cells are in a steady state, and old biofilm parts slough off may take 2–4 weeks (Zottola and Sasahara, 1994). Cell proliferation, especially for spores following germination, is typically accompanied by the production of extracellular materials, primarily extracellular polymeric substance (EPS), which may contain one or more anionic uronic acids (Costerton et al., 1994). As proliferation continues, microcolonies develop with concentrated areas of cells and matrix. Within the matrix, a dynamic environment develops with water channels, delivering oxygen and nutrients and facilitating the movement of metabolic products (De Beer et al., 1994; Sauer et al., 2007).

The structure of mixed biofilms is different from those comprised of single species. As the development of microcolonies occurs, those cells that are located adjacent to others, where a mutually cooperative relationship can develop, have a selective advantage. There may be substrate exchange or the metabolic end products of one species may be removed by a second species. Costerton et al. (1994) have suggested that in mixed biofilms, the cells present are phenotypically different from those in batch culture. The factors they claim are important in this respect are a fixed cell location; the presence of neighbouring cells of different types; the presence of a surface interface; and an EPS matrix.

12.8 Effect of antimicrobials on organisms within biofilms

The effect of antimicrobials on bacteria in biofilms is of particular relevance to the food industry because ineffective surface decontamination can directly or indirectly lead to contamination of food (Gibson et al., 1999). Bacteria that are attached to surfaces as well as those in biofilms are generally more resilient to cleaning and disinfection than those in suspension (Frank and Koffi, 1990; Lee and Frank, 1991; Wong and Ronner, 1993). The precise mechanisms for the resilience are not yet fully understood.

For individual cells in suspension, the entire surface is exposed to toxic cleaning agents; in contrast, surface-attached cells or those in biofilms are protected to a certain extent by the physical presence of the surface, other cells and extracellular polysaccharides and the associated diffusion barrier that it presents. Thus, the use of antimicrobials at increased concentrations compared to those effective against suspended cells may be necessary, unless the biofilm is disturbed (Hood and Zottola, 1995).

12.9 Organisation of cleaning and disinfection and its documentation

12.9.1 *What is cleaning and disinfection?*

Cleaning and disinfection is often referred to as sanitation. It is both a direct and indirect legal requirement. Food businesses and equipment must be clean. Further, food must not be placed on the market if it is unsafe. Poor cleaning and/or disinfection can cause contamination and/or cross-contamination, which can make food unsafe. The process should be designed to limit (and in most cases eliminate) contaminants: physical, microbiological, chemical and allergenic. In addition to controlling hygiene, these can impact on the safety (in terms of absence of physical contaminants such as packaging), organoleptic qualities and shelf life of a product.

12.9.2 *Expected outcomes from cleaning and disinfection*

Effective day-to-day cleaning and disinfection routines will remove soil and microorganisms. Over time, however, some soil deposits and microorganisms are retained on surfaces and build. Holah (2014a) suggests that this periodic build-up may take place over hours, but more typically occurs over several days or weeks. For this reason, in addition to 'clean as you go' and daily or end of shift cleaning, periodic 'deep' cleaning is required.

In terms of the bacterial log reduction for the cleaning and disinfection process, recent research conducted by Campden BRI in factories with good standards and systems suggests a combined 3–4 log reaction of microorganisms per unit area is achievable. This is made up of a c. 2 log reduction in the cleaning phase and a further c. 2 log reduction after disinfection.

12.10 Cost of cleaning and disinfection

Holah (2014a) suggests that the cost of chemical/cleaning products alone is approximately 5% of the overall cost of sanitation in food factories. This leaves 95% of the cost accounted for by labour, energy, water, etc. It is therefore crucial that the relatively modest cost of the products is not seen as disproportionate, leading to false savings on this vital commodity.

12.11 Establishing effective methods

Seeking expert advice in chemical choice and sanitation methodology is recommended. In any new situation, or where a nonconformance triggers a review, it will be my choice to ask a respected cleaning chemical supplier to send a representative to give advice on cleaning and disinfection prior to setting up the routines. The company's knowledge of soils, accessing hard-to-reach parts of intricate food equipment, effective chemicals and appropriate cleaning kit are invaluable. They can also account for the particular chemistry of local source water (for instance hardness, acidity, total dissolved solids) and this can affect chemical suitability and choice. Companies may offer to write the documentation, or leave this up to the food factory. It is my experience that documentation provided by respected suppliers can be very well designed and produced on paper/plastics that do not themselves pose a risk of contamination. They come in various formats, for example, laminated cleaning information cards, posters, etc. and can impress EHOs and certification auditors. With my expert witness hat on though, I must point out that over time processes, equipment and chemicals can change and sometimes the documentation does not keep pace. There is potential for risk and a lack of due diligence in such a situation. Regular and thorough reviews of the systems and documentation are necessary.

When writing documentation, if a business is working to a certification standard, it can be very helpful to write the documentation with an eye on the wording within the standard. This way, during audits, it is much easier for the auditor to identify compliance at a glance.

There are essentially two elements to the control of sanitation documentation: the 'how to' part, often referred to as a cleaning schedule, and the 'recording' part, which might be a cleaning record, or a document covering both the checking of cleaning completion along with any additional production checks.

Schedules can be displayed throughout the factory and particularly at cleaning stations as visual reminders but it is staff training in the execution of the tasks that is key.

12.11.1 Cleaning schedules

Common criteria on cleaning schedules include chemical, concentration, contact time, water temperature, method and tools of application, whether a potable water rinse is necessary, who the task is allocated to (actual person or capacity) and frequency. It is usual for a particular item or areas to be treated in different ways and sometimes with different chemicals depending on the type of cleaning activity. Mid-shift cleaning, for instance, will possibly not be as comprehensive as end of shift or periodic cleaning. It may be that the use of a sanitiser (combining detergent and disinfectant properties in one chemical) is fine for an hourly treatment but a full clean and disinfect with separate detergent followed by a rinse and then a disinfectant step occurs at the end of the shift.

It is important that the information detailed in cleaning schedules is displayed in such a way and with sufficient clarity that the users can fully understand requirements. Writing chemical names with concentrations in percentages or v/v (volume to volumes) may not be clear to some operatives.

12.11.2 Cleaning records

Records indicate that a sanitation task has been completed to a particular, defined standard. From an EHO and auditor's perspective sanitation records ideally need to have an identifiable set of signatures. The first, by the person completing the task – this shows responsibility for its execution – and second, by the person checking their work – this is usually a supervisor or someone from the quality assurance department.

12.11.3 Staff training in sanitation

Staff training in how to clean should have the status of critical control point (CCP) training. Sanitation is often defined as a prerequisite, or an operational prerequisite depending on how the HACCP plan and food safety management system is constructed, but this should not diminish focus on how critical it is to ensure effective training.

Jobs that involve the execution of cleaning in a factory are not generally given high status or high pay. Employers can be loath to pay for general food hygiene training for dedicated cleaners who do not specifically handle food. This can result in a lack of understanding and awareness of the dangers cleaning staff pose particularly if cleaning various areas in low-risk and high-care/high-risk areas of a factory. It is my recommendation that cleaners, if not factory line workers, receive general hygiene training alongside food handlers. They need to understand their own capacity to cross-contaminate.

In the United Kingdom, English as a second language and literacy issues can be a barrier to learning during training. These need to be genuinely accommodated. Factories that excel in this area automatically book translators for training courses. For all hygiene operatives, they also carry out and document an assessment of learning. This could be where a supervisor accompanies the learner onto the factory floor and observes the completion of tasks correctly along with the learner's explanation and a dialogue demonstrating understanding. I have argued in several cases that the absence of such verification of the effectiveness of training has left companies open to the suggestion of a lack of due diligence.

12.12 Cleaning arrangements and management

Sanitation can be effective through a variety of arrangements. It may be completed by:

1. factory line staff typically after production runs and at the end of a shift;
2. an in-house hygiene team and
3. external contractors.

or any combination of the above. Where external contractors are used, arrangements need to be in place to ensure operational collaboration with induction and training of operatives in the particular systems defined in the factory. Further, where new staff are recruited, these too must be trained and competent. The use of agency (typically temporary) staff can be a tricky area to manage and present similar issues to untrained regular staff.

12.13 Getting the chemical concentration right

This can be one of the most challenging areas of hygiene management. Chemicals may be subject to dispensing either automatically, via a calibrated pump system, or manually by the operative pouring or pumping to dilute the chemical.

Auto dosing, in theory, is the safer and more assured option. Chemical drawn directly from the manufacturer's container goes through a pump and is mixed with a calibrated amount of water, emerging at the right concentration. There is very limited potential for exposed chemical or spills.

It is important that the correct chemical is placed in association with its intended pump and line. For example, if substitution of the line for detergent with the line for disinfectant occurs and the two are used at different concentrations, sanitation will be affected.

If the pump is not properly calibrated, or recalibrated, it may dispense the wrong concentration of chemical. This can be critical in high-risk and high-care areas where antimicrobial properties are relied upon to keep food safe. In practice, business owners may not always fully appreciate the implications of this system not being optimal. I once had a client with resident *Listeria* in his factory where the dosing kit was old and unreliable. He had allocated a budget to replace all the sanitation chemical pumps and lines at the time of a forthcoming refurbishment. Luckily, and it was only luck, during the interim months the strains were nonpathogenic but the micro results from routine product sampling were not a pretty sight. Had an issue arisen, the owner would not have had a legal leg to stand on.

Where chemicals are mixed manually there is a strong human temptation to think the stronger the better. Provision of a calibrated measuring jug or cylinder with funnel and clear instructions is needed.

12.14 Changing chemicals

There can be a disconnect between individuals with the budget for sanitation products and those concerned with its effective execution. For example, head office order cleaning chemicals and products, and staff in the factory receive and use them. I had a client once where head office discovered that they could get the same particular chemical formulations more than one-third cheaper, so switched supplier. The new chemicals, however, had different names and different colours. Such a change might not have seemed significant for the office team ordering; indeed the annual saving was compelling. However, unbeknown to head office, the factory had – at huge expense –commissioned training videos showing the old chemicals. Staff had been trained and all cleaning equipment was colour coded. To change to the new regime meant that any saving in chemical costs were quickly wiped out by having to remake the videos, retrain staff and replace colour coded equipment.

Changing cleaning chemicals can be a significant step in hygiene management. Revised documents, systems, cleaning equipment and staff retraining should all be accounted for.

12.15 Records

In practice records help staff know what jobs have been done and what are still to do. They are critical for due diligence, and the sign-offs need to be honest and in real time. I have been involved as an expert witness in cases where, in addition to other factors, the sign-off is dated weeks after the sanitation task has taken place. In other words, the factory records have been completed and stockpiled and sent to the quality assurance or technical department at the end of the month to be signed off in a batch. I have successfully argued that this delay in verifying means that the standard of sanitation was never actually checked in person and thus significant doubt is cast on whether the job was completed to a sufficient standard. This has been enough to prove, on the balance of probability in a civil setting, breach in the duty of care to produce safe food as a consequence of a surface – and subsequently food – being contaminated, so causing illness.

Records may be paper or electronic, and each have their advantages. The crucial point is that they have to be genuine and show that the tasks took place according to the correct method. Any corrective actions should be included. So important is the integrity of recording that in one criminal case I worked on, the food handler and his supervisor both were imprisoned for perverting the course of justice – an offence borne of falsifying records.

12.16 Assessing the effectiveness of cleaning and disinfection

Any cleaning and disinfection programme, be it a prerequisite plan or a requirement as an operational prerequisite, will need to be effective. Testing will help provide confirmation that the system is working. Testing, however, can be an expensive business so asking the right questions about what information is being sought will help target resources effectively.

12.17 Overview of sampling methodologies

The simplest form of sampling of surfaces is visual. There is little point in going further, using expensive technology, if a visual test indicates deposits. Deeper surface sampling involves the removal and quantification of *something* from a surface. That *something* has traditionally been bacteria, however within the last 30 years we have benefited from rapid technologies that also measure adenosine triphosphate (ATP), protein, sucrose and allergens.

12.17.1 Use of swab, wipes and sponges in analysis

For hard environmental or equipment surfaces, swabs, wipes and sponges can be used to dislodge material and transfer microorganisms (usually into a buffer solution to retain in an unchanged state) until analysis. Sponges and wipes are generally used on larger areas; swabs on smaller areas. Typically this method looks for bacteria. These can be identified as either general counts or individual counts for pathogens, spoilage bacteria, etc., depending on the hygiene manager's specific requirements in the test. Confusingly, different labs and references give different names to the general count of all bacteria grown in air on plates: total viable count (TVC) aerobic colony count (ACC) and aerobic plate count (APC). There are technical definitions for each, as well as accredited protocols, but for the hygiene manager it can be a confusing start because in fact they are all pretty much the same.

It should be borne in mind that the number of bacteria recovered by swab for enumeration will be an underestimate of the total on the surface. Barnes et al. (1996) suggest only approximately one-fifth of the total on the surface are recovered and grow on plates probably because some fail to desorb and some may be lethally injured in the recovery process.

As swabbing and analysis can be expensive, it is important to ensure the person taking the swab is properly trained in the use of aseptic techniques.

12.17.2 Contact plates

Contact plates are simple to use. They are a plate with a specified growth medium (agar) – designed to look for all general aerobic bacteria, or a particular bacterial species. The agar is pressed onto the surface and subsequently incubated for the requisite time. The plates may be used in-house and do not always need to be sent to a lab for reading. The difficulty in using a contact plate is that it may simply be down to luck as to whether the number of colonies that grow are countable. If there is a large number, it will not be countable and can only be recorded as greater than X. In contrast, with swabs, the recovery medium is normally serially diluted and plated to enable the right dilution to capture the readable plate.

12.17.3 Rinses

Rinses are primarily used to assess decontamination of inaccessible surfaces such as where cleaning in place systems exist, and pipework does not lend itself to swabbing. As with swabbing, recovery of bacteria is unlikely to be complete.

12.17.4 Air plates

Air plates, sometimes called settle plates, are not a direct measure of surface cleanliness. They are more a measure of microbes within the air, an indirect measure of the effectiveness of general prerequisite controls.

12.17.5 Adenosine triphosphate testing

ATP testing uses a swab to wipe a surface and dislodge residues to analyse with a substrate and enzyme. ATP is found in live and dead cells (including but not exclusively bacteria) and thus in food. A reaction with luciferin and luciferase causes light to be emitted. The more contaminated a surface is, the more light is produced. It is measured in arbitrary 'relative light units' (RLUs).

In practice, ATP testing is relatively cheap and provides results within seconds or minutes. In the majority of systems, it will not differentiate between microbial and nonmicrobial sources.

Most food factories use ATP testing very successfully for determining whether a surface has been cleaned to a determined, adequate standard. Initial set-up trials are needed to establish the pass and fail standard. The trials may resemble the following: clean down on 10 occasions where the process has been carefully controlled and is known to be correct. The results for ATP swabbing are averaged and taken as the standard for a pass. Any results greater than a considered amount, for instance 25% higher, may be considered a fail with a corrective action likely to include a reclean and staff reassessment of competence. Setting the pass fail standard needs to be done carefully with full consideration of the implications.

12.17.6 Protein and carbohydrate swabs

Rapid method test swabs are commercially available for confirming the presence of surface residues. Protein or carbohydrate swabs are commonly used in food manufacturing environments. Some are stand-alone swabs with reagents in the swab cap. These cause a colour change to compare on a chart to indicate relative contamination levels. Others use a unit for reading the relative contamination level. The 'all in one' swabs are more expensive but there is no investment in the reader unit, those that are used with the reader are cheaper but the business needs to buy or lease the reading unit.

The protein swabs lend themselves to product of animal origin types of food businesses and the carbohydrate ones to more bakery-type sectors. For a hygiene manager, the results provide real-time confirmation of the effectiveness of cleaning at removing residues; they do not indicate a microbial status of a surface. They can be a helpful tool for the hygiene team supervisor to use daily after clean down to confirm effective cleaning. It is common to rotate the location of swab sites around a standard set of food contact sites post clean and this then gives data for trend analysis in the longer term.

12.17.7 Surface testing for allergen-free status

Although virtually all allergens are proteins of one sort or another, they can be present in various forms, liquid, solid, airborne particles, etc. Additionally, their ability to adhere and thus be removed varies with the chemical make-up of the food matrix, for instance it may be sticky, baked-on, etc. It is beyond the scope of this chapter to examine allergen removal by cleaning in detail or its verification; however, a greater understanding has developed in the last few years and readers are referred to specialist texts such as Nikoleiski (2015) or Holah (2014,b).

Rapid technology does offer hygiene managers test kits that can detect specific allergens down to 5–10 ppm on a surface. Risk assessment should be conducted to determine the most appropriate method for confirmation of a surface being freed from an allergen by a cleaning process.

In some food factories, the pass/fail standard for ATP testing has been calibrated against tests for specific allergen absence. From an expert witness point of view, this is potentially dangerous territory. Every time an allergen-containing run finishes and a nonallergen-containing run commences using the same equipment, a food business operator needs to know that the line is free from the allergen. In theory, one can calibrate and link a cleaning standard to a consistent negative allergen test, and this to an RLU measure using ATP. In practice, if I were a food business owner, I'd factor in the price of greater certainty and use specific allergen detection tests rather than an indirect method.

12.18 Reasons for sampling

There are likely to be four reasons for making surface sampling checks:

- *Monitoring* (e.g., CCP or equivalent real-time check at a step critical to food safety)
- *System validation* (e.g., demonstrating that a sanitation programme or part of it works)
- *System verification* (e.g., demonstrating that a sanitation programme or part of it is continuing to work on a day-to-day basis)
- *Special projects* (an in-depth study for various reasons, e.g., to establish the optimal interim in shift line cleaning frequency)

12.18.1 Monitoring

Monitoring, being HACCP Principle 4, is needed at a CCP to check that a point critical to food safety is under control. It can only happen in real time, that is, while the factory is operational so will inevitably use rapid technology. It may be that monitoring happens each time the food being prepared on a surface changes, to check the hygiene status of the surface. For instance, on a sandwich line where egg filling is handled followed by a nonegg-containing filling. Because of the allergen status of the egg, it is important to ensure that all egg residues have been effectively removed. Monitoring would therefore check to confirm the absence of the allergen after a clean-down following the egg batch, prior to starting a new batch of nonegg-containing sandwiches.

Depending on how the food safety management system, HACCP and prerequisites are set up, monitoring may be conducted in respect of CCPs or operational prerequisites. Notwithstanding how the point is defined, the important principle is that monitoring takes place during real operational time and its purpose is to determine the difference between safety (cleanliness – however it is defined) and non-safety (contamination).

12.18.2 System validation

When implementing a new sanitation process or chemical, the HACCP or prerequisite plan(s) would normally guide a hygiene manager to validate the system. In short, this is a check that a new cleaning system or chemical is going to work effectively. If the data shows it to be ineffective, further investigations will be needed to ask why. It may be the cleaning system, chemical concentration, contact time, solution temperature or it may be that a wider consideration of transmission routes of contamination and risk assessment will be needed. Once the results demonstrate that the system is robust, the evidence of validation should be retained. The validation process does not then need to be repeated unless there are changes – the principle of review in HACCP, although it could be helpful for due diligence to repeat the validation process occasionally, for example, every couple of years.

Validation is a key element in ensuring correct calibration of an automatic dosing system where sanitation chemicals are used. Being sure that a known volume of chemical is being delivered to a known volume of water is critical in ensuring that the concentration is correct. Wearing my auditor or expert witness hat

– a chemical won't work optimally if it is too dilute, and where is the evidence that the concentration is right?

In respect of chemical concentration validation, this can be done directly by litmus-type papers or a titration-type test by the supplier. Indirectly validation might be a series of surface bacterial counts following the cleaning process, though these would reflect the *process* as opposed to just the chemical *concentration*.

12.18.3 System verification

Once a sanitation programme has been put in place and validated as being effective there is a need simply to reassure oneself that it is operating according to the protocols and with the expected results. There are various ways to verify cleanliness and disinfection and it is important to define the standard: for instance, a particular target bacterial count, a measure of RLUs if an ATP system is being used, or the absence of indicator or pathogenic organisms, etc.

This monitoring or verification is normally carried out after sanitation. Its purpose is not to provide assurance of the safety of food, it is to provide reassurance that the system is working effectively. It might also be used as a step between cleaning and disinfection to demonstrate that the cleaning has achieved a set standard prior to the application of disinfection: a level of cleanliness of a surface is needed in order for the disinfection to work effectively.

12.18.4 Special projects

From time to time there may be a necessity to carry out a special project to answer specific hygiene management questions. It is likely to fall to the hygiene manager either to design the project or specify its form for a laboratory to become involved and be responsible for its execution. Such special projects can form important validation evidence within an HACCP system.

It is clear from academic sources that a surface, be it equipment or structural, can be modified following contact with a food matrix. Further, that bacteria can start to build on this modified surface within minutes. It can therefore be helpful to conduct studies to map the build-up of microbial numbers, particularly on high-care and high-risk product contact equipment or lines to understand where best to carry out interim cleaning. An example may be a processing line cutting and packing smoked salmon. An approach could be to take a sample every 15 min and compare bacterial levels. There is likely to be a point where levels are steady and don't increase despite continuing volumes of product passing over the surface (within one shift). If this point were at, say, 90 min, it could be prudent to introduce cleaning hourly or at least every 90 min.

12.19 Approaches to sampling strategies

Sampling strategies are about what to look for and when. Does one look for organisms that are indicators of hygiene (the *Enterobacteriaceae*), specific pathogens, spoilage bacteria, or how about indirect semiquantitate measures of hygiene such as RLUs using an ATP kit? Strategies should be developed to address the reasons for sampling and select suitable sites and frequencies.

Sampling strategies will link to the risk level of food. For instance a wet (raw) fish factory will require far less sampling compared with one that makes ready-to-eat products. Strategies will also be linked to the method of high-risk/high-care containment. Focus will be on checks to verify that pathogens are being kept out of the high-risk/high-care areas rather than to look for say pathogens in an area handling raw product, where one would expect to find them.

A 'collector site' approach is often helpful where ready-to-eat products are handled. One would not expect pathogens to be present in such an area. However, there are sites that over time are likely to collect pathogens that may occur very intermittently; these include floor drains, trolley wheels and fins on chilled air fans. A positive result indicates a need for a corrective action, but is not necessarily cause for immediate alarm. If the product contact sites are clear and the product is clear, then it probably indicates better attention to cleaning and disinfection of the measures for keeping pathogens out – physical barriers as well as systems and protocols.

Over time banks of results can be analysed and assist in validation and verification of sanitation systems. Triangulating product and environmental samples can also help provide reassurance about systems of sanitation and microbial quality of intermediate and finished products.

Sampling strategies will link to budget. The hygiene manager needs to pitch hard for the right sampling budget and ensure that the money is spent where it will have the greatest benefit. Samples can be pooled to give economies of scale, for instance, pooling swabs looking for indicator organisms on four identical lines; however if the results indicate a problem, it is more difficult to pinpoint the specific area for default, in this case the particular line.

12.20 Conclusion

As small food manufacturing businesses grow it is essential to fully account for hygiene management during the expansion. This includes physical elements of design, layout and equipment provision as well as system development and management. Expensive mistakes and risks can be easily averted with regular review and verification of products, volumes and HACCP plans. Failures in one area can have detrimental knock-on effects on others.

It is clear from the research that microbial attachment to surfaces is influenced by a number of factors. Further, attachment is not entirely predictable. This is partly because of gaps in knowledge, as well as the fact that experimental methodologies may not always recreate the reality of factors in a real, dynamic factory environment. Notwithstanding, there is good evidence that bacterial attachment occurs within minutes of exposure and that levels build over time and get to a point where there is little if any further increase in cell number as a result of attachment. Over time, if attached microorganisms remain undisturbed, their numbers can grow by multiplication within biofilms. Surface-attached bacteria and those enveloped within biofilms are more resistant to removal and destruction from cleaning and disinfection.

The systematic organisation of cleaning and disinfection and its documentation in terms of 'how to', plus the evidence of it having been completed correctly will assist in the effective maintenance of food safety. Professionals external to the food manufacturer can be very helpful in advising on appropriate routines and chemicals. Proper management of processes including completion of records is vital to show due diligence.

Sanitation verification is an important part of a food businesses HACCP-based system. It forms part of due diligence in demonstrating that systems to avoid contamination of the final product are effective. With limited resources, the hygiene manager should establish the purpose of sampling, be it monitoring (real-time checking of a cleaning standard), validation (proving that the cleaning and disinfection system works) or verification (confirming that cleaning and disinfection systems continue to work). The type of samples should be considered (general counts, indicators, pathogens, RLUs, etc.) and targets and trend analysis established. Sanitation performance can then be reviewed and any necessary improvements effected.

References

Allen, R.E. (Ed.), 1990. The Concise Oxford Dictionary, eigth ed. Clarendon Press, Oxford.

Barnes, B.I., Cassar, C.A., Halablab, M.A., Parkinson, N.H., Miles, R.J., 1996. An in situ method for determining bacterial survival on food preparation surfaces using a redox dye. Lett. Appl. Microbiol. 23 (5), 325–328. Available from: https://doi.org/10.1111/j.1472-765X.1996.tb00200.x.

Bower, C.K., McGuire, J., Daeschel, M.A., 1995. Suppression of *Listeria monocytogenes* colonisation following adsorption of nisin onto silica surfaces. Appl. Environ. Microbiol. 61, 992–997.

Bower, C.K., McGuire, J., Daeschel, M.A., 1996. The adhesion and detachment of bacteria and spores on food-contact surfaces. Trends Food Sci. Technol. 7, 152–157.

Busscher, H.J., Weerkamp, A.H., 1987. Specific and non-specific interactions in bacterial adhesion to solid substrata. FEMS Microbiol. Lett. 46 (2), 165–173. Available from: https://doi.org/10.1016/0378-1097(87)90062-0.

Chakrabarti, B.K., Banerjee, P.C., 1991. Surface hydrophobicity of acidophilic heterotrophic bacterial cells in realtion to adhesion on minerals. Can. J. Microbiol. 37, 692–696.

Costerton, J.W., 1999. The role of bacterial exopolysaccharides in nature and disease, (Volume 26). J. Ind. Microbiol. Biotechnol. 22 (4–5), 551–563. Available from: https://doi.org/10.1038/sj.jim.2900665.

Costerton, J.W., Lewandowski, Z., Debeer, D., Caldwell, D., Korber, D., James, F., 1994. Biofilms, the customized microniche. J. Bacteriol. 176, 2137–2142.

De Beer, D., Stoodley, P., Roe, F., Lewandowski, Z., 1994. Effects of biofilm structures on oxygen distribution and mass transport. Biotechnol. Bioeng. 43 (11), 1131–1138. Available from: https://doi.org/10.1002/bit.260431118.

EU Commission, 2004. Regulation (EC) No 852/2004. Official Journal of the European Communities, 2002 (May 2002), 1–54. Retrieved from <http://eur-lex.europa.eu/LexUriServ/LexUriServ.do?uri = OJ:L:2004:139:0001:0054:en:PDF>.

Fletcher, M., 1987. How do bacteria attach to solid surfaces? Microbiol. Sci. 4, 133–136.

Fletcher, M., 1991. The physiological activity of bacteria attached to solid surfaces. Adv. Microb. Physiol. 32, 54–85.

Frank, J.F., Koffi, R.A., 1990. Surface-adherent growth of *Listeria monocytogenes* is associated with increased resistance to surfactant sanitizers and heat. J. Food Prot. 53, 550–554.

Gehin, A., Gelhaye, E., Petitdemange, H., 1996. Adhesion of *Clostridium cellulolyticum* spores to filter paper. J. Appl. Microbiol. 80, 187–190.

Gelinas, P., Goulet, J., 1983. Efficacité de huit désinfectants sur trois types de surfaces contaminées par *Pseudomonas aeruginosa*. Can. J. Microbiol. 29, 1715–1730.

Gibson, H., Taylor, J.H., Hall, K.E., Holah, J.T., 1999. Effectiveness of cleaning techniques used in the food industry in terms of the removal of bacterial biofilms. J. Appl. Microbiol. 87 (1), 41–48. Available from: https://doi.org/10.1046/j.1365-2672.1999.00790.x.

Gilbert, R.J., Watson, H.M., 1971. Some laboratory experiments on various meat preparation surfaces with regard to surface contamination and cleaning. Int. J. Food Sci. Technol. 6 (2), 163–170. Available from: https://doi.org/10.1111/j.1365-2621.1971.tb01604.x.

Holah, J.T., 2014a. Cleaning and disinfection practices in food processing. In: Leliveld, M.L. M., Holah, J.T., Napper, D. (Eds.), Hygiene in Food Processing Principles and Practice, second ed. Woodhead Publishing Limited, Cambridge, pp. 259–304.

Holah, J.T., 2014b. Microbiological environmental sampling, records and record interpretation. In: Leliveld, H.L.M., Holah, J.T., Napper, D. (Eds.), Hygiene in Food Processing Principles and Practice, second ed. Woodhead Publishing Limited, Cambridge.

Hood, S.K., Zottola, E.A., 1995. Biofilms in food processing. Food Control 6 (1), 9–18.

Hood, S.K., Zottola, E.A., 1997. Growth media and surface conditioning influence the adherence of *Pseudomonas fragi, Salmonella typhimurium*, and *Listera monocytogenes* cells to stainless steel. J. Food Prot. 60 (9), 1034–1037. Retrieved from <Go to ISI>://CABI.19980402337.

Lee, S.-H., Frank, J.F., 1991. Inactivation of surface-adherent *Listeria monocytogenes* hypochlorite and heat. J. Food Prot. 54 (1), 4–11. Retrieved from <http://www.ingentaconnect.com/content/iafp/jfp/1991/00000054/00000001/art00001>.

Mafu, A.A., Roy, D., Goulet, J., Magny, P., 1990. Attachment of *Listeria monocytogenes* to stainless steel, glass, polypropylene, and rubber surfaces after short contact times. J. Food Prot. 53, 742–746.

Marshall, K.C., Stout, R., Mitchell, R., 1971. Mechanism of the initial events in the sorption of marine bacteria to surfaces. J. Gen. Microbiol. 68 (3), 337–348. Available from: https://doi.org/10.1099/00221287-68-3-337.

Marshall, K.C., 1985. Mechanisms of bacterial adhesion at solid-water interfaces. In: Savage, D.C., Fletcher, M. (Eds.), Bacterial Adhesion: Mechanisms and Physiological Significance. Plenum Press, New York, US, pp. 133–161.

Mosteller, T.M., Bishop, J.R., 1993. Sanitizer efficacy against attached bacteria in a milk biofilm. J. Food Prot. 56 (I), 34–41.

Nikoleiski, D., 2015. Hygienic design and cleaning as an allergen control measure. In: Flanagan, S. (Ed.), Handbook of Food Allergen Detection and Control. Elsevier, pp. 90–102.

Piette, J.P., Idziak, E.S., 1992. A model study of factors involved in adhesion of *Pseudomonas fluorescens* to meat. Appl. Environ. Microbiol. 58, 2783–2791.

Sauer, K., Rickard, A.H., Davies, D.G., 2007. Biofilms and biocomplexity. Microbe Mag. 2 (7), 347–353.

Suárez, B., Ferreirós, C.M., Criado, M.T., 1992. Adherence of psychrotrophic bacteria to dairy equipment surfaces. J. Dairy Res. 59 (3), 381–388. Available from: https://doi.org/10.1017/S002202990003065X.

van Loosdrecht, M.C., Lyklema, J., Norde, W., Zehnder, A.J., 1990. Influence of interfaces on microbial activity. Microbiol. Rev. 54 (1), 75–87. Retrieved from <http://www.pubmedcentral.nih.gov/articlerender.fcgi?artid = 372760&tool = pmcentrez&rendertype = abstract>.

Warren, T.M., Williams, V., Fletcher, M., 1992. Influence of solid surface, adhesive ability, and inoculum size on bacterial colonization in microcosm studies. Appl. Environ. Microbiol. 58 (9), 2954–2959.

Whitekettle, W.K., 1991. Effects of surface-active chemicals on microbial adhesion. J. Indust. Microbiol. Biotechnol. 7, 105–116.

Wolfaardt, G.M., Cloete, T.E., 1992. The effect of some environmental parameters on surface colonization by microorganisms. Water Res. 26 (4), 527–537. Available from: https://doi.org/10.1016/0043-1354(92)90054-8.

Wong, A.C., Ronner, A.B., 1993. Biofilm development and sanitizer inactivation of *Listeria monocytogenes* and *Salmonella*... – Abstract – UK PubMed Central. J. Food Prot. 56 (9), 750–758.

Zottola, E.A., Sasahara, K.C., 1994. Microbial biofilms in the food processing industry – Should they be a concern? Int. J. Food Microbiol. 23 (2), 125–148.

Third-party audit schemes

13

Kevin Swoffer
NSF International Global Food Division, Hanborough Business Park, Long Hanborough, Oxon, United Kingdom

13.1 Introduction: the development of third-party audit schemes

The development of accredited third-party audit schemes can be traced back to the early 1990s in the United Kingdom. With the introduction of the UK Food Safety Act 1990 (UK Legislation, n.d.), there were significant changes in the way in which companies could defend themselves in the event of a legal challenge relating to food safety. This Act removed the injustice of strict liability by allowing a defence to be established if a person(s) were held responsible for committing a food safety offence for which he had no responsibility at all, or because of an accident or some cause beyond their immediate control. This provision is known as the due diligence defence, and was responsible for the changes in the way in which companies worked with each other. Prior to the introduction of the UK Food Safety Act 1990, legislation contained the warranty defence, whereby a person accused of an offence would escape conviction, if he could prove that, when he procured the product he obtained a written warranty from the supplier stating that the product was safe and could be legally sold; the UK Food Safety Act removed this provision.

The UK Food Safety Act 1990, with respect to the UK food industry, had far-reaching implications, particularly for the suppliers of retailers' own brand product. The requirements for the due diligence defence are defined under section 21 of the UK Food Safety Act 1990 as follows:

> *...it shall... be a defence for the person charged to prove that he took all reasonable precautions and exercised all due diligence to avoid the commission of the offence by himself or by a person under his control.*

The situation was complicated by organisations that were neither the manufacturer nor the importer of the food, for example, retailers, food service and wholesalers. In such circumstances, the statutory defence shall be deemed to be established if the organisation can prove any of these points in section 21, subsection (3) of the Act:

a) 'that the commission of the offence was due to an act or default of another person who was not under his control, or to reliance on information supplied by such a person;
b) that he had carried out all such checks of the food in question as were reasonable in all the circumstances for him to rely on checks carried out by the person who supplied the food to him, and

Swainson's Handbook of Technical and Quality Management for the Food Manufacturing Sector.
DOI: https://doi.org/10.1016/B978-1-78242-275-4.00013-7

c) that he did not know and had no reason to suspect at the time of the commission of the alleged offence that his act or omission would amount to an offence under the relevant provision.'

OR as contained in subsection (4) of the Act:

a) 'that the commission of the offence was due to an act or default of another person who was not under his control, or to reliance on information supplied by such a person:
b) that the sale or intended sale of which the alleged offence consisted was not a sale or intended sale or intended sale under his name or mark; and
c) that he did not know, and could not reasonably have been expected to know, at the time of the commission of the alleged offence that his act or omission would amount to an offence under the relevant provision.'

Therefore in the event of an offence, in accordance with section 21 of the UK Food Safety Act, the defendant is required to prove that he took all reasonable precautions *and* exercised all due diligence. In reality, the UK courts would have to consider whether the steps taken by a defendant are sufficient to demonstrate that a person(s) has taken all reasonable precautions and exercised all due diligence. Therefore an organisation wishing to use a due diligence defence must take positive steps to ensure that it has appropriate systems and procedures in place and that these are fully operational.

With the introduction of the UK Food Safety Act 1990, there was a great deal of debate within the UK food industry concerning the specific sectors of the supply chain and what would be regarded as being reasonable due diligence; for example, a small retailer who purchases purely branded products would not be expected to undertake the same control measures as a large retailer who has a range of its own retailer branded products.

In February 1991, a number of UK food industry trade bodies (the National Consumer Council, Local Authorities Co-ordinating Body on Trading Standards, Institution of Environmental Health Officers, National Farmers Union, Retail Consortium and the Food and Drink Federation) published a guidance document, Food Safety Act 1990: Guidelines on the Statutory Defence of Due Diligence (Food and Drink Federation & National Consumer Council, 1991).

This document provided clear guidance and emphasised the need for businesses to operate control systems based on hazard analysis critical control points (HACCP), quality assurance principles and codes of best practice. The document broke the food industry down into specific sectors: farmers and growers, food processors, wholesalers, importers and distributors, small-scale retailing, larger scale retailing and retailers' own brands. Each sector had guidance on the controls that would be expected to be in place to ensure a due diligence defence could be used if an offence were to occur.

The guidelines advised that large retailers' control systems should operate 'downwards' through systems within their own organisation and 'upwards' through the control of their suppliers. The large retailers should be able to demonstrate that they use reliable suppliers who:

1. follow accepted codes of practice and have effective quality and quantity control procedures and
2. provide written confirmation that the food meets the quality required. This written confirmation should be as specific as possible.

In addition, the guidelines suggested that large retailers purchasing product from new suppliers or suppliers outside the EU may require details to be provided of the suppliers' own control systems or the results of any testing.

Large retailers who sold product under their own labels and whose products were produced or processed by other companies were considered to have influence over these suppliers and were regarded as having a high level of responsibility for the product. In addition to the requirements of being a large retailer, further guidelines were applied to retailers that sold own brand products as they could not transfer their liability to their suppliers.

There were five additional requirements for retailers selling own brand products:

1. To ensure the product has a detailed specification which is not unlawful or inconsistent with compositional standards and good manufacturing practice.
2. *To satisfy themselves that the supplier is competent to produce and/or process the product specified, complies with all relevant legal requirements and operates systems of production control in accordance with good manufacturing practice or agricultural practice.*
3. *From time to time to make visits to suppliers, wherever practical to verify point 2 or to receive the result of any other audit of the suppliers systems for that purpose.*
4. To devise a programme, having regard to the relative hazards of the product, for random samples to be taken for examination, testing or analysis to check continuing compliance with specifications.
5. To monitor any customer complaints about own brand products and to take any necessary action when problems are found.

Since the late 1940s, many UK retailers have employed teams of food scientists and food technologists to work closely with their suppliers of retailers' own brand product, assuring high levels of safety and quality of these products. As the commercial relevance of own brand product grew, so did the importance of the brand reputation to UK retailers and significant numbers of food scientists, food technologists, home economists and laboratory staff were directly employed by the major UK retailers.

With the introduction of the due diligence defence within the UK Food Safety Act 1990 and its relevance to brand protection, the larger UK retailers enhanced their control systems and developed their own codes of practice and supplier auditing programmes in line with the guidelines agreed by the food industry. Therefore, particularly in the early part of the 1990s, second-party audits of the suppliers of retailer brand products became prevalent, with the corresponding increase in resources at the larger UK retailers. These audits varied in scope, covering product safety, product quality or both, and systems were unique to the retailer. The expectations of ISO 9000 (then BS5750) in the UK food industry were largely unfulfilled, due mainly to the fact that the standard left individual companies to identify risks and establish their own quality standard that they believe meets legislative requirements. The standard was also seen by the UK food industry as being broad and not necessarily focusing on specific requirements such as HACCP and food safety prerequisite programmes. The outcome was the development and implementation of individual standards and codes of practice owned by retailers or developed on their behalf.

In 1995, commercial pressures and market forces on the UK food retail industry meant that retailers reviewed their control mechanisms in relation to risk and started to look at the redeployment of technical resources into areas of commercial benefit such as new product development, product quality and new technologies. However, senior technical managers at UK retailers were mindful of their food safety obligations and the 'reasonable' level of activity expected within the due diligence guidelines. Some retailers had made some changes towards a third-party audit approach a number of years before the introduction of the Food Safety Act 1990, but only to a very limited degree.

In mid-1995, the major UK retailers met to discuss how they could work together under the leadership of the British Retail Consortium (BRC) to develop a third-party audit standard for manufactured products and how this could be implemented as a scheme which they could utilise. The culmination of the work of the BRC working groups was a standard, the BRC Technical Standard for Companies Supplying Retailer Branded Products (September 1998) and an associated protocol document, Technical Standard for Companies Supplying Retailer Branded Products, The Inspection Protocol (October 1998) (Swoffer, 2012).

Similarly several UK retailers, working closely with other retailers in Europe, were instrumental in developing a good agricultural practice scheme and took this forward to a Euro-Retailer Produce Working Group in 1997 and the EUREPGAP scheme was formed (GLOBALG.A.P. n.d.a).

Elsewhere in the world, where there was no statutory due diligence defence requirement, nor the substantial growth in the sale of retailer brand products, many progressive organisations did carry out second-party audits to maintain an appropriate level of understanding and control of their suppliers, but not at the intensity of the UK retail environment.

13.2 The growth of third-party audit schemes

Initially the utilisation of third-party schemes was very slow due primarily to the need for confidence to be gained by the users of the schemes. Whereas second-party audits are undertaken by parties that have an interest in the organisation (e.g., a customer's technologist or auditors that are acting on the behalf of that party, e.g., a customer instructs the auditor), a third-party audit is an objective and independent assessment undertaken by an organisation that is neither an employee of a customer nor directly acting on the behalf of any other company.

Because the auditee commissions an organisation to undertake a third-party audit, problems were initially experienced in relation to the ownership of the audit report. The report is owned by the auditee and there are confidentiality issues regarding the sharing of information with third parties. Another crucial factor, which still generates a great deal of debate today, is the level of competence of the individual auditor or auditing team. A great deal of work has been done over the last 20 years to define and validate the qualifications, skills, knowledge and experience of food safety auditors. However, in the United Kingdom, when such schemes as the BRC Food Standard were being introduced, the auditor would invariably be called by the courts as an expert witness in the event of a food safety offence. Given the evidence that an auditor may give is that of an opinion, as it is highly unlikely that the offence took place while the auditor was undertaking the audit, even though the audit is impartial and objective the retailer may feel unease in entrusting their brand reputation to a third party.

Third-party auditing schemes have worked hard to build confidence with users of the schemes, and the schemes that are successful today have had to meet the demands of the users. These schemes have changed beyond recognition from their inception.

The development of the schemes has taken place over a number of years, but can be summarised as follows:

- the implementation of constitution and appropriate governance structures;
- the development and implementation of supporting systems and procedures;
- the use of databases to manage and control systems;
- the continual revision of HACCP-based standards, reflecting best practice and putting in place requirements to meet new food safety issues, for example, food allergen control;
- highly focused on continual compliance to their requirements by certification bodies and accreditation bodies, for example, auditor competence and consistency;
- the development of supporting guidance, codes of practice and training programmes;
- the development of relationships with key stakeholders in the supply chain;
- the implementation of integrity programs to monitor certification activities and
- continually increased internal resources to meet industry demands.

Despite these early implementation issues, third-party auditing schemes have continued to develop and are widely used across all sectors of the food industry, for example, food packaging manufacture, storage and distribution and agents.

The third-party auditing approach differs from the traditional ISO standards approach as the owners of the scheme strive to meet the demands of the food industry by utilising their organisation's resources, which provides increased assurance of continuing and consistent compliance to defined standards. In other words, there is another level of management and control which sits above the traditional ISO standards process. For users of schemes, such as multinational manufacturers and retailers, food is traded on a global basis with highly complex supply chains and this added assurance of consistent and continued supplier monitoring is a highly important factor within their own food safety management systems.

There have also been a number of benefits of the introduction of third-party auditing schemes which are not well appreciated. For example, as many of the scheme standards are reviewed and revised on a regular basis, the standards requirements complement changes in legislative requirements and in many cases translate regulatory requirements into meaningful operational requirements. Given the globalisation of the food supply, the schemes, many of which are used in a large number of countries, have emphasised the need for compliance with legislative requirements in both the country of supply and the country of usage. This must be seen by the legislators and government enforcers as being beneficial and in line with self-regulatory principles which are actively encouraged by a number of governments.

If one closely reviews the format and content of the major international third-party auditing schemes, notably BRC and Safe Quality Food (SQF), over the last 17 years the changes are substantive (particularly the BRC scheme, where originally there were two levels of certification and a series of recommendations). It could be argued that third-party auditing schemes have had a significant influence on the global food supply chain in relation to safety. A note of caution: Many believe third-party auditing is the panacea to solve the issue of supplier management; it is not. Referring back to the UK Food Safety Act 1990, Guidelines on the Statutory Defence of Due Diligence, third-party auditing is one of a number of tools that an organisation should use to monitor supplier performance and product safety. Third-party auditing is a systematic, independent and documented process but it cannot predict nonconformance at all times; it is still the organisation that has direct influence and ultimate responsibility for food safety. Progressive organisations have utilised third-party auditing as one of the tools in the 'toolbox' of their food safety management system, and realise there are certain limitations within the auditing approach and that by the use of third-party auditing they cannot divorce themselves from their corporate legal obligations.

There are numerous organisations which own and manage third-party auditing schemes and these vary greatly in relation to the scope of their operation and the geographic location(s) where certification is awarded. The key scheme owners, who are international in their operation, are those which are recognised by the GFSI (n.d.a), a scheme that has become very familiar to those working in the food industry over the last 15 years.

These organisations are:

- British Retail Consortium (BRC) (BRC Global Standards, 2014)
- CanadaGAP (CanadaGAP, 2015)
- FSSC22000 (Food Safety Certification System, 22000, 2015)
- Global Aquaculture Alliance (GAA) (General Global Aquaculture Alliance, 2011)
- GLOBALG.A.P. (GLOBALG.A.P., n.d.b)
- Global Red Meat Standard (Global Red Meat Standard, n.d.)
- International Featured Standards (IFS) (IFS, n.d.)
- PrimusGFS (PrimusGFS, 2014)
- Safe Quality Food (SQF) (SQF Institute, 2015)

To provide some perspective as to the success of these schemes, they represent over 77,000 certificates, which are issued annually in over 150 countries.

In addition to these international schemes, there are third-party auditing schemes which are very specific in their scope and only operate with one country or region. These have limited recognition outside the country where they operate, but are important within a single market or as an industry norm. These schemes still require the level of control provided by standards such as ISO/IEC 17021 (ISO/IEC 17021:2011, Conformity assessment 2011) and ISO/IEC 17065 (ISO/IEC 17065:2012, Conformity assessment 2012), using national accreditation bodies to assess the competence of the certification bodies involved in the scheme. Such organisations are numerous, but good examples are:

- Red Tractor Farm Assurance Standard (Assured Food Standards, 2015)
- Genesis Quality Assurance Scheme (Genesis Standards, n.d.)
- British Egg Industry Council Code of Practice for Lion Eggs (British Lion Quality, 2015)
- Canadian Grain Commission – Harvest Sample Program (Canadian Grain Commission, 2014)

In recent years there have been schemes developed that follow the principles of third-party auditing schemes, but that deliberately have not used any formal accreditation by national accreditation bodies. The reason for this is because it allows the ability to mentor and advise companies within the scheme, which is not possible with accreditation; however these schemes would still be regarded as third-party auditing schemes and are used as such by sectors of the food industry. A good example is the Safe and Local Supplier Approval (SALSA) scheme (SALSA, 2011), which was developed in 2007 by four trade bodies within the United Kingdom to promote best practice food safety requirements in small and medium-sized enterprises. The need for such a scheme arose from consumer demand for local food and the appreciation that these suppliers of artisan locally produced food could not meet the full management requirements of international schemes such as the BRC Food Standard. The scheme was originally sponsored by the UK Government, the ownership of the scheme rests with the British Hospitality Association, the BRC, the Food and Drink Federation and the National Farmers Union, and the management of the scheme is undertaken under contract on behalf of the Institute of Food Science and Technology (IFST, 2015).

The SALSA scheme standard and scheme management structure does however meet the food safety requirements necessary for a due diligence defence. Such a scheme differs from the traditional accredited schemes in that the individuals who mentor and audit suppliers can advise these smaller suppliers, whereas certification bodies under ISO standard accreditation cannot provide consultative advice or guidance to the auditee. It must be pointed out that the management of the SALSA scheme mirrors the principles applied within the ISO certification standards and fulfils an important niche in the food supply chain, and that some SALSA certified companies have moved forward to formally accredited certification under one of the larger international schemes.

These types of third-party auditing schemes have their place in the food supply chain and users of the schemes have to decide the type(s) of scheme most appropriate for their own food safety management system.

An interesting development concerning formal recognition by government authorities was the commissioning of research by the UK Food Standards Agency in 2010 (Food Standards Agency, 2013). The project was entitled *Assessment and comparison of the third party assessment schemes in the food sector: Towards a common framework* (project no. 245006) and culminated in a final report of the same name in October 2013. The project objective was to identify commonly used UK-based third-party assurance schemes, establish criteria for acceptability and assess a number of current schemes against these criteria. The UK Food Standards Agency had recognised that in recent years third-party assurance schemes had been developed in response to demands from customers, for example, supermarkets, for independent verification of food safety, rather than performance of the functions associated with inspections and other interventions by regulators.

The final report concluded that many of the third-party assurance schemes meet the criteria for the UK Food Standards Agency and they should consider advising enforcement bodies, such as local authorities, to take the assurance scheme's audit results into account when scheduling routine surveillance inspections (Wright et al., 2013).

13.3 The relationship of certification and accreditation with third-party auditing schemes

The processes of certification and accreditation are critically important for scheme owners to provide confidence in the competence of companies undertaking audits and issuing certificates to the scope of their scheme, that is, third-party certification bodies. Notwithstanding the importance of certification and accreditation, there are some schemes which do not recognise formal accreditation of the certification bodies within their scheme, but the scope of these schemes are not normally related to food safety, for example, product quality marks and sustainability. As a general rule, virtually all major food safety schemes have accreditation of their certification bodies and the formal certification of the auditee as a fundamental requirement.

The rationale for this requirement sits with the due diligence and incident mitigation approach, where there is an additional level of control and management exercised by organisations, namely national accreditation bodies, who are recognised by governments, and in some cases owned by governments. The accreditation and certification processes are based on internationally recognised conventions and ISO standards, such as ISO/IEC 17021 (ISO/IEC 17021:2011, Conformity assessment 2011), ISO/IEC 17065 (ISO/IEC 17065:2012, Conformity assessment 2012) and ISO/IEC 17011 (ISO/IEC 17011:2004, Conformity assessment 2004), which provide the principles to inspire confidence to business.

These principles are:

- impartiality,
- competence,
- responsibility,
- openness,
- confidentiality,
- responsiveness to complaints.

Therefore the value of accreditation and certification is the degree of public confidence and trust in product and services that is established and gained by an impartial and competent assessment. This confidence has led to the facilitation of trade, greater acceptance of product through open market access and minimisation of costs associated with multiple certifications.

Accreditation is a process by which an authoritative body gives formal recognition of the competence of certification bodies services against an international standard. The principals regarding conformity assessment are set out in ISO/IEC 17000:2004, *Conformity assessment – vocabulary and general principles* (ISO/IEC 17000:2004, 2004).

The standard used by national accreditation bodies to accredit certification bodies in the food industry is ISO/IEC 17011:2004, *Conformity assessment – general requirements for accreditation bodies accrediting conformity assessment bodies* (ISO/IEC 17011:2004, 2004).

Certification is a process by which accredited certification bodies, based upon an audit, provide written or equivalent assurance that fulfilment of specified requirements has been demonstrated; requirements can relate to products, processes, systems or persons.

Certification in the food industry utilises the standard ISO/IEC 17021:2011, *Conformity assessment – requirements for bodies providing audit and certification of management sys*tems (ISO/IEC 17021:2011, 2011), or the standard ISO/IEC 17065:2012, *Conformity assessment – Requirements for bodies certifying products, process and services* (ISO/IEC 17065:2012, 2012).

The ISO/IEC accreditation and certification standards are detailed documents which clearly and concisely specify the requirements of each process and, like all ISO standards, are subject to review and revision on a regular basis.

The vocabulary used in the food industry has confused matters considerably and in the past the word 'accreditation' has been used incorrectly and out of context. In summary:

- Accreditation bodies *ACCREDIT* certification bodies to an international accreditation standard (e.g., ISO/IEC 17011:2011).
- Certification bodies *CERTIFY* companies to the scope of a scheme standard (e.g., ISO/IEC 17065:2012, 2012 for the scope of the BRC Food Standard or ISO/IEC 17021:2011 for the scope of FSSC 22000).

Within both processes, the status of an organisation can change if continued conformance is not maintained; a certification body can have its accredited status suspended or withdrawn by a national accreditation body and a certified organisation can have its certificate suspended or withdrawn by the awarding certification body. In the case of the certified company, the certificate is owned by the certification body and not by the certified company, which is not always fully appreciated by the certified company.

There is a global dimension to accreditation and considering the globalisation and complexity of the food supply chain, the consistent control and management of certification bodies is vitally important for the international scheme owners.

The International Accreditation Forum (IAF) (IAF, n.d.) provides this global dimension and is the world association of conformity assessment accreditation bodies and other organisations interested in conformity assessment such as certification body associations, industry associations and other key stakeholder organisations.

The IAF coordinates accreditation bodies, who undertake the evaluation of activities of conformity assessment bodies against recognised standards; the IAF thus provides assurance of their impartiality and competence. The IAF is made up of six regional accreditation associations:

- African Accreditation Cooperation (AFRAC) (AFRAC, 2014)
- Inter-American Accreditation Cooperation (IAAC) (IAAC, 2014)
- ARAB Accreditation Cooperation (ARAC) (ARAC, n.d.)
- European Co-operation for Accreditation (EA) (EA, 2015)
- Pacific Accreditation Cooperation (PAC) (PAC, 2009)
- Southern African Development Community in Accreditation Service (SADCAS) (SADCAS, 2015)

One of primary activities of the IAF is the development and maintenance of the multilateral recognition arrangements (MLAs). An MLA is an arrangement whereby signatory accreditation bodies are evaluated by peers as competent, and sign arrangements that enhance the acceptance of products and services across national borders. Signatories to IAF MLA's must recognise the certificates issued by certification bodies accredited by all other members of the MLA. This network of accreditation bodies therefore creates a framework to support international trade through the removal of technical barriers to international trade.

The IAF has a number of committees that develop and publish a series of documents which provide the constitutional framework by which the IAF and its membership operate. These include policy documents (PL series), MLA documents (ML series), IAF guidance documents (GD series), IAF information documents (ID series), IAF mandatory documents (MD series) and procedure documents (PR series).

Given the globalisation of the food supply chain and the need for international third-party auditing schemes to assure the competence and impartiality of certification bodies operating within their schemes, the IAF plays a key role in the higher-level assurance mechanism.

13.4 The development of the Global Food Safety Initiative and its relationship with third-party auditing schemes

The GFSI was established in 2000 by the CIES – The Food Business Forum, which is now known as the Consumer Goods Forum (CGF). It is a business-driven initiative that is managed by the CGF and was founded by a group of European-based retail organisations following a number of food safety issues in Europe that had significantly impacted consumer confidence.

Through international collaboration in 2000, requirements for benchmarking food safety schemes were developed that would allow the evaluation of applicant schemes to a defined level of equivalence and that would be formally recognised by GFSI.

Benchmarking is a term that is widely used but equally misunderstood. The term benchmark can be defined as:

- Criterion, standard, touchstone or measure; a basis for comparison.
- A reference point against which other things can be evaluated.
- A standard or a set of standards used as a point of reference for evaluating performance or level of quality.

In early 2000, the Technical Working Group formed by GFSI was asked to consider a number of options, namely the recognition of a scheme which existed at that time, the development of one global scheme owned by the GFSI, the mutual recognition of schemes through facilitation by GFSI and lastly the development of a benchmarking process owned by GFSI to evaluate the equivalence of schemes. At a meeting of the GFSI Technical Committee in late 2000, it was unanimously agreed that the benchmarking option was by far the preferred approach, given the status of food safety schemes and the politics that existed at that time.

Therefore in late 2000 the GFSI began the development of the GFSI Guidance Document. This document was developed to define the process of benchmarking, which in turn fulfils one of the main objectives of GFSI, that of determining equivalency between food safety management systems and the formal recognition of these schemes. The first edition of the GFSI Guidance Document was published in August 2001. GFSI is responsible for the production and maintenance of the document, which is readily available on the GFSI website (GFSI, n.d.b). The editions of the GFSI Guidance Document have always been developed in consultation with stakeholders and been formally approved by the GFSI Board.

The term equivalence is widely used, but it is important to understand it means a condition of being similar or identical or essentially equal. It will become evident from those using GFSI recognised schemes that the scheme's standards may differ in format, and to an extent content, but that over a number of editions of the GFSI Guidance Document there has been a marked convergence of scheme standards and scheme management which proves a level of consistent outcomes.

There have been a number of editions of the GFSI Guidance Document, but it remains subject to review and revision by a specific GFSI Technical Working Group. Since 2000, a full review of the GFSI Guidance Document has been undertaken at least every 4 years, although addenda may be added at any time with the approval of the GFSI Board. There is a clearly defined process of review and revision which includes full stakeholder consultation.

One of the stated objectives of the GFSI is to reduce the number of audits that suppliers were required to undergo to meet an increasing number of different customer requirements. The GFSI, like organisations such as BRC and IFS, had recognised the significant 'audit fatigue' in the food industry, which resulted in high costs, waste of valuable time and unnecessary duplication. The ultimate goal of the

GFSI from its inception has been 'once certified, accepted everywhere'. The intention of determining equivalence and recognition was for retailers and other sectors of the food industry to accept food supplied by a supplier certified to one of the GFSI recognised schemes in lieu of the preferred scheme or standards of that company.

In this way, a level of flexibility was introduced into the marketplace, especially for large multinational companies and their suppliers who could decide on which GFSI recognised food safety scheme to implement, while reducing food safety audits without compromising the safety of the food supplied. This approach also offered a level of confidence to the customer because these recognised schemes had undergone a rigorous and independent evaluation.

With the introduction of Edition 6 of the GFSI Guidance Document in 2011, there were a number of significant changes to criteria demanded by the GFSI. Primarily the GFSI had strengthened the requirement of scheme management and opened the scope to a whole food supply chain approach, which would include schemes representing specific areas of the food supply chain that were not previously considered for recognition, such as the manufacture of food packaging, storage and distribution, and animal feed. To widen this scope of recognition and to put in place criteria that would be used to evaluate equivalence, GFSI formed Technical Working Groups whose membership included key stakeholders to establish sector-specific requirements. As and when these sector-specific Technical Working Groups complete their work, they are disbanded and the criteria determined for that sector are published in a subissue of the current edition. There have been three versions of Edition 6 of the GFSI Guidance Document, each adopting new scopes of recognition.

The GFSI does not have any direct involvement with any certification or accreditation activities; however, it does stipulate that third-party certification audits carried out against a GFSI recognised scheme must be conducted by certification bodies that are accredited by a national accreditation body. The GFSI also recognises the significance of the control and management of national accreditation bodies and in turn requires all national accreditation bodies working with GFSI recognised schemes to be a member of the IAF. Through this membership in the IAF and as previously described in this chapter, this recognition provides a level of assurance to users of the competence and impartiality of the accredited certification bodies and hence the certification provided.

The GFSI has regularly engaged with key stakeholders in meetings prior to its annual Food Safety Conference. In these meetings, delegates are asked about their food safety concerns and which issues in the food industry they feel require resolution. Over many years, major concerns were expressed about effectiveness of third-party audits and particularly auditor competence. These concerns exist despite the number of processes in place to provide assurance that third-party audits are carried out by technically competent and professional auditors.

The GFSI Guidance Document has, in all its editions, defined specific requirements for all recognised schemes to have in place criteria for the appointment, management and continued integrity of any food safety auditor operating within the framework of its schemes. This, in line with ISO standards, includes requirements in relation to qualification, sector knowledge, training and work experience; in addition, there are also requirements regarding continuous professional development and the maintenance of legislative and sector-specific knowledge.

In version 6 of the GFSI Guidance Document, in addition to the controls undertaken by accreditation bodies, schemes recognised by the GFSI are also required to have in place a series of measures to monitor certification body performance, which includes a risk-based programme of certification body office audits and supplier audits to review the performance of the certification body's auditors.

The competence and performance of food safety auditors is critically important to the integrity of the GFSI and GFSI recognised schemes. Therefore in 2011, the GFSI Board mandated that a Technical Working Group be formed to specifically focus on food safety auditor competence. Following work carried out by this Group during 2011, the Auditor Competence Scheme Committee was formed in 2012. The purpose and role of this committee is to continue to develop competence assessment methodology and credentialing of GFSI recognised scheme auditors, investigate the possible personal certification of food safety auditors and to work with other interested parties involved in this area of work. In 2013 the committee was responsible for the publication of the first edition of the GFSI Food Safety Auditor Competencies covering auditors working in the food manufacturing sector. Competencies for other food industry sectors will continue to be developed, published and, as part of the GFSI Guidance Document process, reviewed and revised as required.

The GFSI has grown over a period of 15 years to become a globally known brand and continues to develop through the principles laid down in the augural meetings held in 1999 of a collaborative approach to achieve its stated vision of 'safe food for consumers everywhere'.

13.5 The future of third-party schemes

Particularly in the past 20 years there has been significant impact by third-party schemes on the food industry. The food industry is not alone in recognising the benefits of such schemes and there are comparable developments in other industries such as automotive and telecommunications. The food industry is unique in the way in which it has approached the development and adoption of third-party schemes. There are a number of reasons for this, primarily relating to the poor industry recognition of ISO standards, the vast diversity of food products and associated services, the establishment of regional and now global schemes, and the competitive market that exists in this sector.

As a consequence we have a myriad of schemes at global, regional, national and organisational levels, and due to a greater understanding of the principles of good scheme development and management, these schemes continue to be successful in delivering safe food to the consumer.

Accredited third-party schemes provide a framework for improving food safety compliance through a number of measures that position themselves with the problems faced by the food industry. This is achieved by:

- the use of well-defined HACCP-based standards that are regularly reviewed and revised to meet the demands of the food industry;
- focus on competent application of food safety management systems and continuous improvement principles;
- compliance with regulatory requirements;
- a checks and balances approach at all steps in the process of accreditation and certification and
- involvement and collaboration of stakeholders in the complete supply chain

Despite the prevalence of third-party schemes there is little quantitative data to establish if indeed these schemes enhance and improve food safety standards. From the review of foodborne illness data, it would seem we still have significant food safety issues to address and third-party schemes may not have made the impact that is commonly ascribed. The GFSI undertook an efficacy study in partnership with Sealed Air's Diversey Consulting during 2013 to understand the impact that GFSI recognised schemes have had on the food manufacturing sector.

The results of this study were presented the GFSI Food Safety Conference in Anaheim, United States, in February 2014. The study involved interviewing over 830 food manufacturing companies across 21 countries and the research indicated a number of interesting results, notably that a large majority of the companies surveyed stated that certification has enhanced their ability to produce consistently safe food, that there has been a marked improvement regarding compliance with regulatory requirements and the preparation for legislative change and that there has been a marked change in corporate culture (GFSI, n.d.c).

There was, however, indication that the level of duplication of food safety audits still remains at a relatively high level and the 'once certified, accepted everywhere' vision of GFSI still remains an elusive goal.

Over the last 30 years, it would be true to say third-party schemes have had a significant impact on the way the food industry self-regulates and therefore takes direct responsibility for the safety of food within the supply chain. These schemes are a means to ensure good and, in most cases, best practice is in place to move industry forward. There are, however, some downsides to this approach, particularly with the larger global third-party schemes, which by their very nature are regularly reviewed and revised. Within developing and emerging markets this will mean industry is forever trying to play catch-up and there is a level of frustration and despair in these situations. This has partly been addressed by the GFSI Global Markets Programme (GFSI, n.d.d) in the case of multinational companies.

Scheme owners face the challenge of commoditisation of third-party auditing in a low margin industry and the potential loss of integrity of forcing price down. Given the current checks and balances approach imposed by organisations such as GFSI, scheme owners and accreditation bodies, the integrity so far has remained unchanged, but the need for continued vigilance cannot be stressed enough.

The pressure to continually improve, particularly with respect to auditor competence, will mean increased costs which must be borne by the industry.

The industry also faces a liability challenge following the cantaloupe *Listeria* outbreak in the United States in 2011 (Flynn, 2013). The legal debate still continues regarding reliance on audit reports and the 'duty of care' of the auditor. True third-party schemes are yet to be challenged, but this is probably just a matter of time. If such a challenge were to come forward in the US courts and negligence of an accredited third-party certification body was found, the whole question of liability could move up the chain from the certification body to the national accreditation body and the principle of third-party schemes would come under scrutiny. Given the efforts made by the food industry to develop, implement and support third-party schemes in the last 20 years, it will be interesting to see the reaction that a guilty verdict would bring, especially for national accreditation bodies who have a direct relationship with governments.

Despite the liability issue, third-party schemes will remain in place, but it is doubtful that the number of schemes will continue to increase. Schemes that are now in place will mature and become more sophisticated in their operation and management. With this maturity there is the distinct probability that governments will recognise the significance of third-party auditing schemes and industry will start to work closer together and proactively collaborate with government agencies; the industry-owned schemes are now ideally positioned to provide data sharing between the public and private sectors, if complex relationship issues can be overcome.

References

AFRAC, 2014. Available from: <http://www.intra-afrac.com/> (accessed 24.02.15).

ARAC, n.d. Available from: <http://www.arabarac.org/index.php/en/> (accessed 24.02.15).

Assured Food Standards, 2015. About us. Available from: <http://www.redtractor.org.uk/about-us> (accessed 24.02.15).

BRC Global Standards, 2014. Available from: <http://www.brcglobalstandards.com/> (accessed 24.02.15).

British Lion Quality, 2015. Available from: <http://www.lioneggfarms.co.uk/default.aspx> (accessed 24.02.15).

CanadaGAP, 2015. Overview of CanadaGAP. Available from: <http://www.canadagap.ca/canadagap-program/overview/> (accessed 24.02.15).

Canadian Grain Commission, 2014. Available from: <http://www.grainscanada.gc.ca/quality-qualite/hsp-per/hspm-mper-eng.htm> (accessed 24.02.15).

EA, 2015. Available from: <http://www.european-accreditation.org/> (accessed 24.02.15).

Flynn, D., 2013. Jensen Brothers take responsibility but blame PrimusLabs. Available from: <http://www.foodsafetynews.com/2013/10/jensen-brothers-sue-primus-over-third-party-audit-they-say-was-faulty/#.VO33V8t0wdV> (accessed 24.02.15).

Food and Drink Federation & National Consumer Council, 1991. Food Safety Act 1990: Guidelines on the Statutory Defence of Due Diligence. National Consumer Council, London.

Food Safety Certification System 22000, 2015. About us. Available from: <http://www.fssc22000.com/documents/about-us.xml?lang = en> (accessed 24.02.15).

Food Standards Agency, 2013. Assessment and comparison of third party assurance schemes in the food sector: towards a common framework. Available from: <http://www.food.gov.uk/science/research/choiceandstandardsresearch/fs245006#toc-2> (accessed 24.02.15).

Genesis Standards, n.d. Available from: <http://www.genesisstandards.com/> (accessed 24.02.15).

GFSI, n.d.a. Schemes and certification. Available from: <http://www.mygfsi.com/schemes-certification/overview.html> (accessed 24.02.15).

GFSI, n.d.b. GFSI Guidance Document. Available from: <http://www.mygfsi.com/schemes-certification/benchmarking/gfsi-guidance-document.html> (accessed 24.02.15).

GFSI, n.d.c. GFSI efficacy study. Available from: <http://www.mygfsi.com/about-us/about-gfsi/benefits.html> (accessed 24.02.15).

GFSI, n.d.d. Global markets programme overview. Available from: <http://www.mygfsi.com/market-access/global-markets-programme/overview.html> (accessed 24.02.15).

Global Aquaculture Alliance, 2011. About GAA. Available from: <http://www.gaalliance.org/about/> (accessed 24.02.15).

Global Red Meat Standard, n.d. About GRMS. Available from: <http://www.grms.org/About_GRMS.aspx> (accessed 24.02.15).

GLOBALG.A.P., n.d.a. GLOBALG.A.P. History. Available from: <http://www.globalgap.org/uk_en/who-we-are/about-us/history/> (accessed 24.02.15).

GLOBALG.A.P., n.d.b. Who we are. Available from: <http://www.globalgap.org/uk_en/who-we-are/> (accessed 24.02.15).

IAAC, 2014. Available from: <http://www.iaac.org.mx/English/Index.php> (accessed 24.02.15).

IAF, n.d. Available from: <http://www.iaf.nu/> (accessed 24.02.15).

IFS, n.d. Available from: <http://www.ifs-certification.com/index.php/en/> (accessed 24.02.15).

IFST, 2015. Available from: <http://www.ifst.org/> (accessed 24.02.15).

ISO/IEC 17000:2004, 2004. Conformity assessment – vocabulary and general principles. International Standards Organization, Geneva.

ISO/IEC 17011:2004, 2004 Conformity assessment – general requirements for accreditation bodies accrediting conformity assessment bodies. International Standards Organization, Geneva.

ISO/IEC 17021:2011, 2011. Conformity assessment – requirements for bodies providing audit and certification of management systems. International Standards Organization, Geneva.

ISO/IEC 17065:2012, 2012 Conformity assessment – requirements for bodies certifying products, processes and services. International Standards Organization, Geneva.

PAC, 2009. Available from: <http://www.apec-pac.org/> (accessed 24.02.15).

PrimusGFS, 2014. Available from: <http://www.primusgfs.com/> (accessed 24.02.15).

SADCAS, 2015. Available from: <http://www.sadcas.org/> (accessed 24.02.15).

SALSA, 2011. Available from: <http://www.salsafood.co.uk/> (accessed 24.02.15).

SQF Institute, 2015, About SQF. Available from: <http://www.sqfi.com/about-sqf/> (accessed 24.02.15).

Swoffer, K., 2012, A history of audits: How we got to the BRC standard. Available from: <https://www.sfdf.org.uk/events/Kevin%20Swoffer.pdf> (accessed 24.02.15).

UK Legislation, n.d. Food Safety Act 1990. Available from: <http://www.legislation.gov.uk/ukpga/1990/16> (accessed 24.02.15).

Wright, M., Palmer, G., Shahriyer, A., Williams, R., Smith, R., 2013. Assessment and comparison of third party assurance schemes in the food sector: towards a common framework. Greenstreet Berman, Reading, UK. Available from: <http://www.food.gov.uk/sites/default/files/835-1-1534_GSB_CR2435_3rd_Party_Assurance_Scheme_R2_V8_FCA.pdf> (accessed 24.02.15).

Supplying the supermarkets: good technical and quality management practices from a retailer perspective

14

Fiona Houston
Independent Researcher, United Kingdom

14.1 Own-label versus branded

There are two types of product sold in store; own-label and branded. There is a noticeable difference between own-label and branded supply from a retailer's (supermarket) point of view:

> A branded product, being sold in a retail store, has a contact address of the manufacturer on the product allowing for enquiries to be directed straight to the manufacturer (even though in some cases customers may contact the store they bought the product in first should they have a query or complaint). The manufacturer is responsible for the quality, safety and legality of the product.
> Retailers look at own-label products differently in that as it is their own name and address printed on the packaging the retailer is clearly far more accountable for the quality, safety and legality of the product. As a result they will take care and appropriate 'due diligence' to ensure the product is safe to eat, meets all relevant legislation, conforms to all industry requirements and has been made to the agreed quality standards.

The focus of this chapter will be on supplying own-label products, as to do so well requires a great deal of time, knowledge and understanding of manufacturing processes, raw materials and legislation. It will cover the following areas:

1. understanding the development process and presenting a product to a retailer;
2. manufacturing site standards;
3. retailers own technical standards;
4. the role of a technologist in the development process; and
5. what can go wrong.

14.2 Understanding the development process

Let us assume that the businesses marketing, new product development (NPD) and/or commercial team have managed to persuade a retailer that you have a good idea for

Swainson's Handbook of Technical and Quality Management for the Food Manufacturing Sector.
DOI: https://doi.org/10.1016/B978-1-78242-275-4.00014-9

a new product. This could have been as a result of an opportunity or indeed a product brief from a retailer. What happens next?

14.2.1 Commercial viability

How soon can your commercial colleagues cost the product? Before presenting a product to a retailer for the first time, indicative costs should be considered based on existing facts and product information, for example the quality or tier within which the product is going to be sold in, i.e. 'value' or 'good' or 'better' or 'best' (top tier). Does the cost price deliver the margin expectations of the retailer and still allow them to sell a competitively priced product? If not, is there a marketing message than can demonstrate why this is the case? For example, 'Madagascan Vanilla' (provenance) or 'handmade' (suggestion of greater care and attention to detail).

14.2.2 The team

As a team, ideally including operations, NPD, technical, engineering and HR, appreciating that not all businesses have individuals to fulfil these roles but need to wear several hats if required, it would be beneficial to discuss on a regular basis how you will deliver this project. A decision on investment is required early on as this could drive the time to launch. It is worth noting that some pieces of equipment or packaging do require a long lead-time and so this should be considered to avoid disappointment later.

14.2.3 Hazard analysis and critical control points and quality assurance requirements

Hazard analysis and critical control points (HACCP) is a process by which businesses monitor the biological, chemical and physical hazards surrounding raw material production, handling, manufacturing and distribution to ensure food safety. Important questions need to be discussed with the retailer while looking to fulfil HACCP requirements.

- Is there any 'provenance' required for this product, for example, Sicilian Lemons?
- What are the volumes being discussed?
- Can you procure enough of the key product ingredients to meet the demand?
- Are there any seasonal issues to manufacture this product, for example, change of harvest when manufacturing chips or crisps or having to order products seasons in advance, for example, canned tomatoes?
- What might the impact be?
- When do these changes occur?

With the concerns raised from the UK Elliott report (July 2014) on the integrity of raw material following incidents of beef products being contaminated with horsemeat, it should be considered which raw materials might cause an issue.

- Do you understand ALL of your supply chain?
- What are your new raw materials for this project? Have you audited new suppliers or requested completion of a self-audit questionnaire?
- What further questions have to be asked?
- Have you asked for HACCP plans, allergen statements, etc.?
- What would you expect the critical control points (CCPs) to be and are these adequately covered?
- Do you need to ask questions of the raw materials from an expert or a scientific organisation?
- Are there any new processes involved?
- How will these be validated and verified?

To view the full report here is the link:

https://www.gov.uk/government/uploads/system/uploads/attachment_data/file/350726/elliot-review-final-report-july2014.pdf

It is important to understand that certain ranges at some supermarkets will have brand values. This can cover a number of areas and, for example, may include the controlled use of additives. In this case all raw materials and processes need to be reviewed in depth to ensure they do not contain additives, not forgetting release agents, processing aids and full breakdowns of subcomponents. Remember colours can contain milk products; you may know and have asked the question but what do you not know? Keep asking questions.

- Does the product use existing processes?
- Who is the target customer?
- How will the process be validated, for example, heat treatment for a dairy product?
- Has it received a pasteurisation or other heat process and will it be cooled within the required time?
- If you change equipment part of the way through the project do not forget to review the process controls.

Raw materials also include the product packaging:

- Is the packaging format an existing or a novel concept?
- If it is a current format is it being used for similar products?
- What effect might product properties such as a higher fat content have?
- Does it need to be able to withstand extreme temperatures, for example, freezing or high cooking temperatures or both?
- How will you validate a new idea?
- Who are the industry experts?
- What trials will need to take place?

14.2.4 Shelf life assessment

Shelf life assessment tends to be made up of two parts, safety and quality. Good technologists will know what organisms the product is at risk from and therefore know what to test for. Think about the medium it is packed in – a protective atmosphere will change the microtesting requirements. When tasting the product, review what happens over life, take photos and consider all areas of flavour, texture, appearance and aroma. It should be considered that for many products the customer is likely to first eat the product at day 2 (of its shelf life) or more and therefore you should start here for a more accurate assessment. Do not be afraid of saying a product changes over shelf life.

14.2.5 Allergens

Allergens are a great cause for concern. According to the UK National Health Service (NHS) who reported in 2014.

- One in fourteen children under the age of three in the United Kingdom have a food allergy.
- Two percent of population and eight percent of children aged under three are affected.

To give clear messages and allow choice it is really important that factory practices are in control every day to avoid any potential for product contamination. Some retailers do not want to 'alibi label' (i.e., 'May contain' labelling) every food for every allergen, while others will add a statement such as the one below after considering a manufacturer's risk assessment:

> *Made in an environment that uses xxxxx (where the xxxxx may be "peanuts", "nuts" or one of the other recognised allergen groups).*

How can you demonstrate that your process is in control and that you are following good manufacturing practices (GMP) which would allow more choice to be given to the retailer's customer, the consumer?

If you are manufacturing a specifically 'allergen free' product, what controls and testing procedures have you put in place to absolutely guarantee the product is safe for the allergen intolerant customer to consume? Such controls must start with raw materials and the environment they are manufactured/prepared in as well as your own environment. Ensure you have spoken to the laboratories to confirm you are using the correct testing methods to validate your product and processes, which will demonstrate good control should there be a customer query or industry challenge in the future.

14.2.6 Nutrition

A theoretical nutritional calculation should be made on the proposed product, which should be assessed against reference intake (RI) and front of pack (FOP) colour charts if applicable (UK 'FIR' Food Information Regulations 2014).

- Do any values come to more than 100%?
- Are any values very close to a lower category?

What are the consequences of the answer to these two questions and is there anything generally that could be amended to improve the recipe to bring the nutrition into a lower category?

Has the product been designed to meet any nutritional claims, 'less than' or 'allergen free' claims? If the answer is yes, then a discussion should be had with the team around how this is going to be achieved every day, focusing on what validation and verification is required.

All product nutritional claims will need to be confirmed by analysing the final product before launch and then on a regular basis after launch.

14.2.7 *Testing/tasting*

Within your organisation do you have trained personnel who can describe the characteristics of a product correctly? Will they know what the product developer/chef means when they say this product is, for example:

- sharp
- rancid
- sweet
- bitter
- burnt
- buttery
- fruity
- farmyard

Do you carry out daily taste panels? Are all disciplines of the team involved and at the same panels do you check packaging integrity, coding accuracy and legibility? If not, this should be considered.

When proposing a product, have you benchmarked your product against similar products already being sold? Knowing what makes your product different/better than the competition will allow you to persuade a retailer why they should stock your product.

Customer panels are often carried out during the development of a new range or when major changes are being made to an established range. This can be sensory tests or consumer panels where there are preferences and likelihood to purchase questions asked to gain a more accurate response from the customer prior to the launch. For more information see Campden Sensory course at www.campdenbri.co.uk or UK National Centre for Food Manufacturing http://www.lincoln.ac.uk/NCFM.

14.2.8 *Training requirements of employees, factory and product*

Training is key to delivering a safe and good quality product every day. All members of staff should receive their basic food hygiene requirements when they start work. What training happens after that is key to delivering consistently to the retailers. Can you demonstrate that personnel have been trained for all of the procedures they are expected to complete at work every day?

Consider what is new for the product you are about to launch for the retailer. What extra training is required, for example, sensory and visual standards not only of finished products but also of the raw materials from intake through to 'work in progress' and then to the finished product.

14.2.9 Trials

During the development of a new product you will conduct many trials. This process in general starts off in the development kitchen and then the product is transferred to the factory for small-scale trial building up to a large-scale 'pre-production' to agree and sign off the final product prior to launch and validate HACCP and process flow.

In all cases, you need to be very aware of what information you are going to find out/confirm during a specific trial. This could include:

- Raw material quality and specification.
- Seasonality issues.
- Ordering lead-time or 'minimum order' quantities.
- Manufacturing procedure.
- Process losses.
- Assessment of shelf life of finished product and also WIP – 'work in progress' of components (the worst case should be tested during the trial period).
- Equipment assessment (especially if using new kit).
- Trial sign off if planning to use different production lines on occasions – all potential production options should be reviewed to ensure that a consistent finished product is going to be produced over time.
- Finished product assessment for safety and quality over the intended shelf life.
- If the product may be frozen as a production aid or by the customer this will also require full assessment.
- Cooking instructions.
- Storage instructions.
- Packing line set-up to achieve the agreed quality standards, that is, packing rate, number of people, component flow.
- Finished product weight control. This information is used to generate average or minimum weight declarations. You need to know what the retailer requires with respect to weight declaration as different controls will be required if a 'minimum' or 'average weight' declaration is being made.
- Microbiological assessment taking into consideration raw materials, components, packing order and therefore environment changes, to confirm product safety and shelf life.
- Packaging barcode 'reads correctly' in the retailer's store – suppliers to retailers should always ask when this can be completed as the product may need to be active in stores for the barcode to read.

14.2.10 Travel tests

Travel/'transit' tests may need to be completed. This should take place as soon as possible and use the proposed packing format with the right number of packs in the outer case or tray. A useful tip would be to consider whether the product will only be sold in one format or could it change for promotion or at certain times of the year, for example, Christmas or summer peaks. If there is more than one format then a company is well advised to travel test them all.

If the format is totally new, then travel test it within your own business first. This will identify issues that you can then resolve or discuss with the retailer before wasting their time finding something you should have already known about and addressed. Retailers have their own travel test requirements so it is important to ensure you have found out what is expected before you waste money doing the wrong thing. Some require product to be sent to them with specific forms and photos whereas others will ask you to send it to a store via a specific depot or transport route where you (or they) will review it and then send a report.

14.2.11 Writing specifications

A specification is essentially a contract between manufacturer and retailer. As a manufacturer you are responsible for making sure the information is correct. In a lot of cases the specification will be used to generate the information and 'pack copy' for artwork. The information required is detailed and will include the following:

- Name or title.
- Subtitle or description of the product. This should allow the customer to know what they are buying.
- Legal name which must be on the legal face of the product (this legal face is not necessarily always the front of the pack).
- Weight.
- Storage conditions, including 'once opened' if appropriate.
- Reference to shelf life indication location.
- Instructions for how the customer should use the product, for example, cooking instructions.
- Name and address so that a customer can make contact if they have a query.
- Ingredients – including in the EU 'Quantitative Ingredient Declarations' (QUID) declaration based on title, subtitle and legal name and sometimes photography.
- Nutrition per 100 g as a minimum requirement.
- All raw materials including their additives and processing aids need to be described even if not declared, if legislation allows, in the final pack. Countries of origin of all components that is the whole supply chain should also be known so that in the case of a crisis this can be reviewed quickly.

Ensure that you keep up to date with legislation changes, for example, FIR (UK Food Information Regulations).

Some suggested UK sources of information are listed below:

Foods Standards Agency – www.food.gov.uk
Campden Research – www.campdenbri.co.uk
Leatherhead – www.leatherheadfood.com
Reading Scientific Services – www.rssl.com
Institute of Food Science and Technology – www.IFST.com
Universities
Local authorities
Other local businesses

14.2.12 Launch

Now the product has been agreed and all safety and quality measures have been signed off. The packaging is printed and waiting in the store. Do not forget to order the raw materials on time. You are ready to launch.

Retailers will often attend preproduction runs and/or product launches to ensure that everything is going well. This involvement should be embraced, as it is much easier to resolve issues on site than have to accept a subsequent rejection from a retailer if something has gone wrong. For more details on such matters please turn to the end of this chapter.

Once the product is in your warehouse do not breathe a sigh of relief just yet. It still has to reach the retailer's distribution system. Ensure that distribution lorries will turn up on time to collect it, the correct quantity is dispatched at the right temperature and the pallets have been constructed correctly and safely. You do not want a rejection at the customer's depot having gone to all of this production effort.

If this is the first time you have delivered product to the retailer it is worth double checking that it has arrived where you expected it to and has been accepted – by this time you will have established a relationship with all personnel in the supply chain. Do not forget to send in samples of the end product to the retailers for their review at the same time as you first dispatch product for launch or beforehand if it is a frozen or long-life product.

14.2.13 Postlaunch

What happens after launch? It may all go quiet and you will receive orders at the agreed frequency and deliver as planned. However, if the product sales are better or not to plan then you will have discussions very quickly with the buying team as to the next actions to take. You will need to be very aware of raw material lead-times, the raw material (including packaging) stocks you have and what the options are to react to the revised sales figures.

14.2.14 Assessment by retailers and customer panels

Retailers will review new lines or lines with significant complaints more frequently than existing lines. Therefore be prepared for feedback and ensure that your taste panel results are clearly recorded. Also some retailers will send product to customer panel either during the trial phase, after launch or following press/magazine reviews of the product. The comments from these will drive analysis and conversation about any required actions.

After launch, maybe 6 – 8 weeks later, a full review of the development and launch process should take place. This may involve the retailer as part of their process but you should certainly carry out a similar review and record any lessons that would help for future projects. This review could include:

- What went well?
- What could have been done better?

- What issues occurred during the development phase and how could these be prevented next time?
- Initial sales review – is there investment required to maximise sales?
- Production planning – should this be reviewed to minimise factory waste?
- Does the case quantity need to be reviewed to reduce waste in stores or maximise sales potential?
- What customer comments have been received? (e.g., technologist feedback or customer complaints), Have they driven changes in design of the product already? What can be learnt from this for future projects?
- Are there any raw materials that need to be discussed from a viewpoint of quality, minimum order quantities (MOQ), seasonality issues following this launch?
- Review supply chain – is all in place correctly?
- Are there going to be any product promotions not previously discussed that need to be considered after this initial launch phase?
- Is the product going to roll-out to more stores and what are the implications of this? Has the product delivered to its success criteria and if not, why?

Once the product is launched ensure that the product is added to all the relevant testing and internal auditing programs at the correct frequency. This could include:

- microbiology assessment;
- taste panels;
- nutritional checks;
- integrity review if you are making a claim on a raw material source;
- quantitative ingredient declarations (QUID);
- manufacturing method; and
- HACCP and process flow.

A review of the whole process should be carried out if this is a new or complex process to see how the output can be maximised without affecting the agreed safety and quality standards. Every little detail should be challenged and often by involving team members who have not been so close to the original project, new ideas and efficiencies will be identified. Remember though; if you change the process the new product will often have to be checked and signed off again with the retailer.

14.3 Manufacturing site standards

There is an expectation from all retailers that sites will follow good manufacturing practices (GMP). If they are to manufacture own-label products, in general, in the United Kingdom this requires a site to be accredited to standards, such as BRC or SALSA, although other accreditations may be accepted.

14.3.1 British Retail Consortium

The British Retail Consortium (BRC) was established in 1992 and has since grown to become a leading trade association recognised across the United Kingdom and worldwide.

BRC have a number of standards to cover different points of the food supply chain including manufacturing, packaging, storage and distribution. Current versions in 2018 are:

- BRC Global Standard for Food Safety, Issue 8;
- BRC Global Standard for Packaging and Packaging Materials, Issue 5;
- BRC Global Standard for Agents and Brokers, Issue 2; and
- BRC Global Standard for Storage and Distribution, Issue 3.

These documents are regularly updated and BRC hold meetings to inform companies of the changes and new requirements which will be additionally reviewed when audits take place. Approved auditors carry out audits; they are not employees of BRC and are independent. You can find out more information on the BRC website at http://www.brc.org.uk.

14.3.2 Safe and local supplier approval

Safe and local supplier approval (SALSA) was set up particularly for small UK businesses which may not have the same extent of resources as larger operations. SALSA states its aim of the accreditation as:

> *It is seen as an approval scheme that helps local and regional food and drink producers supply their products to regional and national buyers.*

Although its requirements are still very much safety and legal first, how you demonstrate control is generally less onerous than the BRC requirements. In the United Kingdom it is recognised by a wide range of organisations in both retail and food service, for example, ASDA, Waitrose, Booths, 3663, as an acceptable accreditation so that business conversations can progress. You can find out more information on their website at http://www.salsafood.co.uk.

The current accreditation is based on the following documents:

- SALSA audit standard issue 5;
- SALSA plus cheese audit standard issue 4; and
- SALSA plus beer audit standard issue 2.

SALSA also offers a mentoring service to suppliers and regular courses to help train personnel.

14.4 Retailer's own technical standards

It is important to note that retailers (supermarkets) also have their own technical standards that you are expected to follow when supplying own-label product. These standards cover many areas and are not limited to:

- Raw material controls throughout the supply chain and for certain risk raw materials define suppliers that must be used.
- Raw materials also include specific packaging controls.

- Hygiene of the site – cleaning standards and procedures that may also specify the chemicals to be used.
- Foreign body control – glass, metal, wood, plastic, bone, etc.
- People:
  - What to ask at interview.
  - What happens if employees are ill at work or at home.
  - Work expectations.
  - Training – allergens, critical control points (CCPs) and personal protective equipment (PPE) which includes protective clothing.
- Processing controls:
  - HACCP.
  - Manufacturing procedure prerequisites.
  - Traceability and mass balance.
  - Packing and coding standards.
- Site standards including:
  - Waste control.
  - Building materials.
  - People flow.
  - Product flow.
  - Waste flow.

As a new supplier the standards may cover top level requirements; however, you should ask what the requirements will be in the future so that you are fully aware of the ongoing expectations and what investment you will have to make to supply a particular retailer in the long term.

Also the requirements vary between retailers, so if you are supplying more than one retailer wherever possible it is better to work to one standard (the most comprehensive and all-encompassing standard) to avoid confusion. You will need to agree as a team which requirement you are going to follow on a day-to-day basis so that your employees have a consistent procedure to follow; this avoids confusion for staff and also means you will be well-prepared if a retailer turns up for an unannounced audit, which is becoming a more common practice.

14.5 The role that the technical manager, technologist, quality assurance manager (QA) or NPD technologist plays

Technologists are often expected to know the answer, or have raised the question and sought out the answer from someone who does know, for every element of safety throughout the process. When we come across new concepts it is important to challenge everything, even if we think we know. You may hear someone explain that 'It is only a change in equipment or one ingredient' or that only the 'pack size or line needs changing' and therefore you may not feel as though you need to question things. However, the danger of doing so is that an important point could be missed which would result in a costly return to manufacturer (RTM) or emergency product withdrawal (EPW). By treating every new, upgrade or size variation with the same respect some may turn out to be easy, but on all of the others you will know that you have thought of everything that you could have to help avoid issues arising.

Depending on the size of your business you may have a quality/technical manager or also an NPD technologist if you are a bit bigger business. The role of either is to ensure that all areas of safety and quality from the start of the process to the end have been considered and the right procedures have been put in place to ensure control. They will be an important member of the development team or indeed lead the development project, depending on the size of the team. The areas that they will cover should include:

- Process flow for the new product.
- HACCP review and add any new processes.
- Add the new product to allergen risk assessments.
- Establish which raw materials are new and if there are any new suppliers (which would then need taking through new ingredient/new supplier approval processes).
- Start to discuss with the new suppliers or existing suppliers with new raw materials:
  - HACCP
  - Process
  - Allergens on site
  - Certification/accreditation
  - Specifications
  - Quality standards
  - Ordering lead-times including minimum ordering quantities and seasonal considerations
- Complete a risk analysis for the microbiological testing procedures required for this product.
- Arrange time for factory trials and ensure raw materials are available.
- Check product during trials for component weights, quality and finished pack weights.
- Train staff in the following areas:
  - Any new processes
  - New raw materials
  - Quality standards
  - Allergen awareness
- Discuss packaging options and agree the proposal.
- Complete travel tests using the proposed packaging.
- Complete shelf life trials for quality changes and microbiological changes.
- Legislation considerations to discuss with the retailers.
- Write specifications so that artwork can be generated.

14.5.1 Handling of complaints

In the United Kingdom many consumers are thought to be quite likely to complain, less so elsewhere in the European Union, which can lead to interesting discussions with manufacturers or raw material suppliers when complaints are received. A complaint system needs to be in place, that should record all of the details of the person complaining, the product they are complaining about and the date you received the complaint and responded to it. Each complaint needs to be investigated and actions agreed, and ideally sign off as completed.

Trends of complaints should be identified at monthly/quarterly reviews. Complaints can be split into categories, such as quality, foreign body, packaging, dissatisfaction plus others that are appropriate to your business. For example, you may want to specifically target a product or complaint-type like bones in fish fingers destined for children's tea plates.

If a foreign body is found it is often the line operators who will know what it is (as they see the product and process all of the time) so it would be reasonable to ask them for help if you do not know what an unexpected object is. Also ensure that senior management discusses complaint trends so that the right investment can be made to resolve the issue (perhaps additional control equipment is required or investment in better suppliers). Some complaints do not arise from the manufacturing site and so investigation needs to be taken further down the supply chain; by working with your raw material suppliers early on in the process and agreeing standards and a specification the frequency of complaints should reduce and the raw material you receive will be better as they will realise that the specification you have agreed is important to you.

14.6 What can go wrong

Too short a time to correctly review new raw materials

- Resulting in complaints or rejections and in worst case RTM:
- Need to know allergens/HACCP/visit factory/initial supplier review/integrity.
- Ensure raw materials are available in the quantity and quality required.

Too short a time to complete the right trials in factory

- Resulting in quality expectations not being met consistently.
- Resulting in nutritional claims not being met.
- Resulting in the product unexpectedly performing inconsistently over shelf life.
- Resulting in unexpected costs due to factory line speed requirements not being met.
- Missing investment lead-times or not assessing kit correctly for use, hygiene requirements, cleaning time and commissioning time.

Too short a time to test products and environment if making a claim either for health or allergen avoidance

- Lack of knowledge of testing requirements.
- Using the wrong test leading to false results.

Investment lead-times

- What new manufacturing kit is required? Where is it coming from? Is it bespoke?
- What requirements brief has to go to the equipment manufacturer?
- When can equipment be available for trials and where in the world are these trials to be held?
- Was there a project manager assigned or was a random team expected to deliver the project?

Lack of understanding of the critical path

- Ensure that a team (even if it is only two people), discuss the whole project early on and also at regular intervals.
- There will be key dates where weekly meetings will be necessary to ensure that all plans stay on track.

Raw material sourcing issues

- Not completing the new supplier review early enough.
- What allergen assurance/information do your final customers require?
- Do you need nut-free or gluten-free components? What tests are going to be required both internally and externally?
- Which organisations do you need to get in touch with to ensure that all requirements are met?
- Do new testing methods need to be considered, for example, DNA testing for integrity claims?
- Not knowing the raw material well enough or being aware of the process in enough detail to ask the right questions, for example, safety considerations.
- Not knowing the raw material well enough and missing key ordering or volume confirmation dates due to seasonality, for example, for bulk processed tomato products you will typically need to order your annual volume requirement before the tomato season even starts for the whole year ahead (as such crops are only grown and processed during a limited amount of the year).

Further information and reading

BRC (British Retail Consortium) Global Standards – http://www.brcglobalstandards.com/
IFS (International Featured Standards) – https://www.ifs-certification.com/index.php/en/
IGD (Institute of Grocery Distribution) – http://www.igd.com/
SALSA (Safe and Local Supplier Approval) – http://www.salsafood.co.uk

Enforcement authority perspective on the food manufacturing sector (UK EHO) 15

Andy Bowles
Specialist Food Law Solicitor, ABC Food Law, Innovation Centre, Norwich Research Park, Norwich, Norfolk, England

15.1 Introduction

Food business operators are required to comply with a wide range of legal obligations relating to the production, handling and marketing of food. This chapter will consider the nature and objectives of food law and describe the arrangements in place in England for the performance of official controls by food authorities to ensure compliance. 'Official control' is defined as 'any form of control that the competent authority or the Community performs for the verification of compliance with feed and food law, animal health and animal welfare rules'.[1] Such controls include audits, inspections, sampling visits, surveillance activities, monitoring of premises and verification visits performed by the 'competent authorities'. In England these authorities include local authorities, port health authorities (PHAs), the Food Standards Agency (FSA), government departments such as Defra and independent control bodies. This chapter will discuss the roles of the main agencies involved in the performance of official controls and will consider the issue of compliance with food law from their perspective. There are a number of enforcement sanctions available to food authorities to secure compliance with food law and these are described. The nature of UK food law enforcement is changing with a shift away from the use of criminal prosecutions to deal with noncompliance with the use of civil sanctions, such as compliance notices, becoming more common. This chapter will describe how food enforcement officers identify breaches of food law through the use of interventions and the factors that they consider when choosing the appropriate sanction to achieve compliance.

[1] Article 1, Regulation (EC) No. 882/2004 of the European Parliament and of the Council on official controls performed to ensure the verification of compliance with feed and food law, animal health and animal welfare rules. Official Journal of the European Union L 191/1 28.5.2004. Note that this regulation is due to be replaced by Regulation (EU) 2017/625 in December 2019.

Swainson's Handbook of Technical and Quality Management for the Food Manufacturing Sector.
DOI: https://doi.org/10.1016/B978-1-78242-275-4.00015-0

15.2 The nature and objectives of food law

What is the purpose of food law and what does it seek to achieve? Indeed is it necessary at all? (Jackman, 2004) One function of food law is to set standards, based on science, which are adequate to ensure that the consumer is protected from harm and this is certainly the case with respect to food contaminants, additives, residues and contact materials legislation. However, food law also provides for minimum standards to meet with consumer expectations which may not actually be required to protect consumers but offer reassurances regarding the manner in which the food that they buy has been produced, processed and distributed. For example there is a legal requirement for 'food premises to be kept clean and maintained in good repair and condition'[2] but is this really necessary? It could be argued by the food industry that such a legal requirement is unnecessary because consumers would only purchase foods from premises that met their own expectations of cleanliness. In other words, clean businesses would succeed while dirty businesses would fail to attract custom and would therefore go out of business. However, we have learnt from history that a failure to set standards by legislators can lead to unacceptable industry practices and as such one objective of food law is to clearly establish minimum criteria that would be deemed to be acceptable by consumers, regulators and the trade.

The principal objectives of modern European food law are to protect and inform the consumer and to facilitate free trade by the setting of goal-based legislation. Regulation (EC) 178/2002,[3] Article 5 states:

> *Food law shall pursue one or more of the general objectives of a high level of protection of human life and health and the protection of consumers' interests, including fair practices in food trade, taking account of, where appropriate, the protection of animal health and welfare, plant health and the environment.*

In terms of protecting consumer's interests, Article 8 states:

> *Food law shall aim at the protection of the interests of consumers and shall provide a basis for consumers to make informed choices in relation to the foods they consume. It shall aim at the prevention of: (a) fraudulent or deceptive practices; (b) the adulteration of food; and (c) any other practices which may mislead the consumer.*

These objectives are set out by the legislators but the means of achieving these objectives are left to the food business operators. For example, Article 5(1) of

[2] Regulation (EC) No. 852/2004 of the European Parliament and of the Council of 29 April 2004 on the hygiene of foodstuffs as amended L 226/20.

[3] Regulation 178/2002 of the European Parliament and of the Council laying down the general principles and requirements of food law, establishing the European Food Safety Authority and laying down procedures in matters of food safety [2002] O.J. L31/1, as amended.

Regulation (EC) 852/2004[4] requires that food business operators develop, implement and maintain documented food safety management systems based on the principles of HACCP (hazard analysis and critical control point) but the Regulation is not prescriptive on how this goal is achieved. Similarly, Article 18 of Regulation (EC) 178/2002 requires that food business operators put in place 'systems and procedures' to demonstrate the traceability of the foods both supplied to and by them. Once again the legislation is silent on how this goal is achieved. The role of the regulators in this flexible legislative framework is to assess whether the steps taken by food business operators are sufficient to meet the objectives of food law. There is, therefore, an element of self-regulation built into the food law system as the regulations require the active involvement of both the food business operator and the regulators to ensure that the stated objectives of food law are achieved.

15.2.1 Criminal law and civil sanctions

While the legal obligations placed on food business operators and the agencies involved in undertaking official controls with respect to food law are discussed elsewhere, it is important to review the nature of sanctions that may be imposed on those who breach food law requirements. In England offences created in food law are criminal in nature[5] but the legal framework provides for the use of administrative,[6] civil[7] and criminal sanctions[8] by the regulatory agencies (Atwood, 2000). For example, on detection of a breach of food legislation, enforcement officers may, depending on the circumstances, serve notices requiring works to be completed; detain food pending further enquiries; seize food for condemnation; or prohibit processes, equipment or premises. Enforcement officers may also decide to instigate criminal proceedings against those who fail to meet the obligations of food law. In practice, the key objectives of food enforcement officers are to promote compliance with food law and take steps to change the behaviour of those businesses which fall short of the legal requirements. As such the use of criminal sanctions is normally reserved for the worst cases or for persistent offenders.

[4] Regulation (EC) No. 852/2004 of the European Parliament and of the Council of 29 April 2004 on the hygiene of foodstuffs as amended L 226/20.

[5] That is to say, breaches of the requirements of food legislation may result in the persons responsible being prosecuted in the criminal courts. Most offences under the Food Safety Act 1990 and the Food Safety and Hygiene (England) Regulations 2013 are triable either way and as such can be heard before Magistrates or at the Crown Court. Some offences are Summary in nature and may only be heard at the Magistrates Court. For example, obstructing an officer contrary to Section 33 Food Safety Act 1990.

[6] For example, certain businesses wishing to place products of animal origin such as meat, fish and dairy products on the wholesale market require the prior consent or 'approval' of the food authority to do so. Failure to comply with the conditions of approval may lead the food authority to take the administrative step of suspending or withdrawing the approval to trade in this manner.

[7] Civil sanctions include the service of Hygiene Improvement Notices requiring remedial works to be completed by the Food Business Operator within a specified time frame or the use of Hygiene Emergency Powers to prohibit a process, treatment piece of equipment or premises.

[8] For a definition of "crime" see Lord Esher M.R. at *Mogul SS Co v McGregor* 23 QBD 598 at 606.

15.2.2 Strict liability offences

The criminal offences created by food law are largely 'strict liability' in nature, there being no requirement to prove intention to commit the offence (*mens rea*). On one hand this reduces some of the burden from the prosecutor of food offences since they do not have to demonstrate a 'guilty mind' on the part of the accused.[9] However some authors have suggested that regulatory strict liability offences are not 'true crimes' and as such there has been reluctance on the part of the criminal courts to treat breaches of food law seriously (Pointing, 2005). It could be argued, for example, that food law, like many other forms of regulatory law,[10] fall within a category of 'crimes' defined by Wright J. as those which: 'are not criminal in any real sense, but are acts which in the public interest are prohibited under a penalty'.[11]

But what of food offences, what is the significance of the strict liability nature of breaches of food law? One important issue to consider here is the likelihood of evidence of *mens rea* being available to food enforcement officers investigating, for example, an outbreak of food poisoning or the contamination of a food with glass. In this context, *mens rea* could be considered to include a negligent act as well as an intentional one and in the case of a criminal prosecution it would be incumbent on the prosecution to lead with evidence of such matters 'beyond all reasonable doubt'. As such, it would place an impossible burden on the enforcement officers

[9] See Lord Reid in *Sweet v. Parsley* [1970]: 'Parliament did not intend to make criminals of persons who were in no way blameworthy in what they did' *Sweet v. Parsley* [1970] A.C. 132, 148G.

[10] In *R v Jackson* (2007), Hooper L.J. commented on the nature of strict liability offences: 'The rationale behind the creation of such offences is generally that they cover conduct which of itself is potentially dangerous to other members of the public and accordingly the public interest overrides the need to prove knowledge on the part of the alleged offender that he was in fact committing an offence or did the act complained of with any particular mens rea. Guilt is to be found in the commission of the act – no particular frame of mind has to be established before guilt is established. This reflects the fact that some conduct, often not truly criminal in the way most people would understand that expression, carries with it such grave risk of endangering public safety, or is so heavily the cause of public concern, that it can properly be punished without the need for the establishment of any degree of mens rea.' This was not a food law case, but instead related to an RAF pilot accused of flying below the minimum level prescribed in law. Clearly the risks to public safety of a pilot ignoring basic safety provisions are self-evident but should the same approach be adopted in terms of food hygiene breaches that may or may not cause actual harm? *R v Jackson* [2007] 1 Cr. App. R. 28.

[11] In *Sherras v De Rutzen* (1895) the defendant was convicted of selling alcohol to a police officer while on duty, contrary to section 16(2) of the Licensing Act 1872. He had reasonably believed the constable to be off duty as he had removed his armband, which was the acknowledged method of signifying off duty. The Divisional Court held that the conviction should be quashed, despite the absence from section 16(2) of any words requiring proof of mens rea as an element of the offence. Wright J. expressed the view that the presumption in favour of mens rea would only be displaced by the wording of the statute itself, or its subject matter. In this case the latter factor was significant, in that no amount of reasonable care by the defendant would have prevented the offence from being committed. Wright J. stated: 'It is plain that if guilty knowledge is not necessary, no care on the part of the publican could save him from a conviction under section 16, subsection (2), since it would be as easy for the constable to deny that he was on duty when asked, or to produce a forged permission from his superior officer, as to remove his armlet before entering the public house. I am, therefore, of opinion that this conviction ought to be quashed.' See *Sherraz v. De Rutzen* [1895] 1 Q.B. 918.

which would result in many cases of serious food offences going unpunished and public protection potentially undermined by the removal of the deterrent of prosecution.[12] There are occasions, however, where the mental element of a food-related crime is required to be demonstrated by the inclusion in the legislation of terms such as 'knowingly' and 'wilfully'. Such terms can be found, for example, in Regulation 15 of the Food Safety and Hygiene (England) Regulations 2013.[13]

15.2.3 Defences

Like other areas of regulatory law, UK food law provides for a number of strict liability offences which are tempered by statutory defences. The intention here is to strike a balance between breaches of legal obligations due to negligent or wilful behaviour and those which might occur despite the best efforts of the well-intentioned business. In the United Kingdom, there are defences available to those charged with food law offences under the Food Safety Act 1990, the Food Safety and Hygiene (England) Regulations 2013[14] or the Trade in Animals and Related Products Regulations 2011.[15] These are offences due to the fault of another person whereby an enforcement authority is able to bypass the immediate offender and to prosecute the real offender where an offence was in fact due to the fault of that other person. This is possible whether or not proceedings are taken against the immediate offender. Where a caterer buys in bread rolls, for example, and these are found to contain a foreign body, the manufacturer may be prosecuted directly by the enforcement authority without involving the caterer in any proceedings and; defence of due diligence whereby a statutory defence is provided for persons charged with food offences whereby the defendant can prove that he or she took all reasonable precautions and exercised all due diligence to avoid committing an offence. For offences under section 14 Food Safety Act 1990 (selling food not of the nature, substance or quality demanded) and section 15 (falsely describing or presenting food) special provision is made for persons who neither prepared the food themselves nor imported it into the United Kingdom.[16] If the defendant alleges that the offence was due to some other person, he or she must provide information to the prosecutor concerning that other person. This information must be given in writing and normally at least 7 days before any court hearing. There is

[12] For a discussion on a similar matter, see *Alphacell Ltd v Woodward* [1972] AC 824.

[13] Regulation 15 Food Safety and Hygiene (England) Regulations 2013 SI 2013 No. 2996.

[14] SI 2013 No. 2996.

[15] SI 2011 No. 1197.

[16] Such persons will have exercised all due diligence if they can prove: (1) the offence was committed by some other person not under the defendant's control, (2) the offence was a result of information provided by some other person not under the defendant's control, (3) checks on the food were carried out or it was reasonable to rely on checks carried out by the supplier, (4) there was no knowledge of the offence and no reason to suspect that the defendant's actions amounted to an offence, or (5) the offending product was not sold under the defendant's brand name and the above points (1), (2) and (4) can be proved.

also a defence available for advertising, printing or publishing businesses who publish an offending advertisement in good faith during the normal course of their business.

In *Wings ltd v Ellis*, Lord Scarman commented on the nature of the due diligence defence found in the Trades Description Act 1968 which was a precursor to the one later incorporated in the Food Safety Act 1990:

> *The employer or principal who has exercised due diligence and taken reasonable precautions and who is not morally blameworthy should not be found guilty of an offence. The net of offences is drawn wide but there is a defence for those who can prove that they are not blameworthy. It is reasonable to put the onus on the defence because (i) the facts are within the peculiar knowledge of the defence, and (ii) to put the onus on the prosecutor would make the Act difficult to enforce.*[17]

As such it is a matter for a defendant, when charged with an offence, to raise and prove that a due diligence defence existed at the time of the alleged offence. The burden of proof placed on the defendant to prove this defence is 'on the balance of probabilities' which is a lower level of burden to that placed on the prosecutor who must prove the offence 'beyond all reasonable doubt'. In *Coupe v Guyett*[18] the owner of a garage was charged with an offence under the Trades Description Act 1968 section 24. The court decided that where the owner of the garage demonstrated the existence of a general defence, an employee could not be convicted in relation to an 'act or default of another person' provision. In *Tesco Supermarkets Ltd v Nattrass* it was held that a manager could be 'another person' for the purposes of the Act. As such a company can only be liable for the defaulting act of its employee if that employee is one of the directing minds or represents the will of the company.[19]

15.3 Arrangements for delivery of official food controls

15.3.1 Who has responsibilities under food law?

The legal responsibility for compliance with food law falls to the relevant Food Business Operator[20] but in practice the responsibility for ensuring that the requirements and objectives of food law are met is shared between the food industry and

[17] *Wings Ltd. v. Ellis* [1985] A.C. 272 House of Lords 1984 June 25, 26; October 25.

[18] *Coupe v. Guyett* [1973] 1 W.L.R. 669, 675.

[19] *Tesco Supermarkets Ltd. v. Nattrass* [1972] A.C. 153.

[20] Food Business Operator is defined in Article 3 of Regulation 178/2002 of the European Parliament and of the Council laying down the general principles and requirements of food law, establishing the European Food Safety Authority and laying down procedures in matters of food safety [2002] O.J. L31/1, as amended as 'the natural or legal persons responsible for ensuring that the requirements of food law are met within the food business under their control'.

the regulators.[21] The former is required to ensure that food placed on the market is safe and that the handling of food is hygienic; the latter is required to perform official controls to verify that these are achieved. The roles and responsibilities of the agencies involved in the regulation of the food industry are set out in the UK National Control Plan.[22] Each Member State of the European Union (EU) is required to establish a central competent authority for food matters and the FSA is the central competent authority for the United Kingdom and is responsible for policy making and the supervision of local authority food service provision.[23] The FSA was created by the Food Standards Act 1999, as a response to public health concerns such as bovine spongiform encephalopathy (BSE), also known as "mad cow disease", *Listeria* and an *Escherichia coli* O157 outbreak in Lanarkshire[24] which resulted in several deaths. While the FSA is the central competent authority for food, it devolves the responsibility for the implementation of official controls programmes in England to food authorities which include County Councils, District Councils, London Boroughs, Unitary Councils and PHAs. Decisions on enforcement matters with respect to food law outside of slaughterhouses, meat cutting plants and approved game handling establishments fall to the food authorities.

15.3.2 Organisations involved in food law enforcement in England

Food authorities are local authorities and PHAs who are responsible for undertaking the majority of official controls with respect to the food industry. There are 388 food authorities in England comprising County Councils, District Councils,[25] London Boroughs, Unitary authorities and PHAs each under the control of locally elected Councillors or 'Members'. The current organisation of local authorities was established in 1974 and since that time has undergone a series of reforms with an initial trend towards unitisation of authorities, but more recently the sharing of services between authorities. Typically food standards official controls are carried out by Trading Standards Officers (TSOs) working for County Councils and some Unitary Councils, whereas food hygiene official controls are carried out by Environmental Health Officers (EHOs) working for District Councils, London Boroughs and Unitary Councils. In London and some Unitary Councils, EHOs are

[21] For a discussion on the shared responsibility for compliance with food law, see Griffith, C.J. 'Food safety in catering establishments' (2000) in *Safe Handling of Foods* (Farber, T. (Ed.), Marcel Dekker, Oshawa) 235.

[22] Details of the National Control Plan can be found at http://www.food.gov.uk/foodindustry/regulation/europeleg/feedandfood/ncpuk.

[23] Note that in Scotland this role is performed by Food Standards Scotland.

[24] See fatal accident inquiry into the deaths of Sarah Kerr Cameron and others. Determination by Graham Cox QC, Sheriff Principal of Sheriffdom of South StrathClyde, Dumfries and Galloway (1998).

[25] Together, County Councils and District Councils form a two tier system of government. The Districts are located within a County Council, the latter taking responsibility for strategic matters such as education, social services and highways management, whereas the District Councils' work is more locally focused and includes Planning Control and Environmental Health.

also responsible for food standards official controls and at County Council level, some TSOs have recently taken on the responsibility for undertaking food hygiene official controls at primary production level.[26] Local authorities have been responsible for the enforcement of food law since 1860 (Paulus, 1974) with TSOs having responsibility for 'nature, substance and quality' matters in two tier authorities until 1990 when EHOs took on the responsibility for foreign object in food matters which were deemed not to be of the substance demanded. However, it could be argued that the distinction between food standards law and food hygiene law is an artificial one which has been created over time to fit in with the organisation of the inspectorate. Similarly, the use of TSOs in two tier authorities appears arbitrary when EHOs in London are deemed to be competent to undertake official controls in both areas of law. The demarcation between functions of the two tier authorities is set out in the Food Law Code of Practice[27] with EHOs taking responsibility for 'microbiological quality and contamination by microorganisms or foreign matter' and TSOs responsible for 'composition, chemical contamination, adulteration and labelling'.

15.3.3 Arrangements at the ports

PHAs are constituted under public health legislation to undertake a wide range of public health checks and controls at seaports. These functions at airports are carried out by the local authority in which the airport is situated and these authorities are referred to as 'local authorities with port health function'. In some cases the PHA is under the direct control of one local authority[28] who appoints a board of Councillors to oversee its functions. In some areas, the PHA is operated by a group of riparian Councils who appoint a management board. In both cases, the funding for food official controls at the ports is primarily provided by the local authorities themselves although some of the functions at some ports[29] may be charged to the food importer and some functions are supported by centrally awarded grants from the FSA and Defra. Officers working at the PHAs include Official Veterinary Surgeons employed by Defra and Port Health Officers (PHOs)[30] and assistants employed by the food authority. The PHOs undertake a range of functions including noise and nuisance control, pest control and infectious disease control as well as food safety work. In addition to these staff, officers from Her Majesty's Revenues and Customs (HMRC) and Border Force are responsible for the detection and destruction of illegally imported food at ports.

[26] Primary production includes farming and fishing.

[27] Available at http://www.food.gov.uk/multimedia/pdfs/codeofpracticeeng.pdf.

[28] Such is the case with the Port of Felixstowe which is under the control of Suffolk Coastal District Council.

[29] The official controls carried out on products of animal origin imported from third countries must be paid for by the importer. However, these imports are only permitted through certain ports called 'Border Inspection Posts'.

[30] PHOs are EHOs with additional training in meat inspection and import controls.

15.3.4 The Food Standards Agency

Proposals for the creation of the FSA were drawn up following the James Report in 1997 (James, 1997). In that report, Prof. James identified a number of factors which he considered to be eroding public confidence in food safety and control. A key factor was the dual role played by the Ministry of Agriculture, Fisheries and Farming, which was responsible for protecting public health but at the same time was the main protagonist for the interests of the food and agricultural industries. Further factors identified in the James report were the lack of coordination of government bodies involved in food safety, and also the uneven enforcement in food law generally. In accepting its findings, the government agreed that the functions of promoting safe food for the consumer and the promotion of the interests of food businesses should be separated. The Government proposed the FSA in its White Paper 'The Food Standards Agency – A Force for Change'[31] in 1998 (The Food Standards Agency – A Force for Change, January, 1998). The White Paper's proposals were generally endorsed and so the Food Standards Bill was produced in response to this, and resulted in the introduction of the Food Standards Act in 2000.[32] The FSA is led by an independent board the members of which are drawn from a variety of backgrounds including enforcement, food industry and consumers.

15.3.5 Communication

Communication between the FSA and food authorities is an important element of food law enforcement. For example, the Agency is responsible for the issue of Food Alerts[33] to food authorities whenever they consider that it is appropriate which may require the withdrawal of foods from the market in circumstances where a risk assessment has identified they may pose a significant risk to health. Local authorities are required to notify the Agency of any food incidents that they are aware of in their area which may have significant or widespread implications.[34] Similarly food business operators are required to notify the competent authority

[31] The white paper proposed several guiding principles under which the proposed agency should operate, which included the following: The protection of public health in relation to food; to provide unbiased expert advice in relation to food standards and safety; to act independently of vested interests; to give information so that the public could make informed choices; to be transparent in its decision-making process and to consult all interested parties in that process; to have a clear and consistent approach to food safety issues.

[32] The main purpose of the Act was as follows: The creation of the Food Standards Agency, and to lay down its functions and powers, as well as its organisational and accountability arrangements; to transfer certain functions in relation to food safety and standards, previously dealt with by the Department of Health and the Ministry of Agriculture, Fisheries and Food, to the Agency; to set out the Agency's main objective in protecting public health in relation to food, and to give it powers to act in this aim at all stages in the food production and supply chain.

[33] Communications issued by the FSA regarding specific food incidents. For more information, see http://www.food.gov.uk/enforcement/alerts/.

[34] Food Law Code of Practice (England) (April 2014) Section 4.1.1. Available at http://www.food.gov.uk/multimedia/pdfs/codeofpracticeeng.pdf.

where food that is unsafe has left their initial control and has been placed on the market.[35]

15.3.6 Monitoring of food authorities and issue of guidance

A significant role of the FSA is the monitoring of food authorities and this is specified in 'The Framework Agreement on Local Authority Enforcement'.[36] The main ways that the FSA achieve this function is through the collection of enforcement data through the LAEMS[37] scheme together with the audit of food authorities. In addition to the monitoring of food authorities, an important function of the FSA is the commissioning of research and the provision of guidance and information to Ministers, the trade and the general public. Initially this was achieved by the use of a series of committees but many of these have now been replaced by ad hoc expert working groups. One surviving committee, the Advisory Committee on the Microbiological Safety of Food has, for example, contributed to guidance on the safe use of vacuum packing as a food preservation method.[38] Such guidance is intended to assist the trade in their understanding of food safety requirements and provide enforcement officers with a method for exerting effective control. The FSA has a significant role in the identification and commissioning of research and also provides a training programme for enforcement officers of food authorities.

15.3.7 Official control bodies

In addition to the FSA and food authorities, a number of official control bodies[39] are involved in the regulatory system in England, most notably in the field of organic food regulation where the authenticity of organic foods is demonstrated throughout the food chain by means of a certification process. In the United Kingdom the Soil Association is one of the recognised control bodies to undertake this function.

[35] Article 19, Regulation 178/2002 of the European Parliament and of the Council laying down the general principles and requirements of food law, establishing the European Food Safety Authority and laying down procedures in matters of food safety [2002] O.J. L31/1, as amended.

[36] 'The Framework Agreement on Local Authority Enforcement' Food Standards Agency 2010. Available at http://www.food.gov.uk/enforcement/enforcework/frameagree/.

[37] Local Authority Enforcement Management System. Further details available at http://www.food.gov.uk/enforcement/auditandmonitoring/laems/.

[38] Food Standards Agency guidance on the safety and shelf-life of vacuum and modified atmosphere packed chilled foods with respect to non-proteolytic Clostridium botulinum June 2016. Available at: https://acmsf.food.gov.uk/sites/default/files/acm_1230a_guidancedoc.pdf.

[39] 'Control bodies' are defined in Regulation (EC) No. 882/2004 of the European Parliament and of the Council on official controls performed to ensure the verification of compliance with feed and food law, animal health and animal welfare rules. Official Journal of the European Union L 191/1 28.5.2004 as 'means an independent third party to which the competent authority has delegated certain control tasks'.

15.3.8 The European and international perspective

The food industry is truly global in nature and as such an international framework has been established to facilitate fair and free trade. The Joint Food and Agriculture Organisation/World Health Organisation Codex Alimentarius Commission is an intergovernmental body set up in 1962 to promote international trade in food by the adoption of standards. The aim of the Codex Alimentarius Commission is to set standards to promote international fair trade practices and the protection of consumer health. Codex standards have become increasingly important since the establishment of the World Trade Organisation (WTO) (Victor, 2000) in 1995 to facilitate and administer international trade agreements and it currently has approximately 155 member countries. The main functions of the WTO are to provide a forum for trade negotiations; handle international trade disputes; monitor national trade policies and provide technical assistance and training for developing countries. The General Agreement on Tariffs and Trade is an international agreement which permits governments to introduce restrictions to free trade where such measures are necessary to protect human, animal or plant health. However, such safeguard measures must not provide artificial barriers to free trade and should not be used as a form of protectionism. There are two WTO agreements in place which provide a framework whereby governments may introduce protective measures without breaching the principles of free trade: the Agreement on Sanitary and Phytosanitary Measures[40] and the Technical Barriers to Trade Agreement.[41] In 2001 the EU became a full member of the Codex Alimentarius Commission and hence be in a position to directly influence international food standards.[42] EU Food Law has its roots in Codex standards and references to Codex standards can often be found in European Case Law.[43]

15.3.9 DG Health and Food Safety

The Directorate F of the DG Health and Food Safety[44] (formerly known as the Food and Veterinary Office) is the food inspectorate of the European Commission. It has a complement of approximately 200 inspectors based in the Republic of Ireland and its functions include the approval of third countries who wish to export

[40] The World Trade Organisation Agreement on the Application of Sanitary and Phytosanitary Measures (SPS Agreement). Available at https://www.wto.org/english/Tratop_e/sps_e/spsagr_e.htm.

[41] World Trade Organisation Agreement on Technical Barriers to Trade (TBT). Available at: https://www.wto.org/english/docs_e/legal_e/17-tbt_e.htm.

[42] The application for accession by the EU to the Codex Commission was opposed by the United States, primarily on the principle that there should be a 'one nation, one vote' approach to Codex, just as there is in the United Nations.

[43] For example in *Pfizer Animal Health SA v Council of the European Union* [2002], Codex was referred to clarify the meaning of key terms such as 'hazard'. Case T-13/99 *Pfizer Animal Health SA v Council of the European Union* [2002] ECR II-3305.

[44] https://ec.europa.eu/food/safety.

live animals and products of animal origin (POAO) to the EU; the approval of the designated ports of entry within the EU permitted to accept such imports[45] and the verification of food official controls and systems in EU Member States. It carries out its functions through a series of audits or 'missions' and produces reports on its findings.[46]

15.3.10 The European Food Safety Authority

In addition to Directorate F of the DG Health and Food Safety, the European Food Safety Authority (EFSA) was created by the adoption of Regulation (EC) 178/2002[47] to provide guidance and advice to the EU legislators and to convene expert committees and commission research to evaluate new and emerging food safety risks. It is an independent organisation and according to Shears et al. (2004) 'Building and maintaining consumer confidence is the business of politicians. Their decisions must be informed by the appliance of science. The scientists must be independent of politics', hence the separation between the scientists and the legislators. Referring to the creation of the EFSA, David Byrne, European Commissioner for Health and Consumer Protection, stated.[48]

> *One of the key features of EFSA is its independence its capacity to inform and advise, free from political interference or industry influence. I sincerely hope it will prove to be a major breakthrough in improving the relationship between public institutions and European citizens.*

15.3.11 Mechanisms for assessing compliance with food law

The traditional method for assessing compliance of food businesses with the requirements of food law is through the implementation of official controls using interventions such as the inspection and audit of premises and the collection of food samples. These remain the main techniques used by food authorities and guidance on their appropriate use is provided in the Food Law Code of Practice.[49] In practice, food authorities use a range of strategies[50] in food businesses using

[45] Known as 'Border Inspection Posts'.

[46] More details of the workings of the Directorate F together with copies of Mission Reports on Member State and third country activities can be found at https://ec.europa.eu/food/audits_analysis/audit_programmes_en.

[47] Chapter III of Regulation 178/2002 of the European Parliament and of the Council laying down the general principles and requirements of food law, establishing the European Food Safety Authority and laying down procedures in matters of food safety [2002] O.J. L31/1, as amended by Regulation 1642/2003 [2003] O.J. L245/4.

[48] Byrne D. Speech/03/593 'Irrational Fears or Legitimate Concerns' – Risk Perception in Perspective Risk Perception: Science, Public Debate and Policy Making Conference Brussels, 4 December 2003.

[49] Food Law Code of Practice (March 2017) Section 4. Available at https://www.food.gov.uk/enforcement/codes-of-practice/food-law-code-of-practice.

[50] For a discussion see Ayres, I., Braithwaite, J, *Responsive Regulation* (Oxford University Press, Oxford, 1992).

both formal and informal approaches to achieve the desired outcome of compliance by the business with food law (Reiss, 1984). These approaches include the dissemination of advice and guidance (Winter, 1985) as well as the requirement for businesses to comply with legal obligations.[51]

15.3.12 Food Interventions

During the last few years there has been a move towards the introduction of new techniques to obtain compliance with food law, for example, by the publication of inspection findings on the Internet (Griffith, 2005). In June 2008, in response to the Hampton Review, a revised Food Law Code of Practice was published which provided for the adoption of a suite of food interventions by local authorities. The intention of this measure was to encourage local authorities to target their resources towards identifying and rectifying breaches of food law in food businesses. Food establishments, including ships, fishing vessels and aeroplanes, are subject to a programme of food hygiene and food standards interventions conducted by the relevant food authority.[52] Food interventions are defined as: 'activities that are designed to monitor, support and increase food law compliance within a food establishment'.[53] The Food Law Code of Practice identifies two types of food intervention: 'official controls' and 'other interventions'. Official controls include inspection and audit of food businesses; monitoring, surveillance and sampling visits. 'Other interventions' include the provision of advice, education and coaching by an enforcement officer together with the gathering of information and intelligence. Food authorities are required to put into place a programme of food interventions for food businesses located within their area based on a risk assessment carried out by their officers using a standard risk rating assessment scheme included in the Food Law Code of Practice.

The food hygiene risk rating scheme is divided into three sections and food officers are required to allocate a numerical score for every business based on how they meet the requirements of descriptors provided in the Code.[54] Risk rating of a

[51] For a description of the inspection work of Environmental Health Officers see: Hutter, B. *The Reasonable Arm of the Law? The Law Enforcement Procedures of Environmental Health Officers* (Clarendon Press, Oxford, 1988)

[52] Food authorities include local authorities, PHAs and local authorities with port health function.

[53] Food Law Code of Practice (England) (March 2017) Section 4.1.1. Available at http://www.food.gov.uk/multimedia/pdfs/codeofpracticeeng.pdf.

[54] The sections of the risk rating scheme are: Part 1: Potential hazard: This section is subdivided into three parts and rates establishments according to their inherent risks due to their size, processing methods and customer base. A large manufacturer of cooked meats for national distribution will attract a high score in this section, a small grocer selling only tinned and packaged food would attract a low score. This part of the rating scheme does not take management controls or levels of compliance into account. The three subsections are: A: Type of food method of handling; B: Method of processing and C: Consumers at risk. Part 2: Level of current compliance: This assesses the current level of compliance with food law in terms of hygiene practices and structural requirements which are scored separately. Part 3: Confidence in management: This part of the scheme assesses the business's previous history of compliance, its current food safety management controls and the ability of the management to comply with food law in the future. It also includes a 'significant risk' score which should be added if the inspecting officer is of the opinion that food safety controls may not be maintained in future resulting in a significant risk of contamination of the food.

Table 15.1 Summary of food hygiene intervention frequencies[a]

Category	Score	Minimum intervention frequency
A	92 or higher	At least every 6 months
B	72–91	At least every 12 months
C	52–71	At least every 18 months
D	31–51	At least every 24 months
E	0–30	A programme of alternative enforcement strategies or interventions every 3 years

[a]Chapter 5.6 Food Law Code of Practice (England) (March 2017).

food business may only take place following a full or partial inspection or audit of a food business where sufficient information has been gathered. Establishments that have received a risk rating are then placed into a risk category depending on the numerical score that they have received and this risk category dictates the minimum frequency and the nature of the next food hygiene intervention. Once an authorised officer has completed a risk rating for a food business, this is recorded on a database and the next intervention date for that business will be generated according to the risk category assigned to it. A summary of the risk rating scores, risk categories and the resultant frequency of intervention is provided in Table 15.1. In addition to the above, scores achieved in parts of this risk rating scheme contribute to a 'broadly compliant' score. Those businesses that score no more than 10 in each of the categories: current compliance with hygiene; current compliance with structure and confidence in management will be regarded as being 'broadly compliant' with food law. While food businesses in category A or B must always receive an inspection or audit as their planned intervention, businesses which are within risk category C and are deemed to be broadly compliant with food law may receive a surveillance/verification visit instead once every other 18 months. Category D businesses need only have alternative official controls and other interventions as their planned hygiene interventions and Category E businesses are deemed to present such a minimal risk to the public that their assessment may be by means of an 'alternative strategy' to interventions. This normally takes the form of a brief surveillance questionnaire or telephone interview.

A similar approach is adopted for the risk rating of businesses for food standards purposes.

Three elements of this risk rating score (relating to hygiene, structure and the confidence that the inspector has in the food business operator and their food safety procedures to maintain/improve standards) are used to provide a food hygiene rating score which can be displayed by the food business operator on their premises and is provided on a national website.[55]

[55] More of the Food Hygiene Rating scheme can be found here: https://www.food.gov.uk/business-industry/hygieneratings.

15.3.13 Food sampling

Food authorities are required to perform food official controls which include the development and implementation of food sampling programmes. Most food authorities are required by the Food Safety Act 1990 to appoint a Public Analyst (PA) to undertake chemical analysis of such food samples whereas samples for microbiological examination are submitted to Public Health England (PHE). The laboratories operated by PAs and the PHE must be listed by the European Commission as official control laboratories and those engaged in the analysis or examination of foods must meet specified qualification and competence standards. Each food authority is required to produce and implement a food sampling policy in their area and identify a food sampling programme which, in practice, many local authorities coordinate with neighbouring authorities through joint food liaison groups. The results of official controls sampling are recorded on the UK Food Surveillance System national database which is operated by the FSA.

15.4 Administrative issues

15.4.1 Registration and approval of food establishments

Food businesses are required to be either registered with or approved by the relevant food authority which in most cases is the local authority Environmental Health department. To register, a food business operator is required to complete a standard form, at least 28 days before commencing food operations, which can be found at https://www.gov.uk/food-business-registration.

Alternatively, establishments which supply other food establishments with POAO may be required to seek approval from their food authority before commencing wholesale trade. Approval is not required in certain circumstances including for retail activities, supply of small amounts of products direct to the final consumer or establishment supplying the final consumer, transport operations or for supply to retailers by retailers as long such supply is local and marginal. The FSA have provided informal definitions for both local and marginal supply as well as suggested definitions for 'small quantities' in their guidance on approved establishments. On receipt of an application for approval, the relevant competent authority will undertake an inspection of the establishment to determine whether it complies with all relevant requirements of food law. In this respect the requirements relate to the hygiene conditions set out in Regulation (EC) 852/2004, the additional requirements provided for in Regulation 853/2004, any relevant microbiological criteria identified in Regulation (EC) 2073/2005, animal by-products requirements required by Regulation (EC) 1069/2009 as well as the general food requirements of Regulation (EC) 178/2002. If these requirements are all satisfied, the food authority will grant an initial conditional approval for a period of 3 months with provision to extend this period to 6 month if necessary. Once this conditional approval period has been completed the food authority will grant the establishment full approval if satisfied of the ongoing compliance with food law.

Once approved all food produced by the establishment must be marked with either a *health mark* or an *identification mark*. A health mark is applied by official inspectors to fresh meat carcasses and primal cuts of meat only following inspection at the approved slaughterhouse or cutting plant. The health mark is therefore an indication that the specific carcass or cut of meat has been examined and declared fit for consumption by the authorised officer. An identification mark, on the other hand, is applied by the food business operator to foods produced within an approved establishment with a primary purpose of providing enhanced traceability for the product. The format of the identification mark is set out in Annex II of Regulation (EC) 853/2004 and comprises an oval containing 'United Kingdom' and 'EC' together with a unique reference number which either identifies the local authority responsible for official controls, or, in the case of meat cuts and offal, the licence number of the approved slaughterhouse or cutting plant.

15.4.2 Imported food control

The United Kingdom imports a significant proportion of its food from countries around the world and as such imported food can be found in virtually every food business. The majority of foods that are imported into this country are perfectly legal, wholesome and safe but it is important to have effective imported food controls in place to ensure that the consumer and food businesses are protected from contaminated products, unfair trading practices and fraud.

Food is imported into the United Kingdom via seaports, airports and by road and rail through the channel tunnel. PHAs have been set up at seaports to undertake checks on certain imported foods and at airports local authorities have this function. Certain ports, which receive POAO such as meat, fishery products, eggs, shellfish, dairy products and honey, are designated as Border Inspection Posts (BIPs)[56] which are approved by the DG Health and Food Safety to carry out veterinary checks on imports. For the purpose of import controls, foods can be placed into one of two categories: POAO and food not of animal origin (FNAO) and can be imported commercially either from countries within the European Economic Area (EEA) or from third countries. The rules regarding the importation of food are different depending on the nature of the food (POAO or FNAO) and the country of origin. The principle of a single market within the EU was established by the EC Treaty[57] and permits the free movement of foods between Member States. This means that POAO produced or processed in another Member State may be imported into the United Kingdom without any specific checks being carried out at the point of entry. Similarly, POAO from the United Kingdom may be freely exported to other Member States without being subjected to any specific border controls. Food that is lawfully imported into a Member State from a third country may subsequently be distributed into other Member States without the need for further checks to be made. This is an important principle and requires that Member States have mutual

[56] These will be known as Border Control Posts from December 2019.

[57] Now the Treaty for the Formation of the European Union (TFEU).

respect for each other's border controls the adequacy of which is policed by DG Health and Food Safety. There are occasions, however, whereby foods produced by Member States are subjected to specific restrictions or 'safeguard measures' mainly due to animal, public or plant health concerns.

POAO may only be imported from third countries that have received prior approval by the European Commission for the relevant type of POAO. In addition, the establishments involved in the handling of such products must also receive prior approval before the food can be lawfully exported to the EU. The prior approval of prospective suppliers of POAO to the EU is undertaken by inspectors from Directorate F of DG Health and Food Safety who assess the ability of the establishments in the third country to produce food of an equivalent standard to those situated in the EU and the adequacy of the official controls performed by the third country's competent authority. POAO, except honey, produced in approved establishments must be marked with the country of origin and a unique 'establishment number'.

When exported from third countries, POAO are normally required to be accompanied by a health certificate signed by the competent authority of the third country to verify that the food meets the requirements of EU Law. The POAO then must enter the EU via an approved port known as a BIP where the food will be subjected to veterinary checks by the relevant authorised officer. These veterinary checks will comprise, as appropriate, a documentary check, an identity check and where necessary a physical check. Veterinary checks are carried out by Official Veterinarians (OV) or Official Fish Inspectors (OFI). Once the veterinary checks have been successfully completed, the OV or OFI will complete, sign and stamp a Common Veterinary Entry Document (CVED)[58] to demonstrate that the food has been subjected to the appropriate official controls and has been released for free circulation within the EEA. This CVED must accompany the POAO to its first destination inland where it must be retained for a period of 1 year. Should the products fail the veterinary checks, the POAO may be destroyed or reexported to the country of origin, depending on the reason for the failure. There may be occasions where the POAO enters the EU via a BIP but its destination is outside of the EU. In these circumstances, special provisions apply. Once cleared by the BIP, food may be sent through to an external temporary storage facilities (ETSF) prior to HMRC checks. The European Commission operates the TRACES website[59] that contains up-to-date information on the currently approved third countries and establishments that might legitimately trade with the EU in addition, the FSA provides the 'GRAIL' website[60] to help authorised officers and businesses to keep abreast of developments in imported food legislation.

In addition to the veterinary checks performed at the BIPs on POAO from third countries, there are a number of 'high-risk' FNAO from third countries which are

[58] This will be known as a "Common Health Entry Document" (CHED) from December 2019.
[59] See https://sanco.ec.europa.eu/traces/output/listsPerActivity_en.htm#.
[60] For Guidance and Regulatory Advice on Import Legislation, see https://grail.foodapps.co.uk/grail/general/home.aspx.

subject to enhanced checks at the ports. These are mainly set out in Annex I of Regulation (EC) 669/2009 which is updated quarterly by the European Commission although there are some additional products which are also subject to these enhanced import checks. Any product subject to these checks must only be imported into the EU via a designated point of entry and be accompanied by a Common Entry Document (CED) which is endorsed at the port to verify legal import.

In December 2019, the current arrangements for the importation of food will be streamlined with the introduction of a Common Health Entry Document which will replace both the CVED and CED. BIPs will become Border Control Posts and will be responsible for official controls on both POAO and FNAO.[61]

15.5 Enforcement sanctions

15.5.1 Approaches to enforcement

Yapp and Fairman (2006) describe two strategies to deal with noncompliance of food law: compliance and deterrence strategies. They quote Reiss (1984) in this respect:

> *The principal objective of a compliance law enforcement system is to secure conformity with the law by means of insuring compliance or by taking action to prevent potential law violations without the necessity to detect, process and penalize violators. The principal objective of deterrence law enforcement systems is to secure conformity with law by detecting violations of law, determining who is responsible for their violation, and penalizing violators to deter violations in the future.*

The United Kingdom has a compliance-based enforcement system, the emphasis is on preventing harm from occurring, generally achieved through a range of cooperative strategies which is in contrast to a deterrent-based strategy which aims to merely detect breaches of the law and punish offenders. Within this compliance strategy, some authorities adopt highly educational approaches by, for example, undertaking advisory visits, running food hygiene courses and subject-specific

[61] These changes will be introduced by Regulation (EU) 2017/625 of the European Parliament and of the Council of 15 March 2017 on official controls and other official activities performed to ensure the application of food and feed law, rules on animal health and welfare, plant health and plant protection products, amending Regulations (EC) No. 999/2001, (EC) No. 396/2005, (EC) No. 1069/2009, (EC) No. 1107/2009, (EU) No. 1151/2012, (EU) No. 652/2014, (EU) 2016/429 and (EU) 2016/2031 of the European Parliament and of the Council, Council Regulations (EC) No. 1/2005 and (EC) No. 1099/2009 and Council Directives 98/58/EC, 1999/74/EC, 2007/43/EC, 2008/119/EC and 2008/120/EC, and repealing Regulations (EC) No. 854/2004 and (EC) No. 882/2004 of the European Parliament and of the Council, Council Directives 89/608/EEC, 89/662/EEC, 90/425/EEC, 91/496/EEC, 96/23/EC, 96/93/EC and 97/78/EC and Council Decision 92/438/EEC (Official Controls Regulation) Text with EEA relevance.

seminars. Others adopt a more formal approach, prosecuting individual businesses and serving notices more frequently where noncompliance is not remedied (Yapp and Fairman, 2005). The relative effectiveness of an authority's strategy can be measured of how well they meet the objective of improving food safety compliance within food businesses in their area (Yapp and Fairman, 2003). In April 2014 the statutory Regulators Code,[62] issued under the Legislative and Regulatory Reform Act 2006, came into force which requires that regulators, including food authorities, should:

- carry out their activities in a way that supports those they regulate to comply and grow;
- provide simple and straightforward ways to engage with those they regulate and hear their views;
- base their regulatory activities on risk;
- share information about compliance and risk;
- ensure clear information, guidance and advice is available to help those they regulate meet their responsibilities to comply and
- ensure that their approach to their regulatory activities is transparent.

15.5.2 The precautionary principle in food law

The precautionary principle was introduced into food law by Regulation (EC) 178/2002[63] although it has been well established in other areas of law, notably environmental law, and is provided for in Article 191 TFEU[64] (ex Article 174 of the EC Treaty). In essence it provides for the intervention of Member States to protect public health in situations where the scientific evidence to do so is less than conclusive. However, Article 7 of Regulation (EC) 178/2002 requires that the principle is applied having regard to the established techniques of risk assessment, risk management and risk communication. Risk assessment techniques require the initial identification of hazards[65] and the subsequent evaluation of the likelihood and severity of these hazards being realised. Scientific guidance is often sought during this evaluation process but in cases such as BSE the science was unable to provide definitive answers in order to properly inform the process of risk assessment (Millstone, 2009). Therein lays the difficulty with a risk assessment-based food control system: it cannot respond to emerging issues of food contamination until there is some scientific justification to do so. The precautionary principle is available, by virtue of Article 7 of Regulation (EC) 178/2002,[66] to all competent authorities, which in England would include local authority enforcement officers.

[62] Regulator's Code, Department of Business, Innovation and Skills, Better Regulation Delivery Office.

[63] Regulation 178/2002 of the European Parliament and of the Council laying down the general principles and requirements of food law, establishing the European Food Safety Authority and laying down procedures in matters of food safety [2002] O.J. L31/1, as amended.

[64] Consolidated version of the Treaty on the Functioning of the European Union [2008] OJ C115/47.

[65] A hazard is something that has the potential to cause harm.

[66] Ibid footnote 54.

15.5.3 Use of the precautionary principle

In 2000 the European Commission provided guidance on the use of the precautionary principle[67] with the aims to:

> *outline the Commission's approach to using the precautionary principle, establish Commission guidelines for applying it, build a common understanding of how to assess, appraise, manage and communicate risks that science is not yet able to evaluate fully, and avoid unwarranted recourse to the precautionary principle, as a disguised form of protectionism.*

15.5.4 The precautionary principle and official food controls

So how does the precautionary principle apply to the implementation of food official controls by food authorities who are encouraged to make risk-based decisions when dealing with breaches of food law? It is important for the enforcement officer to recognise circumstances where they are unable to make informed risk-based decisions but instead, in the absence of clear evidence, are required in the public interest to act decisively. An example was discussed during the public enquiry into an outbreak of *E. coli* O157 food poisoning in South Wales in 2005[68] regarding the dilemma faced by an enforcement officer who was undertaking an inspection of a meat products factory prior to the outbreak. The officer noted some hygiene offences and discovered during the course of the inspection that the food business operator was using a vacuum packing machine for packing both raw and cooked meats. The officer questioned the food business operator regarding his procedures for preventing cross-contamination from the raw meat to the cooked but in the absence of direct evidence that the procedures were flawed, accepted that there was no imminent risk to health. The food poisoning outbreak occurred a few weeks after this inspection. The officer later claimed at the public enquiry that there was insufficient evidence available at the time of the inspection to serve an Emergency Prohibition Notice which would have prohibited the vacuum packing activities. This assertion was rejected by the chair of the enquiry, Prof. Hugh Pennington, who stated that the officer should not have accepted verbal reassurances relating to the implementation of such important hygiene procedures when there was physical evidence available to the officer that other hygiene procedures in the factory were not being followed. This was a classic example where the use of a risk-based decision-making process took precedence over a precautionary principle approach.

[67] Communication from the Commission on the precautionary principle Brussels, 02.02.2000 COM (2000) 1.

[68] The Public Inquiry into the September 2005 Outbreak of *E. coli* O157 in South Wales Chairman: Prof. Hugh Pennington, March 2009. HMSO Press.

15.5.5 Tools of enforcement

Regulation (EC) 882/2004,[69] Article 55 requires that: 'Member States shall lay down the rules on sanctions applicable to infringements of feed and food law and other Community provisions relating to the protection of animal health and welfare and shall take all measures necessary to ensure that they are implemented. The sanctions provided for must be effective, proportionate and dissuasive'. In addition to prosecution, there are a number of different formal enforcement options available to food safety officers and these can be broadly classified into three groups: detention and seizure procedures; compliance notices and prohibition procedures.

15.5.6 Detention and seizure of food

Food can be deemed to fail food safety requirements if it is 'unsafe' as defined in Article 14 of Regulation (EC) 178/2002; if it fails to meet the requirements of certain statutory instruments such as the Food Additives, Flavourings, Enzymes and Extraction Solvents (England) Regulations 2013[70] or the Contaminants in Food (England) Regulations 2013[71] or; if the authorised officer believes that the food has not been produced, processed or distributed in accordance with the Hygiene Regulations and he certifies that the food fails food safety requirements as a consequence using the provisions of Regulation 29 of the Food Safety and Hygiene (England) Regulations 2013.[72] Where an authorised officer *suspects* that a food 'fails to meet food safety requirements' he may serve a Detention Notice on the person in charge of the food using the provisions of section 9 of the Food Safety Act 1990. The officer may then direct that the food is detained at the premises where it was discovered or at another place as specified in the notice. Once a section 9 Food Safety Act 1990 detention notice has been served, the officer must decide whether the food in question fails to meet food safety requirements within 21 days of service. If during this period the officer becomes satisfied that the food fails to meet food safety requirements, he may seize the food and make an application to a Magistrate to have the food condemned. If, however, during the 21-day period the officer becomes satisfied that the food is safe for human consumption, he/she must release the food from detention and compensate the owner of the food for any losses incurred as a result of the detention.

Where an officer is *satisfied* that a food fails food safety requirements, he/she may seize the food and, normally within 2 days, apply to a Magistrate to have the food condemned. The application is by means of a written complaint and the person who is culpable in law for the food has the right to attend the court and make representations. The Magistrate may decide to hear whatever evidence he/she chooses

[69] Regulation (EC) No. 882/2004 of the European Parliament and of the Council on official controls performed to ensure the verification of compliance with feed and food law, animal health and animal welfare rules. Official Journal of the European Union L 191/1 28.5.2004.

[70] SI 2013 No. 2210.

[71] SI 2013 No. 2196.

[72] SI 2013 No. 2996.

and if after hearing the relevant evidence is satisfied that the food fails to meet food safety requirements, must condemn the food. If the Magistrate does not condemn the food, the owner of the food is entitled to compensation for any losses incurred as a result of the enforcement action. While a prerequisite for the use of Food Safety Act powers of seizure is normally that the contamination or decomposition of the food has rendered it unfit for human consumption, the Food Safety and Hygiene (England) Regulations 2013 permit an authorised officer to certify that a specified food fails to meet food safety requirements if it has not been produced, processed or distributed in accordance with the hygiene regulations. The hygiene regulations in this context are Regulation (EC) 178/2002,[73] Regulation (EC) 852/2004,[74] Regulation (EC) 853/2004[75] and Regulation (EC) 2073/2005.[76] As such, if a food is found in an establishment which is not operating to the hygiene standards specified in Regulation (EC) 852/2004, an authorised officer may use Regulation 29 of the Food Safety and Hygiene (England) Regulations 2013 to certify that it fails food safety requirements and subsequently seize the food using section 9 of the Food Safety Act 1990.[77]

15.5.7 Compliance notices

Regulation 6 of the Food Safety and Hygiene (England) Regulations 2013 provides for the use of a Hygiene Improvement Notice (HIN) by an authorised officer to rectify any breaches of hygiene law within a food business. The HIN must be served on the Food Business Operator and must give the recipient at least 14 days to comply with its requirements. HINs may be used to require, for example, structural improvements in food businesses, the training of staff in hygiene matters or the development and implementation of a food safety management system. They are not suitable for use to rectify transient breaches of food law or for ongoing offences such as a lack of routine cleaning. Other compliance notices are available under a number of food regulations including the Fish Labelling (England Regulations 2013 and the Food Information Regulations 2014).

15.5.8 Prohibition procedures

Regulation 7 of the Food Safety and Hygiene (England) Regulations 2013 provides the courts with prohibition powers which may be used to prohibit a process or

[73] Regulation 178/2002 of the European Parliament and of the Council laying down the general principles and requirements of food law, establishing the European Food Safety Authority and laying down procedures in matters of food safety [2002] O.J. L31/1, as amended.

[74] Regulation (EC) No. 852/2004 of the European Parliament and of the Council of 29 April 2004 on the hygiene of foodstuffs as amended L 226/20.

[75] Regulation (EC) 853/2004 of the European Parliament and of the Council of 29 April 2004 laying down specific hygiene rules for food of animal origin L 226/22.

[76] Council Regulation (EC) 2073/2005 on microbiological criteria for foodstuffs OJ L 338/1.

[77] For a discussion on the appropriate use of this 'Regulation 29 certification', see *R (on the application of Allan Rich Seafoods Ltd) v West Lindsey DC* [2010] WL 2131265.

treatment, piece of equipment, premises or part premises following conviction of a Food Business Operator under relevant hygiene offences or the service of a Hygiene Emergency Prohibition Notice (HEPB) by an enforcement officer. The courts may also prohibit a person from participating in the management of a food business if, following a conviction for a relevant hygiene offence, they consider that this is an appropriate course of action. Regulation 8 of the Food Safety and Hygiene (England) Regulations 2013 provides for an HEPN which may be served by a suitably qualified and experienced enforcement officer where he/she is satisfied that an imminent risk to health exists. The service of the HEPN has the effect of immediately prohibiting the piece of equipment, process or treatment or premises specified in the notice. Once an HEPN has been served on the Food Business Operator, the authorised officer must make an application to the Magistrates' court for a Hygiene Emergency Prohibition Order (HEPO) but only once the Food Business Operator has been given at least one day's notice of the officer's intention to apply for the HEPO. On application to the court, if the Magistrates are satisfied that an imminent risk to health exists, they shall issue the HEPO. If they are not satisfied that the imminent risk to health exists, they will not issue the HEPO and the HEPN no longer has effect. In addition, the food business operator should be compensated by the food authority for any losses incurred as a result of their enforcement action. Where an HEPO is issued by the court, the prohibition remains in effect until the HEPO is lifted in writing by the enforcement officer. This will be achieved once the officer is satisfied that an imminent risk to health no longer exists.

Regulation 9 of the Food Safety and Hygiene (England) Regulations 2013 provides for a Remedial Action Notice (RAN) and a Detention Notice but the use of both of these enforcement notices is restricted, in England, to establishments which require approval under Regulation (EC) 853/2004[78] such as meat products factories, dairies and other establishments approved by the food authority for the handling and wholesale supply of POAO. The RAN can require the immediate prohibition of a piece of equipment or an establishment, the imposition of conditions or the reduction in the rate of production of the food business. It may only be served where the authorised officer has evidence that the hygiene regulations have been breached or their inspection has been hampered.

In addition to the above enforcement procedures, authorised officers may seize any POAO which has been illegally imported from a third country using the provisions of Regulation 19 of the Trade in Animals and related products Regulations 2011.[79] Where a noncompliant FNAO, such as fruit or vegetables, imported from a third country is found by an enforcement officer, he may detain or seize the food using Regulation 32 of the Official Feed and Food Controls (England) Regulations 2009.[80]

[78] Regulation (EC) 853/2004 of the European Parliament and of the Council of 29 April 2004 laying down specific hygiene rules for food of animal origin L 226/22.

[79] SI 2011 No. 1197.

[80] SI 2006 No. 15.

15.5.9 Food Information Regulation

Regulation 1169/2011 came into force in three main stages: the first stage which required information regarding the placing on the market of minced meat came into force on 1 January 2014; the second stage, which provides for the provision of information and the labelling of both prepacked and non-prepacked food came into force on the 13 December 2014; and the final stage, which requires the mandatory nutritional labelling of most prepacked foods came into force on 13 December 2016. The Food Information Regulation is largely a consolidation of the existing rules although it does introduce some new requirements, for example, a minimum font size for mandatory information and rules relating origin and provenance of food. However, the most significant change in England, from the perspective of the enforcement officer, is the nature of the sanctions that may be used to rectify non-compliance. Prior to the introduction of the Food Information Regulation, the only enforcement sanction available to food authorities in England who had identified a breach of food labelling legislation was prosecution. Under the new regulations, food authorities will only be able to initiate legal proceedings under a limited set of circumstances and for most food labelling breaches the only sanction available is the Improvement Notice. Legal proceedings may still be brought by the food authority for failure to declare food allergens or irradiated ingredients in food and for noncompliance with the requirements of Improvement Notices. As such it is very likely that enforcement activity with respect to identified breaches of food labelling law will increase as food enforcement officers are likely to view the issue of Improvement Notices a more proportionate response, in accordance with the Regulator's Code, than prosecution.

15.6 Future developments

There are two key policies which will have a marked, but as yet, unknown effect on food law enforcement in the England.

1. BREXIT – In June 2016 the United Kingdom voted in a referendum to leave the EU and formal notice of its intention to do so was served by the UK government on the EU in March 2017. As such, the United Kingdom will leave the EU at the end of March 2019. At the time of writing there has been no agreement between the United Kingdom and the remaining EU states (EU 27) on the nature of the relationship between the parties. It is not clear whether the United Kingdom will become a third country (the so-called hard BREXIT) or will remain in the single market and under some jurisdiction of the EU courts (the so-called soft BREXIT). It is also unclear as to the nature and extent of any transitional arrangement between full UK membership of the EU and the final relationship.
2. Regulating our future – In February 2016 the FSA began a review of the delivery of food official controls in England, Wales and Northern Ireland. Food Standards Scotland are undertaking a similar review. At the time of writing no firm proposals have been published but it is understood that there is likely to be a new form of food registration scheme

introduced to risk assess businesses and place them under an appropriate level of official control. The FSA have also discussed ways of exploiting data and services from the private sector to enhance the effectiveness of food official controls.

15.6.1 Review of Official Controls Regulations

The review of Regulation (EC) 882/2004, the Official Controls Regulation, has been undertaken in accordance with the European Commission's better regulation initiative 'Smarter rules for safer food'.

The revised Official Controls Regulation (Regulation (EU) 2017/625[81]) consolidates and simplifies the arrangements for the importation of food from third countries. It also places greater emphasis on the prevention and detection of food fraud. This Regulation comes into effect in December 2019.

References

Atwood, B., 2000. Butterworths Food Law, second ed. Butterworths, Reed Elsevier, UK (Chapter 2).

Griffith, C.J., 2005. Are we making the most of food safety inspections? A glimpse into the future. Br. Food J. 107 (3), 132.

Jackman, D., 2004. Does regulation make it worse? J. Finan. Regul. Compl. 12 (2), 106.

James, P. The Food Standards Agency: an interim proposal by Professor Philip James, 30th April 1997. Available at <http://archive.food.gov.uk/maff/archive/food/james/part2a.htm#(l)>.

Millstone, E., 2009. Science, risk and governance: radical rhetorics and the realities of reform in food safety governance. Res. Pol. 38 (4), 624.

Paulus, I., 1974. The Search for Pure Food. Martin Robertson, London.

Pointing, J., 2005. Food crime and food safety: trading in bushmeat – Is new legislation needed? J. Crim. L. 69 (1), 42.

Reiss, A., 1984. Selecting strategies of social control over organisational life. In: Hawkins, J., Thomas, J. (Eds.), Enforcing Regulation. Kluwer-Nijhoff, London.

Shears, P., Frances, E., Zollers, Hurd, Sandra N., 2004. The European Food Safety Authority: towards coherence in food safety policy and practice. Br. Food J. 106 (4), 336.

The Food Standards Agency – A Force for Change, 1998. Cm 3830. Available at: <http://www.archive.official-documents.co.uk/document/maffdh/fsa/fsa.htm>.

[81] Regulation (EU) 2017/625 of the European Parliament and of the Council of 15 March 2017 on official controls and other official activities performed to ensure the application of food and feed law, rules on animal health and welfare, plant health and plant protection products, amending Regulations (EC) No. 999/2001, (EC) No. 396/2005, (EC) No. 1069/2009, (EC) No. 1107/2009, (EU) No. 1151/2012, (EU) No. 652/2014, (EU) 2016/429 and (EU) 2016/2031 of the European Parliament and of the Council, Council Regulations (EC) No. 1/2005 and (EC) No. 1099/2009 and Council Directives 98/58/EC, 1999/74/EC, 2007/43/EC, 2008/119/EC and 2008/120/EC, and repealing Regulations (EC) No. 854/2004 and (EC) No. 882/2004 of the European Parliament and of the Council, Council Directives 89/608/EEC, 89/662/EEC, 90/425/EEC, 91/496/EEC, 96/23/EC, 96/93/EC and 97/78/EC and Council Decision 92/438/EEC (Official Controls Regulation) Text with EEA relevance.

Victor, D.G., 2000. The sanitary and phytosanitary of the World Trade Organisation: an assessment after five years. N.Y. Univ. J. Int. Law Pol. 32, 865.
Winter, G., 1985. Bartering rationality in regulation. Law Soc. Rev. 19 (2), 219.
Yapp, C., Fairman, R., 2005. Assessing compliance with food safety legislation in small businesses. Br. Food J. 107 (3), 150.
Yapp, C., Fairman, R., 2003. Performance assessment in local authority food safety services. J. Environ. Health Res. 2 (1), 2.
Yapp, C., Fairman, R., 2006. Factors affecting food safety compliance within small and medium-sized enterprises: implications for regulatory and enforcement strategies. Food Contr. 17 (1), 42.

Consumers and consumption of fruits and vegetables: Who wants more of a good thing?

16

Wojciech J. Florkowski
Department of Agricultural and Applied Economics, University of Georgia, Griffin, GA, United States

16.1 Introduction

This chapter focuses on quality and safety of fruits and vegetables, recognising that production, distribution and consumption are fundamentally dependent on consumer preferences and choices. The exercised choice is conditional on the budget constraint, but the number of factors other than the budget appears to influence the ultimate choice. Motives behind the consumption of fruits and vegetables can be realised if they are affordable, accessible and of desired quality, including product safety. The specific fruit or vegetable has to possess other characteristics compatible with the modern lifestyles, especially congruent with convenience. Last, but not least, fruits and vegetables must be wholesome and safe to consumers. Concerns about safety are reflected in increasing domestic and international regulations that guide fruit and vegetable production, processing and distribution.

This chapter reviews various aspects affecting fruit and vegetable consumption with the emphasis on fresh consumption. The commonly presented argument for such consumption is the overriding evidence that fresh produce maintains health and prevents numerous diseases. Numerous nutritional and epidemiological studies stress the benefits of eating fresh produce, although some, for example, broccoli, provides more benefits after cooking as compared to its fresh consumption. The objective of this review is not to exhaust all relevant issues, but to examine those that are of particular research interest to the author.

16.2 Consumption patterns

Fruit and vegetable consumption shows different patterns in recent years. Both categories are large and represent a rich diversity of plants. Only a small portion of fruit-bearing plants, tubers, roots and other plant parts considered as vegetables have been subject to cultivation and breeding. A major force driving cultivation has been natural conditions for production, but over time, as cultural

Swainson's Handbook of Technical and Quality Management for the Food Manufacturing Sector.
DOI: https://doi.org/10.1016/B978-1-78242-275-4.00016-2

practices have improved, such conditions have become less limiting in determining growers' choices. Currently, fruit and vegetable growers respond primarily to consumer preferences reflected in the purchase of specific items at large retail outlets. Preferences are shaped by numerous factors, but of increasing in importance are nutritional attributes and disease prevention effects associated with eating fruits and vegetables. Rapidly expanding knowledge has increased awareness of these benefits of fruit and vegetable consumption promoting the desire to consume fruits and vegetables across continents and climate zones. The world's primary fruit production increased 55.6% between 1993 and 2009 and reached 610 million tons (FAO, 2016). The 3.13% growth rate resulted from the annual 1.42% growth of the area harvested and 1.7% growth of the yield increase, respectively, during that period.

16.3 Affordability

Although preferences are a key driving force, the type, quantity, and form a fruit or vegetable in which it is consumed, is ultimately determined by affordability. For example, a given fruit may be available in fresh, dry, frozen or canned form, but affordability is the deciding factor with regard to how much of a fruit is ultimately consumed and in what form.

Fruit consumption is a very good indicator of prosperous households. An example of affordability is provided by Florkowski (2012) with the illustration of fresh apple consumption by households from different income strata in Poland following the recent global financial crisis. Apples are the most common fruit in Poland because of large domestic production and suitability for long-term storage. Consequently, domestically produced apples are available year-round. With the decreasing role of self-supply of food in general, and specifically apples, consumers depend on the commercial retail outlets for fresh apple purchase. In the years immediately following the occurrence of the financial crisis, the pattern of expenditure on apple purchase between Polish households classified as poor and those above the poverty level showed differences. Low-income households were much more dependent on apple consumption than well-off households as reflected in higher expenditure on apples than citrus fruit (Florkowski, 2012) making low-income households more sensitive to price changes. Price sensitivity might have led to increased apple consumption by low-income households as a result of the embargo imposed by the Russian Federation on apple imports from Poland in August 2014. The embargo led to a precipitous decline in fresh apple prices in the 2014/15 marketing season and reversed the declining trend in fresh apple consumption in Poland (Klepacka and Florkowski, 2016).

Processed fruit, including frozen, canned and dry fruit, is often not more affordable than fresh fruit and it typically has a different taste. The decreasing home processing of fruit also reduces this type of fruit consumption by households at all income levels, but a decline in home processing was not associated with an increase in demand for commercially processed fruit. In general, the demand for processed fruit has declined, not only because such products are not price-competitive, but also because of changing preferences. The change in preferences is associated, for example, with increased commuting time to work or school, smaller household size and less frequent meals eaten by all household members. Instead, fresh fruit consumption is a very good indicator of prosperous households, which tend to spend considerably more on fresh fruit purchase than household with less income. In addition, affordability seems to be of particular importance to fruit purchase and relatively less important to vegetable purchase (Florkowski et al., 2014).

16.4 Accessibility

Consumption of many fruits and vegetables is also influenced by accessibility. The diversity of fruits and vegetables consumed is reflected by their source, that is, from wild (not cultivated) or from cultivated plots on a limited scale. Limited scale and quantity supplied often results from specific attributes of a vegetable or a fruit, and prevents distribution through commercial channels. Therefore, the increasing international trade in fruits and vegetables, although encompassing a gradually expanding variety, involves a small portion of all consumed fruits and vegetables.

However, from an individual household perspective, a key factor influencing accessibility in the developed economy is accessibility of a retail outlet. Many transition economies across the globe have experienced a rapid growth of the modern retail outlet types such as chain hyper- or supermarkets. Changes in the food retail sector altered consumer choices. An illustration in Klepacka et al. (2014) shows the change in the fresh fruit consumption pattern in Poland in recent years. Estimation results for five income quantiles show that income increase will lead to increased expenditure on apples in lower income quantiles among farm households, and upper income quantiles for pensioner households. Moreover, households in regions with large apple production spent more on apples, while those in regions with small production spent less in quantiles with incomes at or below poverty level likely reflecting the local supply and accessibility. Also, low income households spent less on apple purchase after Poland's accession to the EU in 2004 because Polish apple export increased lowering the domestically available volume.

The restructuring of food retail sector further fundamentally alters consumer accessibility to fruit and vegetables. The global expansion of supermarket chains in the past two decades reduced the presence of independently owned retail stores offering produce for sale. As a result, only organisations capable of supplying large volume of uniform quality of fruits or vegetables are capable of entering their products in the distribution channels and reach consumers. On one hand, small, independent growers, who could compete on price but not on volume, have been limited to negotiating sales with alike grocers, that is, small and independent, while the supermarket chain chooses and decides about the consumer accessibility to variety of fruits and vegetables. The restructuring of the retail sector changed the fruit and vegetable selection, influenced the supply channels, and production practices and location. Some traditional, often locally well accepted and produced, fruits, common several decades ago, have been replaced by new species originating from other countries or climate zones.

Fundamentally, accessibility is a precondition of consumption and the efficient supply chain of tropical fruit (e.g., bananas) offsets the advantage of locally produced fruit with fragmented distribution system. The differences are ultimately captured in changes in fruit and vegetable consumption (e.g., the decreasing consumption of cabbage in medium and large urban centres). Consumers face different choices and may change their dietary patterns.

16.5 Budget constraint

The importance of fruits and vegetables, from a consumer viewpoint, captures their share in household budgets. Food expenditure data are available for many countries, but the shares of food categories, including fruits and vegetables, are unavailable. It is commonly thought that expenditures on fruits and vegetables represent a relatively small portion of total food expenditures. Obviously, price and quality influence the expenditure share, so households buying low-quality fruits and vegetables may actually eat a relatively large quantity. However, in general, it is plausible to expect that households with high income consume more fruits and vegetables as predicted by Engel's law. This relationship holds regardless of the country's income classification (e.g., Florkowski et al., 2014, pp. 156–157; Revoredo-Giha and Florkowski, 2013). A study of urban Ghana consumers showed that income was an important factor influencing the expenditure on vegetables and, especially, fruit.

Besides differences across countries, there are differences across regions within a country. The already mentioned study by Meng et al. (2014) shows that urban households in various cities in Ghana spent different amounts on vegetables. To some extent, this is a reflection of varied incomes across regions or difference in locally produced items resulting from difference in natural resource endowment, but such regional differences also reflect regionally shaped preferences. In the aforementioned study, the authors reported that income gains would increase vegetable expenditure in the less wealthy region as compared to relatively well-off areas. In contrast, the study did not find differences in expenditure on fruit. It appears that factors other than location determined fruit expenditure in urban Ghana, in particular income, marital status and the number of adults in a household.

But an income increase leads to an increase in fruit and vegetable consumption only to a certain level. In some countries or regions, an increase in income eventually has a negligible effect on fruit and vegetable expenditure and consumption, that is the income elasticity with regard to fruits and vegetables increases at a decreasing rate. What changes are the type of fruit and vegetable that is being eaten (e.g., Cook, 2011) and the degree of processing or value added in fruits and vegetables purchased by consumers.

16.6 Contemporary consumer lifestyle demands

To a large extent, both the change in the type of fruit and vegetable eaten and its form are influenced by the demand for convenience. Although an apple is more convenient to eat than an orange, a banana is more convenient than an apple: an orange is somewhat difficult to peel, while an apple needs to be washed before eating, while a banana is easy to peel and does not have to be washed at least as far as prevailing perception is concerned. Convenience explains the observed consumption changes in fruit: banana is consumed in the largest volume and surpassed the previously most commonly eaten apple, while citrus fruit consumption has been decreasing (Cook, 2011). A similar trend towards eating imported tropical and subtropical fruit in larger quantities than, for example, abundant domestic apples, has been observed in Poland, but among the residents of large cities (Klepacka et al., 2014). Such tendency creates new challenge for all domestic, temperate zone fruit growers, who have to increase efforts to expand consumption, protect quality and reduce production and handling costs to remain competitive and counteract the declining consumption.

In response to convenience driven by changing lifestyles and time allocation preferences, clementine consumption is increasing among citrus, while many vegetables, and to a lesser degree fruits, are consumed in the form of freshcut. Freshcut popularity reflects demand for convenience, but also an inclination to transfer vegetable handling to commercial providers. Purchase of a freshcut vegetable or fruit salad saves labour in preparation in the home as well as assures a high degree of safety. Shifting preparation time away from the household to freshcut plants transfers the responsibility for safety to processors and distributors who now bear the risk and have to protect the freshcut product integrity in storage, transportation and handling.

Incentives to add value, for example, by developing freshcut salads, are driven by entrepreneurs who seek higher returns than offered by existing forms of processing. Such activities create new opportunities and new risks. Commercial freshcut operations require implementation and continued vigilance of strict safety and hygiene procedures to prevent the growth of microbiological contaminants if the cold chain is compromised, most likely at the point of consumption (e.g., at home, in a restaurant). Procedures safeguarding freshcut produce involve application of various sanitisers (see, for example, Beuchat, 2000), which are subject to verification and approval by regulatory agencies. The burden of meeting the regulatory requirements concerning the composition of sanitisers falls on the manufacturer, but the user is responsible for its judicious application to produce.

The growth of the freshcut sector appears sensitive to changes in disposable income. The recent financial crisis may have tempered the sector's growth. It also appears that the consumer is only ready to purchase certain volume of freshcut at retail food outlets and a major user of freshcut is the hospitality industry and institutional buyers, for example, university cafeterias, schools. The type of buyer exerts additional pressure on freshcut processors and distributors to follow stringent safety procedures. An example of school children sickened with norovirus in German schools a few years ago serves as an illustration.

16.7 Factors driving consumption and quality of fruits and vegetables: energy

Fruits and vegetables are a dietary food group and a source of energy. Their share in terms of the diet is less than that of grains, but in some cases, starchy vegetables play the role of a staple. Potatoes or sweet potatoes are examples of vegetables that in many countries are viewed as a staple. Interestingly, potato consumption has been falling as incomes increase (an inferior good in the context of preferences in economics), but in high-income countries the potato has shifted to a more prominent position reflecting ever more diversified consumption forms. For example, baking potatoes and fingerlings (small, often of various colours, narrow, but elongated tubers) are viewed as specialty vegetables.

Fruits and vegetables are not energy-dense food with few exceptions, for example, avocado. They supply the bulk of dietary fibre. For example, vegetables supply 16% and fruits 12% of daily recommended fibre intake in the United States (Hoy and Goldman, 2014). Indeed, their importance as a source of fibre has increased with the increase in consumption of highly processed foods and fats.

Additionally, the relatively low calorie content of fruits and vegetables enhances their role in weight management (Obesity and Overweight, 2015). The recommendation of consumption of fruits and vegetables increases the importance of assuring that they are safe and wholesome. Among developed countries there are differences related dietary patterns across continents, including the consumption of fruits and vegetables. For example in Asia, the Republic of Korea (ROK) reports relatively low shares of men and women having weight management problems. The OECD (2010) reported an average obesity rate of 3.5% in South Korea with 3.3% and 3.7% among adult men and women, respectively. The obesity rates in the ROK are low among the Organization for Economic Co-operation and Development (OECD) member countries. The rates of overweight men and women are reported by OECD as 23.7% and 30.3%, respectively. A recent study of factors affecting the body mass index (BMI) among urban female residents in Korea identified higher expenditures on fruits and vegetables are associated with lower BMI. However, urban women engaged in processing Chinese cabbage into the traditional condiment kimchi tended to have BMI exceeding the upper range associated with healthy body weight (Madhaven Nambiar et al., 2015). Because kimchi is a condiment, although made primarily of vegetables, the homemaking of kimchi captures a lifestyle and overall dietary pattern rather than only vegetable consumption. The upside of kimchi consumption is that it is a relatively good source of fibre needed for maintaining health.

16.8 Factors driving consumption and quality of fruits and vegetables: nutrition and health benefits

The caloric content of fruits and vegetables does not reflect their crucial role in supplying vitamins, minerals, dietary fibre and antioxidants (e.g., Li et al., 2011). These components (a minimal energy value) provide numerous health promoting and disease prevention benefits. Fruits and vegetables are good sources of water-soluble vitamins. Green leafy vegetables have been found to offset the negative effects of foods contaminated with aflatoxin, a metabolite of *Aspergillus* spp. (Alpsoy et al., 2009). In lesser developed countries located at low latitudes, leafy greens are commonly cooked or sautéed potentially eliminating microbial contamination, but in temperate zone countries consumers are encouraged to eat fresh (uncooked) vegetables. Leafy greens are particularly difficult to clean during postharvest handling and have been a cause of recurring outbreaks of foodborne illnesses. Other outbreaks have been linked to contaminated sprouts, including those grown on organic farms (Mohel-Boetani et al., 2009; Marler, 2011; Bayer et al., 2014) which are added to salads to improve taste and also because of their perceived healthfulness.

The expanding importation of fresh fruits and vegetables reflects an inadequate volume supplied by domestic producers. For example, the United States imports large volume of fruits, vegetables, nuts and spices (Florkowski, 2008). The importation reflects the erosion of competitiveness of the domestic fruit and vegetable growers and the termination of domestic production. Another factor is the seasonality of domestic production and the need for imports in the effort to assure year-round accessibility. Ultimately, the growing imports create a threat to consumer safety resulting from the use of pesticides and microbiological contamination. Whereas the risk of pesticide contamination has been reduced by the gradual withdrawal of a number of pesticides, the risk of microbiological contamination has persisted as evidenced in publicised periodic outbreaks of poisonings due to consumption of imported fresh or processed fruits or vegetables in North America or Europe.

Importation of fruits and vegetables also results from differences in natural conditions that prevent production of produce in various regions, for example, tropical versus temperate fruits. The increasing dependence on imports mandates the involvement of government agencies in testing imported produce for safety, as a demonstration of sovereignty and to protect consumer well-being. Typically, regulations vary with regard to the degree of processing and may be enforced by different agencies. Information about self-sufficiency rates is sparse, but is now considered an important measure of a country's food security. In the case of fruits and vegetables, governments do not view this category as particularly important and as such the adequacy of domestic production and domestic consumption is often ignored. The European Union (EU) self-sufficiency ratio is high in the case of vegetables (99%), but substantially lower in the case of fruits (77%) (Noleppa and Cartsburg, 2013). Furthermore, self-sufficiency ratios vary among countries within the EU and across the globe, causing the volume of imports to vary as well.

16.9 What limits fruit and vegetable consumption in developed countries?

Increasing incomes, decreasing share of food expenditures, including expenditures for fruits and vegetables in total household expenditures, demand for convenience reflecting different values for time and preferences in time allocation, concerns about health maintenance and disease prevention, including weight management, varying self-sufficiency ratios, desire for variety, and expanding international trade in fruits and vegetables suggest possible patterns across countries. Figs 16.1–16.5 are based on a group of selected European countries. Such comparisons on an international scale are difficult because of variation of methodologies applied in estimating per capita consumption of fruits and vegetables as well as measures against which such consumption is compared. Nevertheless, even such graphic comparisons provide some insights, even though a discernable pattern does not always clearly emerge from this exercise.

Figs 16.1 through 16.5 illustrate the complexity of forces affecting the importance of fruit and vegetable safety. The illustration is based on selected member

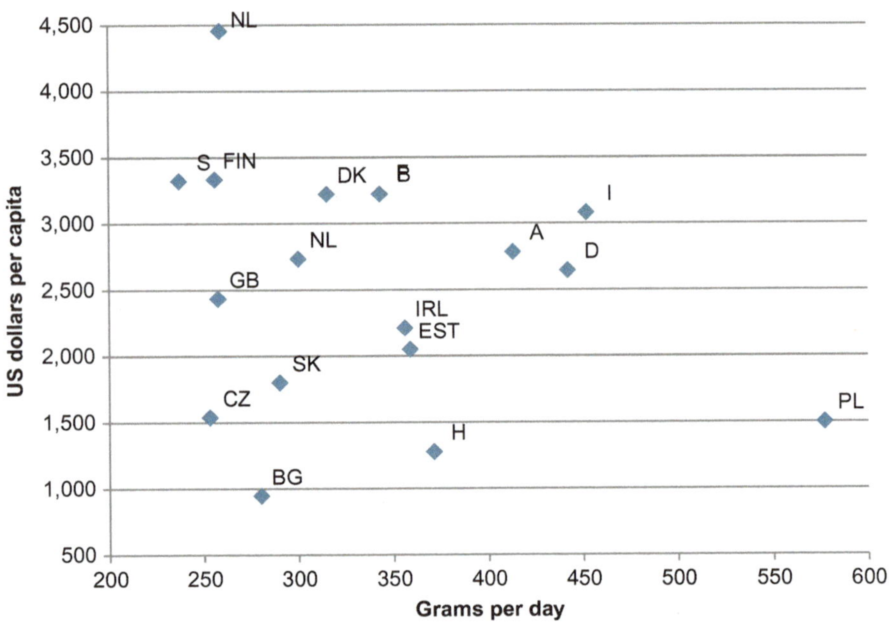

Figure 16.1 Expenditures on food in US dollars and consumption of fruit and vegetables in grams per day for selected EU countries and Norway.
Note: Country designations are A – Austria, B – Belgium, BG – Bulgaria, CZ – Czech Republic, D – Germany, DK – Denmark, EST – Estonia, F – France, FIN – Finland, GB – United Kingdom, H – Hungary, I – Italy, IRL – Ireland, N – Norway, NL – Netherlands, PL – Poland, S – Sweden, SK – Slovakia.

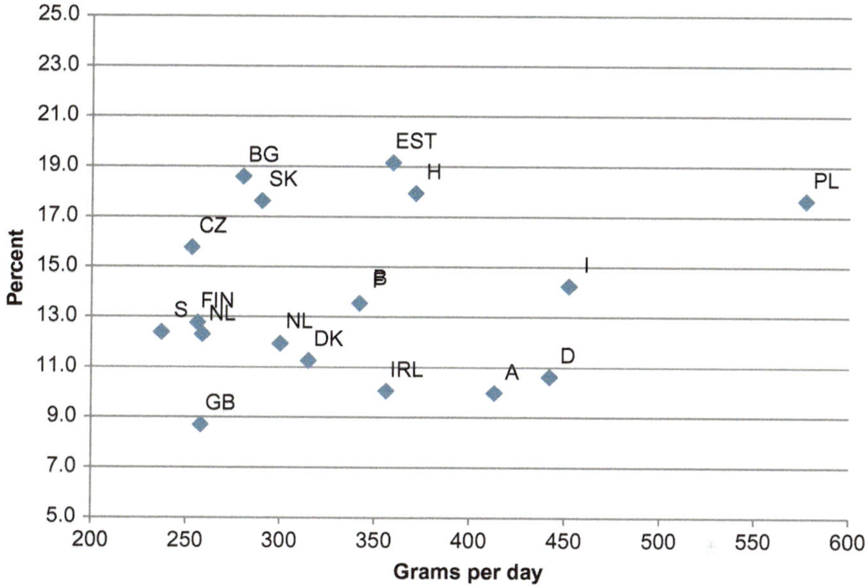

Figure 16.2 Share of expenditure on food in per cent and consumption of fruit and vegetables in grams per day in selected EU countries and Norway.
Note: Country designations are A – Austria, B – Belgium, BG – Bulgaria, CZ – Czech Republic, D – Germany, DK – Denmark, EST – Estonia, F – France, FIN – Finland, GB – United Kingdom, H – Hungary, I – Italy, IRL – Ireland, N – Norway, NL – Netherlands, PL – Poland, S – Sweden, SK – Slovakia.

countries of the EU and non-EU countries. Although the observed patterns apply to other geographic areas, specific conditions prevailing in individual countries may require further scrutiny. Fig. 16.1 shows a plot of per capita income against the daily consumption of fruits and vegetables in grams. The pattern seems somewhat counterintuitive if fitted visually and suggests a downward sloping curve, implying that countries with relatively lower per capita income consume more fruits and vegetables.

A plot of the share of expenditures on fruits and vegetables in total food expenditures and the daily consumption of fruits and vegetables does not reveal any particular pattern (Fig. 16.2). Countries with a low-to-high share of expenditures consume a relatively small quantity of fruits and vegetables. Similarly, there is a large diversity in terms of the expenditure share among countries consuming a medium-to-large daily amount of fruits and vegetables.

Finally, Fig. 16.3 shows the plot of daily wage rates against the daily consumption of fruits and vegetables, indicating the absence of a distinct pattern. An attempt to fit a linear trend results in very low *R*-square values, which is consistent with the visual assessment. However, from the point of view of cost of a potential disease outbreak caused by eating contaminated produce, such costs are likely to be higher in countries with higher wages and higher incomes than in their counterparts. These

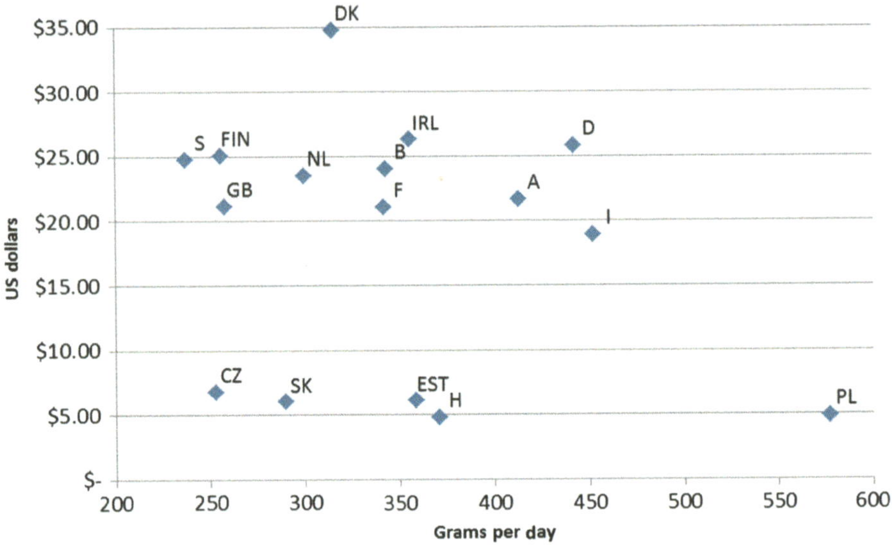

Figure 16.3 Direct hourly pay and consumption of fruit and vegetables in grams per day in selected EU countries and Norway.
Note: Country designations are A – Austria, B – Belgium, BG – Bulgaria, CZ – Czech Republic, D – Germany, DK – Denmark, EST – Estonia, F – France, FIN – Finland, GB – United Kingdom, H – Hungary, I – Italy, IRL – Ireland, N – Norway, NL – Netherlands, PL – Poland, S – Sweden, SK – Slovakia.

potential costs encourage the risk avoidance driving the regulations. Countries with a higher per capita income tend to impose regulations to minimise or substantially reduce risks associated with consumption of unsafe fruits and vegetables whether fresh, dried, or processed. This is, among others, a reaction to the costs of an outbreak of disease caused by microbiologically unsafe fruits or vegetables. Employees who are sick cause losses of productivity, increasing employer costs. The loss is higher in countries where wages are higher, but in those countries health care costs also tend to be higher than in countries where low wages prevail, causing the treatment costs in the former to be high as well. Although lost productivity affects private companies, health care services affect the public sector, especially where health care services are supplied by the government. Not surprisingly, safety requirements are initiated, formulated, and enforced in countries with high per capita incomes, often the primary markets for imported fruits and vegetables. However, domestic regulations regarding fruits and vegetables no longer limit their effects to the country that introduced them, but force the change in production, packaging, transportation and storage outside its boundaries as well.

Within a national economy, the wage rates vary according to a number of factors, but education seems to be one of major reasons. Overall, it is reasonable to expect that those with more formal education receive higher wages and have more disposable income. Therefore, the cost of getting sick due to contaminated fruit or

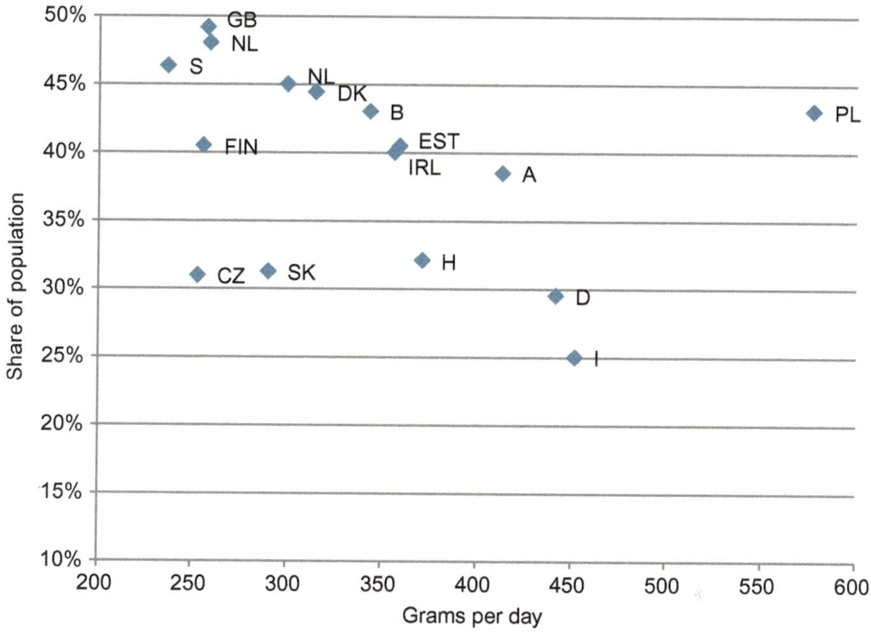

Figure 16.4 Share of 24- to 35-year-old citizens with tertiary education and consumption of fruit and vegetables in grams per day for selected EU countries and Norway.
Note: Country designations are A – Austria, B – Belgium, CZ – Czech Republic, D – Germany, DK – Denmark, EST – Estonia, FIN – Finland, GB – United Kingdom, H – Hungary, I – Italy, IRL – Ireland, N – Norway, NL – Netherlands, PL – Poland, S – Sweden, SK – Slovakia.

vegetable for those with more education is higher than for those with less education. Also, the risk of sickness increases (or decreases) as the volume of eaten fruits and vegetables increases (or decreases). Figs 16.4 and 16.5 show plots of the shares of college-educated population in two age groups and the per capita fruit and vegetable consumption. Both plots show that the countries with larger share of college-educated population tend to eat less fruits and vegetables per capita than countries with a smaller share of highly educated residents. The observed tendencies appear to contradict the notion that better educated tend to eat more fruits and vegetables, but those empirical studies confirming such link use individual data rather than the national aggregates. Nevertheless, Figs 16.4 and 16.5 signal a potential problem that requires scrutiny using time series data. Individual countries (e.g., the United States) already show a rather troubling phenomenon of fruit and vegetable per capita consumption levelling off, while the share of college-educated population increases.

Overall, the common measures applied in the estimation of empirical models of fruit and vegetable consumption or purchase may become increasingly inadequate in predicting consumer behaviour. Issues such as data requirements, data

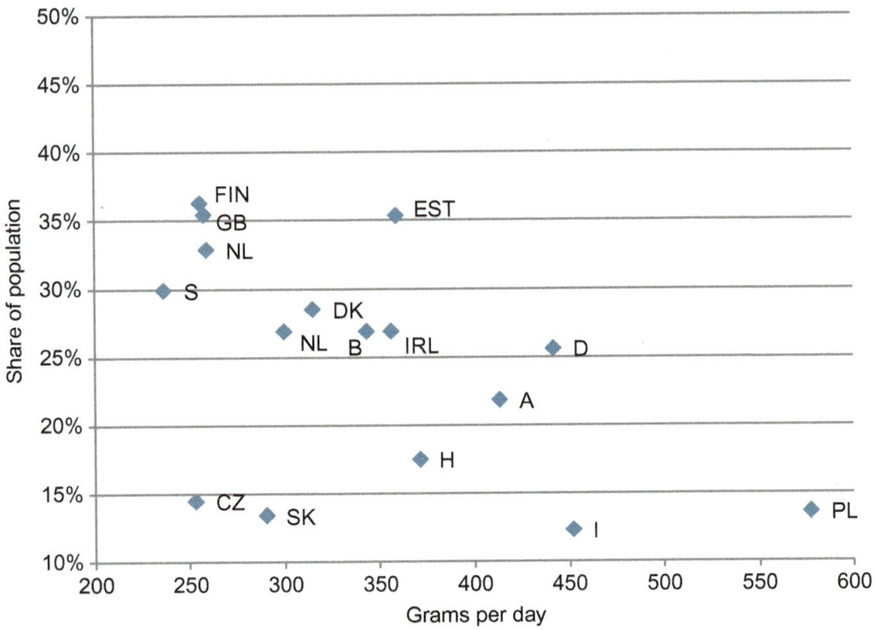

Figure 16.5 Share of 36- to 64-year-old citizens with tertiary education and consumption of fruit and vegetables in grams per day for selected EU countries and Norway.
Note: Country designations are A – Austria, B – Belgium, CZ – Czech Republic, D – Germany, DK – Denmark, EST – Estonia, FIN – Finland, GB – United Kingdom, H – Hungary, I – Italy, IRL – Ireland, N – Norway, NL – Netherlands, PL – Poland, S – Sweden, SK – Slovakia.

measurement and estimation techniques reflecting the perceived consumer behaviour become increasingly challenging.

16.10 Safety of fruits and vegetables

The increasing fruit and vegetable consumption and the associated expanding distribution of fruits, vegetables and their products is governed by a growing number of laws and regulations originating in the public and private sectors. Fruits and vegetables are traded and consumed in several fresh and processed forms. The degree of processing changes the relative importance of sensorial attributes, while safety is an overriding requirement. Safety attribute requirements are subject to regulations, some of which are voluntary, while others are mandatory. Their scope is determined by the regulatory institution and can have national or international impact. They can apply to a single subsector or distribution channel as well as have sector-wide applications. Recent decades have witnessed the development and application of numerous regulations aimed at controlling fruit and vegetable production practices, postharvest handling and distribution as potential sources of foodborne pathogens. Numerous initiatives have substantially reduced and in some cases eliminated a number of risk sources. Information presented in this chapter describes a complex set of causes that stimulated efforts to assure fruit and vegetable safety undertaken by private and public institutions on national and international scale culminating in the recently implemented Food Safety and Modernization Act. Ultimately, consumer choices reflect preferences. Consequently, as consumer preferences and choices change in the future, they will modify quality interpretation, revise perception of safety and alter regulations.

Safety issues associated with growing, processing and distributing fruits and vegetables have changed over time. Until recently, the safety of fruits and vegetables has been focused on the presence of pesticides on finished products. However, through regulation, many pesticides, especially insecticides and fungicides, have been banned or limited to use on fewer crops. Testing for pesticide residues by processors and regulatory agencies continues on both domestically grown and imported fruits and vegetables. Surprisingly, the type of produce and countries from which the shipments originate and continue to be associated with relatively frequent violations remain the same. For example, among imported vegetables, peas, string beans, leafy greens and taro have been found more often to exceed allowable pesticide residue (FDA, 2009, 2012; Galt, 2009), while among fruits, papaya has 'warranted special attention' (FDA, 2009, 2012). Similarly, there is a difference among exporting countries, where some register none or relatively few pesticide residue violations, while others often reappear on lists of violators. Overall, the per cent of sampled produce showing residues above the permitted threshold is small and generally less in domestic produce as compared to imported shipments. As such, countries that import relatively more fruits and vegetables may expose their populations to residue levels that possibly exceed the regulated threshold.

16.11 Cost of regulation

Encouraging fruit and vegetable consumption is expected to result in health benefits and prevent diseases and is increasingly important as health care costs rapidly grow across the world. However, consumption benefits cannot materialise if they are associated with produce containing pesticide residues in excess of allowable limits or are contaminated with harmful microbes. To assure produce safety, governments introduce regulations or revise existing regulations to further reduce the risk of illness. Regulations, however, are costly because they force changes in postharvest handling or production techniques and may in the long term determine land-use decisions (Galt, 2009, p. 338; Zhang, 2016). Higher costs of regulations are eventually incorporated in prices and the ultimate consumer will pay higher prices than prior to the introduction of safety regulations. Safer produce implies higher quality justifying higher prices in economic terms if consumers are willing to pay a new price. In countries with strong regulatory institutions, consumers may have little or no access to fruits and vegetables that are produced outside the regulatory regime. But in countries with lesser developed institutional environment, consumer purchase of fruits and vegetables is price-sensitive and they can obtain them at a lower price from a parallel distribution system that trades produce outside regulated channels. Affordability segments consumers.

16.12 Regulatory focus

In recent decades, the focus of government regulatory agencies was the use of pesticides. The direct cause was evidence of detrimental health effects in people who consumed fruits and vegetables. Additionally, the use of pesticides was linked to a number of negative environmental effects and an indirect effect on the health of living organisms. The use of pesticides was also linked to harmful effects on applicators leading to strict rules guiding their use. Efforts in eliminating and reducing the use of pesticides resulted in narrowing the number of pesticides, their permitted use, and imposed specific rules on distribution and disposal of pesticides. Consumer exposure to synthetic pesticides has been reduced.

As the risk associated with the harmful effects of synthetic pesticides has been reduced, microbiological contamination, a new source of risk associated with fruit and vegetable consumption, has gained visibility. Microbiological risks are more complex and diversified than synthetic chemicals. Microbes as living organisms have the ability to adapt to various conditions and can mutate over time. Therefore, the reduction of microbiological risks requires substantial resources, continuing scientific research and monitoring. Fruits and vegetables are produced in different natural conditions, multiple climatic zones, handled at multiple distribution points, and consumed in various forms at different venues. Regulation changes the competitive advantage of fruit and vegetable suppliers altering regional economies and international flows of fresh produce. The multiple private and public entities that regulate the fruit and vegetable sector with the goal of microbiological risk reduction sometimes compete and sometimes collude in their efforts, affecting long-term competitiveness and altering consumer responsibility for handling fruits and vegetables.

16.13 Internationally sanctioned standard quality assurance procedures: Codex Alimentarius

Among numerous quality assurance systems and inspection procedures, one with a long history of continuing development and improvement is Codex Alimentarius. Initiated by Food and Agricultural Organization (FAO) in the early 1960s and soon joined by WHO (World Health Organization), Codex is a reference in international food safety disputes recognised by WTO (World Trade Organization). Codex results from the work of the Codex Alimentarius Commission, which is an intergovernmental body of 186 countries and the EU. A number of other organisations participate in the Commission as observers.

Codex has been an attempt to develop uniform standards for foods, including fresh fruits and vegetables, influenced by the desire to facilitate trade by lowering the costs of quality verification through inspection, sample collection, testing, etc. The origin of Codex is reflected in specific rules and procedures that it recommends for a number of fruits and vegetables (Codex Alimentarius, 2007). Many of the rules directly specify tolerance level or presence of various defects.

For example, Codex Alimentarius rule of fresh asparagus apply only to asparagus intended for sale to consumers and exclude industrial processors. The rules involve a number of requirements regarding the presence of foreign matter, rotten or deteriorating parts, without visible pest damage, without foreign smell or taste, among others. The rules specify three classes and distinguish between green and white asparagus. Specific provisions define tolerances for size, diameter, uniformity, packaging and presentation (Codex Alimentarius, pp. 4–5, 2007). However, the rules go beyond typical description of standards and include labelling, contaminants such as heavy metals and pesticide residue, and hygiene. In practice the trading parties adhering to the Codex Alimentarius rules for specific produce follow the inspection procedures that verify a shipment of a particular class meets all listed tolerance thresholds (defects are listed either in percentage terms or in square centimeters in the case of an area) as well as the general applicable requirements. Some requirements are obviously specific to fresh produce at hand, for example, in the case of baby corn, "Extra" class lot must be free of silk, while the great variation across avocado cultivars is accommodated by 15 size codes. Additionally, the length of a stem or presence of a stem is defined for various fresh fruits and vegetables.

Codex Alimentarius provisions require the use of proper labelling and packaging. Such requirements are included in the set of rules specific to the produce at hand. Also, included are references to separate documents addressing specific aspects of hygiene.

Hygiene has been the focus of Codex Alimentarius Commission in early years of its existence. Food hygiene (Codex Alimentarius, 2009) encompasses several separate documents adopted by the Commission over time. General principles of food hygiene developed initially in 1969 have since been modified and currently include HACCP (hazard analysis and critical control point) principles. Two recent documents cover microbiological risk management and principles for control of *Listeria monocytogenes* in food. Both apply to foods in general, but the latter explicitly lists raw vegetables as a type of food that was a source of contamination and foodborne illnesses.

Overall, the efforts of the Codex Alimentarius Commission combined efforts of the FAO, WHO and individual countries to secure safe supply of fruits and vegetables on an international scale. The adoption of numerous rules for particular produce has been accomplished over several decades. The length of the process is characteristic to negotiations involving the development and approval of regulations applicable on international scale. Each set of regulations is subject not only to discussions within the Commission, but has to undergo the approval process of sovereign states. The latter is a potential source of delay in adopting even the simplest rules, but respects the individual country's contribution to the process.

16.14 Food Safety Modernization Act

Recent developments in the United States' internal regulations appear to have fundamentally changed the safety procedures applied by the fruit and vegetable sector. The goal of the law passed in the United States in 2011 (Public Law, 2011) is to prevent microbial contamination incidents by developing a comprehensive set of rules guiding a number of issues associated with production, processing and distribution of food including fruits and vegetables. The provisions of the law have to be implemented at different dates giving the industry time to undertake the necessary steps to comply. For example, small or medium-sized firms were obligated to apply the new safety rules on a date later than that for large firms. Many fruit and vegetable production and processing entities are small compared to large processed food manufacturing corporations. The law became mandatory on 1 January 2016.

The Food Safety Modernization Act (FSMA) is an example of the rivalry between the private sector and regulators from the public sector. In recent decades, as government authorities placed new requirements on the fruit and vegetable sector, private entities responded by accepting them and subsequently creating their own specifications that sometimes exceed government-imposed measures. Stricter produce quality requirements were a source of competitive advantage in the fruit and vegetable market. But, the FSMA is so broad and encompasses so many aspects of production, processing and distribution that it has become an umbrella regulation relegating the existing quality control and safety assurance programmes to the role of tools that may be used by the private sector to accommodate the new requirements.

Although FSMA provisions apply to firms in the United States, fruit and vegetable exporters must also obey the rules. It is, therefore, quite possible that a number of countries will adopt at least some of the new rules when exporters from those countries adopt them for their internal operations. Such a process would be similar to the widening adoption of good agricultural practices (GAPs) or good manufacturing practices in the past.

16.14.1 FSMA key requirements

The scope and reach of FSMA is unprecedented with regard to fruit and vegetable growing, processing and distribution. Implementation of the Act results in the immediate cost of learning what needs to be done to comply with new regulations. This one-time cost is followed by the need to permanently reorganise the procedures within the fresh fruit and vegetable supply chain. The purpose is to reduce the possibility of consumers purchasing produce with substandard quality. If FSMA is successfully implemented, the primary reason for illness caused by consumption of contaminated fruit and vegetable consumption will result from improper consumer behaviour, for example, not washing hands when eating fresh fruit. It could be expected that under new conditions, any outbreaks of foodborne disease will be spatially contained and involve a small number of people. How the added costs absorbed by the suppliers will transfer into higher market prices and change the demand for fruits and vegetables will eventually be revealed in the marketplace and in structural changes of the supply industry itself.

Covered facilities are not uncommon in the fruit and vegetable sector. For example, harvested produce is subject to sorting and grading in packinghouses. According to the new law, each such facility must establish and implement a food safety system that includes an analysis of hazards and risk-based preventive controls as well as a written food safety plan. A food safety system in a facility comprises hazard analysis and preventive control. Both are relevant in the case of fresh and processed fruits and vegetables, but the new regulation stresses a need for a 'system', implying an intertwined process leaving no gaps in monitoring produce with regard to safety. This is a challenge in the fresh fruit and vegetable sector, where the transfer of ownership is more frequent than in many other food industries.

Hazard analysis involves identification of spots, procedures or handling of fruits and vegetables. Such hazards are 'reasonably foreseeable' and encompass biological, chemical and physical hazards. In the case of fruits and vegetables, hazards may occur naturally. For example, a fruit becomes contaminated by fungi that produce mycotoxins while still on a tree, collected from the ground, or infected in a collection bin; chemical hazards are commonly associated with the detectable presence of synthetic pesticides; and physical hazards may result from the presence of pieces of metal, rocks or inedible plant parts. Hazards are also unintentionally introduced, but the processor is responsible for undertaking *a reasonable effort* to guard against them. Intentionally introduced hazards for economic gain that pose a risk to safety of the food contradict the new law. An example would be the application of a banned pesticide to fruits and vegetables in the field or in a short-term storage area to harvested and sorted produce.

Preventive controls of hazards require attention in the case of fruits and vegetables at all steps to minimise safety risks. Such controls include processing and sanitation controls as well as supply-chain controls and a recall plan. It is likely that such controls and recall plans require a set of written procedures and familiarity with them on the part of all personnel involved in handling produce at any pertinent stage of the supply system.

16.14.1.1 Oversight and management of preventive controls

Monitoring production, harvesting and handling procedures provide assurance that preventive controls are consistently performed. For example, monitoring of a washing process to kill pathogens would include checking the actual temperature as well as the testing the sanitiser solution. Although such procedures are applied at packinghouse sorting, the new regulations may impose additional procedures to record and document the actual performance of each step. There likely will be varying frequency in the measured phenomenon, for example, tests for strength of sanitiser solution may be more frequent than monitoring preventive activities used to minimise metal hazards on the packinghouse sorting line. Each test, however, requires a verifiable record.

Departure from the required procedures must be corrected. Corrections may involve timely identification and remediation of a minor, isolated problem that occurs during production, harvesting or postharvest handling of fresh produce or its further processing. Corrective actions must be documented with records and retained on file. Verification encompasses activities required to ensure that preventive controls are consistently implemented and effective. They include validating with scientific evidence that a preventive control is indeed capable of effectively controlling an identified hazard. An example of a validation is a record of accuracy checks of process monitoring and verification of instruments such as thermometers. Performing any corrective actions must be documented.

Product testing and environmental monitoring are verification activities required as appropriate to the form in which fruits or vegetables are to be eaten, and differ for fresh and processed items. The form determines the preventive control and the role of that control in the facility's safety system. Environmental monitoring generally would be required when contamination of a ready-to-eat food, such as whole fresh fruit or freshcut salad, with a pathogen is a hazard that requires a preventive control. As fruits and vegetables enter the distribution chain, companies purchasing produce for resale, which may involve a transfer of ownership, must have a risk-based supply-chain programme. This requirement applies not only to fruits and vegetables, but also to ingredients used in the final product if those ingredients have been identified by a hazard requiring a supply-chain applied control, unless the entity controls a hazard using preventive controls. Alternatively, the entity that relies on a supplier to control hazards does not need a supply-chain control.

Fruit and vegetable postharvest handling starts as it is being transported from the field or orchard to a packinghouse. Procedures of receiving the incoming produce must ensure tracking back to originating fields or orchards. Produce may originate from approved suppliers who follow a hazard analysis and control for that hazard. For example, cultural practices, including the use of pesticides, follow an approved protocol with a detailed record of each application. Unapproved supplier deliveries must be verified in terms of following an identical protocol prior to acceptance and processing of produce. An unapproved supplier cannot continue deliveries. However, if the preventive control of an identified hazard, for example, microbiological or pesticide residue, is implemented by the organisation that takes over the delivery, the supplier must identify such delivery as 'not processed to control' the hazard and issue a written assurance (from such organisation like a broker or distributor) that it agrees to take certain actions to protect produce safety.

The FSMA imposes a new requirement on growers and handlers of fresh fruits and vegetables as well as those processing fruits and vegetables into canned, frozen or dried products, namely the requirement of education and training. This new function creates a new fixed cost that will affect competitiveness of each company and intensify the search for new cost-cutting measures. The industry needs to reevaluate to what extent such initiatives as the GAP meets the new requirements. GAPs combined with the good hygienic practices programme is administered by USDA, but another GAP is the initiative of the FAO. It appears that both will become at least linked with other procedures in the food safety system expected from companies involved in fruit and vegetable handling.

Each company must ensure that employees handling fruits and vegetables are 'qualified' to perform assigned duties as demonstrated in their education, training and experience necessary to process the produce. Training includes principles in food safety and hygiene. Also, the employees' health becomes a factor. Such requirements will pose new challenges in human resource management, including recruitment and retention.

16.14.2 The cost of new regulations to growers

The economic effects of introduction of such broad reaching regulations like those associated with the FSMA are commonly impossible to fully anticipate. Even though the new law has been introduced gradually, due to its nature characterised by plant biology and required conditions for growth as well as the production organisation (e.g., produce originating from numerous small farms), the effect of the regulation takes time to reveal itself. The intent of the regulator and the anticipated effects at the aggregate, national or international level are estimated to generate substantial gains, and the distributional effects at the level of a specific sector, such as the fresh market tomato, are nearly impossible to anticipate.

A study by Zhang (2016) illustrates the effects of the FSMA on the tomato growers in the United States and Mexico. The latter exports a large volume of fresh tomatoes to the United States in winter months. The results suggest major adjustments in the tomato sectors in both countries and ultimately negative effects on the small growers. Consumers may expect safer product, but a number of small producers are forced to make major adjustments due to new regulations or forced out of the business of growing tomatoes; American growers lose their domestic market and Mexico growers are banned from the export market. It is these types of micro- or sector-level effects that have the most serious effects on growers, and their households. The policy-makers formulating the new regulations tend to ignore such effects drive by other factors than the recognition of the full cost of their actions.

The distributional effects of trade liberalisation and the common imbalance in regulations between two trading countries or regions are of major consequences for what consumers will have access to, whether they will be able to afford it and what will be the product quality. Consumers may be easier convinced of the benefits of a new trade arrangement than growers and distributors, but for a national economy those groups are by far more important because the adjustment costs will burden local and national economies. The lack of accurate cost–benefit estimates of trade agreements in terms of their effects on specific sectors is reflected in the scepticism and opposition to the two trade agreements negotiated by the EU, that is, Transatlantic Trade and Investment Partnership (TTIP) and a similar agreement with Canada.

16.15 Future trends

Elimination of the human factor in fresh fruit and vegetable handling could become an attractive proposition if the cost of managing a food safety system continues to increase. In the long run, many vegetables and some fruits will be produced in a controlled environment if destined to large distribution networks. The already tested automatic production of some leafy vegetables harvested by robots may, over time, become price-competitive because reprogramming a robotic system of production, harvesting and handling is easier and faster than retraining a number of employees, especially if they are working only part-time or seasonally. Robots can be kept clean and assure a level of consistency that is impossible to replicate in large-scale fresh produce production. Such automatisation, combined with the newest greenhouse technology that is energy-efficient and environmentally friendly, will supply large quantities of safe, fresh produce. Still, such ultra-modern systems will not be able to deliver the full variety of fruits or vegetables, but may be embraced by large supermarket chains requiring a large volume of consistent quality, limited in variety produce with assurance of safety.

In the short run, the requirement of education and training of employees that becomes a new added fixed cost will increase produce prices to consumers. Consumers, who are not informed about the FSMA regulations intended to benefit them, are less likely to appreciate such developments, and also unlikely to link prices to new regulations. There appears to be conflict between the desire of the public sector to assure that fruits and vegetables are safe, the cost of a food safety system at the company level and the budget constraint faced by consumers, which affects the composition of the food basket in a household. If price increases resulting from new regulations is noticed, it may lead to decreased consumption. At present, few studies have attempted to project the overall outcomes of the FSMA but have focus instead on a single item, tomatoes (Zhang, 2016).

Internationally, suppliers to the United States already have made or will soon make the necessary adjustments. But, some suppliers not able to compete under new conditions will seek markets elsewhere. Because price is a major factor in determining sales, it is possible that a separate distribution system will persist with an alternative safety standard. Consequently, the safety of fresh fruits and vegetables becomes a rationed good, segmenting consumers into alternative supply systems offering different levels of quality.

References

Alpsoy, L., Yildirim, A., Agar, G., 2009. The antioxidant effects of vitamin A, C, and E on aflatoxin B1-induced oxidative stress in human lymphocytes. Toxicol. Indus. Health 25 (2), 121–127.

Bayer, C., Bernard, H., Prager, R., Rabsch, W., Hiller, P., Malorny, R., et al., 2014. An outbreak of *Salmonella* Newport associated with mung bean sprouts in Germany and the Netherlands, October to November 2011. Euro Surveill. 19 (1), pii/20676.

Beuchat, L.R., 2000. Use of sanitizers in raw fruit and vegetable processing. In: Alzamora, S. M., Tapia, M.S., Lopez-Malo, A. (Eds.), Minimally Processed Fruits and Vegetables: Fundamental Aspects and Applications. Aspen Publ., Inc., Gaithersburg, MD, pp. 63–78. (Chapter 4).

Codex Alimentarius, 2007. Fresh Fruits and Vegetables, first ed. WHO & FAO, Rome, 195 pp.

Codex Alimentarius, 2009. Food Hygiene – Basic Texts, fourth ed. WHO & FAO, Rome, 125 pp.

Cook, R. 2011. Tracking demographics and U.S. fruit and vegetable consumption patterns. Available online at <http://www.fao.org/3/a-a1389e.pdf>.

Florkowski, W.J., 2008. Status and projections for foods imported into the United States. In: Doyle, M.P., Erickson, M.C. (Eds.), Imported Foods: Microbiological Issues and Challenges. ASM Press, Washington, DC, pp. 1–43.

Florkowski, W.J., 2012. Evidence of household consumption changes associated with the global economic slowdown. Poznań Univ. Econ. Rev. 12 (3), 42–80.

Florkowski, W.J., Klepacka, A.M., Nambiar, P.M., Meng, T., Fu, S., Sheremenko, G., et al., 2014. Consumer expenditures on fresh fruit and vegetables. In: Florkowski, W.J., Shewfelt, R.L., Brueckner, B., Prussia, S.E. (Eds.), Postharvest Handling: A Systems Approach. Elsevier, New York, pp. 147–166.

Food and Agriculture Organization of the United Nations. 2016 FAOSTAT, Food and agriculture data.

Galt, R.E., 2009. "It just goes to kill Ticos": national market regulation and the political ecology of farmers' pesticide use in Costa Rica. J. Political Ecol. 16, 1–33.

Hoy, M.K., Goldman, J.D., 2014. Fiber Intake of the U.S. Population: What We Eat in America, NHANES 2009–2010. Food Survey Research Group, Dietary Data Brief No. 12, U.S. Department of Agriculture, September, 6 pp.

Klepacka, A.M., Florkowski, W.J., 2016. Poland's apple sector and the embargo on fruit exports to Russia. Acta Hortic 1132, 31–38. Available from: https://doi.org/10.17660/ActaHortic.2016.1132.5.

Klepacka, A.M., Meng, T., Florkowski, W.J., 2014. Apples or oranges? Recent fruit consumption in Poland. Zeszyty Naukowe SERiA 16 (6), 255–259.

Li, H., Tsao, R., Deng, Z., 2011. Factors affecting the antioxidant potential and health benefits of plant foods. Can. J. Plant Sci. 92, 1101–1111.

Madhaven Nambiar, P., Florkowski, W.J., Suh, D., 2015. Measuring vulnerability of urban Korean women to weight management problems. Food Nutr. Sci. Available from: https://doi.org/10.4236/fns.2015.

Marler, B., 2011. Local organic sprouts cause of German *E. coli* O104:H4 outbreak. Food Poison J. . Available online at http//:www.foodpoisonjournal.com/foodborne-ilness-outbreaks/orga... (accessed 01.02.16.).

Meng, T., Florkowski, W.J., Sarpong, D., Resurreccion, A.V.A., Chinnan, M., 2014. Expenditure on fresh vegetables, fresh fruits, and peanut products in urban Ghana: Does location matter? Paper presented at the Southern Agricultural Economics Association Annual Meeting, Dallas, TX, February 1–4.

Mohel-Boetani, J.C., Farrar, J., Bradley, P., Barak, J.D., Miller, M., Mandrell, M., et al., 2009. *Salmonella* infections associated with mung bean sprouts: epidemiological and environmental investigations. Epidemiol. Infect. 1337 (3), 357–366.

Noleppa, S., Cartsburg, M., 2013. Agricultural self-sufficiency of the European Union: statistical evidence. Agripol Research Paper 2013-02, 19 pp.

Obesity and Overweight, 2015. WHO. Fact sheet No. 311, updated January 2015. Available online at <http://www.who.int/mediacentre/factsheets/fs311/en/> (accessed 30.01.16.).

Organization for Economic Co-operation and Development (2010). Obesity and the economics of prevention: fit not fat – Korea key facts. <http://www.oecd.org/document/20/0,3746,en_33873108_33873245_46038740_1_1_1_1,00.html> (accessed 20.03.14.).

Revoredo-Giha, C., Florkowski, W.J., 2013. An analysis of the demand for fresh fruit in Scotland. Zeszyty Naukowe SERiA (Annals of SERiA) 15 (3), 279–287.

U.S. Food and Drug Administration, Pesticide Monitoring Program, Fiscal year 2009 pesticide report. www.fda.gov/Food/FoodborneIllnessContaminants/Pesticides/default.htm.

U.S. Food and Drug Administration. Pesticide Monitoring Program, Fiscal year 2012 pesticide report www.fda.gov/Food/FoodborneIllnessContaminants/Pesticides/default.htm.

U.S. Public Law 111–353, Food Safety Modernization Act (FSMA), Public Law 111–353—January 4, 2011 124 Stat. 3932.

Zhang, L., 2016. Domestic production in import demand analyses and implications of the Food Safety Modernization Act on the fresh tomato industry. Ph.D. dissertation, University of Florida, Gainesville.

Herbs and Spices

17

Steve Clemenson
Group QA & Regulatory Manager, Nedspice EMEA BV, Rotterdam, The Netherlands

17.1 Introduction

Herbs and spices have played an essential part in traditional medicine and cuisine for centuries and their use can be traced back to the ancient Egyptians. These natural ingredients possess many qualities that have been used to protect health, preserve foods and even embalm the dead. The first documented use of spices dates back to the pyramidal age in Egypt, where onion and garlic were used to protect the health of the workers and cinnamon was used to embalm the dead. Many of our famed explorers such as Vasco da Gama, Christopher Columbus and Marco Polo brought back exotic spices from their travels to the new worlds. Spices, such as black pepper, were used as a currency in Medieval times – hence the term 'peppercorn rent', which was also used in legal documents to explain a ground rent of nominal value. The first culinary use of spices was to preserve meat due to their antimicrobial properties. Herbs and spices now play an important role in modern food production as they are able to mask salt reduction, reduce the use of artificial flavourings, provide natural colour to food and act as natural antioxidants. The use of herbs and spices has more recently moved into the hot beverage sector and into 'healthy' food products where they are being used in 'soft claim' applications, where the medical benefits are implied. EU Regulation 1924/2006 Article 12 states that health claims which make reference to recommendations of individual doctors or health professionals cannot be made on food.

17.2 Definitions

Before we move onto more detail in this chapter, we must first define what we mean by herbs and spices. The following definitions have been accepted in the trade and by the European Regulators to describe the products:

European Spice Association (ESA) Definition of Culinary Herbs and Spices

'Culinary Herbs and Spices are edible parts of plants that are traditionally added to foodstuffs for either their natural flavouring, aromatic and/or visual properties'.

Swainson's Handbook of Technical and Quality Management for the Food Manufacturing Sector.
DOI: https://doi.org/10.1016/B978-1-78242-275-4.00017-4

17.3 Types of herbs and spices

According to the ESA list of culinary herbs and spices the details are listed in Table 17.1, including their botanical names. Table 17.1 shows the items of commerce that its membership use or trade on a frequent basis, although there are some notable omissions, such as sumac (*Rhus aromatic and coriaria*) and black cardamom (*Amomum subulatum*).

17.4 Formats

This section deals with the formats in which herbs and spices are commercially available for use by food manufacturers and their benefits/drawbacks.

17.4.1 Rubbed herb or whole spice

The basic agricultural product that will be processed immediately postharvest to produce a dried herb with a small leaf size and a minimum of residual stalky material or a whole dried spice with a minimum of residual extraneous matter. The product can be used in this format providing the manufacturer has risk assessed the implications of the use. Typically, the rubbed herb or whole spice will be re-cleaned and treated to reduce the inherent microbial loading prior to use. Permitted treatments include steam sterilisation, steam washing and irradiation (labelling required). Treatment with ethylene oxide is not permitted within the European Union but is permissible in the United States of America (USA).

17.4.2 Ground herb and spice

The product obtained from the above, which can be milled into various particle sizes from very fine powder (100 microns) to kibbled pieces (7000 microns). Different terminologies are used to describe the particle size; fine ground (<300 microns), ground (300–700 microns), coarse ground (800–1200 microns), cracked (1300–3500 microns) and kibbled (>4000 microns). The end product use usually determines the format applied.

17.4.3 Herb or spice blend

A standardised blend of herbs and/or spices designed to provide flavour (e.g., garam masala) or visual impact (mixed herbs) to a food product.

Table 17.1 **The ESA list of culinary herbs and spices**

Herbs and spices (Common name)	Botanical name	Part of plant[a] (Botanical definition)	Climate zone[b]	Herb or spice
Aniseed	*Pimpinella anisum*	Fruit[c]	TE	Spice
Asafoetida	*Ferula asafoetida*	Rhizome	TE, SU	Spice
Basil	*Ocimum basilicum*	[d]	TE	Herb
Borage leaf	*Borago officinalis*	Leaf	TE	Herb
Caraway	*Carum carvi*	Fruit[c]	TE	Spice
Cardamom seeds	*Elettaria cardamomum*	Seed	TR	Spice
Cardamom, whole pods	*Elettaria cardamomum*	Fruit	TR	Spice
Celery leaves	*Apium graveolens dulce*	[d]	TE	Herb
Celery seed	*Apium graveolens*	Seed	TE	Spice
Chervil	*Anthriscus cerefolium*	[d]	TE	Herb
Chillies	*Capsicum frutescens or annuum*	Fruit	TR, TE	Spice
Chives	*Allium schoenoprasum*	Leaf	TE	Herb
Cinnamon	*Cinnamomum zeylanicum*	Inner bark	TR	Spice
	Cinnamomum burmanii[e]			Spice
	Cinnamomum loureirii[e]			Spice
	Cinnamomum aromaticum[e]			Spice
Cloves	*Syzygium aromaticum*	Flower bud	TR	Spice
Coriander leaves	*Coriandrum sativum*	[d]	TE	Herb
Coriander seed	*Coriandrum sativum*	Fruit[c]	TE	Spice
Cumin seed	*Cuminum cyminum*	Fruit[c]	TE	Spice
Curry leaf	*Murraya koenigii*	Leaf	TR, SU	Herb
Dill seed	*Anethum graveolens, Anethum sowa*	Fruit[c]	TE, SU	Spice
Dill tops	*Anethum graveolens*	[d]	TE	Herb
Fennel seed	*Foeniculum vulgare*	Fruit[c]	TE	Spice
Fenugreek	*Trigonella foenum-graecum*	Seed	TE	Spice
Galangal	*Alpinia officinalis, Alpinia officinarum, Kaempferia galangal*	Rhizome	TR	Spice

(*Continued*)

Table 17.1 **(Continued)**

Herbs and spices (Common name)	**Botanical name**	**Part of plant[a] (Botanical definition)**	**Climate zone[b]**	**Herb or spice**
Garlic	*Allium sativum*	Bulb	TE	Spice
Ginger	*Zingiber officinale*	Rhizome	TR	Spice
Grains of paradise	*Aframomum melegueta*	Seed	TR	Spice
Juniper berries	*Juniperus communis*	Fruit	TE	Spice
Kaffir Lime Leaves	*Citrus hystrix*	Leaf	TR	Herb
Laurel (Bay) leaves	*Laurus nobilis*	Leaf	TE	Herb
Lavender flower	*Lavandula officinalis*	Flower	TE	Herb
Lavender leaf	*Lavandula officinalis*	Leaf	TE	Herb
Lemon Grass	*Cymbopogon citratus*	Leaf	TR	Herb
Lovage	*Levisticum officinalis*	Leaf	TE	Herb
Lovage	*Levisticum officinalis*	Rhizome	TE	Spice
Mace	*Myristica fragrans, Myristica argentea*	Aril	TR	Spice
Marjoram	*Majorana hortensis, Syn. Origanum majorana*	[d]	TE	Herb
Mexican oregano	*Lippia graveolens*	Leaf	TE, SU	Herb
Mustard seed	*Sinapis alba, Sinapis nigra* *Brassica nigra, Brassica juncea*	Seed	TE	Spice
Nigella seeds (kalonji seed, or improperly 'onion seeds', or 'black cumin seeds', or 'black caraway seeds')	*Nigella sativa*	Seed	TR, SU	Spice
Nutmeg	*Myristica fragrans*	Seed	TR	Spice
Onion and shallot, dried	*Allium cepa and Allium cepa var. aggregatum*	Bulb	TE	Spice
Oregano	*Origanum vulgare* *Origanum onites*	[d]	TE	Herb
Paprika	*Capsicum annuum or frutescens*	Fruit	TE	Spice
Parsley	*Petroselinum crispum, Petroselinum sativum*	[d]	TE	Herb

Pepper, green[f]	*Piper nigrum*	Fruit	TR	Spice
Pepper, black	*Piper nigrum*	Fruit	TR	Spice
Pepper, white	*Piper nigrum*	Seed[c]	TR	Spice
Pink pepper	*Schinus terebinthifolius* *Schinus molle*	Fruit	TR	Spice
Cubeb pepper, Cubebs, Java pepper	*Piper cubeba L.*	Fruit	TR	Spice
Peppermint	*Mentha piperita*	Leaf	TE	Herb
Pimento	*Pimenta dioica*	Fruit	TR	Spice
Pimento (Allspice)	*Pimenta dioica*	Fruit	TR	Spice
Rosemary	*Rosmarinus officinalis*	Leaf[d]	TE	Herb
Saffron	*Crocus sativus*	Parts of pistil	TE	Spice
Sage	*Salvia officinalis, Salvia triloba*	[d]	TE	Herb
Savory	*Satureja hortensis*	[d]	TE	Herb
Summer savory, winter savory	*Satureja Montana*	[d]	TE	Herb
Sichuan pepper	*Zanthoxylum piperitum*	Fruit	TE	Spice
Spearmint	*Mentha spicata*	[d]	TE	Herb
Star anise	*Illicium verum*	Fruit	TE	Spice
Tarragon	*Arteminisia dracuncuklus*	[b]	TE	Herb
Thyme	*Thymus vulgaris, Thymus zygis, Thymus serpyllum*	[d]	TE	Herb
Turmeric	*Curcuma longa*	Rhizome	TR	Spice
Vanilla	*Vanilla planifolia* *Vanilla tahitensis*	Fruit	TR	Spice

[a]Part of plant used (botanical definition).
[b]Spices are classified in various ways of which the most used are as per part of the plant or as per indigenous climate zone. *TR*, tropical zone; *TE*, temperate zone; *SU*, subtropical.
[c]The traditional names and definitions developed over time sometimes deviate from the botanical definitions, for example aniseed and other seeds which are botanically fruits and white pepper which is sometimes defined as fruit while the botanical definition is seed.
[d]For culinary herbs part of plant is normally leaves or whole, soft part of the plant.
[e]The 3 botanical names are sometimes referred to as cassia.
[f]Pepper, green, red, and capers in brine are generally defined as condiment.
Source: From http://www.esa-spices.org/download/esa-list-of-culinary-herbs-and-spices.pdf.

17.4.4 Seasoning (UK Seasoning and Spice Association – [SSA] definition)

A blend of permitted food ingredients containing one or more, whole or ground spices/herbs and/or other flavour-enhancing or -imparting ingredients. They often contain added functional ingredients to achieve the technological purpose for which the seasoning in question is designed.

17.4.5 Essential (volatile) oil

The product is obtained by steam distillation of the herb or spice. The herb or spice is rough ground (if necessary) and placed in a vessel, water is added which is heated and the volatile flavouring components (essential oil) is liberated from the herb or spice in the form of a gas mixed with the steam and separated from the steam by a cold-water condenser. The essential oil is cleaned after distillation and the quality is assessed before use.

17.4.6 Oleoresin

The product obtained by the solvent extraction of the herb or spice. The herb or spice is rough ground (if necessary) and placed in an extraction vessel where a polar solvent (acetone, hexane, ethyl acetate, methanol or alcohol) is added to the vessel. The mixture is left to soak or percolate for a fixed time, which causes the main flavouring components (e.g., piperine in pepper or capsaicin in chillies) plus the waxes, fats, gums and fixed oils to be separated from the herb or spice into the solvent. The solvent is subsequently removed by distillation to leave behind a viscous mobile liquid or oleoresin. The oleoresin is normally standardised by adding back a portion of the essential oil and carrier oil (sunflower oil) or permitted food grade solvent (propylene glycol).

17.4.7 Plated herbs and spices

A plated product can be produced by dry blending the extracts (essential oil and/or oleoresin) from the herb or spice onto a carrier (salt, dextrose, rusk, etc.), to produce a standardised soluble herb or spice. These products are typically used in canned foods or processed comminuted meat products.

17.4.8 Encapsulated herbs and spices

An encapsulated or spray dried product is produced by diluting the extract (essential oil and/or oleoresin) from the herb or spice into water and microencapsulating the droplets into a starch or gum base. Bases used are typically maltodextrin, vegetable fat, gum arabic or a mixture of carriers, dependant on the intended final use. High fat containing carriers are used in bakery applications where the encapsulating material is required to retain the extract until the end of the baking process.

17.5 Major origins

In global terms herbs and spices are classified as minor crops; however, production quantities can range from 1 to 200 metric ton per annum for saffron, through to 1.5–2 million metric tons per annum for chillies. Spices are either produced as cash crops, i.e., mainly grown for export or as part of the indigenous regions' staple diet. Global production of black pepper is approximately 575,000 metric ton, of which approximately 400,000 metric ton are exported from the source countries, whereas for chillies production it is approximately 1.5–2 million metric tons and global exports stand at about 500,000 metric ton. The major production origins are listed in Table 17.2.

17.6 International quality standards

Herbs and particularly spices have always been highly priced commodities and vulnerable to adulteration. In consequence simple standards evolved early. As an example, in 1180 in the United Kingdom in the reign of Henry II, a 'peppers' guild was established in London to set and enforce standards for spices. In 1429 it was incorporated into the Grocers Company that is still in existence. This guild was granted a charter by Henry VI to manage the trade in spices. This organisation was given exclusive power to garble (e.g., cleanse and separate) spices. The term is still in use today, for example, in classifying types of pepper such as Telicherry Garbled Extra Bold Black Pepper (TGEB). Today the two major international standards are those set by the United States and those set by the European Union. Standards relying on the same general parameters also exist in those countries responsible for growing herbs and spices, for example, the Indian Spices Board (ISB) and the Pepper Marketing Board (Malaysia). These standards are influenced by those set by the major importing countries.

There are various types of test that make up the range of international standards:

Cleanliness. This is a measure of the amount of foreign and extraneous matter, for example insect contamination, excreta or foreign bodies. Measurement is by physical determination (using microscopic analysis (30 ×)) of contamination within aliquots (samples) of the product.

Ash level. This is a measure of the level of impurities in a product, obtained by burning off the organic matter and measuring the residue of ash. This measurement is carried out by incinerating the herb or spice at 550°C to constant weight. Characteristic maximum figures exist for most herbs and spices.

Acid insoluble ash (AIA) (or sand content). This is a classic determination of the cleanliness of the herb or spice. The measure is usually made in conjunction with the ash content by boiling the ash in 2 N HCl and incinerating the residue (again at 550°C) to a constant weight. Again maximum figures exist for most herbs and spices.

Prosecutions have in the past been based on high AIA levels within Europe, which are seen as indicating an unacceptably high level of gross contamination.

Volatile oil (V/O) determination. This measure helps to identify whether the herb or spice has been adulterated, perhaps by addition of foreign materials, low quality materials or

Table 17.2 **The major production origins**

Herbs and spices (Common name)	Major origins
Aniseed	Turkey, Syria
Asafoetida	India
Basil	Egypt
Borage leaf	Germany, United Kingdom
Caraway	Finland, Holland
Cardamom seeds	Guatemala
Cardamom, whole pods	Guatemala, India
Celery leaves	Germany, Poland, United Kingdom
Celery seed	India
Chervil	Germany, United Kingdom
Chillies	India, Mexico, Thailand, China, Malawi, Uganda, Kenya
Chives	Germany, United Kingdom
Cinnamon	Sri Lanka, Seychelles, Madagascar – Cinnamomum zeylanicum China, Vietnam, Indonesia – Cinnamomum burmanii, loureirii, aromaticum
Cloves	Indonesia, Madagascar, Zanzibar, Comores, Brazil
Coriander leaves	Germany, United Kingdom
Coriander seed	India, Russia, Romania, Bulgaria, Canada, Egypt, Morocco
Cumin seed	India, Syria, Iran, Turkey, China
Curry leaf	India
Dill seed	India
Dill tops	Poland, Germany
Fennel seed	India, Egypt
Fenugreek	India, Egypt, Australia
Galangal	China, Thailand, Indonesia
Garlic	China, India, Egypt
Ginger	India, China, Nigeria, Ethiopia, Vietnam
Grains of paradise	Ghana
Juniper berries	Croatia, Italy
Kaffir lime leaves	Thailand
Laurel (Bay) leaves	Turkey

Lavender flower	United Kingdom
Lavender leaf	United Kingdom
Lemon Grass	India, Thailand
Lovage	Poland
Mace	Indonesia, Grenada, Papua New Guinea
Marjoram	Egypt
Mexican oregano	Mexico
Mustard seed	India, Canada, Russia
Nigella seeds (kaljoni seed, or improperly 'onion seeds', or 'black cumin seeds', or 'black caraway seeds')	India
Nutmeg	Indonesia, Grenada, India, Papua New Guinea
Onion and shallot, dried	India, China, France, Holland, Egypt
Oregano	Turkey, Chile, Peru
Paprika	Peru, China, Zimbabwe, Morocco, Spain
Parsley	United Kingdom, Germany, China, Egypt
Pepper, green	India
Pepper, black	Vietnam, India, Indonesia, Malaysia, Brazil, Ecuador, Thailand, Cambodia, Madagascar, Sri Lanka
Pepper, white	Vietnam, Indonesia, Malaysia, Brazil, China
Pink pepper	Brazil, Madagascar, Reunion Islands
Cubeb pepper, Cubebs, Java pepper	Indonesia
Peppermint	United Kingdom
Pimento	Mexico, Jamaica, Honduras, Guatemala
Rosemary	Spain, Morocco, Turkey
Saffron	Iran, Spain
Sage	Turkey, Albania, Italy, Greece
Summer savory, winter savory	Macedonia, Albania, Italy, Spain
Sichuan pepper	China
Spearmint	Egypt
Star anise	China, Vietnam
Tarragon	France, Russia
Thyme	Spain, Morocco, Turkey
Turmeric	India, Vietnam, Indonesia
Vanilla	Madagascar, Uganda, India, Papua New Guinea

spent herb or spices (byproduct of essential oil or oleoresin production). The herb or spice is boiled with water under reflux conditions. The oil separates on top of the water and is reported as a proportion of the mass of the product under test. Minimum percentage levels of oil exist for most major herbs and spices.

Moisture content. This measure is important since moisture content determines weight, and weight is used in pricing. With highly priced commodities traded on weight, a 1% moisture increase in the product as shipped can result in increased weight and increased profits for the original exporter.

Maximum moisture contents are set for all herbs and spices, based on the maximum allowable amount of moisture for the product to remain stable. Moisture content is generally determined within the herb and spice industry using the Dean & Stark methodology. This involves re-fluxing a known weight of the herb or spice in petroleum spirit and measuring the water that condenses at the bottom of the reflux chamber from the known weight of herb or spice. Generally the level is 12% max.

Water activity. In recent years moisture content has been related to the water activity (a_w) of the herb or spice. A level of 0.65 a_w is generally accepted as a figure below, which microbial growth cannot occur. However, this figure does not take into account other natural antimicrobial factors and several herbs and spices may be stored at significantly higher water activity levels without problem due to the preservative effect of other components, especially of the oils they contain.

Microbiological measures. There are a range of techniques available for enumerating microorganisms (both spoilage species and pathogens) in herbs and spices. While new technologies for microbiological analysis are continually being developed and improved the standard methods are slow to change (usually described in ISO standards) and it is these standard methods that tend to be used in cases of dispute and which are quoted on specifications.

Pesticide levels. Pesticide levels are not seen as a major health problem given the (low) average daily intakes of herbs and spices by consumers. EC Regulation 396/2005 that came into force on 21 September 2008, details the maximum residue levels (MRLs) permitted. A list of dehydration factors for fresh herbs and capsicums can be accessed from the ESA website (www.esa-spices.org).

Legislation is different in the United States, as in general they do not accept tolerance levels; however, limits exist for certain pesticides in specific herbs and spices. In the meantime, Codex limits for the nearest equivalent commodity may be a useful guide. Pesticide levels are usually assessed either by gas chromatography (GC) or high-performance liquid chromatography (HPLC), linked to tandem mass spectrometry (MS/MS), depending on the pesticide in question.

Mycotoxin levels. Mycotoxins, particularly aflatoxin and ochratoxin A, have been of increasing concern within the industry. EC Regulation 1881/2006 governing the maximum levels of aflatoxins and EU Regulation 105/2010 governing the maximum levels of ochratoxin A in capsicum species, piper species, nutmeg, mace, ginger and turmeric have been implemented in recent years. Within the United States the maximum limit for total aflatoxins is currently 20 μg/kg and there are no levels for ochratoxin A.

HPLC linked to immunoaffinity columns is the reference methodology employed in these determinations.

Bulk density/bulk index. This is an important measure, particularly in filling retail containers of herbs and spices. The herb or spice must be sifted or ground to give a certain density so that retail units not only appear satisfactorily full but also comply with the declared weight. Densities may be measured packed down, e.g., after tapping the product so that it assumes a minimum density, or untapped (as it falls into the container without compression). This measure is usually reported in grams/Litre or mL/100 g.

Mesh/particle size. Many spices and herbs are ground to give easier dispersion in the final food product. This process also aids the dispersion of flavour. Particle size is generally specified and is carried out using standardised sieves. Products are ground to pass through a specified sieve, while coarse matter is recycled through the mill until it finally passes through the sieve. Sieves are often characterised in 'micron sizes' and typical specification requirements will be for a 95% pass on a specified size of sieve. The older method of measuring sieve mesh (hole) sizes was related to the number of holes per inch. However, confusing differences exist between American and British mesh sizes. The mesh size depends on the diameter of the wire making up the sieves and this differs between nations. Thus a 25-mesh US sieve is equivalent to a 30-mesh BS (United Kingdom) sieve and both are equivalent to a 500-micron aperture size. Tables are available giving the relationships between national sieve sizes and micron sizes. There are a number of internationally approved standards for testing procedures, established by the International Standards Organisation (ISO) as referenced in Appendix I.

Adulteration (ESA definition). Adulteration is the commercially motivated deliberate and intentional inclusion in herbs and spices of constituents whose presence is not declared. The blending together of different qualities of the same material to reduce the variation in the flavour profile is not considered adulteration. In addition, naturally occurring contaminants such as heavy metals, mycotoxins, pathogens, etc. are not considered adulterants.

Product may be adulterated with:

- ingredients, additives or constituents not approved for use in food;
- ingredients, additives or constituents approved for use in food but not declared; and
- material that has had any valuable constituent omitted or removed.

Prevention. Responsible companies are aware of the risk of adulteration and closely monitor the products they handle. They will use risk assessment to determine the most appropriate tools to combat this problem, including:

- Evaluation of the supply chain
  - History of supply
  - Supplier capability of meeting legal and other requirements
  - Adherence to GAP (Good Agricultural Practice) and GMP (Good Manufacturing Practice)
  - Adherence to HACCP (Hazard Analysis Critical Control Point) principles
  - Traceability
- Evaluation of the raw material
  - Form (whole or ground)
  - Price
  - Species
- Evaluation of raw material control through appropriate product testing.

Responsible companies are constantly working with agencies and suppliers around the world to increase the awareness of this issue and eliminate the practice of adulteration of herbs and spices.

Standardisation. Is the process by which various component parts of the same botanical species are separated and then blended back together in a different ratio, to give the desired physical characteristics (colour/flavour/aroma), e.g., separation of black peppercorns, light berries and pinheads. These are then blended back together in different proportions to standardise the piperine content.

17.7 Product-specific quality parameters

There are several other tests used in the industry, some of which are for specific herbs or spices. Some of the best known and widely used are:

Piperine levels. The test is specifically for peppers of the *Piper* genus. This involves extraction and measurement of the characterising heat portion of the pepper (the piperine content). After re-fluxing in alcohol to extract the piperine, absorbency is compared to a standard using a spectrophotometer at 342–345 nm.

American Spice Trade Association (ASTA) Colour values. This is a measurement of the extractable colour of products of the *Capsicum* genus and is used as a quality indicator for paprika. Extraction is in acetone (16-h ambient extraction) and is followed by spectrophotometric analysis (against a standard at 460 nm). The methodology was developed by the ASTA and results are still often expressed as the ASTA colour value.

Capsaicin content. Capsaicin is the pungent principle that gives heat to species of *Capsicum.* Extraction of capsaicin by re-fluxing with alcohol is followed by the determination of HPLC using acetonitrile/water as the carrier. Results can be related to Scoville test (see below).

Scoville heat units. The Scoville heat unit is a measure of the capsaicin content of *Capsicum* species. The test involves extraction of the capsaicin in alcohol and tasting of successively stronger dilutions in sugar syrup until the chilli heat (burning sensation) is detected. It gives a compatible result to capsaicin content but obviates a need for sophisticated laboratory equipment. A trained tasting panel is required (Scoville units divided by 150,000 = percent capsaicin).

Curcumin content. This is a test specific to the measurement of the extractive colour of turmeric. This is carried out by reflux extraction in acetone followed by measurement using a spectrophotometer at 415–425 nm.

Nonpermitted colours. These colours are sometimes added illegally to spices such as chillies, cassia, fennel, paprika, saffron and star anise to enhance their physical appearance and therefore their value. Nonpermitted colours are assessed by HPLC with a diode array detector (DAD) or by liquid chromatography, tandem mass spectrometry (LC/MS/MS) dependent on the level of detection (LOD) required.

17.8 Trade bodies

The global trade in herbs and spices is controlled by National Trade Associations who report to or are members of International Trade Associations. Examples of National Trade Associations are as follows:

Spices Board of India
World Spice Organisation
Vietnam Pepper Association
Canadian Spice Association

Examples of the Global Trade Associations, who set quality specifications/standards are as follows:

The American Spice Trade Association (ASTA): The ASTA was established at the beginning of the 20th century. Given its long involvement in regulating the quality of herbs and spices entering the United States, ASTA standards are recognised and endorsed by the United States Food & Drug Administration (US FDA).

Cleanliness: Specifications exist for all major herbs and spices, in terms of permitted amounts of extraneous matter or filth, mould (visible), excreta, insects and insect damaged material. The amount of contamination is measured by microscopic analysis (30 ×) of aliquots of the material. The ASTA cleanliness specifications are shown in Table 17.1.

For the purposes of these specifications, extraneous matter is defined as everything foreign to the product itself, including, but not restricted to: stones, dirt, wire, string, stems, sticks, excreta, and other animal contamination, nontoxic foreign seeds and other plant material (in some cases) such as foreign leaves.

The level of contaminants must fall below those shown in Table 17.4, except for the column 'Whole insects, dead' which must not exceed the limit mentioned. Herbs and spices not meeting this standard must be re-cleaned/re-conditioned before distribution and sale within the United States.

ASTA also sets a range of other standards. These are broadly comparable to those set by the ESA, which are discussed in Section 17.8.1.

Microbiological standards now play an increasingly important role in determining the quality of herbs and spices. They are becoming a crucial quality parameter due to the increasingly varied uses of herbs and spices in the developed world.

Increased travel has led to a society demanding multicultural foods. This coupled with ready meals, cook–chill products, etc., has meant that herbs and spices are not 'always cooked' as was assumed in the past.

A certain level of microbiological contamination is unavoidable on plant materials grown outdoors, this coupled with low levels of technology employed in harvesting, handling and storage in the countries of origin of many herbs and spices plus the concentration that occurs due to drying means these products can have high microbiological counts. Total counts in excess of 10^6 colony forming units (cfu) per gram are common and food poisoning bacteria such as *Salmonella* are estimated to be present in approximately 10% of consignments.

There are currently three major methods of microbiological control:

1. Within the United States (US), the principal method of microbiological control is fumigation with ethylene oxide (a biocidal gas). Sometimes multiple fumigations are used to achieve a satisfactory microbiological reduction. In recent years concern about residues left by ethylene oxide (ETO) and ethylene chlorhydrin (ECH), has led to bans on its use (within the EU for example) and recently the United States has set MRLs for ETO and ECH.
2. This has led to the use of heat treatment for decontamination, generally using high pressure steam in highly specialised equipment. ASTA has published guidelines for the Validation of thermal processes.
3. Irradiation is permitted for microbiological control of herbs and spices in many countries of the world. However, the use of irradiation must be declared on the consumer packaging, and public concern about the safety of irradiated foods has prevented the widespread application of this undeniably efficient process in many areas where its use is permitted by law (Table 17.3).

17.8.1 ASTA Cleanliness specifications for spices, seeds and herbs revised 2011

Table 17.3 ASTA cleanliness specifications for spices, seeds and herbs

Name of spice, seed or herb	ΔWhole insects, dead[a]	Excreta, mammalian	Excreta, other	Mould	Insect defiled/ infested	Extraneous/ foreign matter
	By count	By mg/lb	By mg/ lb	% By weight	% By weight	% By weight
Allspice	2	5	5.0	2.00	1.00	0.50
Anise	4	3	5.0	1.00	1.00	1.00
Sweet Basil	2	1	2.0	1.00	1.00	0.50[b]
Caraway	4	3	1.0	1.00	1.00	0.50
Cardamom	4	3	10.0	1.00	1.00	0.50
Cassia	2	1	1.0	5.00	2.50	0.50
Cinnamon	2	1	2.0	1.00	1.00	0.50
Celery seeds	4	3	3.0	1.00	1.00	0.50
Chillies	4	1	8.0	3.00	2.50	0.50
Cloves[c]	4	5	8.0	1.00	1.00	1.00
Coriander	4	3	10.0	1.00	1.00	0.50
Cumin seed	4	3	5.0	1.00	1.00	0.50
Dill seed	4	3	2.0	1.00	1.00	0.50
Fennel seed	SF[d]	SF[d]	SF[d]	1.00	1.00	0.50
Ginger	4	3	3.0	SF[e]	SF[e]	1.00
Laurel leaves[f]	2	1	10.0	2.00	2.50	0.50
Mace	4	3	1.0	2.00	1.00	0.50
Marjoram	3	1	10.0	1.00	1.00	1.00[b]
Nutmeg (broken)	4	5	1.0	SF[g]	SF[g]	0.50
Nutmeg (whole)	4	0	0.0	SF[h]	SF[h]	0.00
Oregano[i]	3	1	10.0	1.00	1.00	1.00[b]
Black pepper	2	1	5.0	SF[j]	SF[j]	1.00
White pepper[k]	2	1	1.0	SF[l]	SF[l]	0.50
Poppy seed	2	3	3.0	1.00	1.00	0.50
Rosemary leaves	2	1	4.0	1.00	1.00	0.50[b]
Sage[f]	2	1	4.0	1.00	1.00	0.50
Savory	2	1	10.0	1.00	1.00	0.50[b]
Sesame seed	4	5	10.0	1.00	1.00	0.50
Sesame seed, hulled	4	5	1.0	1.00	1.00	0.50

(Continued)

Table 17.3 (Continued)

Name of spice, seed or herb	ΔWhole insects, dead[a]	Excreta, mammalian	Excreta, other	Mould	Insect defiled/ infested	Extraneous/ foreign matter
	By count	By mg/lb	By mg/ lb	% By weight	% By weight	% By weight
Tarragon	2	1	1.0	1.00	1.00	0.50[b]
Thyme	4	1	5.0	1.00	1.00	0.50[b]
Turmeric	3	5	5.0	3.00	2.50	0.50

Ground processed spice[m] (Cannot exceed limit shown)

Spices	Whole equivalent insects	Insect fragments	Mites	Other insects	Rats/mouse hairs	Animal hairs
Ground Paprika		Average of more than 75 fragments/ 25 g			Average of more than 11 rodent hairs/25 g	

[a]Δ Whole insects, dead: Cannot exceed the limits shown.
[b]Extraneous matter: Includes other plant material, e.g., foreign leaves.
[c]Clove stems: Less than (<) 3% allowance by weight for unattached clove stems over and above the tolerance for other extraneous matter is permitted.
[d]Fennel seed: In the case of fennel seed, if 20% or more of the subsamples contain any rodent, other excreta or whole insects, or an average of 3 mg/lb or more of mammalian excreta, the lot must be reconditioned.
[e]Ginger: More than 3% mouldy pieces and/or insect pieces by weight.
[f]Laurel leaves and sage: 'Stems' will be reported separately for economic purposes and will not represent a pass/fail criteria.
[g]Broken nutmeg: More than 5% mould/insect defiled combined by weight.
[h]Whole nutmeg: More than 10% insect defiled and/or mouldy pieces, with a maximum of 5% insect defiled pieces by count.
[i]Oregano: Sumac negative. Analysis for sumac shall not be mandatory if samples are marked 'Product of Mexico'.
[j]Black pepper: 1% mouldy and/or infested pieces by weight.
[k]White pepper: 'Percent black pepper' will be reported separately for economic purposes and will not represent a pass/fail criteria.
[l]White pepper: 1% mouldy and/or infested pieces by weight.
[m]Microanalytical methods for paprika and ground capsicums can be found in the 'Analytical Procedures' section of the ASTA cleanliness specifications for spices, seeds and herbs.

17.8.1.1 The European Spice Association

ESA represents national seasoning and spice trade associations within Europe. ESA members are involved in the processing, marketing, distribution and trading of herbs, spices and seasoning mixes.

Its objectives are:

- to promote and protect the interests of its membership;
- to be the leading organisation of the European Seasoning and Spice Industry;
- to ensure members are fully aware of all current and future legislation;
- to act as central medium of communication with the EU Commission, European Parliament and FoodDrinkEurope (FDE); and
- to strengthen links with worldwide spice trade associations and nongovernmental organisations (NGO's) including the American Spice Trade Association (ASTA), International Organisation of Spice Trade Associations (IOSTA), Indian Spices Board (ISB), Codex Alimentarius and ISO.

Quality standards in Europe are typified by the standards set by ESA, which draw both on national standards such as those issued by AFNOR (the French standards authority) and BSI (British Standards Institute), and the international standards issued by ISO.

The ESA minimum general quality standards for all herbs and spices are summarised in Table 17.4, while ESA quality standards for specific herbs and spices are shown in Table 17.5.

The ESA general standards are more relaxed and represent *minimum* standards allowable for trade.

They do not preclude buyer and seller setting additional, stricter standards fit for the final purpose for which the herb and spice is to be used.

17.8.1.2 International Organisation of Spice Trade Associations

The IOSTAs is made up of spice associations from around the world whose members meet regularly to discuss issues relevant to the spice industry.

The IOSTA mission statement reads:

> *IOSTA brings together spice associations from around the world to address common issues and seek sensible solutions to ensure the sustainability of the spice industry.*

IOSTA has developed a GAPs Guide to be used as a resource in the growing and harvesting of spices (General Guidelines for Good Agricultural Practices Spices, April 2008: Issue 1. Produced by IOSTA with assistance from the International Trade Centre, Geneva).

The current IOSTA membership comprises: All Nippon Spice Association (ANSA), American Spice Trade Association (ASTA), Canadian Spice Association, European Spice Association (ESA), International General Produce Association (IGPA), Spices Board of India, All India Spices Exporters Forum, The Spice Council of Sri Lanka and Vietnam Pepper Association.

17.8.1.3 All Nippon Spice Trade Association

The ANSA membership comprises leading Japanese spice, flavour and fragrance and seasoning companies.

The ANSA mission statement is:

> *to increase the consumption of spices in Japan through stable supply of good quality spices and to contribute to the better, healthier and safer food life of consumers.*

ANSA's activities include providing information on food safety issues to consumers and communicating with the relevant governmental bureaus to establish realistic and adequate food safety regulations for spices. The most recent activities were to request the Ministry of Health, Labour and Welfare to undertake a risk assessment on the irradiation of spices, which is currently banned in Japan, and to provide information on approved chemical residue levels in spices.

Table 17.4 European Spice Association (ESA) specifications of quality minima for herbs and spices

Subject	
Sampling	ISO 948 For mycotoxins: See the relevant Commission Regulation at: www.esa-spices.org.
Chemical/physical analysis	
Ash	For values see Appendix I.
Acid insoluble ash	For values see Appendix I.
Moisture	For values see Appendix I.
Volatile oil	For values see Appendix I.
Water activity	Water activity is a key parameter that affects microbial growth. Therefore ESA recommends a target value of maximum 0.65.
Bulk density	Due to methodology variability both method and value should be agreed between buyer and seller.
Microbiology	The product shall be free from micro-organisms at such levels which may represent a hazard to health. If the product is treated to reduce microbial loads before being imported into destination country the treatment will be such as to render/ensure the microbiological safety of consumers. Specific requirements to be agreed between buyer and seller.
Contaminants/residues	
Pesticides	Shall be utilised in accordance with good agricultural practice. Application and residue limits must comply with existing EU and/or national legislation.
Heavy metals	Must comply with national/EU legislation (e.g., cadmium, lead).
Mycotoxins	Herbs and spices must be grown, harvested, handled, and stored in such a manner as to prevent the occurrence of mycotoxins. If found, levels must comply with existing national and/or EU legislation.
Allergens	Refer to ESA Position Statement
Treatments	Only approved legal processing procedures may be applied in any treatment used for product quality or safety. EC approved fumigants may be used in accordance with manufacturers' instructions, but this must be indicated on the accompanying documents. Ethylene oxide (ETO) treatment has been banned under European legislation. This ban covers both the materials treated within and outside of the EU (i.e., the use of material that has been ETO treated before importation is also illegal). Irradiation, at present, does not have full consumer acceptability, so the treatment has to be agreed between buyer and seller. If it is agreed,

(*Continued*)

Table 17.4 (Continued)

Subject	
	irradiation is only permitted in EU-approved irradiation plants. However, EU legislation requires that the irradiated product is declared at all levels within the food chain. Members of ESA support the use of environmental friendly fumigants (Montreal protocol) and nontoxic processes (e.g., microbial reduction under pressure, steam treatment). All products subject to processing (for example grinding, microbial reduction) are not in the scope of this document, unless otherwise stated.
Purity	
Species	In compliance with Food Law Regulations. If not regulated: ESA list of culinary herbs and spices, or to be agreed between buyer and seller.
Adulteration	Must be free from.
Infestation	Should be free in practical terms from live and/or dead insects, insect fragments and rodent contamination visible to the naked eye (corrected if necessary for abnormal vision).
Extraneous matter	Herbs max. 2%, Spices max. 1% by weight
Foreign matter	The European food operator has to evaluate if products fully comply with safety requirements before selling them to the final consumer. If not, additional processing will be necessary.
Sensory properties	Must be free from off odour or off flavour.
Packaging	The packaging must not be a source of contamination or migration, should be food grade and must protect the product quality during transportation and storage.

Source: Courtesy of ESA.

17.8.1.4 International Pepper Community

The IPC is an intergovernmental organisation of pepper producing countries. The community now includes Brazil, India, Indonesia, Malaysia, Sri Lanka and Vietnam as permanent members and Papua New Guinea as an associate member.

The IPC was established in 1972 under the auspices of the United Nations Economic and Social Commission for Asia and the Pacific (UN-ESCAP). The IPC Secretariat is located in Jakarta, Indonesia and is headed by an executive director.

Membership to the community is open to all pepper producing countries, with the unanimous consent of the existing members and by acceding to the agreement establishing the Pepper Community.

Table 17.5 **Appendix I chemical/physical parameters; dry base for ASH, AIA, V/O**

Product	Ash (% w/w) Max*	AIA (% w/w) Max*	H_2O (% w/w) Max*	V/O (mL/100g) Min	Notes
Anise	9.0	2.5	12	1.0	
Basil	16	2.0	12	0.5	
Caraway	8.0	1.5	13	2.5	
Cardamom	9.0	2.5	12	4.0	
Celery seed	12	3.0	11	1.5	
Celery leaves	20	1.0	8.0	Traces**	
Chervil	17	2.0	8.0	Traces**	
Chilli	10	1.6	11		
Chives	13	2.0	8.0	Traces**	
Cinnamon (Ceylon)	7.0	2.0	14	0.7–1.0	*The use of SO_2 is only permitted for Ceylon cinnamon, Annex III part B Directive 95/2/EC*
(Cassia)				(ISO 6539 ISO 6538) Depending on botanical species	Styrene off notes can be prevented through control of moisture content throughout the supply chain
Cloves	7.0	0.5	12	14	
Coriander seed	7.0	1.5	12		
Microcarpum				0.6	
Macrocarpum				Traces**	
Coriander leaves	15	1.0	8.0	Traces**	
Cumin	14	3.0	13	1.5	
Dill seed	10	2.5	12	1.0	
Dill tops	15	2.0	8.0	Traces**	
Fennel	9.0	2.0	12	1.5	
Fenugreek	7.0	1.5	11	Traces**	
Galangal (ground)	9.0	4.0	10	Traces**	
Garlic products	6.0	0.5	6.5	–	Due to the hygroscopic nature of these products, lower moisture content may be required.

(Continued)

Table 17.5 **(Continued)**

Product	Ash (% w/w) Max*	AIA (% w/w) Max*	H_2O (% w/w) Max*	V/O (mL/100g) Min	Notes
Ginger	8.0	2.0	12	1.5	
Juniper berries	5.0	1.0	16	1.4	
Laurel leaves	7.0	2.0	8.0	1.0	
Lemon grass	8.0	2.5	10	Traces**	
Mace	4.0	0.5	10	5.0	
Marjoram	10	2.0	12	0.7	
Mustard	6.5	1.0	10	–	
Nutmeg	3.0	0.5	10	5.0–6.5 Depending on grade	
Onion products *Allium cepa*	5.0	0.5	6.0–8.0	–	Due to the hygroscopic nature of these products lower moisture content may be required.
Oregano	10	2.0	12	1.5	
Paprika powder	10	2.0	11	–	
Parsley	14	1.5	7.5	Traces**	English origin is not covered.
Pepper black	7.0	1.5	12	2.0	
Pepper white	3.5	0.3	12	1.5	
Pepper green (dried)	3.0	0.3	13*	1.0	Min* If freeze dried: 8%
Pink pepper (Schinus)	7.0	1.8	14*	2.0	* If freeze dried: 8%
Pimento					
Jamaica	4.5	0.4	12	3.0	
Other origins	5.0	1.0	12	2.0	
Poppy seeds	8.0	1.0	8.0	N.A.	
Rosemary	8.0	1.0	10	1.0	
Saffron (whole)	8.0	1.0	12	–	

Saffron (ground)	8.0	1.5	10	–	
Sage	12	2.0	12	1.5	
Savoury	12	1.0	12	0.5	
Mountain type (wild grown)					
Savoury	12	4	12	1.5	
Garden type (cultivated)					
Spearmint	12	2.5	13	0.5	
Star anise	3.0	0.5	8.0	7.0	
Tarragon	12	1.5	8.0	0.5	
Thyme	12	3.5	12	1.0	
Turmeric					
Whole	8.0	2.0	12	2.5	
Ground	9.0	2.5	10	1.5	

**Traces—low levels of volatiles (< 0.5%) for which analytical quantification by using ISO 6571 is not accurate and reliable. The sensorial flavouring properties should be agreed between buyer and seller.

The functions of the community are to promote, coordinate and harmonise all activities relating to the pepper economy with a view to achieving the following broad objectives:

- To coordinate and stimulate research on technical and economic aspects of production, including research on diseases affecting the pepper plant, research on development of disease-resistant and high-yielding varieties.
- To facilitate the exchange of information on programmes and policies and on any other aspects relating to production.
- To develop programmes for increasing consumption in traditional and new markets, including programmes of cooperation in promotion activities.
- To intensify and coordinate research on new uses of pepper.
- To further joint action for the relaxation of tariff and nontariff barriers and for the removal of other obstacles in trade.
- To coordinate standards of quality so as to facilitate international marketing.
- To keep under constant review developments relating to supply, demand and prices of pepper.
- To carry out investigations into the causes and consequences of fluctuations in the price of pepper and suggest appropriate solutions.
- To provide statistical and other information on pepper production, consumption, trade and prices including techniques of production and consumption forecasting.
- To undertake other activities and functions that may be considered desirable in the interests of the world pepper economy.

Appendix I References

ASTA *Clean Spices – A handbook for American Spice Trade Association Members.*

ASTA *Official Analytical Methods of the American Spice Trade Association.*

ASTA *Official Microbiological Methods of the American Spice Trade Association.*

ASTA *General Protocol for the Validation of Microbiological Processes on Pathogen Contaminated Spices and Culinary Herbs.*

BS4547 *Specification for Herbs and Spices*, British Standards Institute.

BS7087 *Methodology for Analysis of Herbs and Spice*, British Standards Institute.

ISO series: *Specifications for Herbs and Spices. International Standards Organisation.*

ESA (2010), *European Spice Association: Quality Minima for Herbs and Spices.*

ESA (2008), *European Spice Association: Dehydration Factors for Products of the Spice Industry.*

Appendix II Recommended analytical methods

Unless otherwise agreed between buyer and seller, ESA recommends the following methods:

1. Spices and condiments – Sampling EN ISO 948 – 2009.
2. Spices and condiments – Preparation of a ground sample for analysis ISO 2825 – 1981.
3. Spices and condiments – Determination of extraneous matter and foreign matter content ISO 927 – 2009 (*see definition* Chapter 5).
4. Spices and condiments – Determination of total ash ISO 928 – 1997.
5. Spices and condiments – Determination of acid insoluble ash ISO 930 – 1997.
6. Spices and condiments – Determination of moisture content – (Entrainment method) ISO 939 – 1980.
7. Spices and condiments – Determination of volatile oil EN ISO 6571 2009.
8. Analysis of spices and condiments – Determination of loss in mass of capsicum and allium species and of dried vegetables by vacuum oven drying – DIN 10236 (German standard).

Cereals

Colin W. Wrigley
University of Queensland, Brisbane, QLD, Australia

18

18.1 Introduction

The grain-bearing plants constitute the most important food resource for mankind. These plants are the direct source of much of our food, fats, fuel, fibre, fluids (beverages) and the indirect source of animal-based foods via feed, forage and fodder. These plants use photosynthesis, carbon dioxide and water, with the sun's energy, to generate carbohydrates and the many chemical compounds required by all forms of life on earth, thus providing:

- *Food*. Grains directly supply over half of humanity's energy and protein.
- *Fluids* (beverages). Grains are used to provide various beers and spirits.
- *Fats*. Oils are derived from the oilseed species and from the germ of cereal grains.
- *Fuel*. Biofuels, increasingly a renewable source of fuel, include oils and ethanol from the starch of cereal grains.
- *Fibre*. Linen from flax plants and cotton from cottonseed species.
- *Feed*. Much of the feed for animals, birds and fish is based on grains and grain by-products.
- *Forage and fodder*. The vegetable matter from grain production (before and after grain harvesting) is feed for our food animals.

This chapter concentrates on the food uses of the cereal grains, especially wheat. However, the multitude of foods from grains is ever before our eyes and upon our plates, so we are liable to be oblivious of the wider range of ways in which the family of grains impacts our lives and contributes to our sustenance.

18.2 The family of grains worldwide

Grains used as food sources are grouped:

- as pulses and oilseeds (Table 18.1); both are from dicotyledonous plants (dicots) (see a food example of lentils in Fig. 18.1A);
- as cereals (Table 18.2), whose plants are monocotyledonous (monocots).

The annual production of cereal grains (about two and a half billion tonnes in 2014/15) is much greater than that of the pulses and oilseeds (over a quarter of a billion tonnes in 2014).

Swainson's Handbook of Technical and Quality Management for the Food Manufacturing Sector.
DOI: https://doi.org/10.1016/B978-1-78242-275-4.00018-6

Table 18.1 World production in 2014 of grains classified as pulses and as oilseeds in millions of tonnes

***Pulses* (grain legumes)**		**Oilseeds**	
Beans, dry	25	*Soybeans*[a]	*308*
Chick peas	14		
Peas, dry	11	Canola (rape seed)	71
Cow peas, dry	6	Cottonseed	45
Lentils	5	Groundnuts, with shell	42
Pigeon peas	5	Sunflower seed	41
Broad beans, horse beans, dry	3	Sesame seed	5
Lupins	1	Linseed	3
Vetches	1	Safflower seed	1
Total production	73	Total production	208

[a]The soybean is classified as an oilseed by FAO, but it is strictly a legume.
Source: FAOSTAT data, accessed in March 2016 (faostat3.fao.org).

Figure 18.1 Many grains are eaten as the cooked whole grain, often after removal of the outer bran or glume layers. Examples shown are (A) dhal, a stew of lentil grains or other legumes; (B) boiled rice, a complement to many savoury dishes; and (C) a salad containing barley grains that have been boiled after pearling to expose the starchy endosperm.

Table 18.2 World production in 2014 of the most significant cereal grains in millions of tonnes

Common name	Genus and species	Millions of tonnes
Maize	*Zea mays*	1,022
Rice, paddy	*Oryza sativa*	741
Wheat	*Triticum aestivum*	729
Barley	*Hordeum vulgare*	144
Sorghum	*Sorghum bicolor*	67
Millet	*Pennisetum glaucum*[a]	28
Oats	*Avena sativa*	23
Rye	*Secale cereale*	15
Triticale	× *Triticosecale* sp.	17
Buckwheat[b]	*Fagopyrum esculentum*	2
Fonio	*Digitaria exilis*	0.7
Canary seed	*Phalaris canariensis*	0.2
Quinoa[b]	*Chenopodium quinoa*	0.2
Total		2,800

[a]One of several millets.
[b]Pseudocereals.
Source: FAOSTAT data accessed in 2016 (faostat3.fao.org).

Grain production from all these species (Tables 18.1 and 18.2) totals over three billion tonnes. Although world population is approaching eight billion people, this great volume of grain is equivalent to 1.1 kg of grain per day for each person on earth! For cereal grain production alone, the notional distribution is 1 kg/day/person. Why, then, is there any hunger on earth? Major sites of production are distant from major sites of need. Furthermore, significant volumes of grain are used for animal feed and for industrial purposes.

18.3 Cereal grains: a range of species

The cereal grains cover a wide range of species (even more than in Table 18.2) with many uses, as described by Arendt and Zannini (2013) and Wrigley and Batey (2010).

Maize heads the list of cereal grains, based on the volume of production in 2014 (Table 18.2). This cereal is known as 'corn' in North America, but in the United Kingdom the word 'corn' is also used as a general name for grains. Maize production has increased in recent years, being 1018 million tonnes in 2013, 872 million tonnes in 2012, 888 million tonnes in 2011 and 851 million tonnes in 2010, all well up from about 700 million tonnes in earlier seasons. Maize is used for animal feed, breakfast cereals, tortilla, cornbread, porridge, snack foods, a wide range of industrial uses (e.g. corn syrup, food ingredients) (White and Johnson, 2003; Orthoefer and Eastman, 2004; Lawton, 2016).

Rice is second in volume, if the production volume (as in Table 18.2) is for paddy rice – the grain as harvested before the outer husks are removed by milling.

Rice is the staple diet of a large proportion of the world's population, being taken as cooked and fried milled rice (Fig. 18.1B); other uses include sake and extruded snack foods (Champagne, 2004).

Wheat is the third of 'The Big Three' cereal species. It is also the grain with the widest range of uses and thus with the most exacting range of quality traits (Khan and Shewry, 2009; Carver, 2009). Accordingly, wheat has been selected as the grain for more extensive description in this chapter. Rye and triticale are close relatives of wheat. As such, they can be used for bread baking, but the relationship is not close enough for their dough-forming qualities to be as good as those of wheat.

Barley has the specialty uses for malting and beer production; also for spirits (Shewry and Ullrich, 2014). It is pearled (removal of husk and bran layers) for food use in porridge, and as an ingredient in soups and salads (Fig. 18.1C). Animal feed is a major use of barley.

Buckwheat and quinoa are classed as pseudocereals (Fletcher, 2016). They are dicots and thus not strictly cereal grains, but they have been combined with the cereal species in Food and Agriculture Organization's (FAO's) statistics. Fonio is a type of millet, there being several species classed as millets.

All the cereal grains and grain-based foods are highly recommended in nutritional guidelines disseminated by government agencies around the world (e.g. Fig. 18.2). In some cases, these recommendations are presented graphically in the form of a pyramid, with the base representing grain-based foods as a major part of food intake.

Figure 18.2 Cereal grains and grain-based foods are recommended as the main source of food in these Australian nutritional guidelines, cereal-based foods occupying the largest sector of the 'plate'.

18.4 The grain chain: the sequence from genes to products

The production and processing of all grain species follow the same path of progressive value addition shown in Fig. 18.3, and this sequence is followed in this chapter. The small item of pastry shown may cost a dollar or two in a patisserie. The main component, flour, came from harvested grain worth a cent or two; the value addition is thus one hundred-fold. Before and between those items are many stages. Cross-breeding leads to the production of new varieties with better grain yield and quality. The farmer manages grain production within the bounds of climatic uncertainties (Gooding, 2009; Carver, 2009).

Quality evaluation and preservation during transport is especially important for wheat, so that the miller can provide flour of the appropriate type for each of the various uses. Ultimately, the customer determines which retailer (and those upstream) has done best in providing the desired product. There is thus a feedback chain back upstream, from one stage to the next, and ultimately to the breeder who plans the breeding programme according to the requirements throughout the chain (Wrigley, 2016).

The sequence stops half way for food products consumed as the whole grain, as illustrated in the examples in Fig. 18.1; such examples for wheat are bulgur and couscous. The complete chain may extend over great distances and significant time periods, given the great extent of international trade. At another extreme, the whole sequence may occur in one village for a family on subsistence farming. This sequence is most detailed in the case of wheat, so that is the grain to be described in the remainder of this chapter.

Figure 18.3 The sequence of events from breeder to the consumer, who is the final arbiter of product quality at the culmination of the grain chain.

18.5 Wheat: a major cereal with unique attributes and diverse uses

The unique dough-forming properties of wheat gluten are the reason for its wide appeal as a grain (Faridi and Faubion, 1995). Only wheat is suitable for the production of leavened bread (Fig. 18.4) and the many other wheat-based products (Uthayakumaran and Wrigley, 2010). The need for cohesive dough properties is shown in the rolling out of dough for a pizza base (Fig. 18.5).

A further variety of wheat-based foods is illustrated by breakfast cereals (Fig. 18.6), the ice cream cone (Fig. 18.7), a fancy cake (Fig. 18.8) and instant noodles (Fig. 18.9). The papadum bread in Fig. 18.1A is used as a spoon to eat the dhal. Animal feed and industrial applications account for major uses of wheat. One of these industrial applications is the separation of gluten from starch. The gluten is dried and used to supplement low-protein flour in bread-making. Wheat starch is used in food manufacture, as well as for the production of bioethanol fuel (Guragain et al., 2016).

Figure 18.4 Loaves of leavened bread from wheat, two made from whole-meal flour and two from white flour, both of these flours require high-protein content. For either flour type, a strong extensible dough is needed to retain the gas bubbles produced by the action of yeast on the sugars released from digested starch. The flour was milled from a hard wheat so that a good degree of 'damaged starch' is available for digestion by amylase action and for good water absorption.

Figure 18.5 Dough extensibility is important for a pizza dough made from white flour from a hard wheat of moderate protein content; thus the dough can be rolled out in a thin circle that does not change shape during cooking.

Figure 18.6 Breakfast cereals take many shapes (and even many colours!). In this case, a dough of wheat flour has been extruded and expanded due to the sudden release of high pressure and temperature. The global market for breakfast cereals is over 30 billion US dollars.

All these products have very different requirements for combinations of grain hardness, dough properties and the particle-size distribution for the respective flours, having been milled from grain of specific varieties (strictly 'cultivars') of hexaploid wheat, *Triticum aestivum.* About 30 million tonnes of annual wheat production worldwide is *Triticum durum* (tetraploid) which is mainly used for pasta manufacture (Fig. 18.10).

The major cereal-producing countries (Table 18.3) are also major producers of wheat. The differences between these two columns of figures (cereals vs wheat) for each country are mainly due to the added production of maize and/or rice. Nearly half of the wheat production in the European Union is due to France (39 million tonnes) and Germany (25 million tonnes). Both countries achieve the highest wheat yields of 7–8 tonnes per hectare.

Figure 18.7 The ice cream cone is one of many forms of the wafer, which is made from a liquid batter of soft-wheat flour. The batter is heated quickly between a pair of heated cast-iron moulds. The finished wafer is then released as an aerated solid-starch foam.

Figure 18.8 Flour for cakes and cookies (biscuits) is milled from a low-protein soft wheat to provide low water absorption. The wheat type needs to provide weak dough strength and good extensibility. Grocery flour for home baking is often of a similar type, unless it is designed specifically for bread-making, such as in a domestic bread machine.

Figure 18.9 Noodles are made from common wheat (*T. aestivum*). These instant noodles come in a disposable container ready to eat after adding boiling water. They are made from hard wheat of medium protein content and medium dough strength.

Figure 18.10 Pasta, in their many shapes, are made from high-protein durum wheat, a species distinct from common wheat. The very hard durum grains mill into semolina which has a larger particle-size distribution than bread flour. The semolina's strong dough is extruded through a specially shaped orifice and is then dried. Pasta types, such as spaghetti, are thus distinct from noodles, despite some similarity in appearance.

Table 18.3 Major producers of cereal grains, ranked in order, are matched with their wheat production (millions of tonnes (metric tons)) and wheat yield (tonnes per hectare) in the year 2014

Country/region	Production of all cereal grains (million tonnes)	Wheat production (million tonnes)	Wheat yields (tonnes/hectare)
China	480	121	5.0
United States	436	58	3.2
European Union	304	143	5.5
India	240	93	3.1
Russia	88	52	2.2
Ukraine	60	22	3.4
Canada	59	33	3.2
Argentina	47	11	3.0
Australia	37	26	1.9
Pakistan	34	24	2.8
Turkey	32	18	2.3
WORLD	*2800*	*729*	*3.2*

Source: Data from FAO and US Department of Agriculture.

18.6 Wheat: from the breeder to the farm gate

The wheat breeder, at the top of the grain chain (Fig. 18.3), has a responsibility to all of the people downstream. The grain grower needs the potential for high grain yield, resistance to diseases and pests, plus good grain quality that will secure premium payments at harvest. Such premiums are at the discretion of grain buyers representing millers and bakers. All these considerations in the grain produced add up to the phenotype – the result of genotype (G) and growth environment (E). The breeder has control over the genotype – the complement of genes in a variety. The performance of this genetic potential is subject to the growth conditions such as climate and soil type. Thus the quality of harvested grain is the result of $G + E$ and the interaction of $G \times E$.

Over a century ago, genetic improvement of wheat involved the farmer selecting promising plants for propagation. The methodology of cross-breeding between two parent lines provides a wider range of genetic diversity from which to select improved plants. The selection of these parents is obviously critical; the hard work starts when the large number of resulting progeny must be screened to select the best combination of traits provided by this combination of parents.

Many of these desirable traits are evident only after several generations (involving years of re-sowing) when progeny lines have become homozygous and after enough grain is available for quality testing. In recent decades, great advances have been achieved in selection methodologies to identify the presence of specific genes using marker-assisted selection (Carver, 2009; DePauw and O'Brien, 2016).

Further advances have lifted the limitations of sourcing desirable genes, so that breeders are now able to source genetic resistance to a disease, for example, from beyond the target grain species (Privalle, 2016). The improved plant that carries genes from an unrelated genus is termed a 'genetically modified organism'. This technology has been used to good effect in research on wheat, but not in commercial practice, even though the practical advantages shown by the widespread production of other grain species are also possible in wheat.

Intermediate between the breeder and the grower is the pure-seed producer, who has the responsibility of seed multiplication for new varieties, whilst also ensuring that the seed is free from disease and from foreign seeds.

After sowing, comes farm management (M), a third factor to add to G × E (above). Whilst the grower has little control over climate (rainfall, temperatures, heat stress or frost), there are many opportunities to manage grain production. These include fertilizer use, irrigation, sowing time and rate, and the use of insecticides, fungicides and herbicides to control diseases and pests.

The interaction of G × E × M is meaningful, for example, in relation to biotic stresses such as plant diseases. The breeder may not have been able to introduce genes for resistance to a disease that is a greater risk in a specific environment, such as prolonged moist conditions. Without the resistance genes and without the moist environment, the grower may have no disease problem. If the moist condition does become evident, fungicide spraying may be necessary, thus invoking the 'M' of G × E × M in this example, which is one of many such management instances through to harvest.

A distinct approach to farm management is the trend towards organic growing of crops. Many consumers are prepared to pay much more for organic products despite the absence of credible evidence that organic grains differ nutritionally from conventionally grown grain. Organic grains look the same as grain that has been grown by conventional procedures. In general, detailed analyses fail to show any differences as a result of organic methods. It is therefore difficult for the consumer to be certain that the price difference is justified.

However, organic growers have been attempting to provide reassurance and to prevent fraud by organising themselves into cooperatives in many countries (Gelinas and David, 2016). Furthermore, international standards have been set up as Codex Alimentarius Commission Guidelines to regulate the production, processing, labelling and marketing of organically produced foods. Legally registered labelling is based on accredited certification agencies inspecting farms and processing industries. At least organic customers have the assurance that many aspects of organic farm practice are environmentally beneficial. Grain yields are often lower with organic growing. An added cost is the need to segregate this grain and ensure organic principles are continued right through the grain chain into processing (Gelinas and David, 2016).

18.7 Wheat: harvesting receival, storage and transport

Harvest is a critical stage for the farmer. Grain moisture in the standing grain must have decreased to a specified maximum for the grain to be accepted at the local grain elevator (silo) or mill. This threshold, generally about 12.5%–14%, is set to ensure the maintenance of grain quality during storage and transport. Thus, when the crop is sufficiently dry, it must be harvested quickly because of the risks that rain at harvest will cause incipient sprouting of the grain in the head and that the added moisture will encourage fungal infection in the grain. Partly sprouted grain has an elevated level of the starch-degrading enzyme amylase, which causes serious problems during baking and other forms of processing.

Delivery of the harvested grain is the stage at which grain quality must be determined to ensure the segregation of grain of one quality class and grade from another. The range of quality specifications is indicated in Table 18.4, contrasting two classes of Australian wheat – Prime Hard 1 and General Purpose 1. Specifications are naturally stricter for the higher-value class (APH1) in many respects, but not all. In both cases, the maximum moisture content is the same and varietal composition is specified for both (although involving different variety mixes). On the other hand, for the General Purpose class, specifications are more relaxed for test weight, protein content, sprouting damage (indicated by the Falling Number test) and various aspects of defective grains and contaminants. Although these specifications relate to Australian wheat, similar considerations apply in other wheat-growing countries, such as the United States (http://www.gipsa.usda.gov/) and Canada (http://www.grainscanada.gc.ca/). More detailed information is provided by McLean and Wrigley (2016).

It is clearly a difficult task to test for so many aspects of grain quality while the truck driver is waiting to be instructed where the load should be tipped. Analytical procedures at grain receival have been reviewed by Delwiche (2010). Sampling grain from the load (usually a truck) is the first step and this part is critical. It is usual for grain samples to be taken from several places in the grain load, to ensure representative sampling. This sample may then be divided into several lots (using a sample divider). One sub-sample is likely to be tested by near infrared spectroscopy for moisture and protein content. A moisture content above the specified maximum could mean rejection of the whole truckload. Grain protein content often determines the premium classification of a grain load, thus also determining the site for tipping of the grain load.

Another critical quality factor is the plumpness of the grains, determined as bulk density by weighing a specified volume (often one half litre). This quality attribute relates to potential milling quality and flour yield, as pinched grains are unlikely to provide a good yield of white flour. It is thus a quality specification that distinguishes the APH milling wheat in Table 18.4 from the general-purpose wheat.

Sieving of the sample reveals the extent of dockage, i.e. material that is not plump wheat grains. This material is given in Table 18.4 as being material above and below the screen. Such contaminants include husks, stones, sticks, snails and

Table 18.4 Grain receival standards for two contrasting grades of wheat

Grain-quality attribute	A high-quality class (APH1)	A lower-quality class (AGP1)
Moisture	<12.5%	<12.5%
Test weight	>76.0 kg/hL	>68.0 kg/hL
Protein (N × 5.70 at 11% moisture basis)	>14%	No minimum
Falling number	>350 s	>200 s
Sprouted grains	Nil tolerance	Nil tolerance
Varieties specified?	Yes	Yes
Unmillable material retained by a 2 mm screen	<0.6% by mass	<1.2% by mass
Screenings passing through a 2 mm screen	<5% by mass	<10% by mass
***Defective grains* by count in a 300-grain sample**		
Insect damaged	<1%	2%
Frosted grains	<1%	10%
Smuts	Nil	Nil
Pink-stained grains	<2%	<5%
Dry green, sappy grains	<1%	<5%
***Contaminants* in ½ L sample**		
Foreign seeds of 10 types	Nil to 50 seeds, depending on type	Nil to 150 seeds, depending on type
Small foreign seeds	<0.6% by mass	<1.2% by mass
Cereal ergot pieces	Maximum of 1	Maximum of 1
Stored-grain insects	Nil	Nil
Field insects	3–10 max. acc. to species	3–10 max. acc. to species
Snails, dead or alive	1	10
Objectionable matter (stone, sticks, glass, concrete)	Nil for the entire load	Nil for the entire load
Other foreign material (% by weight)	0.1 max.	0.1 max.

These specifications are incomplete; the actual details may vary from season to season and from region to region.
Source: For full specifications, go to http://www.graintrade.org.au/sites/default/files/.

foreign seeds. Special concerns at this stage are the presence of insect-eaten grains or insects relevant to stored grain (nil tolerance).

The presence of foreign seeds is more serious for some species than for others. The ranges of foreign-seed numbers in Table 18.4 depend on whether the seed species is relatively innocuous, such as another cereal grain, or whether the contaminating seed presents serious problems. The propagation of seeds of noxious weeds needs to be discouraged as they threaten cropping and grazing. Of significant concern is the presence of seeds that can impart a taint to the entire grain lot (such as Hexham scent, *Melilotus indicus*) or seeds that are actually toxic. Fragments (ergot bodies) of the fungus *Claviceps purpurea* are in this latter category, as ergot contains toxic alkaloids, which are injurious to cattle at a level as low as 0.05%. The presence of bunt (*Tilletia caries* or *Tilletia foetida*) imparts a pungent unpleasant odour. In this case, the infection replaces the endosperm of the developing grain spores of the stinking smut, which may then spread through the grain shipment.

According to Alldrick (2010): 'Cereal-based foods can be considered to present a lower risk to food safety than many other foods.' [Although there are hazards...] 'once identified, these hazards can be managed and the threat to the consumer minimised'. For example, the European Union's 'Rapid Alert System for Food and Feed' (www.moniqa.org) notified alerts for cereal-based foods were essentially similar in a 5-year period as in 1 year for cheese and poultry (Alldrick, 2010).

Of particular concern are the mycotoxins, which are poisonous metabolites produced by specific filamentous fungi, either during crop development or storage, especially in warm, wet conditions (Alldrick, 2010). Minimising the problem before harvest involves appropriate crop rotation and tillage to reduce the risks of infection. Post harvest, regulation of temperature and moisture during storage are the strategies.

In any specific region, the grain harvest occurs over a limited period so that the resulting grain needs to be stored to be processed during the remainder of the year. In regions of major production, grain must be transported – in many cases across vast distances involving rail and ocean transport. In these cases of storage and transport, grain value must be maintained by ensuring quality and identity preservation. Testing and segregation according to quality type at receival is costly, so the benefits must be preserved.

Many procedures have been devised to avoid predation by insects, microorganisms, rodents and even birds. Temperature and humidity must be maintained at reasonably low levels to discourage insect and microorganism attack. Sealed storages offer the added advantage of permitting chemical fumigation (Annis, 2016b; Fleurat-Lessard, 2016). On the other hand, aeration of stored grain is used to good effect to reduce the temperature and moisture of stored grain (Navarro et al., 2012; Annis, 2016a). The use of carbon dioxide offers the advantage that it is suitable for organically grown grain, but the level of CO_2 must be maintained for over a week (Annis, 2016a, b).

18.8 Wheat: the grain buyer, the miller, the baker and other processors

The marketing of grain is complex, involving direct negotiation between grower and miller in many situations, government intervention in other cases and private grain companies to a large extent. The grower who has on-farm storage is theoretically in a good position to enter into extensive negotiations both long before harvest and afterwards, but grain prices can be volatile.

The grain buyer seeks quality types that match the needs (and financial constraints) of the processor, who is next down the grain chain (Fig. 18.3) – especially flour millers, and also feed formulators and industry processors. In the case of the feed miller, the grains compete with a wide range of energy sources to provide the nutritional needs of the target animals. The appropriate mix of these sources is likely to change from week to week, depending on price fluctuations. A pet-food manufacturer, for example, may have a dozen or more formulations for each target animal, switching between these ingredient mixes according to least-cost programmes.

The separation of starch and gluten is an important industrial use of wheat. There are processes for doing so based on wet milling of the whole grain, but more often the grain is milled to flour before being formed into dough, from which the starch is water-washed. Major uses of the starch include the manufacture of paper, adhesives and sweeteners after hydrolysis to sugars (Lawton, 2016). The co-product, gluten, is dried and used in various food and non-food products, most prominently for improving the dough properties of low-protein flour for bread-making.

The task of the flour miller is to provide the food processor with flour (white or whole meal) that suits the next stage of the production chain (Fig. 18.3). The major parts of the wheat grain are the endosperm, the bran and the germ (Fig. 18.11) (Fardet, 2014). White flour is derived from the endosperm, which accounts for about 80% of the grain mass. The miller must remove as much as possible of the endosperm from the bran and germ, as well as comminuting the endosperm to the specified particle size (Sarkar and Dexter, 2016).

All parts of the wheat grain are rich in a wide diversity of nutritional compounds, as indicated in Fig. 18.11. Despite some contrary publicity, the endosperm (white flour) is a good source of protein, complex carbohydrate, fibre and minor nutrients, but an even wider range of nutrients is available in the other tissues of the grain. Nevertheless, the greatest value for the miller is in the white flour, with much of the bran and germ being relegated as 'offal' for animal feed. This anomaly is largely because the bran and germ fractions have rather negative effects on baking performance compared to white flour. Furthermore, if 'grain protectants' (pesticides) have been used, they will be concentrated in the bran layers.

The miller supplies flour to a variety of customers making a range of products, each with its specific quality requirements. Examples of these specifications may be very exacting, as indicated by the examples (Table 18.5) for flour-protein content and dough properties. To meet these specifications, the miller has the options of blending the wheat lots available and also blending the resulting flour streams, thus to attempt to satisfy the specifications.

The two measures of dough properties in Table 18.5 are explained in Fig. 18.12.

Dough strength can be measured in the Farinograph by mixing the test flour with water (AACC International Method No. 115/1 at http://www.aaccnet.org/

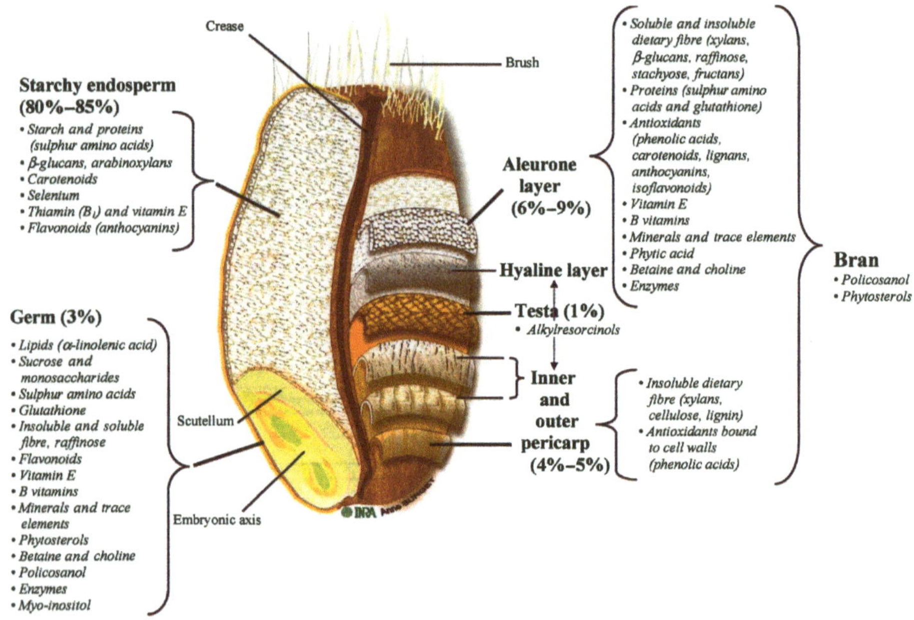

Figure 18.11 The wheat grain, showing the nutritional compounds available in its many anatomical parts.
Source: Reproduced with permission from Fardet, A., 2014. New approaches to studying the potential health benefits of cereals: from reductionism to holism. Cereal Foods World 59, 224–229.

Table 18.5 Examples of flour specifications required by bakery customers for supply by a miller

Product	Customer (fictional)	Protein content (%)	Dough properties	
			Strength, Rmax (BU)	Extensibility (cm)
Sliced bread	Plant bakery #1	11.6–13	250–600	18–24
Sliced bread	Plant bakery #2	>11	350–400	18.5–23
Sliced bread	Plant bakery #3	>11	330–380	>20
Hearth bread	Fred's breads	13–14	350–500	>19
Buns	Fast food	13–14	>350	>20
Pizza	Kwik	11–12	250–500	18–23
Noodles	Oodles	>11	>250	>18
Cookies	The Best	7–8	100–250	>15
Crackers	Savoury snacks	11–12	310–390	Not specified
Starch-gluten	Arch starch	>10.5	250–400	17–25

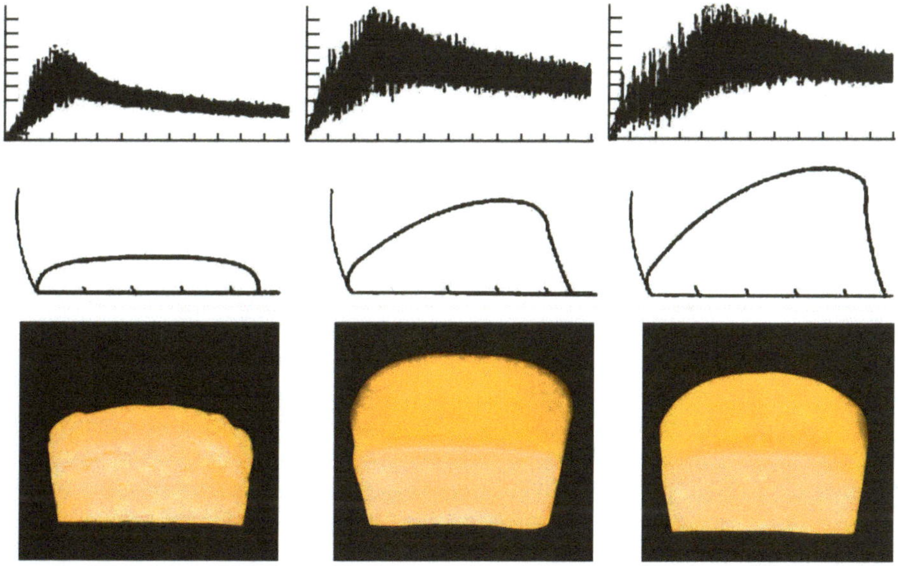

Figure 18.12 Quality characteristics for three flour samples and the resulting bread quality. Poor bread quality results from a weak dough (left), whereas good bread is the result of strong extensible dough (right), based on dough testing with the Farinograph (top row) and the Extensograph (middle row).
Source: Reproduced with permission from Wrigley, C.W., Bekes, F., and Bushuk, W. (Eds.) (2006). Gliadin and Glutenin: The Unique Balance of Wheat Quality. AACC International, St Paul, MN.

Pages/default.aspx). The three traces at the top of Fig. 18.12 compare Farinograms for the doughs resulting from three flour samples – weak, medium and strong. Initially in each trace, there is little resistance to mixing (vertical axis). Then the flour starts to hydrate and resistance increases with time (horizontal axis). A peak of resistance is reached – early for the weak dough and later for the strong dough which also has a higher peak of resistance. The 'breakdown' in resistance after the peak is more rapid for the weak dough than for the strong one.

Traces in the middle row are produced by the Extensograph in which dough from the mixer is stretched. Resistance to extension and time are depicted in the vertical and horizontal axes, respectively. The maximum resistance (Rmax) is reached just as the stretched dough piece breaks. Resistance is measured in arbitrary 'Brabender Units'. Extensibility is the length, in centimetres, of the full trace. See AACC International Method No. 114/1 at http://www.aaccnet.org/Pages/default.aspx.

Flour specifications in Table 18.5 can be compared to the requirements for the several wheat-based foods described in the legends of Figs 18.4–18.10. Bread and buns require flour of higher protein content, giving a stronger dough, demonstrated by the row of test loaves at the bottom of Fig. 18.12. In contrast, cookie and cake manufacture calls for lower protein content and weaker dough (Table 18.5; Fig. 18.8). On the other hand, flour for cracker manufacture should have a moderate protein content and good dough strength.

Additional aspects of flour quality include the level of starch damage, required for bread-making. See legend to Fig. 18.4. The degree of starch damage is greater for hard wheats and is increased by hard grinding with the smooth rolls.

There are many ways in which the skill of the miller is shown; these can be expressed by 'P' for processing, which should be added to the earlier mentioned factors that determine the quality of the final products, namely, G × E × M × P (genotype × growth environment × farm management × processing). It may be possible for deficiencies at an earlier stage to be compensated at processing by intelligent modification of the normal milling process, but the need for such changes in milling is not desirable. On the contrary, the miller requests 'no surprises!' from the grain trader. The miller and baker have the expectation that the grain and flour provided, respectively, will be of uniform quality, according to specification, so that no changes are required to routine operations.

'P', for processing variables, continues on into baking and into the many other functions en route to the final consumer products. The baker, for example, can vary the presence and proportions of the many ingredients and 'processing aids' available to use in achieving the desired product quality. In addition, dough-mixing time, proofing and baking times are some of the other processing variables. If the baker's flour specifications are not met by the miller, the blame continues up the grain chain, with blame to the grain segregation system, through the grower and eventually to the wheat breeder. The breeder, in turn, takes note of the feedback in planning the future of the breeding programme, but the implementation is slow because of the long lead time to produce a new variety.

18.9 Wheat: the biochemistry and genetics of grain quality

The long process of selecting for appropriate grain quality in wheat breeding is currently becoming shorter because of advances in breeding technology and marker-assisted selection, combined with a better understanding of the chemistry of wheat quality. The gluten-forming storage-protein complex is made up of about equal parts of gliadin and glutenin (Fig. 18.13). The gliadin fraction consists of many proteins, all of modest size (less than 100,000 Da), with intra-chain disulphide bonds. The genes for their synthesis are located at the *Gli-1* loci on the short ends of Chromosomes 1A, 1B and 1D and at the *Gli-2* loci on the short ends of Chromosomes 6A, 6B and 6D (Fig. 18.13).

In contrast, the glutenin fraction is made up of a spectrum of proteins with sizes ranging up into the tens of millions of Daltons. These very large proteins are disulphide cross-linked polypeptides (subunits) with varying degrees of polymerisation. Greater dough strength is associated with a larger molecular-weight size distribution for the glutenin polymers, and also with the involvement of specific subunits in these polymers. The subunits of glutenin are coded by the *Glu-1* genes (for the high-molecular-weight subunits) and the *Glu-3* genes (for the low-molecular-weight subunits) on Chromosomes 1A, 1B and 1D (Fig. 18.13).

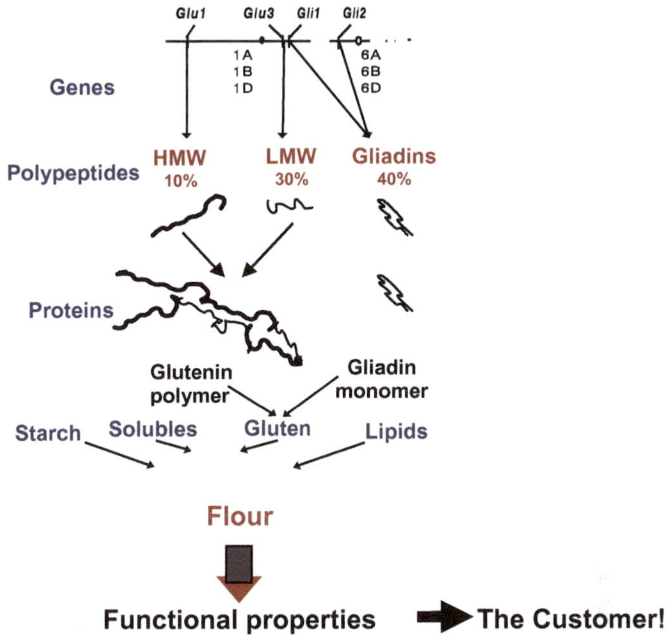

Figure 18.13 The genes, polypeptides and proteins that contribute to the functional properties of dough and wheat quality.

It is now possible to make predictions of the dough strength (as Rmax) and extensibility (Fig. 18.12) of wheat genotypes (at normal protein content) based on the composition of the alleles at the *Glu-1* and *Glu-3* loci (Békés, 2012; Juhasz et al., 2014). A catalogue of gluten alleles for 8500 wheat varieties and breeding lines is available (Bekes and Wrigley, 2013) at www.aaccnet.org/initiatives/definitions/Pages/Gluten.aspx. Estimates of Rmax and Ext are provided for the many genotypes for which there are *Glu-1* and *Glu-3* alleles.

18.10 Cereal grains: the final member of the grain chain – the consumer

Ultimately, the consumer has the final 'say' about the quality of grain-based foods, and this 'say' echoes right back up the supply chain. The consumer's 'say' is buying power. It determines the commercial success or otherwise of one baker against another or of one brand of breakfast cereal versus others. The aspects of quality, as seen by the consumer, include appearance, taste, safety, nutrition as well as perceptions of possible allergic reactions and intolerances to some foods and ingredients.

Nutritional guidelines are published by the government agencies in many countries (e.g. Fig. 18.2). A common recommendation is the consumption of grain-based foods as a major part of food intake. This emphasis on grains in the diet provides major opportunities for the industries that produce and process food grains. Nevertheless, consumers too often ignore such 'official' advice, preferring to rely instead on concepts and popular fads from friends, social media and advertising (Sheats and Jones, 2016).

An example is the trend towards 'gluten-free' foods. In 2010, global sales of gluten-free foods amounted to US$2.5 billion (Sapone et al., 2012). Lifelong gluten avoidance is necessary for people diagnosed with coeliac disease. In this case, 'gluten' includes the storage protein of wheat, rye, triticale, barley and possibly oats. Beyond coeliac disease, there are other conditions for which gluten avoidance is warranted, including gluten ataxia, dermatitis herpetiformis, more general intolerances to gluten and IgE allergic reactions to ingested wheat (Sapone et al., 2012). However, 'going gluten-free' has become a trend that may not be medically justified in many cases.

There are many more cases relating to eating trends, nutrition and dieting where consumers have been duped by the media, advertisements, popular booklets and websites with hidden agendas. Nevertheless, these consumer trends and fads are real, in terms of which products are selling from retail shelves, so that processors and retailers must be aware of such trends and act accordingly.

18.11 The future for grains: still the major source of our food, feed, fuel and fibre

The grain-producing and -processing industries must continue to keep pace with genuine nutritional trends, as well as with other less genuine fads, as all these factors affect the types of food product that sell well. Customer-based pressures will continue to be reasons to supply specific requirements. One of these trends is the demand for foods based on organically grown grains. This is a trend (some would say a fad) that the industry has accepted and capitalised upon. Whereas customers are prepared to pay higher prices, there are considerable costs involved in the whole process of organic growing and processing (Gelinas and David, 2016).

On the other hand, genetically modified (GM) grains offer considerable advantages to the grower and also to the environment in many cases. Corn is the main cereal grain that is GM. Like any special crop, GM grain must be identity-preserved throughout the grain chain; in this case, there are well characterised tests by which to identify GM grain. The future may well see the extension of GM crops to the benefit of growers and consumers with improvements in processing and nutrition (Privalle, 2016).

Increasing affluence for privileged people in developing countries is affecting their food choices with respect to grain-based foods, due to perceived increases in social level for wheat-based foods over rice, maize or sorghum. Again the result of such attitudes provides regional opportunities for growers, processors and retailers.

Each stage of the supply chain (Fig. 18.2) can be seen as an interaction between supplier and customer, with the possibility of disagreement, such as the baker blaming the miller for failing to meet quality specifications, or the grower blaming the breeder for lack of disease resistance. At each stage, the 'customer' (farmer, miller, food processor) has to decide between paying more for raw material that is reliable in quality and supply versus managing with poorer quality material, at a lower cost, while using skill to continue in delivering good product for the next 'customer' down the chain. Thus, the grower debates between seed sources and ability to achieve yield and quality grain by more effective farm management. Thus the grain chain can be summarised by the factors described in this chapter; namely the combined factors that determine the quality of the final products – G × E × M × P (genotype × growth environment × farm management × processing).

18.12 Further information and advice

Scientific societies relevant to the grain sciences are valuable sources of information, advice and support for researchers and industry (Bridges and Wrigley, 2016). For example, AACC International (formerly the American Association of Cereal Chemists), based in St. Paul, MN, United States, has a worldwide membership of scientists involved in grain production and the grain-based food industry. So too does the International Association for Cereal Science and Technology (ICC), based in Vienna, Austria. The members of both organisations cooperate to develop test methods relevant to the situations throughout the grain chain (Fig. 18.3). So too do many other similar organisations, such as those listed below. See details of these and others described by Bridges and Wrigley (2016).

- AACC International, http://www.aaccnet.org
- International Association for Cereal Science and Technology, http://www.icc.or.at
- Standard Methods of the American Oil Chemists' Society, http://www.aocs.org
- International Seed Testing Association, http://www.seedtest.org
- American Society of Agronomy, http://www.agronomy.org
- Grain and Feed Trade Association, http://www.gafta.com
- American Society of Brewing Chemists, http://asbcnet.org

These organisations also produce books that review topics about specific grain species and their uses, such as for barley (Shewry and Ullrich, 2014), corn (White and Johnson, 2003), rice (Champagne, 2004) and wheat (Khan and Shewry, 2009). The full range of grain species at all stages of the grain chain is covered by articles in the Encyclopedia of Food Grains (Wrigley et al., 2016). Some specific articles are cited in the chapter as they expand on topics mentioned.

References

Alldrick, A.J., 2010. Food safety aspects of grain and cereal product quality. In: Wrigley, C. W., Batey, I.L. (Eds.), Cereal Grains: Assessing and Managing Quality. Woodhead Publishing Ltd, Oxford, UK, pp. 342–366.

Annis, P.C., 2016a. Stored grain: invertebrate pests. In: Wrigley, C., Corke, H., Seetharaman, K., Faubion, J. (Eds.), Encyclopedia of Food Grains, Vol. 4. Elsevier Ltd, Oxford, UK, pp. 110–116.

Annis, P.C., 2016b. Chemicals for grain production and protection. In: Wrigley, C., Corke, H., Seetharaman, K., Faubion, J. (Eds.), Encyclopedia of Food Grains, Vol. 4. Elsevier Ltd, Oxford, UK, pp. 99–104.

Arendt, E.K., Zannini, E. (Eds.), 2013. Cereal Grains for the Food and Beverage Industries. Woodhead Publishing Ltd, Oxford, UK.

Békés, F., 2012. New aspects in quality related wheat research: II. New methodologies for better quality wheat. Cereal Res. Commun. 40, 307.

Bekes, F., Wrigley, C.W., 2013. Gluten alleles and predicted dough-quality for wheat varieties world-wide: a great resource – free on the AACC International website. Cereal Foods World 58 (6), 325–328.

Bridges, A.R., Wrigley, C.W., 2016. Standardized test methods for grains and grain-based products. In: Wrigley, C., Corke, H., Seetharaman, K., Faubion, J. (Eds.), Encyclopedia of Food Grains, Vol. 2. Elsevier Ltd, Oxford, UK, pp. 300–307.

Carver, B. (Ed.), 2009. Wheat: Science and Trade. Wiley-Blackwell Publishing, Ames, IA.

Champagne, E.T. (Ed.), 2004. Rice: Chemistry and Technology, third ed. American Association of Cereal Chemists, St Paul, MN.

Delwiche, S.R., 2010. Analysis of grain quality at receival. In: Wrigley, C.W., Batey, I.L. (Eds.), Cereal Grains: Assessing and Managing Quality. Woodhead Publishing Ltd, Oxford, UK, pp. 267–310.

DePauw, R., O'Brien, L., 2016. Wheat breeding: exploiting and fixing genetic variation by selection and evaluation. In: Wrigley, C., Corke, H., Seetharaman, K., Faubion, J. (Eds.), Encyclopedia of Food Grains, Vol. 4. Elsevier Ltd, Oxford, UK, pp. 279–286.

Fardet, A., 2014. New approaches to studying the potential health benefits of cereals: from reductionism to holism. Cereal Foods World 59, 224–229.

Faridi, H., Faubion, J., 1995. Wheat End Uses Around the World. American Association of Cereal Chemists, St. Paul, MN.

Fletcher, R., 2016. Pseudocereals: overview. In: Wrigley, C., Corke, H., Seetharaman, K., Faubion, J. (Eds.), Encyclopedia of Food Grains, Vol. 1. Elsevier Ltd, Oxford, UK, pp. 274–279.

Fleurat-Lessard, F., 2016. Stored-grain pest management. In: Wrigley, C., Corke, H., Seetharaman, K., Faubion, J. (Eds.), Encyclopedia of Food Grains, Vol. 4. Elsevier Ltd, Oxford, UK, pp. 126–139.

Gelinas, P., David, C., 2016. Organic grain production and food processing. In: Wrigley, C., Corke, H., Seetharaman, K., Faubion, J. (Eds.), Encyclopedia of Food Grains, Vol. 4. Elsevier Ltd, Oxford, UK, pp. 154–161.

Gooding, M.J., 2009. The wheat crop, Wheat Chemistry and Technology, fourth ed. AACC International, St Paul, MN, pp. 19–49.

Guragain, Y.N., Probst, K.V., Vladlani, P., 2016. Fuel alcohol production. In: Wrigley, C., Corke, H., Seetharaman, K., Faubion, J. (Eds.), Encyclopedia of Food Grains, Vol. 4. Elsevier Ltd, Oxford, UK, pp. 235–244.

Juhasz, A., Bekes, F., Wrigley, C.W., 2014. Wheat proteins. In: Ustunol, Z. (Ed.), Applied Food Protein Chemistry. Wiley-Blackwell. John Wiley & Sons, West Sussex, UK, pp. 219–304. ISBN 978-1-119-94449-2.

Khan, K., Shewry, P.R. (Eds.), 2009. Wheat: Chemistry and Technology, fourth ed. AACC International, St. Paul, MN.

Lawton, J.W., 2016. Use of native starch. In: Wrigley, C., Corke, H., Seetharaman, K., Faubion, J. (Eds.), Encyclopedia of Food Grains, Vol. 3. Elsevier Ltd, Oxford, UK.

McLean, J., Wrigley, C.W., 2016. Contaminants that down-grade the value of grain deliveries. In: Wrigley, C., Corke, H., Seetharaman, K., Faubion, J. (Eds.), Encyclopedia of Food Grains, Vol. 4. Elsevier Ltd, Oxford, UK, pp. 83–98.

Navarro, S., Noyes, R.T., Casada, M., Arthur, F.H., 2012. Grain aeration. In: Hagstrum, D. W., Phillips, T.W., Cuperus, G. (Eds.), Stored Product Protection. KSU Publisher, Manhattan, KS, pp. 121–134.

Orthoefer, F.T., Eastman, J., 2004. Corn processing and products. Corn: Origin, History, Technology, and Production. John Wiley & Sons, Hoboken, NJ, pp. 867–896.

Privalle, L., 2016. The development of genetically modified grains. In: Wrigley, C., Corke, H., Seetharaman, K., Faubion, J. (Eds.), Encyclopedia of Food Grains, Vol. 4. Elsevier Ltd, Oxford, UK, pp. 343–349.

Sapone, A., Bai, J.C., Ciacci, C., Dolinsek, J., Green, P.H.R., Hadjivassiliou, M., et al., 2012. Spectrum of gluten-related disorders: consensus on new nomenclature and classification. BMC Med. 10, 13. Available from: https://doi.org/10.1186/1741-7015-10-13.

Sarkar, A.K., Dexter, J., 2016. Wheat: dry milling. In: Wrigley, C., Corke, H., Seetharaman, K., Faubion, J. (Eds.), Encyclopedia of Food Grains, Vol. 3. Elsevier Ltd, Oxford, UK, pp. 307–319.

Sheats, D., Jones, J.M., 2016. Grains and health: misinformation and misconceptions. In: Wrigley, C., Corke, H., Seetharaman, K., Faubion, J. (Eds.), Encyclopedia of Food Grains, Vol. 2. Elsevier Ltd, Oxford, UK, pp. 23–28.

Shewry, P.R., Ullrich, S.E., 2014. Barley: Chemistry and Technology, second ed. AACC International, St. Paul, MN.

Uthayakumaran, S., Wrigley, C.W., 2010. Wheat: characteristics and quality requirements. In: Wrigley, C.W., Batey, I.L. (Eds.), Cereal Grains: Assessing and Managing Quality. Woodhead Publishing, Cambridge, UK, pp. 59–111.

White, P.J., Johnson, L.A., 2003. Corn: Chemistry and Technology. AACC International, St. Paul, MN.

Wrigley, C., Corke, H., Seetharaman, K., Faubion, J. (Eds.), 2016. Encyclopedia of Food Grains. Elsevier Ltd, Oxford, UK.

Wrigley, C.W., 2016. The grain chain – the route from genes to grain-based products. In: Wrigley, C., Corke, H., Seetharaman, K., Faubion, J. (Eds.), Encyclopedia of Food Grains, Vol. 1. Elsevier Ltd, Oxford, UK, pp. 22–30.

Wrigley, C.W., Batey, I.L. (Eds.), 2010. Cereal Grains: Assessing and Managing Quality. Woodhead Publishing, Cambridge, UK.

Wrigley, C.W., Bekes, F., Bushuk, W. (Eds.), 2006. Gliadin and Glutenin: The Unique Balance of Wheat Quality. AACC International, St. Paul, MN.

Further reading

Bonjean, A.P., Angus, W.J. (Eds.), 2001. The World Wheat Book: A History of Wheat Breeding. Lavoisier, Paris.

Champagne, E.T., Wood, D.F., Juliano, B.O., Bechtel, D.B., 2004. The rice grain and its gross composition. In: Champagne, E.T. (Ed.), Rice: Chemistry and Technology, third ed. American Association of Cereal Chemists, St Paul, MN, pp. 77–108.

Baked products

19

Colin Jeffery[†]*, Luanne Gounder and Lindy Jeffery*
Johannesburg, South Africa

19.1 Introduction

Intrigue !! Government intervention !! Monopolizing !! Price fixing !! Unionization !!

This is the South African Baking Industry.

From its inception in 1687 to the present day this industry has struggled to survive. But survive it did, despite many obstacles.

The following presentations will endeavour to unbundle this fantastic industry from the perspective of two people who are and have been actively involved in the day-to-day running of large bakeries and with a combined experience of well over 40 years both have a unique perspective of the history and challenges that have been and still have to be faced.

There will be many similarities to other baking industries in the world as we all face the same challenges.

19.2 The baking industry in South Africa

19.2.1 *Where we came from*

To the original people of South Africa, bread and any form of baking was a completely foreign entity. They had planted maize (corn) and millet and other cereals as their main source of carbohydrates, as they still do today, as a complement to a protein source.

The Settlers (who arrived from Holland and France in the mid-1600s and subsequently colonised the southern part of the country which eventually would be called the CAPE, and then the British who settled in the eastern part of the country which would be known as NATAL) brought with them the skills and it is well documented that they also brought with them the first wheat cultivar. It was found that the climate in the Cape was well suited for wheat production.

Up until about 1879 most baking was for home consumption and probably produced as follows:

The wild wheat berry would have been crushed between two round stones – the *now 100% flour* would have been mixed with water, salt and a form of wild yeast, to achieve 'fermentation'. Bakers very quickly learnt that a 'sour dough', that is,

[†] Deceased.

Swainson's Handbook of Technical and Quality Management for the Food Manufacturing Sector.
DOI: https://doi.org/10.1016/B978-1-78242-275-4.00019-8

dough that has fermented for a number of hours, could be used as a source of yeast and when added to a fresh batch of dough would replicate the fermentation process. It was mixed by hand, left to stand overnight and baked in a home built oven. It should be noted that probably only the four compulsory ingredients were used – no fats/sugars/preservatives/enzymes/oxidising agents, etc.!!

But there's a paradox here – some 135 years later bread made using more or less the same method is in great demand and consumers will actively source this type of bread and are prepared to pay a premium to be able to enjoy it. This type of reinvention has helped to spell the demise of the *plant bakery* which by definition is a factory semi- or completely automated, consisting of all the necessary machinery, that is, mixers, dividers, moulders, provers, ovens and packaging machinery, required to produce bread in large quantities, and has been the backbone of the South African industry during the past 100 years.

With the discovery of diamonds in 1868 and gold in 1884 there was an influx of settlers from Europe. Consequently, the baking industry started to gain momentum.

The demand for bread had always been linked to the availability and cost of maize (corn) but slowly the demand for bread increased. It very quickly became apparent that bakeries had to produce more, and faster. Before 1938 the industry was completely unregulated and it became apparent that lower income groups were unable to afford the ever-increasing cost of bread. When maize was unavailable due to scarcity or cost, the lower income groups who made up a large part of the population acted in the only way they could – by protesting.

In 1936 the then government introduced strict measures to control the industry. By 1941 all bakeries had to be registered. Price control was introduced and bakeries were paid a subsidy. The main motivation for this was to ensure that all citizens would have access to a reasonable quality and affordable loaf of bread. However, by the 1990s the South African bread industry was controlled by a handful of companies.

This was about to change and the changes would have a significant effect.

19.2.2 Where we are now

New government – prior to 1980 trade unions only represented the white workers and it was illegal for non-white workers to form or belong to a trade union. This all changed in 1996 when the new constitution for South Africa was put into law. Collective bargaining and the right to strike was now the right of every South African regardless of colour as long as he/she belonged to a recognised trade union. This resulted in a completely decontrolled industry and what would become very clear was that the consumer would require a different type of product from the baking industry.

The plant bakeries could produce vast quantities of products and because of the onset of unions they made them bigger and more automated. No longer would the workers work for a minimum wage, so one way for bakeries to address this was to automate and reduce the amount of workers required. Bakeries often had the backing of their parent companies (milling), so investment in advanced processing equipment was rarely a problem. But the one thing a plant bakery could not give was *freshness*. The very nature of a plant bakery necessitated that the bread would be at least 12 hrs old before it was available to the consumer – in order to understand this, it is necessary to know how a plant bakery works. Generally a plant bakery would supply bread to retail customers only and members of the public would not be allowed to purchase bread directly from them. In South Africa plant bakeries can have a capacity of 4,000–12,000 loaves per hour depending on the size of the bakery. It can take approximately 4 hrs from mixing to slicing of the bread. Each plant bakery has a fleet of delivery vehicles to deliver locally and to other towns.

Hence the statement that plant bakeries cannot compete with 'in-store bakeries' as far as freshness is concerned. Consequently in-store bakeries and artisan bread shops have become extremely popular, and this is what the consumer wanted. This one fact alone reduced the number of plant bakeries (in the space of 20 years) by a third.

Government started to institute measures that ensured that the consumer was aware of which ingredients were present in the form of labelling and the industry adopted recognised food safety programmes such as hazard analysis and critical control points (HACCP) as well as worker safety programmes. Quality control became a major factor.

All of these innovations required the one criteria that the baking industry had neglected for far too long, that being the development of skills. Most countries in the developed world have some form of institutionalised training for bakers and confectioners which would consist of practical and academic training. South Africa to this day has very little. There are basic training courses available but set at entry level only.

For example:

> About 2 years ago I had reason to visit a large 4000 loaves per hour bread plant in Africa – they had two high-speed mixers using the Chorleywood bread process and I started to chat to the operator. He had worked for this company for over 20 years. When I asked him to explain his duties, he very proudly told me that he presses a button. This pressing of the button tilts the mixing bowl which allowed him to add the yeast and premix. He then presses another button which rights the bowl, and yet again presses another button which allows the flour and water to go in and finally starts the mixing process by pressing yet another button. I asked him if he knew how the yeast works and if he knew what ingredients were in the premix and how they work. He replied that he did not know but he thought the yeast made the bread rise and that the premix was for flavour. Good guess work on his part – but shouldn't he know far more? Who is at fault here, the operator or the management? This is an all too familiar scenario in the South African baking sector.

The baking industry in South Africa is currently at a crossroads – we need to find a balance between the ever-insatiable need for profit and concentrate on the greater good of our fellow South Africans who wish to enter this industry but because of the lack of professional training facilities and therefore lack of qualifications, find it difficult to do.

19.2.3 Future challenges

19.2.3.1 Reduction of sodium levels

The South African Government, in keeping with world trends, has asked the baking industry to significantly reduce the amount of sodium chloride (salt) levels in baked products. It is well documented that heavy salt usage is detrimental to one's health and can cause elevated blood pressure which could lead to heart disease. The recommended intake of salt at time of writing is 5 g per adult per day. To understand the effect this legislation will have on the industry one must understand the critical role sodium chloride (salt) plays in the baking process.

Salt is one of the big four baking ingredients and without it, or a substitute, it is not possible to produce a bread-type product as we know it, as the appearance and taste of these products would be drastically changed.

Salt imparts a flavour and without it bread would be bland and tasteless and would almost be impossible to produce because of the following reasons:

- Salt has a major impact and control on fermentation.
- Salt has a toughening effect on the flour proteins.

Bakers have learnt the correct levels of salt and yeast to achieve a balance between flavour and proving times.

Potassium chloride is probably the alternative choice to salt although there are others. However, there are concerns regarding changes in process control and the flavour profile and side effects of some of these alternatives and of course the **COST** of these alternatives. South Africans are generally extremely price sensitive especially the lower income groups and any significant negative change in such a basic and common food item could have major consequences. It is also documented that potassium chloride can leave a bitter aftertaste.

19.2.3.2 Industry image

Generally the South African baking industry is not a profession of choice for new job seekers. It lacks the glamour that other industries have been able to achieve. Low wages and difficult working hours make it hard to attract good quality future recruits. Industry blames the government and government blames industry for these circumstances.

The industry needs to attract committed and able recruits to replace the relatively low number of qualified people that are currently running South African baking operations. To achieve this, the sector needs to be made more attractive via improvements in pay and working conditions.

19.2.3.3 Raw material price fluctuations

In order to maintain the status of products supplied to the market, all manufacturers have to be able to allow for constant raw material price fluctuations. Price increases may occur every quarter or yearly depending on suppliers. These price rises can often range from a 4% to a 12% increase per annum depending on supply and demand, import costs and availability. For businesses to remain profitable, constant reformulation of recipes is one of the biggest factors in the baking industry to allow manufacturers to supply the customer with a good quality product and still secure a sufficient mark-up in order to remain profitable and stay in business.

Such steps can be achieved via routinely searching for suppliers with the lowest costs, striving for additional process efficiencies/savings, or perhaps sometimes via reducing the amounts or quality of some of the ingredients used, where such steps are not detrimental to the overall eating quality and presentation of the product.

19.2.3.4 Product quality consistency

There will always be new businesses striving to make their mark in the market. However, companies able to supply consistent quality will have a stronger chance of retaining the market share.

The following needs to be considered:

- Ensure that the product is manufactured in accordance with product specifications specified by the consumer.
- Ensure that product quality is inspected by quality control before being dispatched to the consumer.
- Ensure service delivery regarding transports and sales in accordance to customer satisfaction.
- Ensure continuous product development and recipe improvements are in line with changes in the market.

19.2.3.5 Innovation

The ability to innovate is an ever-changing mindset of what the market would like to consume. Market research is a key function in this industry and manufacturers need to be versatile in order to be able to adapt their processes constantly to stay on par with the market requirements. Bakeries benefit from remembering that it is rarely what the manufacturer would like, but is what is required by the consumer that will secure additional future sales.

19.2.3.6 Food safety

The application of food safety systems has become far more recognisable in recent years. Consumers have become more aware of the potential food safety risks associated with the baking industry thus further enforcing manufacturers to comply with increasing government regulations. This will be explained more in Section 19.4.

19.3 What the South African consumer expects

19.3.1 The diversity of different income groups and buying power

Prior to 1994 there was a distinct gap in the 'haves' and 'have nots'. Products that targeted the very low income groups tended to be of a lower quality and cheaper to produce. There has been a huge demand for such low-cost products and a whole industry developed to supply this sector of the market. The key factor being 'cost' and the ability of the product to pacify hunger. For such products quality was usually considered a low priority.

After 1994, post-apartheid, a new consumer pattern started to emerge. A new and aspirational middle class developed. This middle-class group has continued to grow and prosper to this day and their expectations are high. Bakeries have always targeted specific income groups and as a result a wider diversity of good quality baked products have become available since 1994.

19.3.2 Rainbow products for a rainbow nation

It is conceivable that a consumer could be sitting in a pavement cafe in Paris admiring the Eiffel tower, enjoying a latte coffee and a fresh Brioche and only 12 hrs later the same consumer could be in downtown Johannesburg looking at the Brixton tower and wishing to enjoy the same products. The latter would not be a problem; however, the request for a *Brioche* would be likely to draw a blank stare from the waiter.

However, South Africa does have a number of products that are unique to the country, for example: koeksusters are a type of plaited doughnut soaked in a sweet syrup. Chakalaka flavoured breadrolls – chakalaka is a spicy relish, and of course, Biltong flavoured products. *Biltong* is similar to American *Jerky* and is considered a national treasure.

Pudding cake is a uniquely South African product and is produced as follows:

Step 1:

All rejected/damaged/burnt/under baked/returned baked products (with the exception of meat products!!) are collected and added to a mixing bowl; apricot jam and caramel colour are added; mixed for 15 min at medium speed.

The mixing of reject/waste products in the manufacture of this product makes any assurance of allergen control very difficult indeed.

Step 2:

Line a baking tray with sweet pastry; fill it with the above; add a layer of pastry; egg wash and sprinkle with white sugar.

Step 3:

Bake at 190°C for 20–25 min; allow to cool and then cut into squares. This process creates the famous South African pudding cake. Due to their nature, such products are extremely profitable, albeit not altogether healthy – and most likely would never be seen in a French Patisserie but it is truly a rainbow product for a rainbow nation!!

19.3.3 Future trends and expectations

The baking industry in South Africa is relatively small compared to other developed nations. Bakeries have learnt to diversify in order to survive. A stand-alone retail bakery in a mall is unlikely to survive even if the raw materials and skilled staff produce the finest of products. However, if that same retail bakery combines baked goods with a coffee bar and offers a limited menu, it can be very successful. The very nature of such turnkey operations will mean that the businesses are unlikely to be able to produce all products under one roof.

To service this sector industrial factories have grown to supply a range of products in a frozen ready for baking/regeneration in the sales outlet to give a 'fresh made' appeal. This 'centralised kitchen' concept has mushroomed in the last 15 years and is likely to continue to do so while the sector grows.

Plant bakeries are likely to decline even when using the latest enzyme technology to extend the shelf life of products. This latest trend to control moisture loss will be insufficient to stop the decline of this sector. As previously stated, a consumer is unlikely to wish to buy a loaf of bread that is 12 hrs old when the consumer can pay a fraction more for a loaf of bread, that is, in some cases, minutes old!

Consumers will continue to have ever-increasing expectations and the two critical factors will be 'freshness' and 'quality'.

Very specialised products, such as – gluten free; low GI; egg free and sugar free, will continue to be popular and the demand will grow. There is also a need for products to assist in weight loss as obesity is becoming a major problem – as it is in most developed countries.

As ever the South African bakeries will have to evolve if they wish to survive.

19.4 Manufacturing

19.4.1 Raw materials

'A raw material of ideal functionality is one which may be processed to give a first quality product while permitting maximum process effectiveness. The functional properties required in a raw material depend on the process to which it is to be subjected' (Sivasankar, 2002).

When it comes to developing a bread product, paying attention to the size, shape and colour affects the finished product quality and processing capabilities.

19.4.2 Factors influencing raw materials

1. **Property of raw materials**
   Bakers and technologists ask themselves these questions:
   a. What am I making?
   b. Why do I need this product?
   c. How am I processing this product?
   d. Is this product able to withstand my manufacturing process?
   By understanding these questions, finding the right specifications of raw materials becomes much easier.
2. **Quality**
   In maintaining quality, bakeries can request product specifications from suppliers. These should be checked to ensure that they are in accordance with what was agreed to during selection of the raw ingredients. The specifications must include name of product, ingredient declaration, allergens, origin of product, grade of product, shelf-life of product, colour, size, shape, aroma and any other important characteristics.
3. **Supplier**
   Manufacturers need to ensure consistent supply of raw ingredients from suppliers. The raw materials need to be available all year round, or seasonally, depending on the production plan of the bakery.
4. **Cost**
   The right quality and consistency comes at a price. South African manufacturers often lean towards low cost as it keeps the product price down and protects profit margins. However, in keeping product prices down, sometimes the quality of these raw materials can suffer. It is beneficial to the company to find the right raw material, at the right quality, available on demand and at the right price.

19.4.2.1 Wheat/flour

The baker or confectioner can use literally 100s of different ingredients, but the principal ingredient has to be flour. To say that all flour comes from wheat is a misconception. There are many types of cereals and flour can be milled from all of them; however, *wheat flour* has particularly unique characteristics that enables us to produce bread and confectionery products.

South Africa is a wheat growing country but because of varied soil conditions and erratic rainfall it is very rarely self-sufficient. It is in the interest of the baker to have a good knowledge of the characteristics of the flour he is receiving and likewise for the miller to have a basic knowledge of baking.

However, the reality is that generally this is not the case.

The purpose of milling is to extract the endosperm from the wheat berry, with as little contamination by the bran and wheat germ as possible.

The *modern way* is by using rollers and sieves. Of course, this is all highly automated and a modern roller mill can produce many tons of flour per hour.

The wheat berry

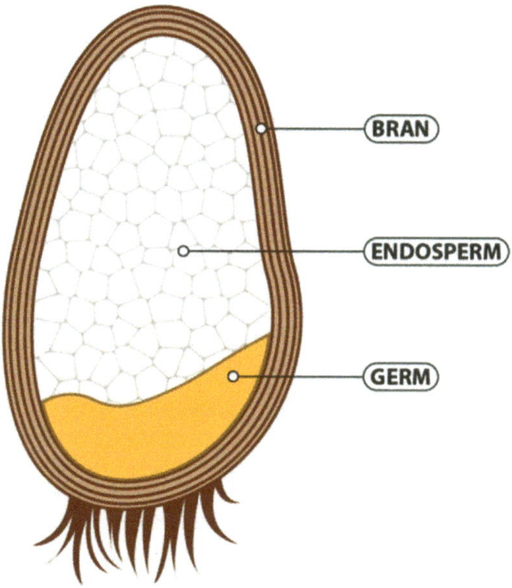

The *old way* is stone milling just as the settlers did hundreds of years ago and there is still a demand for this type of milling today; but to have endosperm with a 100% wheat germ and a large percentage of bran tests the baker's ability because of the proliferation of enzymes in the wheat germ and inert characteristics of bran.

The type of product produced will depend on which type of flour is being used. In South Africa there is significant variation in the quality available and in the rest of Africa such inconsistencies are far worse.

There are several milling companies in South Africa and as competition is quite high, the South African Baker enjoys a relatively good quality and variety of flour. The type of flour required will depend on the type of product produced. If one is producing sponge cakes, one would favour a low protein flour as opposed to, for example, producing hamburger buns where a much higher protein level is required. The reality is that the milling industry is unable and/or unwilling to cater for the needs of the baking industry and it may therefore be necessary to use additives to regulate the flour according to the product being produced. As there are several milling companies in South Africa it is up to the bakeries to decide which one to use.

The situation in some African countries is far worse. There is often a monopoly in place with stiff penalties for imported flour (this is done to protect the local industry),and to ask for a specific type of flour or information regarding the rheological status of the flour is often met with scorn.

The baker's greatest concern is inconsistency in flour quality, low falling numbers (diastatic activity) and fluctuations in protein quality/quantity – these will play havoc with a baker's ability to maintain quality parameters.

The typical composition of South African flour is as follows:

COMPOSITION OF FLOUR	
• STARCH	70%
• PROTEIN	12%
• MOISTURE	14%
• OIL / FAT	1%
• MINERALS / ASH	0.5%
• SUGARS	2.5%

Starch $(C_6H_{10}O_5)n$ is a carbohydrate and when mixed with water forms a glutenous-type substance. When this is subjected to temperature (baked) the starch changes its structure – this is known as gelatinisation – starch needs a temperature of 65°C to achieve this. As most baked products will achieve a core temperature of 90°C it is more than sufficient to reach this state. However, in a normal dough there is insufficient moisture present to achieve 100% gelatinisation and only a partial state will be achieved and it is this partial gelatinisation that forms the characteristics of bread as we know it. From a baker's perspective starch is not very exciting, although during the milling process some of the starch is damaged – this damaged starch can be of great benefit to the baker with regard to the flour's ability to absorb water.

As previously stated, all cereals contain certain amounts of ***protein*** but only flour from wheat (rye could also be included) has specific types of protein. The most important ones are *Gliadin* and *Glutenin*. There are others but these two are the most active in the baking process – they have the remarkable ability to hold the gas which is formed during the fermentation stage. The protein chain or matrix is a series of interlinked molecules which will, during the mixing process, break down, reform and strengthen. This is called *dough development* and it is absolutely critical for the baker to have a good understanding of this very important function.

Flour contains a small amount of natural sugar – this is often overlooked by cereal scientists as being of no importance. However, there are sufficient natural sugars to achieve caramelisation and they are also a source of yeast food which assist in fermentation.

19.4.2.2 Water

To be able to make flour into dough and to achieve gelatinisation of the starch the baker needs to hydrate the flour and this is always done by adding water. How much water will depend on several factors, including

- the moisture content of flour;
- type of product being produced;

- type of process being used and
- protein quantity and quality.

Water is by far the cheapest ingredient used and bakers are mindful that the more water that can be incorporated, the bigger the yield.

Most developed countries have safe water but in some underdeveloped countries this is not the case and water may have to be treated to make it fit for human consumption. One must be mindful of pH levels as this may affect yeast performance. It is a misconception that all bacteria are killed during the baking process.

19.4.2.3 Yeast

There are two types of yeast used for the purpose of fermentation – one for the brewing industry and one for the baking industry.

Saccharomyces cerevisiae (baker's yeast) starts its life, as all living things do, as a single cell.

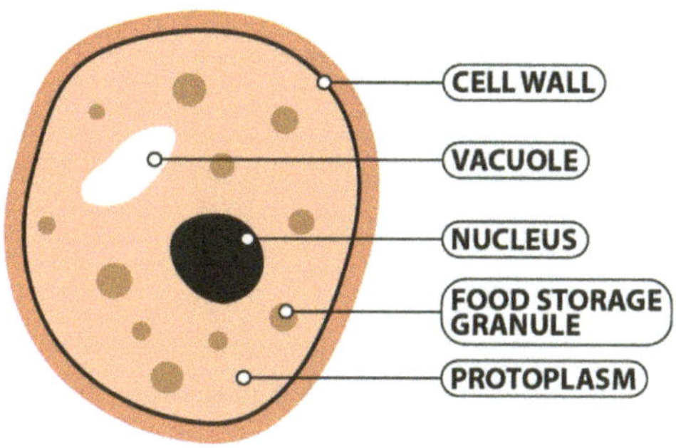

It consists of a cell wall with a liquid substance called the protoplasm. Inside this liquid is the nucleus.

Given the correct conditions it will reproduce by converting sugar into alcohol and in doing so will give off a gas (carbon dioxide). There are other chemicals that will also exhibit the same characteristics namely, *baking powder* or *bicarbonate of soda* and these can be used as aerating agents when producing baked products. But only yeast will impart in the finished product the structure and flavour that has become synonymous with bread products.

In South Africa there is access to three types of yeast – *dried, liquid* and *compressed*. A bakery's geographical position and type of *plant* will dictate which type of yeast to use.

This can be explained as follows:

- Compressed yeast has only 30% active ingredient (the rest being water) and it does require refrigeration.
- Liquid yeast has more active ingredients and requires refrigerated storage tanks as well as an automated dispensing system. It would only be viable for a large bakery within reasonably close proximity to the yeast factory.

- Dried yeast contains no water and is ideal for bakeries that are in remote areas or in countries that do not have fully operational yeast factories. It should be noted that the cost of dried yeast is considerably higher but the usage level can be reduced by a third compared to compressed yeast.

The correct conditions that enable yeast to *grow* are food, moisture, temperature and time.

Like all living things yeast can be killed and under normal circumstances this will occur during the first few minutes of baking. Low temperatures will place yeast in a dormant state.

Certain chemicals will inhibit yeast growth and may even completely destroy it. It has already been mentioned that salt has a profound effect on yeast – to recap – too much salt will actually kill the yeast whilst too little salt will result in uncontrollable yeast activity, that is, excess fermentation.

The amount of yeast (compressed) required will depend on the type of product and the type of process used, but as a general rule South Africa would use 2% for standard bread and as much as 4% for roll type products based on a percentage to flour mass.

19.4.2.4 Salt

Salt (sodium chloride) has previously been discussed in Section 19.2.3.

19.4.2.5 Other ingredients

Bread can be made with the above four ingredients. However other ingredients are beneficial in order to enhance the appearance and taste and thereby create a more viable commercial product.

Potential additional ingredients are numerous and often complicated.

Some examples include the following:

- *Preservatives:*

  All baked products after a certain time will be susceptible to mould infection. It is unavoidable unless certain preventative measures are taken. This can be in the form of packaging or with the addition of mould inhibitors but usually it will be a combination of both.

  The trend to market products as 'preservative-free' is justified if a limited shelf-life is commercially viable. This particularly applies to countries that have warmer climates as this will have an accelerated effect on mould growth. Countries in close proximity to the equator have high humidity and temperature levels all of which advance the occurrence of mould growth unless preservatives are used.

  Having good hygiene standards and employing a strict housekeeping policy will all assist in minimising the likelihood of contamination by airborne bacteria.

  The reality is that most bakeries will use mould inhibitors. These are normally acid based and work by adjusting the pH of the dough making it less attractive to contamination. The quantities used are very small and legislated so care must be taken.

  There are numerous products available but the product of choice in South Africa is typically calcium propionate at a dosage level of 0.2% to flour mass.

It should be noted that it is a misconception that a mould inhibitor will impart softness to the product. It is merely there to inhibit mould growth.

- *Sugar:*
  A distinction must be made between sugar used for yeast-based products and sugar used for confectionery-type products. In 'yeast-based' products there is sufficient natural sugar in flour to achieve caramelisation and as a food for fermentation as well as the taste profile. However, bakers will generally add sugar in one form or another to enhance these characteristics. South Africa is fortunate to have a healthy sugar cane industry so there is an abundance of sucrose available to the baker. Extra sugar will have a beneficial effect on the softness as well as crumb texture. Care should be taken as excessive use of sugar will have a negative effect on fermentation.
- *Fats:*
  Fats can be derived from three main sources – animal, marine and plants. Religious and cultural diversity in South Africa necessitates that most bakery fats will be derived from plants – notably palm and sunflower seeds. The advantages of adding fats are as follows:
  - gas retention;
  - crumb structure and shelf-life and softness;
  - eating quality;
  - improved appearance and
  - machinability.
- *Flour improvers and enzymes:*
  All too frequently bakers experience quality problems. It could be bread collapsing prior to oven entry resulting in poor volume or complaints about inadequate or variable shelf-life. Any product that fails to meet the quality specification needs to be corrected. Root cause could be the result of human error or it could be suspect raw materials. An experienced baker will develop a feel for troubleshooting, and it may be necessary to add something to rectify the problem. That something could be a flour improver or even an enzyme-based improver.

  Wheat gluten/ascorbic acid/data esters/soya flour/malt flour are just a few of the many flour improvers that are available.

  Most bakers will not be cereal chemists so it is very unlikely that they will actually purchase enzymes in their pure form. More often they are purchased from a specialised bakery supplier or will be added to a premix. However, it is a fact that increasingly bakers are relying on enzymes to enhance the volume, softness and shelf-life of baked products.

  Enzymes are proteins that accelerate chemical reaction. They will zero in on either the carbohydrates (starch) or the proteins in the flour. They are very specific and depending on country of origin they can be extremely aggressive.

Premixes:

Premixes are generally expensive and suppliers typically refuse to divulge the contents in detail but more and more bakeries are using premixes as opposed to the so-called 'scratch recipes'.

South Africa has had an explosion of premix factories in the last 10 years, and premixes are available for every conceivable bakery product. There are numerous negative and positive factors in using a prepurchased premix.

Negative:

- Cost – Ready made premixes are always considerably more expensive than those made from a 'scratch recipe'.
- Premix companies are reluctant to divulge the contents of their premixes apart from a mandatory general disclosure.

- Stereo-typing – If numerous bakeries purchase the same premixes and produce products as per the instructions, then all bakeries will be producing similar products – the appearance may be changed but the taste profile will be identical.
- Skill requirements for using scratch recipes is high.

Positive:

- All the ingredients are prepacked.
- Less likelihood of mistakes being made – that is, in a scratch recipe, the person who is responsible for making the premix could omit the salt or another important ingredient. In a prepacked premix this is unlikely to happen.
- Not necessary to stock large quantities of different ingredients.
- The supplying company will render technical advice or assistance.
- Skill requirements for using prepacked premixes is low.

19.4.3 Process

19.4.3.1 Mixing

For a baker or confectioner this is by far the most critical stage in the production process. Both disciplines have scores of different ways to blend the ingredients to achieve the desired product. In bread or roll production a baker will more than likely place all of the ingredients in a machine and start the mixing cycle in order to blend all of the ingredients to form a dough.

What is desirable is to mix until the dough has reached a maximum development stage, no more and no less. Maximum development stage is a breaking down and reforming of the protein chain (matrix) to form a cohesive mass that will give maximum gas retention during the fermentation process. This is particularly relevant when using a 'no time dough process' when long fermentation times are not applied. In some sophisticated bakeries the mixer's *built-in* computer will measure the amount of energy that is inputted into the dough and automatically switch off when maximum dough development has been reached, but in less sophisticated establishments it's left to the skill of the operator. Temperature of the finished dough is critical with 26°C–27°C being the norm.

The mixing times may change depending on fluctuations in flour quality.

A professional flour supplier will supply the bakery with certain information in the form of Chopin alveographs, mixographs, farinographs, falling numbers, etc. The production manager or operator should learn to interpret this information as it can forewarn of any impending fluctuations in flour rheological changes.

There are many different types of methods to mix doughs, that is, *sponge and dough* and *delayed salt*, to mention a couple, but the most commonly used method in South Africa is the *straight dough* method where all the ingredients are placed in the mixing bowl and the mixing cycle is then started.

19.4.3.2 Scaling and panning

At this stage the dough is cut into the correct mass and shaped and placed onto or into baking pans. An experienced baker will take great pride in his ability to mould dough by hand. It is difficult to explain, but an artisan, just by feeling the dough, will know the likelihood of success or failure in the finished product. All too sadly this practice is increasingly being replaced by machine processes.

19.4.3.3 Proving

The prepared dough is then placed into a prover. This is normally a box-type structure that is heated and will have a supply of low pressure steam. This will allow the dough to rise in a controlled fashion – the reason for the steam, apart from the heat, is to stop a skin forming as the dough rises.

Some products do not require a prover. French *pistolet* is a good example, and the dough is covered to prevent skinning. The amount of proving time will depend on a few factors, that is, dough temperature/amount of yeast/type of product. Generally 50–60 min at a temperature of 35°C and a relative humidity of 75% is appropriate.

19.4.3.4 Baking

There are many types and sizes of ovens available which can be heated by the following heat sources, that is, electricity, gas, diesel fuel and/or paraffin and they all have the ultimate goal which is to partially gelatinise the starch and the caramelisation of the sugars.

When this stage is reached, the product is considered to be baked.

The exact time and temperature will depend on the type of oven and product being produced. The latest trend is to minimise moisture loss by baking faster thereby extending shelf-life. Bakers have to remain mindful that as soon as the product is baked it will start to lose moisture (*staling*). Under normal circumstances this staling process cannot be stopped, only slowed down. However, new technology can sometimes be utilised to allow a product to increase moisture during the cooling down stage by exposing it to a more moist environment.

19.4.4 Packaging

Packaging requirements depend on the type of bakery. Some bakeries will sell their products loose and let the customer pack by supplying plastic bags, paper bags or boxes. More automated bakeries will have fully automated packaging machines and at some stage products may well have to undergo the dreaded 'squeeze test' by the customer. The in-store customer is also likely to ask the question, **'IS IT FRESH?'**. (As previously stated, when the product exits the oven it will start to lose moisture and this process will continue until it reaches a stage where it is deemed stale.) The baking industry has invested huge amounts of money to try and reverse this situation and in many cases they have managed to extend the shelf-life of baked products considerably.

Innovations such as gas 'modified atmosphere' packaging and/or vacuum packaging all assist in extending shelf-life. Vacuum packaging is not widely used for bakery products in South Africa as it is quite costly but the technology is available.

- **Functions of packaging**
  1. to protect the product from physical damage;
  2. to prevent the product from losing moisture;
  3. to prevent pest contamination, for example, rodents and cockroaches;
  4. to enhance the shelf-life of the product;
  5. to attract customers to the product being sold and
  6. to educate consumers on their purchase especially if a product has a 'product label'.
- **Labelling**
  In the South African food industry, the government has regulated guidelines regarding product labelling. The purpose of these regulations is to eliminate false advertising from product labelling, media and company branding. Each product is labelled as follows:
  1. product name;
  2. traceability details – batch coding;
  3. ingredient declaration;
  4. allergen declaration;
  5. nutritional table (voluntary);
  6. storage instructions;
  7. heating instructions (if applicable) and
  8. use by and best before dates.

This allows consumers to make informed decisions when purchasing a product and they can evaluate if a product meets their needs.

19.4.5 Food safety

Food safety has become a vital and critical requirement in the South African food industry. Consumers require good, clean and safe food. Due to growing concerns of the consumer and their growing awareness of the risks associated with baked goods, the South African government has put more pressure on the safety of the products being manufactured and sold to the customers. In order to implement a food safety system, companies have had to become more aware of the necessary guidelines that are in place and also be willing to make the effort to continuously maintain such systems of control.

- **Getting started**
  In South Africa, government has a regulation [(R962 – Foodstuffs, Cosmetic and Disinfectant act, 1972 (Act 54 of 1972)]. This regulation states the minimum requirements that a company would need to satisfy in order to manufacture goods to be supplied to the public. There are many widely recognised certifications available such as
  - SABS (South African Bureau of Standards) certification
  - HACCP (Hazard Analysis and Critical Control Points) certification
  - ISO (International Organisation for Standardisation) certification
  - AIB (American Institute of Baking) certification
  - GFSI (Global Food Safety Initiative) certification.

It is up to each manufacturer to select the certification goal they would like to achieve as these certifications also can come with a hefty price tag. Often a particular standard will be most associated with the market being supplied, and perhaps even have to be held and maintained as a 'prerequisite of supply'.

Key fundamentals regarding food safety systems:

1. **Food safety policy**

   This policy outlines the good manufacturing processes that a company must follow.

   It highlights all of the responsible persons to manage and implement the system as well as outline challenges, guidelines and objectives of the company.
2. **Personal hygiene**

   Staff hygiene plays a major role in a food safety system. It is important that all workers understand what is required and that they are also equipped with the right gear in order for them to carry out their tasks. A personal hygiene policy stipulates the expectations and procedures related to the following:
   - **a.** protective clothing – type of uniform, head covering, ear covering, type of shoe, jewellery policy, etc.;
   - **b.** eating, drinking and smoking;
   - **c.** plaster and glove policy and
   - **d.** what to do in the case of injuries and sickness.
3. **Foreign objects**

   Foreign objects are defined as a physical matter found in a food product that was not meant to be there and which can cause harm or concern to the consumer. Some examples of foreign matter present in bakeries include hair, wood, paper, plastic, rubber, metal and glass. These are a potential risk to food preparation areas and can cause harm to the consumer.

   Manufacturers regard foreign matter as one of the very important elements in the food safety system. Controls are usually put in place by the following:
   - **a.** identify the foreign matter, for example, wooden baskets for bread;
   - **b.** implement a daily check, for example, check each wooden basket for wooden splinters;
   - **c.** remove nonconforming equipment (perhaps showing signs of damage) in order to protect the integrity of the product and
   - **d.** metal detectors are more commonly used as opposed to x-ray detectors. The former being more cost-effective.
4. **Cleaning**

   Cleaning plays an important role in ensuring that a bakery reaches a certain hygiene standard. Bakeries are known for having flour dust cover everything. Cleaning procedures are set up by determining the following:
   - **a.** daily, weekly, monthly cleaning schedules;
   - **b.** determining frequency of cleaning and what food grade chemicals are to be used in in cleaning;
   - **c.** what equipment is used in cleaning?
   - **d.** how to clean? – step-by-step procedure on how to clean and
   - **e.** verification of cleaning – this is done through microbiological swabs to test the effectiveness of cleaning.

   It is also required by food safety standards that the use of colour-coded equipment is used in all working areas so as to avoid cross-contamination. For example, use a blue broom in the production area and use a red broom in the toilet facilities thus enabling the segregation of production areas from staff facilities and thereby help avoid cross-contamination. Cleaning equipment may also be segregated for food contact surfaces and nonfood contact surfaces as well as allergenic product lines, etc. Allocation of coloured equipment or type of equipment is a great benefit when seeking to differentiate.

5. **Control of allergens**

   Allergens are any substance in food that triggers an allergic reaction in people. The most common allergens found in the baking industry are gluten, soya, nuts, tree nuts, eggs, milk and milk products. Seeds, although not declared as an official allergen in South Africa, also seem to cause certain allergies amongst people, and as a result require controlling. In creating an allergen control policy, the following needs to be implemented:

   a. Identify what allergens are present in the raw materials.
   b. Make use of separate demarcated storage areas for each of the allergens to avoid contaminating other allergenic ingredients and nonallergenic ingredients.
   c. Identify the production line in which the allergenic product will be manufactured.
   d. Ensure pre-issued ingredients are separated via colour coding as well as colour-coded tools for each allergen present, that is, ingredients containing allergens placed in colour-coded bags in order to avoid cross-contamination.
   e. Once the allergenic product has been manufactured, ensure cleaning takes place on the production line. This will help avoid cross-contamination to other manufactured products.
   f. Allergenic swabs can be used to detect whether any allergen residue remains after cleaning.

6. **Pest control**

   Pest control programmes need to be put in place. Any food factory will attract pests such as rodents, cockroaches and other insects. It is important to have a registered contractor with a monthly contract in place to service the interior and exterior of the facility to prevent pest entry. Bakeries also need to look at their production, toilet and waste areas and take the necessary measures to pest proof entrances, doors and walkways.

7. **Food microbiology**

   Food microbiology forms an important role in the food industry as it helps ensure that safe and clean food is being produced. This prerequisite programme monitors the following through random sampling:

   a. identifies and verifies the shelf-life of finished goods;
   b. monitors hygiene and sanitation in the manufacturing plant through cleaning swabs and
   c. detects food borne illnesses in water, raw materials and finished goods.

   Often, samples are sent to a South African National Accredited System (SANAS) laboratory for testing and results are forwarded to the manufacturer. Some companies have their own internal laboratory and are equipped to do the necessary testing.

8. **Traceability**

   This is one of the most important programmes in a bakery or any food sector. Sometimes unforeseen circumstances occur such as, foreign contamination, microbiological complaints and/or malicious intent from staff. A traceable system allows manufacturers to trace all of the ingredients that have been used to manufacture a product through the baking process, up until the product arrives at the customer's door. Should a food scare arise, a good traceability system will enable quick reaction by the bakery to recall all of the products produced on that date of production.

It is important to remember when implementing a food safety system that, the system needs to be user-friendly, workable and maintained. Training plays a vital role in informing staff of what is expected of them in order to have a functional food safety system.

19.4.6 Quality control

Quality control is a system that checks and monitors the conformance of a product to the required standard (usually defined in a 'specification'). Quality control plays an important role in the food industry as it ensures product consistency throughout the manufacturing and distribution process.

1. Effect of quality control on the business

Customer satisfaction

↓

Incoming orders and increased sales

↓

Business growth and good reputation

↓

Assure jobs for the workers

2. Food quality is typically divided into three groups

a. Appearance factors
This relates to the visual appearance of the product in terms of size, shape and colour.

b. Textural factors
These factors apply to the feel of a product, for example, softness, crispiness, chewiness, firmness, coarseness, etc.

c. Flavour factors
This relates to taste sensations, that is, sweet, salty, sour, bitter as well after taste aromas.

For all the factors above it is important to evaluate each product and develop clearly defined accept/reject criteria.

A quality control process (sometimes also termed 'quality assurance') consists of monitoring raw materials, process, product and service. In order for quality monitoring to be successful, a finished product specification must be developed by the manufacturer. A finished product specification is a document that typically includes the following:

- Raw material specification – type of raw material and evaluation thereof;
- Process specification – recording of batch numbers, recipe batching, water temperature, mixing time, dough temperatures, weights, proving time, baking time and baking temperature, cooling time, date coding etc.
- Final product specification – determining the appearance, textural and flavour factors as well as final product weights.

19.4.6.1 Factors that cause variability in the quality of a finished product

1. People

Each staff member's contribution differs from one employee to another. For example, one person's rounding of a piece of dough will be different to another person's rounding. This can affect the quality of the finished product.

2. **Equipment**
   All equipments – mixers, dividers, moulders, provers, ovens, wrapping machines – work differently to produce a specific quality product. It is important to understand what type of product quality is needed and to ensure that the equipment is capable of achieving this.
3. **Methods or technologies employed in the process**
   Proper process control methods which must be measured are vital for the assurance of product quality.

Every system requires constant monitoring in order to ensure that the correct finished product specifications are adhered to. The more controls there are in place, the more effective a quality control system is.

19.4.7 Distribution and transportation

This is a key function of any bakery that needs to distribute its products for sale. Depending on the size of the bakery it could even have its own transport management structure to ensure the smooth running of this department. Key points would be:

- distribution staff need to have good product knowledge;
- distribution staff need to present themselves well;
- vehicles need to be clean inside and out;
- lugs (crates) need to be kept clean and
- distribution staff needs to show respect and be courteous to customers.

A huge problem in South Africa is the control of lugs (crates). Bakeries can spend 1000s of Rands in replacing lost (stolen) crates. There is a highly controlled illegal industry in recycling crates. It's not impossible that you could buy your own crates back at a fraction of the cost of new ones. The industry seems powerless to stop it. This all adds to the end cost of the product.

19.4.8 Training

Most companies will have at least some sort of rudimentary training programme, be it an induction or safety training – in some cases it is possible that a new recruit will be placed in charge of a machine that is perhaps worth US$ 100,000 after only a few hours of instruction.

By far the biggest challenge facing the baking industry in South Africa is the lack of skills. Employers are often reluctant to thoroughly train their staff due to a preference to save money and a feeling that when the business has invested in thoroughly training individuals they will be head-hunted by another company.

Such reluctant businesses overlook the point that the skills will stay in the industry.

The larger bakeries do offer in-house training courses, and a few raw material companies have basic training schemes. There are currently only two universities in the whole country that offer a BSc (Bachelor of Science Degree) in Food Science. South Africa does have a National Chamber of Baking who to their credit are trying to reverse this trend, seeking to engage the industry and/or government in improvement initiatives.

Sector development may include:

1. improve the image of the baking industry;
2. make it compulsory for a bakery that employs more than 15 workers to operate an accredited training scheme with a qualified mentor on a full-time basis;
3. give tax incentives for companies that comply and
4. create a national bakery college – funded by the baking industry.

The initiative must come from the baking industry. Government's role would be to ensure that it is implemented. But before all this can happen, the industry's business owners/directors will most likely need to realign their priorities to then enable progress to be made.

19.5 Future trends

19.5.1 Impact of skills shortage

Scenario 1

The craftsmen who have many years' experience will not be unaware of the fact that they can command a very high wage for their services. Bakery bosses who are unwilling or cannot meet this expectation usually try and rely on bakery raw material suppliers to cater for this need.

The companies of the last 20 years realised that a fantastic marketing tool to have on their fulltime staff is one or many fully qualified bakers and confectioners who can be 'borrowed' out to a bakery for training or product development work. Of course this service only lasts while the providing company remains a supplier and is only reserved for the larger clients, and in all likelihood only during office hours and definitely not over a weekend.

The scenario is likely to continue; however, there are obvious negative implications. The 'skills' pool is exhaustible and increasingly these companies are relying on more recently qualified food technologists to fulfil this need (therefore overall the system is losing valuable experience over time).

Scenario 2

Use of bakery consultants in recent years has started to increase. However this is a recent phenomenon and is likely to gain momentum once the industry learns to trust them and finally realise that there is a high price to pay for 'skills'. It's interesting to note that they are more popular in African countries north of South Africa where there are even fewer skills and where the use of consultants has been embedded for many years. The South African baking industry is relatively poor and is only now beginning to see the benefit of consultants.

Scenario 3

Frozen doughs and confectionery products are being increasingly utilised by bakeries. Most products can be frozen either in an unbaked or baked form and finalised at point of sale. This is not new technology and the key is slight change in formulations to optimise performance, freezing as fast as possible and then maintaining the cold chain up to the point of use.

19.5.2 Legislation

Production managers in the 1980s would quake with fear when informed that the inspector from trade metrology had arrived to check the weight of the bread being produced. With the power to reject huge amounts of bread if it was found to be underweight it could be 'career suicide' if the business was found wanting. This is not the case today – The National Regulator for Compulsory Specifications (NRCS) advises on the correct mass of bread and it is up to trade metrology to police the industry; however, perpetrators are very rarely dealt with sternly. At most a small fine but there is a growing concern by the consumer that more regular checks should be made and harsher penalties implemented.

Present government has implemented certain mandatory regulations in respect of vitamin enrichment of flour as well as labelling of wrapped bakery products. The required labelling should include the following:

- baked mass of product;
- full list of ingredients used;
- analytical breakdown of ingredients;
- expiry date as well as best before date and
- allergens.

It is believed that currently only a few bakeries strictly enforce these practices. Unfortunately it's the consumer that could suffer in the event of noncompliance. The reality is that small and medium bakeries can often completely ignore any labelling requirements and without accountability this practice could continue.

19.5.3 Manufacturing

Plant bakeries: Will continue to realign themselves with changing market trends. Automation will become a necessity with the ever-increasing demands by workers for a better salary and working conditions. However, this is contrary to government policy regarding job creation, so conflict is likely between management and workers. Margins will continue to be put under pressure and therefore this sector of the industry may struggle to show significant growth in the foreseeable future.

Small-to-medium wholesale bakeries: There is likely to be significant growth in this sector. Increasingly smaller bakeries will be required to provide more specialised products that they are currently reluctant to make or lack the skills to produce. Export of these products to other African countries is likely to become more widespread. As this sector expands they will obviously use more ingredients and could catch the attention of raw material suppliers, especially the flour milling industry which will need to secure their ailing market share by acquiring such ventures. Customers will apply more and more pressure on this sector in the form of internationally recognised food safety programmes as well as worker safety programmes, and such matters could therefore play a more active role in the day-to-day running of these businesses.

Small-to-medium retail bakeries: The reality in South Africa is that to have a retail business based on the 'high street principle' is no longer viable. South Africa currently has one of the highest crime rates in the world and this type of business is extremely vulnerable. However, in a shopping mall with adequate security and linked with a coffee bar and limited menu (not to mention the added attraction of free wifi), if managed properly such ventures will succeed and could even prosper.

19.5.3.1 Thinking out of the box

Consumers will continue to require more specialised products, that is, gluten free/low carb/high protein products, etc. – these will become significant areas of market growth. Businesses and government may wish to consider the benefits of specialised bakery products being publicised (or actually available) in doctors' waiting rooms aimed at people who are gluten intolerant or products for people on particular diets.

The same case could be made for seeing bakery products at the local gymnasiums which have become a high growth area in recent years.

I recently developed a product for a world famous body builder (see Fig. 19.1). His requirement to add 71 % pure protein (to flour mass) initially sounded impossible to develop with the expectation being that such a recipe would completely destructure the finished product and therefore not resemble a loaf of bread.

However a successful recipe and process was found to be possible and it is this kind of alternative thinking that could significantly help to regenerate the South African baking industry.

19.5.4 Availability of raw materials

South Africa is fortunate that it is self-sufficient as far as basic ingredients are concerned and in fact there is currently an over-supply. There are also a number of overseas bakery ingredient companies that are represented in South Africa and for the more specialised products the worst-case scenario is 12 hr delivery time from Europe.

The rest of Africa is not so fortunate. With very strict import duties and almost overwhelming bureaucracy it is sometimes almost impossible to produce a reasonable quality product when forced to use all local ingredients. There are a few

Figure 19.1 71% pure protein to flour mass.

exceptions – Namibia is a good example, often in cases where German business interests produce excellent products. A few of the French speaking countries should also be mentioned as having good quality baked goods, that is, DRC and Senegal.

Overall comparison with other countries across Africa highlights that there is a vibrant bakery ingredient sector in South Africa albeit a sector that is still well set for further development.

19.5.5 Future challenges

Baking in South Africa has evolved a lot over the past years; every year a new trend is set which brings about new challenges in bakeries that aim to deliver and keep up with ever changing market requirements.

The biggest challenges are the following:

- **Ability to adapt**
  The willingness to adapt to the market trends and consumer requirements.
- **Innovation**
  The ability to innovate is crucial. Not all manufacturing companies have the resources and capital to achieve this. Manufacturers need to grow at a constant pace to allow for business expansion.
- **South African economy**
  Companies wishing to expand by procuring machinery from countries abroad may find that the exchange rate as well as the fluctuations in the economy are stumbling blocks.

The *challenges* are many and have all been discussed before. Perhaps the biggest challenge facing the South African baking industry is the one thing we are all frightened of the most, that being the 'fear of change'.

And last, but not least – As a great man once said –

> *There are people in the world so hungry, that God cannot appear to them except in the form of Bread.*
>
> *Mahatma Gandhi*

Acknowledgements

1. South African Chamber of Baking
2. Sivasankar, 2002.
3. Wheat berry and yeast cell images by Theo Poulsen, Graphic Designer, Johannesburg and South Africa.
4. Mahatma Gandhi.

Reference

Sivasankar, B., 2002. Food Processing and Preservation. published by Asoke K Ghosh, Prentice Hall of India, p. 167.

Technical and quality management of canning

20

Richard F. Stier
Consulting Food Scientist, Sonoma, CA, United States

20.1 Introduction to canning

Canning has evolved exponentially since its creation over 200 years ago. The best way to define canning may be to utilise the definition of commercial sterility found in the United States Code of Federal Regulations (CFR) for low-acid canned foods found in Title 21, Part 113, "Thermally Processed Low-Acid Foods Packaged in Hermetically Sealed Containers."

> *(e) Commercial sterility: (1) "Commercial sterility" of thermally processed food means the condition achieved--*
>
> *(i) By the application of heat which renders the food free of--*
>
> *(a) Microorganisms capable of reproducing in the food under normal non-refrigerated conditions of storage and distribution; and*
>
> *(b) Viable microorganisms (including spores) of public health significance; or*
>
> *(ii) By the control of water activity and the application of heat, which renders the food free of microorganisms capable of reproducing in the food under normal non-refrigerated conditions of storage and distribution.*

And, what the definition does not state is that such foods will be packaged in a hermetically sealed container, which is also defined in this regulation as:

> *(j) Hermetically sealed container means a container that is designed and intended to be secure against the entry of microorganisms and thereby to maintain the commercial sterility of its contents after processing.*

When we talk about "canned foods," we are referring to a class of foods that goes way beyond cans. Canned foods are now packed in cans, glass, plastics, composite containers, flexible pouches, paperboard boxes and other containers. If that product can be sterilised and the package has the ability to maintain a hermetic seal, it could become a canned food. It would, perhaps, be better to call these foods

Swainson's Handbook of Technical and Quality Management for the Food Manufacturing Sector.
DOI: https://doi.org/10.1016/B978-1-78242-275-4.00020-4

shelf stable products in hermetically sealed containers, but the world has used and understands that term canning, so let's leave it at that.

20.2 History and evolution of canning and processing

Many people may not like to admit it, but war and strife often lead to significant innovations. And, it was war that led to the development of canning. Napoleons armies dominated the European continent from the late 1700s until his final defeat at Waterloo in 1815. He was a brilliant general who understood the dictum that "an army moves on its stomach." Relying on foraging as his armies moved was inefficient and tended to alienate the locals who were stripped of their crops and animals as the army moved. He offered a prize to anyone who could come up with a means of preserving foods for an army on March. The challenge was answered by Nicolas Appert in 1810, who devised a means of packaging foods in wide mouthed glass bottles, sealing said bottles and heating them in boiling water. He published his findings in *L'Art de conserver les substances animales et végétales* (or *The Art of Preserving Animal and Vegetable Substances*). He was awarded a 12,000 franc prize for his efforts.

So the technology known as canning was born, but the original package was glass not a can. Knowing what we know now about food microorganisms, especially *Clostridium botulinum*, one wonders what happened with low-acid foods such as meats and vegetables at that time.

The English, seeing a good technology, promptly adopted it. Peter Durand, a British merchant, utilised what he had learned of Appert's technology through a French colleague to file a patent on the subject in 1810. His original patent mentioned containers such as glass, clay pots and cans, but he focused on cans only. He sold the patent rights in 1812. He also patented the process in the United States in 1818, but never actually commercialised it himself. With his canning processing, foods were stuffed in cans, boiled and then sealed while hot using corks, soldered metal plugs or some other means.

So, the technology was up and running by the 1820s. Wars helped the canning industry grow throughout the 19th century. Canned foods played a prominent role in feeding soldiers in the Crimean War in the mid-1850s and in the American Civil War between 1861 and 1865. The quality of these products was not uniformly high, but they did provide soldiers with some variety in the daily diets.

With the discovery of gold in California, canned foods helped the pioneers as they crossed the country. As the population grew, so did the industries supporting the expanding population. One of these industries was canning. Beginning in 1864, salmon were canned on the Sacramento River in Sacramento. The migrating salmon were captured, butchered and placed into cans which were then processed in open vats containing salts such as calcium chloride. The salts allowed process temperatures to be increased. The cooked cans were cooled and allowed to sit for a time. Cans which did not spoil were then painted red and sent to market, which is why canned salmon is associated with a red can even today (Lerke, 1979).

Perhaps the most significant advances related to canning that occurred in the 1800s were the confirmation of the germ theory by Luis Pasteur, the invention of the pressure cooker and the development of the sanitary can. Pasteur conducted experiments that clearly demonstrated that food spoilage was caused by microorganisms and said microorganisms could be destroyed by heating (pasteurisation). This explained why canning actually worked.

A.K. Shriver invented the pressure cooker in 1874, which allowed cans to not only be processed at higher temperatures but also allowed the processor to better control process temperatures.

The sanitary can consisted of two ends that were attached to the cylindrical can body. The cylinder was flanged allowing the two ends to be rolled and flattened onto what is now referred to as the can body. The ends contained a sealing compound which would flow under pressure creating the hermetic seal and effectively excluding air and other contaminants. This allowed containers to be sealed quickly and easily which increased production rates and helped enhance quality.

20.3 Canning becomes a science

The canning industry continued to grow and prosper, but ran into some problems at the start of the 20th century. Specifically, there were a number of botulism cases linked to commercially canned low-acid foods. Botulism is caused by the ingestion of a neurotoxin that is produced by the food pathogen, *C. botulinum*. This organism is a sporeforming anaerobe that is ubiquitous throughout the world. The organism was first isolated in 1895 by German scientist, Emile van Ermengem from ham. It was also associated with spoiled sausage products, hence the species name botulinum that is derived from the Latin word for sausage – botulus. Today, botulism is normally associated with improperly processed home canned foods, but outbreaks from commercially canned products can be devastating.

Canned ripe olives produced in California were found to be the source of the most significant of these outbreaks. A 1919 report in Public Health Reports describes an investigation into an outbreak that killed seven in Canton, Ohio and 17 all totalled. This outbreak significantly damaged the California canned olive industry, and provided the impetus for research that led to the development processes that helped turn canning from an art to a science. Hard on the heels of this issue, there were 20 deaths attributed to botulism in canned spinach. These issues and others led to the passage of the Cannery Inspection Act in 1925 by the State of California. This was a joint effort by the State Board of Health and the California canning industry to prevent further outbreaks of botulism attributable to commercially canned foods (CDPH, 2015).

The science behind canning took two distinct paths that were linked. These were developing a better understanding of the microbiology of canning, especially the microbiology of *C. botulinum*, and creating the tools and programs to calculate thermal processes for different food processes.

The microbiology of canning was in large part driven by Dr Karl Meyer from the University of California at San Francisco. The work can be broken out into two basic areas, the ubiquitous nature of *C. botulinum* and the heat resistance of its spore. He teamed with Dubovsky in 1922 to publish two papers on the distribution of *C. botulinum* spores. The work in California clearly showed that spores of the pathogen could be found in the soil. They found that 30% of the isolates were toxic and that Type A *C. botulinum* predominated in California. They collected over 1500 samples across the United States and found Type A to be predominant in the West whereas *C. botulinum* Type B spores were more common in the East. A number of other workers were able to demonstrate that *C. botulinum* Type E was often found in marine sediments in addition to being associated with numerous botulism outbreaks linked to seafood products.

Dr Meyer also collaborated with his colleague at the university, Dr J.R. Esty and Meyer (1922) to publish what is acknowledged as one of the classic studies on *C. botulinum* heat resistance. Their work demonstrated that the death rate of *C. botulinum* spores was logarithmic between 100°C and 121°C.

The other half of the equation was the development of methods to calculate thermal processes for low-acid canned foods utilising heat resistance of the pathogens and heating rate of the different food products. For this work, the industry owes a debt of gratitude to the National Canners Association (NCA). In 1920, Bigelow et al. published an NCA Bulletin which incorporated heat resistance of *C. botulinum* and heat penetration data. The concepts that were developed were known as the general method, which is still used today.

The fourth author on the Bigelow publication, C. Olin Ball, was destined to play a major role in advancement of the canning industry. Ball joined the NCA as a chemist immediately after returning from Europe where he was part of the Allied Expeditionary Force. One of his first tasks at the Association was the refinement of the copper-constantan thermocouple used for heat penetration work (Stier, 2004). While working at NCA, Ball pursued a masters degree at George Washington University. His thesis was entitled "The Electron Theory Applied to Thermo-Electricity." This work focused on the various conductivities of different metals which allow thermocouples to be used to measure temperature (Ball, 1922).

Ball felt that the method for calculating processes could be improved and set out to improve the general method. The new method was published in 1923 and served as the basis for his PhD in 1926. He recognised that the rate of heating in a container and bacterial destruction were both logarithmic which allowed mathematical values to be assigned to each. The end result was the Ball Formula that is still used today. When persons working with canned foods see terms such as Z-Value or Fo, they can thank Ball (see Fig. 20.1). And, what makes this work even more impressive was that it was not only done in his spare time but also without a computer or other tools today's engineers and mathematicians rely so much on. Among the many honours that Drs Meyer and Ball received in their lifetime was the Nicolas Appert Award from the Institute of Food Technologists, awarded for lifelong contributions to advancing the food processing industry.

The research that was conducted during this short stretch of time in the 1920s has held up for almost 100 years. One other issue that evolved from this work is the

D-Value – time in minutes at a given temperature required to reduce the number of organisms by 90% or one log.
Z-Value – number of degrees required for the thermal death time curve to traverse one-log cycle.
F – time in minutes required to destroy a stated number of organisms at a defined temperature, usually 250°F (121°C) and when the Z-value is 18°F. F-Value is a common term employed in the canning industry to express the lethality or sterilising value of a sterilisation process.

Figure 20.1 Key definitions.

12-D concept. To ensure the safety of canned foods, a very conservative approach was taken to ensure complete destruction of the spores of *C. botulinum*. Processes are established to ensure that there is sufficient lethality to achieve a 12-log reduction (Perkins, 1964). To some, this seems excessive, but the safety of commercially canned foods belies that concern. Outbreaks of botulism from commercially canned foods over the past century are few and far between. Many of the botulism outbreaks that were linked to canned foods were not as a result of processing issues, but postprocess failures. This will be addressed later.

20.4 Process determination and validation

The work conducted in the 1920s set the stage for ensuring the production of safe and wholesome canned foods. The NCA and the container companies were the leaders when it came to developing and validating thermal processes for canned foods. With an understanding of microbial death rates, the ability to measure heat penetration into containers and having established the 12-D reduction for low-acid canned foods, the basic elements were now in place to properly determine and validate thermal processes for canned low-acid foods.

There are many different canned food products. The focus thus far has been the low-acid products, which as defined as those products with a pH of 4.6 or above. These are the products that will support the growth of *C. botulinum* and are, therefore, potentially the most hazardous of canned foods. The acidified foods regulation found in 21 CFR Part 114.3 defines acid and acidified foods as:

(a) Acid foods means foods that have a natural pH of 4.6 or below.

(b) Acidified foods means low-acid foods to which acid(s) or acid food(s) are added; these foods include, but are not limited to, beans, cucumbers, cabbage, artichokes, cauliflower, puddings, peppers, tropical fruits, and fish, singly or in any combination. They have a water activity (aw) greater than 0.85 and have a finished equilibrium pH of 4.6 or below. These foods may be called, or may purport to be, "pickles" or "pickled ______," carbonated beverages, jams, jellies, preserves, acid foods (including such foods as standardized and nonstandardized food dressings and condiment sauces) that contain small amounts of low-acid food(s)

and have a resultant finished equilibrium pH that does not significantly differ from that of the predominant acid or acid food, and foods that are stored, distributed, and retailed under refrigeration are excluded from the coverage of this part.

Acid and acidified foods will be discussed later.

There is one more element that needed to be established for each food and that was the resistance of *C. botulinum* or a nontoxic surrogate in different products. For example, the organisms might be more resistant to heat in cream style corn as opposed to canned corn in salt brine, so information of the specific heat resistance in different products needed to be developed (Fig. 20.2).

There were several steps needed to accomplish this. These were:

1. producing spore cultures to do the work;
2. verifying that the cultures had adequate heat resistance to conduct the studies;
3. conducting the inoculated studies in product; and
4. determining heat resistance based on these studies.

When doing thermobacteriology or any other validation studies, it is essential to have not only the proper culture but also to ensure that the culture that is being used has the proper resistance. Depending upon the product and process, the resistance that must be established may be wet heat, dry heat or some other parameter such as hydrogen peroxide. One of the most common mistakes made by researchers doing this kind of work is that they fail to take the time and make the effort to verify the resistance of the cultures used in the study. When conducting studies with canned low-acid foods, the organisms most often used and target heat resistance are:

- 1809 – Appert invents canning (really in glass)
- 1810 – Peter Durant and the soldered can
- 1845 – Crimean War and deadly canned meats
- 1850s – Canning grows; salmon in Sacramento
- 1860 – 1864 – Pasteur and the germ theory
- 1874 – Shriver invents the pressure cooker
- 1888 – Invention of the sanitary can
- 1895 – Emile van Ermengen isolates the organism from home cured sausage
- 1900 – 1920 – Botulism outbreaks from commercially processed foods
- 1920 – Esty and Meyer define resistance and that the *Clostridium botulinum* may be found everywhere
- 1925 – Cannery Inspection Act passed by the State of California
- 1926 – Olin Ball and Ball Formula
- 1930 – 1970 – Science behind canning expanded; Townsend, Ball, Mercer and others
- 1971 – Botulism outbreak from Bon Vivant prompts development of US LACF regulations based on California's Cannery Inspection Act.
- 1981 – Approval of hydrogen peroxide for use in United States as a sterilant; aseptic processing takes off

Figure 20.2 Evolution of canning.

- *C. botulinum* – $D_{250°F/121°C} = 0.1–0.2$ min
- *Clostridium sporogenes* (nontoxic Clostridia) – $D_{250°F/121°C} = 0.1–1.5$ min
- *Geobacillus stearothermophilus* – $D_{250°F/121°C} = 4.0–5.0$ min

So, spore cultures must first be grown and harvested, and then tested for heat resistance in neutral buffer solutions. Testing may be done in glass thermal death time (TDT) tubes (Fig. 20.3). The tubes are filled with the spore culture in buffer, sealed and then heated in TDT retorts. Determining heat resistance usually involves heating the spores at three temperatures and at a minimum of five times at each temperature. Each retort-containing tubes being heated will contain a tube with a thermocouple in it so that the rate of heating and cooling may be monitored. The heating times are selected to destroy all spores at the highest temperature whereas the lower temperatures will show increasing destruction of the spores as heating times increase. After heating the containers in the retorts are immediately cooled and cultured to determine survivors.

Another tool to determine heat resistance of spores is an instrument called a thermoresistomer. This unit heats spores in buffer in direct steam. The spores are inoculated into small aluminium cups and placed in the unit. Because the inoculum size is small (0.1 mL) and it is placed in direct steam, heating is assumed to be instantaneous. At the end of the heating time, the steam is turned off and the cups are immediately dumped into culture media. This cools the inoculum and allows surviving spores to be cultured.

If the spore culture does not have the proper heat resistance, a new culture must be produced.

If the spore culture has good heat resistance, studies will be conducted to evaluate heat resistance of the spores in product. This usually done in TDT tubes as described earlier or in TDT cans. These cans are 208 × 006 containers

Figure 20.3 Early aseptic unit (Dr Olin Ball, far left at Rutgers University).

(2 8/16 in. diameter × 6/16 in. high). The containers are filled with approximately 20 g of product and inoculated with the spore culture which is mixed to evenly distribute the spores. The containers are then sealed and placed in a holder for insertion into the TDT retort. The TDT retorts are heated and cooled as described earlier. The TDT cans are not normally cultured, however. They are allowed to incubate and growth is determined by whether the cans swell. These studies allow the researcher to determine the *D*-value at different temperatures in the product.

Heat penetration studies are conducted to determine not only how the product heats but also the rate of heat penetration. Does the product heat via conduction as would occur in canned tuna fish or pet food, or is it a convection heating product?. There are also products that exhibit what are known as broken heating curves. In these cases, the product will initially exhibit convection heating but will change conduction heating through the process. Examples of products that exhibit such heating curves would be dry beans. As the product is retorted and starches begin to set up in the brine, the heating rate slows.

In some cases, the calculated processes must be confirmed in the processing plant. Of course, when doing this the researchers must utilise the nontoxic species. One never conducts in-plant studies with a food pathogen. These studies are basically expanded versions of what was done on the bench using the TDT cans. The containers used in the study will be clearly marked using coloured engine paints, inoculated and subjected to the process. When utilising the nontoxic clostridia, the inoculated cans are recovered, incubated and monitored over time. Based on swelling and verification that the swelling was caused by the inoculated organism, a process can be confirmed and issued.

20.5 Raw materials quality

As with all foods destined for processing, raw materials quality plays a key role in the overall quality and safety of foods in hermetically sealed containers. Failure to utilise high-quality ingredients can result in failure (spoilage or safety concerns) of the products. Anyone who has worked in the industry has observed such failures. During my time in industry, the following incidents were among the many that were observed:

1. butyric acid anaerobe spoilage of canned pears;
2. canned soup spoilage by putrefactive anaerobes; and
3. thermophilic spoilage of canned vegetable with added starch.

In the first case, a lot of pears experienced almost 100% failure due to contamination of the pears with butyric acid anaerobes. The cause was poor in-plant sanitation which resulted in high loads of *Clostridium butyricum* going into the cans. The normal thermal process was inadequate to eliminate such a high load.

The canned soup incident was caused through the use of the wrong ingredient. Institutional sized cans (603 × 700) were being processed in discontinuous rotary system called an Orbitort (FMC Corporation). The ingredient employed was a quick set starch rather than the one that was supposed to be used. The starch set up too quickly inhibiting heat penetration resulting in an under-processed product which eventually spoiled.

The final example underscores the importance of ingredients such as sugar and starch used in canning. The NCA's Laboratory Manual for Food Canners and Processors (1968) addresses the importance of establishing microbiological standards for sugars and starches for thermophilic bacteria. The thermal processes that are established for low-acid canned foods are designed to achieve a 12-D reduction of *C. botulinum*, but do not have the accumulated lethality to destroy high levels of thermophilic spores. Thermophiles are bacteria that have the ability to germinate and grow at elevated temperatures, usually 50°C (122°F). These organisms are not addressed when establishing thermal processes because they will not grow at "normal conditions of distribution and storage."

20.6 Canned food spoilage and evaluation

The discussion on ingredient quality naturally leads into a discussion of canned food spoilage and its evaluation. Canned foods are one of the safest and most stable products that are manufactured. Their shelf life can be years. In fact, cans of food manufactured during the American Civil War have been recovered, tested and been found to be commercially sterile. The quality may not have been great, but the food remained sterile. Shelf life is a function of the type of product (acidic or low-acid), how it was processed and the type of container. For example, tuna fish in a can may have a shelf life of 2–3 years whereas that in a pouch might be 9–12 months. However, all foods in hermetically sealed containers are prone to spoilage. Spoilage of canned foods may be caused by any of the following situations:

1. Postprocess contamination (leakage) – this type of spoilage when a processed container spoils as a result of microorganisms entering the container after processing. The spoilage organisms enter the container through the seams, seals or caps and begin to grow which usually causes the container to swell. This will be discussed at greater length later.
2. Underprocessing – underprocessing is a type of spoilage that occurs when the container or product is inadequately processed, that is, a failure to properly introduce sufficient heat to the product in question. Low-acid foods have established processes to ensure the destruction of *C. botulinum* spores, a significant public health pathogen. The definition given in the first paragraph of this chapter addresses this concern.
3. Thermophilic spoilage – this type of spoilage is a result of the outgrowth of sporeforming bacteria capable of growing at elevated temperatures, that is, greater than 120°F (49°C). These organisms are called thermophiles or heat loving. They are not pathogenic nor will they grow at normal conditions of distribution and storage. Spoilage will occur when containers are not cooled properly or are exposed to elevated temperatures during storage and distribution.

4. Incipient spoilage – incipient spoilage is a type of spoilage that occurs prior to prior to processing. It can occur in the container or in a hold tank. As an example, a lot of cans might be staged for retort processing when the plant suffers a power outage. The filled and sealed cans might not be processed for several hours. In the interim, microorganism in the can could grow. Their growth could alter the product characteristics and produce gas so cans would deform during processing.
5. Hydrogen swells – this occurs when an acid product reacts with a metal container resulting in the production of hydrogen gas and a swollen container.

The first four kinds of spoilage are caused by microorganisms. Hydrogen swelling is a result of the product reacting with the can causing the generation of hydrogen. Canned foods may appear to be spoiled as a result of excessive fill or loss of vacuum, but are, in fact, not so.

The FDA's Bacteriological Analytical Manual (2001) provides an excellent overview of canned food spoilage and analysis. The method described in the chapter may also be applied to hermetically sealed containers in glass, pouches, cups, trays and any of the other containers that are being used today.

Postprocess contamination or leakage may be caused by a wide range of conditions. The low-acid canned food regulations in the United States (1989a) include a number of clauses designed to minimise the potential for postprocess contamination. These include mandates for:

1. Cooling water – container cooling water shall be chlorinated or otherwise sanitised as necessary for cooling canals and for recirculated water supplies. There should be a measurable residual of the sanitiser employed at the water discharge point of the container cooler.
2. Postprocess handling – when cans are handled on belt conveyors, the conveyors should be so constructed as to minimise contact by the belt with the double seam, that is, cans should not be rolled on the double seam. All worn and frayed belting, can retarders, cushions, etc., should be replaced with new nonporous material. All tracks and belts that come into contact with the can seams should be thoroughly scrubbed and sanitised at intervals of sufficient frequency to avoid product contamination. Automatic equipment used in handling filled containers should be so designed and operated as to preserve the can seam or other container closure integrity.
3. Container closures – regular observations shall be maintained during production runs for gross closure defects. Any such defects shall be recorded and corrective action taken and recorded. At intervals of sufficient frequency to ensure proper closure, the operator, closure supervisor or other qualified container closure inspection person shall visually examine either the top seam of a can randomly selected from each seaming head or the closure of any other type of container being used and shall record the observations made. For double-seam cans, each can should be examined for cutover or sharpness, skidding or deadheading, false seam, droop at the crossover or lap and condition of inside of countersink wall for evidence of broken chuck. Such measurements and recordings should be made at intervals not to exceed 30 min.
4. Teardown examinations – teardown examinations for double-seam cans shall be performed by a qualified individual and the results therefrom shall be recorded at intervals of sufficient frequency on enough containers from each seaming station to ensure maintenance of seam integrity. Such examinations and recordings should be made at intervals not to exceed 4 h. All containers for low-acid foods are subject to this requirement be

they cans, "*paper boxes* like Tetra Pak and Combibloc containers," pouches or trays. The objective of the teardown is to examine the seams or seals to ensure that they have been properly formed.

Food regulations and guidelines from around the world include similar mandates, yet the industry experiences incidents of postprocess contamination on a regular basis. The mandates noted in points 1 and 2 earlier are designed to protect the can, glass jar or other container when they are most sensitive – immediately after processing during cooling. During cooling, a vacuum begins to form in the container. In addition, with cans and jars, the sealing compound is soft and more prone to failure. If the container has a bad seam, pinhole or has the soft compound, it is more prone to failure. Maintaining residual chlorine in cooling water and ensuring that lines used to handle containers after process are clean, sanitary and dry will minimise the potential for postprocess contamination. Those factors noted in points 3 and 4 mandate that the containers themselves be examined on a regular basis, the hope being that failures of the can seamer or sealer will be detected and corrected.

These controls do not work all the time, however. In 1981 there was an outbreak of *C. botulinum* Type E poisoning attributed to canned salmon produced in Alaska. Due to the nature of the industry, can bodies were produced and shipped to the Pacific Northwest (Alaska and Canada) after being flattened. This was done to save space during shipping. When the flattened containers were used at the canneries, they were formed into cylinders, flanged and an end placed on the container. The equipment designed to do this was called a reformer. The flattened containers were unloaded from boxes and placed in a magazine. A "picker fingers" would then remove a can from the bottom of the magazine and place it on a mandrel to be formed into a cylinder. During this unit operation, the picker fingers would occasionally come into contact with the container above the one being picked. The end result was a triangular shaped tear in the container which came to be called the "Index Defect." These damaged cans were not detected and were eventually filled with salmon and processed. During cooling, water containing spores of Type E *C. botulinum* ended up contaminating the cans. The spores grew out and produced toxin but no gas. There were several illnesses as a result of this. In addition, the major market for canned salmon, the United Kingdom, demanded that the industry take steps to prevent such an occurrence in the future.

The industry, under the leadership of the National Food Processors Association, made a commitment to improve operations. They improved sanitation, chlorination practices for cooling water and adopted several new technologies to minimise potential problems. These included the adoption of the one-piece nested can thanks to the work of the Western Can Company and the incorporation of dud detection on all labelling lines. Dud detection measures lid deflection so low-vacuum or overfilled cans can be detected and removed from the production line. The nested cans could be shipped north in the same foot print as the flattened cans and better yet, eliminated the need for the reformers. The incorporation of 100% dud detection ensured that low-vacuum containers, such as those which had the index defects, would be detected and removed from the line.

20.7 Canning technologies over the years

Canning technology has evolved significantly over the years. This evolution has included the adoption of new containers, filling machines, the adoption of technologies such as check weighers, dud detection and x-ray machines and new and improved processing systems. Today's canning lines operate at high speeds and may include a number of on- or at-line systems to monitor fill weights, seam integrity and final vacuums. Let's simply focus on how processing systems have evolved.

All low-acid foods must be processed underpressure in order to ensure the destruction of *C. botulinum* spores. It took some time before the industry learned this fact. Acidified or acid foods may be processed at atmospheric pressures so maximum temperatures that can be reached are the boiling point of water. Today, most acid or acidified products are filled hot ($>$185°F/85°C), held for a short period of time and cooled. In some cases, these processing lines will include a twist which inverts the container to ensure that the lid or cap is in direct contact with the hot product. These products, which range from beverages, to condiments, to pickles and to sauces, are packaged in many different types of containers including metal cans, plastic pouches, high-density plastic bottles and metallised pouches.

Systems for processing low-acid canned and glass packaged foods have evolved considerably over the years. As noted earlier, products were originally processed in boiling water. The industry then evolved to using open tanks containing salts which allowed the boiling point of the water to be increased. The invention of the pressure cooker in 1874 allowed canned low-acid foods to be processed more rapidly and improved quality. These still retort systems are still in use. Containers were loaded into the retort in baskets or plates, the retort sealed and steam introduced to the system. The pressure vessel allowed steam pressure to be increased raising the temperature well above that of boiling water.

When processing in glass containers, the industry learned that it was essential to provide overpressure to the retort in order to keep the lids securely on the jars. In glass processing, critical control points are usually time and temperature plus overpressure. Depending upon the product being packed, other critical factors might include headspace, fill weight, viscosity or others. It is probably more common to see acidic or acidified products such as apple sauce, sauerkraut, pickled cucumber (gherkins), artichoke hearts and condiments in glass than it is low-acid products such as beans, peas or carrots. Glass tends to be viewed by consumers as a higher quality package, especially in developing nations, as the product is visible. It is for this reason that baby foods are still packaged in glass in many parts of the world.

20.8 Still retorts

As alluded to earlier, the still retort is the oldest and simplest of retort systems. These may be vertical or horizontal systems meaning that they are loaded from the top or the side, respectively. These pressure vessels are manufactured of heavy gage metals and must be designed to withstand high pressures. There are some horizontal retort systems that have two doors. Unprocessed product is loaded through one door and the processed product from the door at the opposite end of the vessel. These systems are more expensive but allow easy segregation of unprocessed and processed products. Section 113.40 (a) from the US regulations governing low-acid food processing provides an excellent overview of what is required for still retort systems.

Perhaps the most important device required for retort systems is the temperature indicating device. For many years, the required device was a mercury-in-glass (MIG) thermometer. Today, MIGs are being replaced by other devices of equivalent accuracy. Food processors simply do not want to have either glass or mercury on their plants. These devices must be calibrated regularly to ensure that they are accurate. When installed on a retort, they must be installed in a location where they will be easily visible to the operator of the system. Retorts must also be fitted with a temperature recording device. This instrument allows temperatures to be continuously monitored and recorded. When adjusting the temperature recorder, the instrument shall be adjusted so that it reads a minimum of 1 degree below the temperature indicating device. Both instruments must be fitted with a bleeder to ensure that there is a continuous flow of steam past the sensor on the device.

Other required elements include: temperature controllers/steam controllers, bleeder valves, pressure gages, steam inlet valves, steam spreaders, vents, air valves, water valves and crates or baskets for loading containers into the still retort systems. The crates or baskets are fitted with divider plates to stack the containers and to ensure adequate steam flow to and around the containers being processed.

Processes for low-acid canned foods must be determined by a recognised process authority. They are based on the rate of heating of the food in the container, the death rate of the target organism in the food and the rate of heating in the cold spot within the still retort. The latter is determined by filling the retort with the smallest size containers scheduled to be processed. These containers are usually filled with water. Thermocouples are placed throughout each retort basket or crate in a pattern that allows temperatures to be continuously monitored from when steam is introduced to the retort until it achieves the target operating temperature. The slowest heating thermocouple will denote the cold spot in the retort and form the basis of the process. These heat distribution studies should be conducted at least three times to properly validate the cold spot determination studies. Once a process has been properly developed and validated, it is filed with the appropriate regulatory agency. Actual processing may now be initiated.

To ensure that processes are properly delivered and the products that are manufactured are safe, records from each cook must be reviewed within one working day. In addition, processors build in additional safety factors by using heat sensitive ink when coding containers designed for thermal processing and placing temperature indicators on each retort basket or crate. The inks are designed to change colours when exposed to retort process temperatures. The usual colour change is black or dark blue to red. The temperature indicators in baskets tend to change from green to blue or purple. Many processors will retain the tags with their process records as a means to demonstrate that an effective process was delivered.

Containers in still retort systems may also be processed in water. This is how the glass containers alluded to earlier are processed. The containers must be completely submerged during processing and are processed with overpressure.

20.9 Continuous agitating retorts

There are several different kinds of continuous agitating retorts. Containers are fed continuously into the processing system through a transfer valve. The transfer valve allows the cans to be processed to move from the plant environment which is atmospheric pressure to the pressurised processing shell of the retort without a pressure loss in the retort itself. The cans are then transferred to flights within the retort and roll along the inside of the retort shell. The reel moves the containers through the retort. Continuous agitating retorts consist of either two or three shells. Two-shell units consist of a cooking (pressure cooker) and cooling shell whereas three-shell units consist of a pressure cooker and two cooling shells. Cooling may be done underpressure or at atmospheric conditions. Pressure cooling is done to ensure that the containers are not damaged due to rapid pressure changes (NFPA, 1995).

As with still retorts, these systems must have the proper instrumentation which includes a MIG thermometer or equivalent, a temperature recording device and steam controller. It is imperative that there also be a device to control reel speed as failure to do may adversely affect the process. There are other critical factors that must be addressed when utilising a continuous agitating system. These are product dependent. Containers must have adequate headspace to allow proper heat transfer. In addition, product fill weight, viscosity of the product (consistency) and initial temperature of the containers entering the retort may be mandated as critical factors for the product. Failure to adhere to established critical factors may result in the production of potentially unsafe foods.

Continuous agitating retorts are used for high-volume products such as canned vegetables and cream style corn. These are expensive systems and definitely not for small operators.

20.10 Discontinuous agitating retorts

Discontinuous agitating retorts are batch retort systems. The containers are loaded into the retort, the unit is sealed and the process is initiated. These retort systems can process in steam, water and in steam – air mixtures. There are different systems on the market including one that resembles the continuous agitating system described earlier. FMC (now JBT Food Tech) manufactured a system called an Orbitort. This system works very much like the continuous system described earlier.

Cans are loaded into the retort, the retort sealed and processing is initiated. The cans roll on the shell of the retort. At the end of the process, the containers are cooled and rolled from the system for labelling and casing.

Another system is commonly used for processing pouches and trays. This system, manufactured by Stock, is called a Rotomat. In this unit, the containers are placed in trays which are stacked in baskets or trays. The containers are secured to minimise movement. The retort baskets or stacks of trays are placed in the retort and processing is initiated. The agitation increases the rate of processing and enhances product quality. The process can be conducted in hot water or water sprays. The product may be fully or partially immersed (NFPA, 1995).

The instrumentation required is very similar to what was described for continuous agitating systems. For systems that rely on immersion or partial immersion of the containers in hot water, level indicators are required. Critical factors for processing must be established by process authorities and must be adhered to and properly documented to ensure the production of safe foods.

20.11 Other processing systems and technologies

There are other systems available in the marketplace for producing foods in hermetically sealed containers. Processors packing canned pet foods often employ a system called a hydrostat or hydrostatic retort. This is a continuous processing system in which the containers are loaded into the retort in what are called sticks which are conveyed on flights. A stick is a row of cans on their side. Whereas the containers in the continuous agitating system described earlier entered the retort through a transfer valve, in a hydrostat they enter and exit through water legs. The water legs maintain the pressure and temperature of the steam in the processing chamber. The water level in the processing area must be maintained at a certain level to ensure that the cans remain in steam (NFPA, 1995).

There are many products that are packed in hermetically sealed containers that are processed using a hot fill and hold process. These include most acid products including fruits, acidified foods, many fermented foods, juices and a variety of condiments. Products can be hot filled into glass bottles and jars, cans, plastic bottles and jars and other packages. One of the newest and most popular containers now being used are multilaminate pouches for infant formulae and juices. Hot fill processes may be conducted in various ways. These include:

1. hot fill and hold with no additional processing;
2. hot fill and hold with additional processing; and
3. hot fill and hold following an in-line twist.

Some products are filled hot, usually at 185°F (85°C) or above and held to ensure that the container itself has been adequately sterilised. There are also systems which include an in-line twist or flip in the line. The hot filled containers are flipped so that the hot product hits the underside of the lid or cap. Many processors have hot fill lines that include a walking beam cooker and cooler. The containers are filled and then enter a heating chamber where they are sprayed with hot water to ensure that the container has been fully processed. The containers then enter a cooling chamber where they are sprayed with cool water to bring the temperature of the containers down. Processors should chlorinate their cooling water to minimise the potential for postprocess contamination or leakage. Some processors will utilise a chill water bath to cool their products. The hot products will be transferred to a cold water bath and cooled there. Again, it is imperative that the cooling water be chlorinated to minimise the potential for postprocess contamination.

Hot fill processes that are used for products such as spaghetti and pizza sauces may vary from those used for other acidified foods. These products are usually manufactured with ingredients that increase soluble solids. Products that are produced to have a final brix or soluble solids content of 18% or above will not support the growth of sporeforming bacteria. So, a spaghetti sauce produced this way would not be subject to flat sour spoilage caused by *Bacillus coagulans*.

20.12 The rise of aseptics

If a consumer walks through a supermarket today, he or she will see a wide range of aseptic products. The low-acid canned food regulations from the United States in 21 CFR Part 113.3 (a) define aseptic products as follows:

> *Aseptic processing and packaging means the filling of a commercially sterilized cooled product into pre-sterilized containers, followed by aseptic hermetical sealing, with a pre-sterilized closure, in an atmosphere free of microorganisms.*

Dr C. Olin Ball was preeminent in the development of aseptic processing. In 1926 while at the American Can Company, Dr Ball described a process at a convention of the Illinois Canners in which hot food would be filled into a can under-pressure (Stier, 2004). He led a research team that developed the process known as pressure filling and closing, which evolved into the process known as heat-fill-cool, which is now known as aseptic processing. This was the technology used when the Wander Company began packing aseptic chocolate milk for the US Army in 1938. This year also saw Dr Ball publish an article that reviewed the history of food preservation from Appert to the present, which addressed how aseptic processing had developed thanks in large part to his own efforts (Ball, 1938). Fig. 20.3 shows Dr Ball and his graduate students with an early aseptic system.

The aseptic processing systems that were most utilised prior to 1980 were systems that packaged sterile product into cans. The Martin system was the first commercially successful aseptic processing system. This evolved into the James Dole aseptic canning system (Heldman, 2005). This system sterilised lids and cans in separate chambers using superheated steam. The lids and cans were the conveyed

to a sterile filling chamber that was also sterilised using superheated steam and filled with sterile product. The product was sterilised using a high-temperature short-time or an Ultrahigh-temperature short-time processing system. Research established that heating the lids and cans to a set surface temperature would ensure that they were sterile. The lid and can sterilising chambers were validated by placing temperature indicating tabs on each. The maximum surface temperature reached would be indicated by the tab.

Europe was ahead of the United States when it came to aseptic processing. The 1950s saw the introduction of the tetrahedron package for dairy products by Tetra Pak. The same company introduced the Teta Brik in 1963 (Heldman, 2005).

The world of aseptic processing changed in 1981 when the US Food and Drug Administration (FDA) approved the use of 35% hydrogen peroxide as a sterilising agent for food packaging materials. The European manufacturers of aseptic systems that utilised peroxide as a sterilant for packaging materials immediately sought to enter the US market. There was one problem; these systems did not meet the requirements defined by the US FDA in the low-acid canned food regulations. Specifically, the systems to sterilise the containers and the filling area had not been properly validated and until these equipment manufacturers could demonstrate this to the FDA, these systems could not be used for processing foods in the United States.

The National Food Processor's Association through its wholly owned subsidiary, the Tech S Corporation stepped forward to assist the European and American equipment makers in addressing these gaps. Under the leadership of Mr Keith Ito and Dr Kenneth Stevenson from the Associations Berkeley Laboratory, the organisation worked hand-in-hand with the equipment manufacturers and utilised their contacts in the thermal processing group at the US FDA to develop a road map for what was needed to ensure that these systems were properly validated. Among the client's with whom the group worked to validate these aseptic systems were Tetra Pak, Combibloc, Metal Box, International Paper, Bosch and the Continental Can Company. Upon completion of the work, the association assisted the different clients with the filing of the processes for the different systems, which included the validation data developed for each system. Table 20.1 (Ito and Stevenson, 1984) summarises how different types of packaging materials may be sterilised.

The challenge with the systems utilising hydrogen peroxide as a sterilant for packaging materials and the filling area were many. They included:

1. Verifying a nontoxic surrogate or surrogates for *C. botulinum.*
2. Developing methods to properly validate sterilisation processes for packaging materials.
3. Developing methods to properly evaluate the sterilisation processes for the filling areas.
4. Evaluating the systems themselves to ensure that they could be properly sterilised and that they would remain sterile during operations.

The first phase of the work was to verify which surrogates could be used for *C. botulinum.* Studies confirmed that the best surrogates when it came to peroxide resistance were *Bacillus subtilis A* and *B. subtilis var globygii* (Stevenson and

Table 20.1 Methods for sterilising aseptic packages

Method	Application	Advantages/ disadvantages	References
Superheated steam	Metal containers	High temperature at atmospheric pressure Microorganisms are more resistant than in saturated steam	Collier and Townsend (1956)
Dry hot air	Metal or composite juice and beverage containers	High temperature at atmospheric pressure Microorganisms are more resistant than in saturated steam	Denny and Mathys (1975)
Hot hydrogen peroxide	Plastic containers, laminated foil	Fast and efficient method	Denny et al., (1974)
Hydrogen peroxide/UV light combination	Plastic containers (preformed cartons)	UV increases effectiveness of hydrogen peroxide	Bayliss and Waites (1982)
Ethylene oxide	Glass and plastic containers	Cannot be used where chlorides are present or where residuals would remain	Blake and Stumbo (1970)
Heat from coextrusion process	Plastic containers	No chemical used	—
Radiation	Heat sensitive plastic containers	Can be used to sterilise heat sensitive packaging materials Expensive. Problems with location of radiation source.	—

Shafer, 1983). This work also established minimum values for hydrogen peroxide resistance and how new spore cultures could be tested to ensure that any organisms used for validation work had the proper resistance; a basic element for any validation study. When working with other types of sterilisation systems, the microorganisms of choice are:

- superheated steam – *G. stearothermophilus* and *Bacillus polymyxa*;
- hot air – *G. stearothermophilus* and *B. subtilis*;
- radiation – *B. subtilis*; and
- saturated steam – *G. stearothermophilus* and *C. sporogenes*.

The next phase was to develop methodologies to evaluate whether the different packaging materials could be sterilised using hydrogen peroxide. There was an added challenge here in that the 1981 approval of hydrogen peroxide as a sterilant for packaging materials also specified that there be no more than 0.5 ppm hydrogen peroxide in the finished package. To verify this, packages were filled with sterile water and tested for the presence of hydrogen peroxide.

When evaluating sterilisation of packaging materials, the packaging materials were inoculated with the appropriate spore cultures. Each culture that was produced was tested in the laboratory to ensure that it had the proper resistance to the sterilising media (Ito and Stevenson, 1984). The researchers would inoculate the packaging material with the target organism and allow it to dry. Part of the challenge was to ensure that the inoculum is placed in the proper location to ensure that it is within the container. Utilising a template to ensure that this is done properly is standard (Ito and Stevenson, 1984). The inoculated packaging materials were then passed through the system starting with the lowest inoculum level and filled with sterile growth medium. The packages are then placed into incubation and checked for evidence of growth. The packages that received the highest inoculum levels would be used for count reduction studies since survivors are expected. The objective of these studies is to get an indication of how many spores survive the process, which serves as an indicator of the system's performance.

The next step was to evaluate the sterile zone where filling would be done. The key here is to ensure that all surfaces in this area could be properly sterilised and that sterility was maintained. The latter point is usually accomplished by maintaining a positive pressure in the filling area through the use of sterile air. The air may be heated or passed through a series of microbiological filters. During operation, air pressure is monitored and maintained to ensure that sterility is maintained.

Testing the parameters employed to sterilise the filling area entailed placing disks or strips that were inoculated with the appropriate spores within the filling area. Disks were punched from cans, hollowed with a ball pein hammer and mounted using thin wire that was run through a hole punched into the cans. Using cans allowed a researcher to utilise disks of two colours – silver and bronze/gold. Because the design of sterile zones on the aseptic systems often provided little space to hang the disks, a new technology was developed. Strips were cut from aluminium tape with an adhesive and inoculated with the target organisms. The strips were then placed throughout the sterile zone and the sterilisation cycle initiated. When placing strips into the sterile zone, researchers must clearly document where each strip was placed. This is best done utilising a map or diagram of the filling area with a detailed written description of where the strip was placed. At the end of the sterilisation cycle, the strips were recovered and placed into recovery media. In the case of systems utilising hydrogen peroxide, the recovery media for the *B. subtilis* strains is usually Dextrose Tryptone Agar with Bromecresol Purple indicator. The indicator changes from blue to yellow when acid is produced allowing researchers to tell at a glance if an organism is growing. Based on the results of the trials, adjustments to the sterilisation cycle could be made. For systems using hydrogen peroxide, these changes might include using more peroxide,

adding additional peroxide reservoirs or increasing drying temperatures. Three consecutive trials with no positives were needed to validate a system.

Ito and Stevenson (1984) recommended that aseptic systems must meet the following criteria:

1. Data must be established to indicate that the equipment is adequate to insure and maintain sterility.
2. Location of control and monitoring devices must be adequate to verify that the specified process has been delivered.
3. If a chemical sterilant is used for containers, a means must be provided to maintain and control the proper level and concentration.
4. If the sporicidal effect of the chemical is dependent on time/temperature or any other physical condition, this must be monitored and controlled.
5. Residue limitations of the sterilant must be met.
6. Filled and closed containers must maintain integrity under normal and intensified conditions of commercial warehousing and transportation.
7. On-line monitoring of container seals must be adequate to ensure proper seals.
8. Commercial installations should undergo on-site testing to demonstrate the adequacy of the system.

Aseptic processing and packaging has been a boon to consumers. There are dozens of products and packages on the market today that were not available 40 years ago. Because aseptic products are processed at high temperatures for short time, these items are of excellent quality. Of course, we cannot forget how aseptic processing in bulk containers (drums or totes) has also benefited the processors. Having high-quality bulk ingredients has helped produce many different products for consumers.

20.13 Canning regulations

Canning regulations have been established the world over to protect the public health. The target organism for these HACCP-based regulations (hazard analysis, critical control points) is *C. botulinum*. This organism has the ability to grow in a vacuum as is typical of canned foods but is inhibited by pH values of 4.6 or below. The latter characteristic is the basis for the acidified food regulations in the United States found in 21 CFR Part 114. The key to control is destroying the spores or creating an environment where they cannot germinate and grow.

The regulations governing canning throughout the world all evolved from the Cannery Inspection Program established in California in 1925 as a result of the outbreaks in olives and other products between 1919 and 1924 (CDPH, 2015). These outbreaks were discussed earlier. The regulations included mandated procedures for retort operations, a mandate that processors of low-acid and acidified foods file their processes for review and established a program to inspect canneries within the state and evaluate their products. Following the Bob Vivant botulism outbreak in 1971, the California Cannery Inspection Program served as the model for federal

regulations governing canning operations. These regulations were codified as 21 CFR Part 113, "Thermally Processed Low-Acid Foods Packaged in Hermetically Sealed Containers." It was this regulation that has served as a template for regulations governing canning around the world. These regulations have evolved throughout the years as new processes and new containers have been developed.

One important element in the US regulation governing canning may be found in 21 CFR Part 113.10. The regulation states the following:

> *The operators of processing systems, retorts, aseptic processing and packaging systems and product formulating systems (including systems wherein water activity is used in conjunction with thermal processing) and container closure inspectors shall be under the operating supervision of a person who has attended a school approved by the Commissioner for giving instruction appropriate to the preservation technology involved and who has been identified by that school as having satisfactorily completed the prescribed course of instruction. This person shall supervise only in those areas for which a school approved by the Commissioner identifies the person as having satisfactorily completed training.*

The regulation specifically states that supervisors in cannery operations must attend a program, known as the Better Process Control School, which addresses cannery operations and includes topics such as microbiology, chlorination, container seal integrity and evaluation, monitoring and recordkeeping and operation of different retort systems. This program has spread throughout the world and the manual, entitled "Canned Foods: Principles of Thermal Process Control, Acidification and Container Closure Evaluation" is now in its eighth edition. The manual itself has been translated into different languages and the program is not only taught in the United States but also around the world. The author has participated in schools in Belgium, Morocco and Egypt. The latter programs were done with simultaneous translation into French and Arabic. They have been done in other parts of the world thanks to faculty at American universities taking the program "on the road."

20.14 The future of canning

Canning is not going away. It is an excellent means to preserve food and given the wide variety of packages available to processors will continue to provide consumers with an ever-expanding number of products. There will still be cans and glass jars, but there will be many more options including different types of pouches, trays, plastic containers both aseptic and hot fill, paper boxes and others. How this will evolve depends upon consumer preferences and perceptions of containers. For example, in some parts of the world, infant foods may remain almost exclusively in glass jars simply because glass is deemed to be a high-end package by the people from that area.

In addition, with the population of the globe expected to increase significantly over the years, it will be imperative that foods that are produced are preserved to meet the demands of these people.

References

Ball, C.O., 1922. The Electron Theory Applied to Thermo-Electricity, Master Thesis. George Washington University.

Ball, C.O., 1923. Thermal Process Time for Canned Food: Bulletin of the National Research Council, 37, 7, Part 1, pp. 1–76.

Ball, C.O., 1926. Calculations Involved in the Mathematical Study of the Thermal Processing of Canned Food, Doctoral Thesis. George Washington University.

Ball, C.O., 1938. Advancement of sterilization methods for canned foods. Food Res. 3, 13–55.

Bayliss, C.E., Waites, W.M., 1982. Effect of simultaneous high intensity ultraviolet irradiation and hydrogen peroxide on bacterial spores. J. Fd. Technol. 17 (4), 467–470.

Blake, O.F., Stumbo, C.R., 1970. Ethylene oxide resistance of microorganisms important in spoilage of acid and high-acid foods. J. Food Sci. 35 (1), 26–29.

California Department of Public Health (CDPH), 2015. Significance and History of the California Cannery Inspection Program. https://www.cdph.ca.gov/Programs/CEH/DFDCS/CDPH%20Document%20Library/FDB/FoodSafetyProgram/Cannery/SignificanceHistory.pdf

Collier, C.P., Townsend, C.T., 1956. The resistance of bacterial spores to superheated steam. Food Technol. 10, 477–481.

Denny, C.B., Brown, C.K., Yao, M.G., 1974. NCA Tests of Prime Pak Aseptic Canning Machine Using Hydrogen Peroxide for Equipment and Container Sterilization, Research Report no. 1-74, March 20–23 and 26–28, 1973. National Canners Association, Washington, DC.

Denny, C.B., Mathys, A.W., 1975. NCA Test on Dry Heat as a Means of Sterilization of Containers, Lids and a Closing Unit for Aseptic Canning. Final Report on RF4614, December 1975. National Canners Association Research Foundation, Washington, DC.

Esty, J.R., Meyer, K.F., 1922. The heat resistance of the spores of B. botulinus and allied anaerobes. XI. J. Infect. Dis. 31 (6), 650–663.

Heldman, D., 2005. Asceptic Processing and Packaging of Foods, Conference of Food Engineering, November 4, 2005.

Ito, K.A., Stevenson, K.E., 1984. Sterilization of packaging materials using aseptic systems. Food Technol. 38 (3), 60–62.

Lerke, P., 1979. History of Canning in California, Presentation at NFPA Western Laboratory Annual Meeting, Berkeley, CA.

National Canners Association, 1968. Laboratory Manual for Food Canners and Processors, Volume 1 and 2. The AVI Publishing Company, Westport, Connecticut.

National Food Processors Association, 1995. In: Gavin, A., Weddig, L. (Eds.), Canned Foods: Principles of Thermal Process Control, Acidification and Container Closure Evaluation, sixth ed. National Food Processors Association, Washington, DC.

Perkins, W.E., 1964. Prevention of botulism by thermal processing. In: From Botulism: Proceeding of a Symposium, Public Health Services Publication No. 999-FP-1, pp. 187–204.

Stevenson, K.E., Shafer, B.D., 1983. Bacterial spore resistance to hydrogen peroxide. Food Technol. 37 (11), 111-116 and 126.

Stier, R.F., 2004. C. Olin Ball – Father of Thermal Processing, Volume 2. Pioneers in Food Science, Food & Nutrition Press, Inc., Trumbull, CT.

United States Food and Drug Administration, 1981. Hydrogen Peroxide Solution, Title 21, Part 178.1005.

United States Food & Drug Administration, 1989a. Thermally Processed Low-Acid Foods Packaged in Hermetically Sealed Containers, Title 21, Code of Federal Regulations, Part 113, U.S. Government Printing Office, Washington, DC.

United States Foods and Drug Administration, 2001. Bacteriological Analytical Manual, Chapter 21, Examination of Canned Foods, Landry, W.L, Schwab, A.H. and Lancette, G. http://www.fda.gov/Food/FoodScienceResearch/LaboratoryMethods/ucm109398.htm

Further reading

Anonymous, 1919. Botulism from eating canned ripe olives. Public Health Reports 34 (51), 2877–2905.

Bigelow, W.D., Bohart, G.S., Richardson, A.C., Ball, C.O., 1920. Heat Penetration in Processing Canned Foods. National Canners Association Research Laboratory Bulletin 16-L.

Food Processing and Packaging Association, 2014. Testing Filing Machines of VDMA Hygiene Class V (Aseptic Plants) Sterilizing the Sterile Zone in a Machine Interior, Frankfurt am Main, 2nd Revised Edition. http://nuv.vdma.org/documents/256988/1036554/FS_08_2003_%C3%BCberarbeitung%202014_englisch.pdf/ce447afe-876e-46b5-bf28-078f061cd497 http://nuv.vdma.org/documents/256988/1036554/FS_08_2003_%C3%BCberarbeitung%202014_englisch.pdf/ce447afe-876e-46b5-bf28-078f061cd497

Meyer, K.F., Dibovsky, B.J., 1922. The distribution of spores of *B. botulinus* in United States. J. Infect. Dis. 31, 559–594.

Meyer, K.F., Dubovsky, B.J., 1922. The distribution of spores of *B. botulinus* in California. J. Infect. Dis. 31, 541–555.

United States Food & Drug Administration, 1989b. Acidified Foods, Title 21, Code of Federal Regulations, Part 114, U.S. Government Printing Office, Washington, DC.

'Cook-chill' ready meals, soups and sauces

21

Mark Swainson
National Centre for Food Manufacturing, University of Lincoln, Holbeach, Lincolnshire, United Kingdom

21.1 Introduction

Chilled ready meals, soups and sauces are often purchased for their convenience and quality. Such products tend to be produced by 'cook-chill' processing methods and comprise a diverse range of ingredients. The final quality and safety of the foods produced will typically be heavily influenced by the quality of the raw materials, the consistency of the production processes applied and the effectiveness of 'product temperature control' in the onward food supply chain.

With particular regard to these factors, quality and technical management has a key role to play in assuring product quality, safety and legality at each stage of the manufacturing process. This chapter considers many development and processing stages of typical cook-chill manufacturing operations producing ready meals, soups or sauces, and provides technical considerations with regard to the assurance of quality and safety.

21.2 New product development

Engaging with the customer and sales representatives at the required points, the business product development department will usually lead the recipe creation and development for businesses involved in cook-chill food production.

Development work is often considered to be the lifeblood of a business as the extent of customer acceptance of its new product creations will be reflected in the sales levels achieved. A very high proportion of the customer's engagement with food products is based upon organoleptic factors (e.g., appearance, aroma, taste and texture), therefore it is vital that the new product development (NPD) department utilise a wide range of sensory skills and techniques to create products which will satisfy expectations.

21.2.1 Recipe development

In the early recipe development phase of a new product it is vital that the development team work closely with the technical team to ensure that the product will meet

Swainson's Handbook of Technical and Quality Management for the Food Manufacturing Sector.
DOI: https://doi.org/10.1016/B978-1-78242-275-4.00021-6

any additional 'technical requirements' (usually relating to quality, safety and legality) outside of the development primary focus on 'sensory/organoleptic' profile.

Examples of these 'technical requirements' could be a business or customer requirement for the product to be 'free from' certain allergens, avoidance of additives/E numbers ('suitable for' vegetarians, vegans, kosher, halal), produced to 'organic' standards, etc; Or could be related to product shelf life/safety requirements, such as avoidance of ingredients known to be difficult to process from a microbiological viewpoint; Or perhaps working to achieve a certain level of product acidity (pH) to inhibit microbiological growth and thereby potentially increase product shelf life and further assure product safety.

As this development stage, if the food business is moving into a particular cook-chill category for the first time, or if it is evident that the new product proposed does not fit any of the current business product categories, then the technical management will be expected to lead the required HACCP (hazard analysis and critical control point) study and implementation, including validation of all related processes in advance of the potential to supply end product to the customer.

With regard to NPD processes, development teams sometimes work creatively from scratch. This can be referred to as 'Blue-Sky Creativity' – thinking of ideas for new products, utilising their sensory skills to then create the product which has been envisaged. More often the NPD team will work to a sales or 'customer brief'. This brief being a description of the type of product or product range required. These briefs can be extremely detailed, providing the desired key sensory points (KSPs) for each individual recipe. Other times the 'customer brief' may only outline a general range of products required and just state the product names or target consumer groups, in which case the NPD team has a wide degree of artistic licence to formulate product samples designed to gain the customer's business (but therefore at the same time has a lot more scope to drift 'off track').

With a relatively high proportion in the chilled foods sector of newly developed products either failing to launch or failing to sell well once launched, it is very beneficial during the NPD process to apply methods which can help assure the selection of the best products/recipes. These 'sensory evaluation techniques' include 'customer panels' which can be used to help define the product KSPs (e.g., appearance, aroma, taste, texture, etc.) and thereby assure a business that the products being developed are likely to meet with the approval of the end consumer once launched.

These customer panels can utilise sensory evaluation techniques including 'acceptance/preference testing', 'focus groups', 'central location testing' and 'product placement' – The ultimate question being 'would the panellist purchase the product at the price that it is intended to be sold for'?

OPD – 'old product development' – While most focus of the development department is on NPD, the practice of 'OPD' is often required to make particular improvements or other adjustments to existing 'live' product recipes. These changes may often be related to aspects of product quality or cost.

With regard to 'quality OPD' the business may wish to reformulate or change a particular ingredient in the recipe in order to respond to a complaint trend or general customer preference feedback on a particular sensory attribute of a product.

'Cost OPD' can involve seeking to change the recipe or process in some way that reduces its cost and in doing so secures a sales margin increase on the product. Such work can be an important phase of the product life cycle in terms of protecting product sales/operating margins.

OPD is also often required as businesses seek to review their product ranges to meet the increasing customer requirements for healthier foods, for example, when seeking to develop product nutritional claims such as 'high fibre', 'reduced fat' or 'reduced salt'.

As with NPD it is vitally important that OPD is progressed in close liaison with the technical department in order to ensure that the changes made do not adversely affect product quality, safety or legality.

21.2.2 Prelaunch – 'product scale-up'

Quality assurance achieved via 'sensory evaluation' plays an important role in the product scale-up phase where it is vital to ensure that the product – when manufactured on a full industrial scale – still achieves all of the required KSPs of the 'customer/sales approved' development kitchen sample.

It is a relatively frequent issue in cook-chill operations that during factory trials of a new product there are problems encountered when trying to match the 'factory made product' to the development 'kitchen samples' (which may have already been approved by/promised to the customer).

Aspects of the scale-up process that can cause such variations include:

- Ingredients purchased in large 'industrial volumes' may not be as 'quality consistent' as those purchased in small volumes for the development kitchen product (which are usually 'hand prepared' ingredients). The 'industrial scale supply' of meat/vegetable particulates is therefore likely to contain more 'off-cuts' or 'fines' which can affect the appearance of the end product. To help avoid this issue, when preparing product samples for customer/sales approval the development department should be encouraged to always use 'factory grade' ingredients. If such points are not controlled then the business is likely to regularly 'overpromise and underdeliver' in terms of product quality/consistency.
- In the large-scale factory manufacturing process, factors such as product mixing, blending and cooking are on a large volume industrial scale. Typically, these large volume processes in the factory will take far longer than when making the very small quantity of sample product in the development kitchen. High volumes and consequent longer processing timescales can lead to an increased potential for product texture breakdown, colour deterioration and flavour changes.

  In order to counter these issues the process technologist overseeing the scale-up operation will be tasked with selecting the best factory methods to help minimise these issues. If this approach does not sufficiently address the issues then the process technologist can be asked to make recommendations on potential new equipment which is less prone to the issues encountered. Another common mitigation approach is to place a limit on the maximum factory batch sizes (often used when it is the excessive factory processing time which is causing the adverse impact upon product organoleptic quality).
- Another significant difference is the extent of thermal processing given to the product in the development kitchen compared to the factory process. The development sample may

only have been cooked to a very limited extent in order to preserve texture, colour and flavour – seeking to make the 'best sample possible'. However, in the full industrial process the product is also likely to need to secure a sufficiently long 'commercially viable' end product shelf life, and to achieve this will often involve having to cook the product at higher temperatures and/or for longer periods of time (compared to the kitchen sample) in order to ensure a sufficient level of microbiological reduction – thereby reducing product organoleptic quality while achieving the extent of product safety/shelf life conferred by the increase in thermal processing. As product safety is a nonnegotiable product requirement the process technologist in the scale-up phase must ensure that the process and operating times/temperatures selected achieve the required levels of safety and shelf life while seeking to reduce the impacts/product organoleptic deterioration caused by overprocessing.

21.2.3 Hazard analysis and critical control point considerations and end product shelf life evaluation

HACCP considerations: In line with European Legislation the food safety system approach in UK cook-chill production systems is that of HACCP.

Useful HACCP guidance can be found in the following Codex Alimentarius guidelines: 'Hazard Analysis & Critical Control Point (HACCP) System and Guidelines for its Application'. http://www.fao.org/docrep/005/y1579e/y1579e03.htm

This guidance is underpinned by the Food Hygiene – Basic Texts (fourth edition) Codex Alimentarius (2009). http://www.fao.org/docrep/012/a1552e/a1552e00.htm

When the principles and implementation steps of HACCP are followed and applied to the specific 'cook-chill foods' in question this approach often results in critical control points (CCPs) focused on assuring the effectiveness and consistency of the cooking process.

This white paper, 'Microbiological quality of chilled pasteurised food products' – Campden BRI (2014) (https://www.campdenbri.co.uk/white-papers/microbiology-safety-chilled-foods.php), provides a detailed insight into the target pasteurisation/thermal treatment processes and shelf life settings typically utilised for chilled soups, sauces and ready meals to assure microbiological safety, particularly with regard to *Listeria monocytogenes* and *Clostridium botulinum*. The paper also describes the survival and growth characteristics of relevant bacterial species in order to understand their potential effect on product quality and shelf life of chilled pasteurised food products.

In addition to the cooking CCP, the 'hot-filling' temperature of product can often be considered another temperature-based CCP (e.g., in the high-risk area 'hot-fill packing' of cooked soups and sauces, with this 'hot filling' temperature set sufficiently high to serve to heat-treat the packaging and counter post-cook process microbiological contamination).

With regard to CCPs for product chilling 'times and temperatures' in cook-chill processes, depending on the product (e.g., 'post-hot-fill' spiral chiller cooling for soups and sauces, or 'post-oven cook' blast chilling for ready meal components) this CCP tends to be focused on confirmation that the product cools consistently to below 4°C and usually required to be achieved within 90 min). Another regular CCP in cook-chill food production is the use of metal detection or X-ray units to scan the final sealed end product pack to help assure foreign body control.

Occasionally when product acidity (pH) or water activity (aw) is required to reach a certain level to assure the microbiological safety of the particular product then these too can sometimes be applied as CCPs in cook-chill operations.

End product shelf life evaluation: With regard to the achievement of a consistent, safe product shelf life, in the cook-chill sector a particularly helpful guide for technologists is 'Evaluation of product shelf-life for chilled foods' (Campden BRI 2004) – This Guideline No. 46 guides the reader through measures to establish the shelf life of chilled food products in terms of both safety and quality, on an informed and scientific basis.

The safety of cook-chilled foods is a complex matter as the relatively light thermal processes typical in the sector result in a range of microorganisms still having the potential to be present in the end product. The product shelf life achieved can be influenced by numerous factors including raw material quality, the manufacturing process applied (e.g., thermal processing – cooking and chilling times and temperatures), operational control (e.g., levels of hygiene and temperature control during the process), end product properties (e.g., acidity pH, water activity aw), packaging (e.g., barrier properties) and storage temperatures.

Shelf life testing will often involve holding the end product within storage and handling conditions that reflect both the product supply chain and the holding conditions of the end consumer (most retailers/food service operations will have preset criteria detailing the expected times, temperatures and storage conditions of the shelf life trials required for any products which are to be sold through their operations). Shelf life testing usually requires both microbiological and organoleptic assessments over the course of the intended shelf life period (in some cases nutritional analysis will also be required to ensure that the nutritional performance of the product is as required over shelf life).

Often during the shelf life assessment of a proposed new product, samples of the product will be despatched to a contract microbiology laboratory to be held at the required temperatures for the designated amounts of time between microbiological tests. The storage temperatures during such trials for chilled foods will usually be elevated to reflect the typically higher temperature storage conditions within the distribution chain and consumer storage (compared to relatively low and consistent storage temperatures achievable within a closely controlled business chill store).

The microbiological targets and limits to be used for this testing will typically be set by the technical department utilising a combination of internal standards, customer codes of practice/specifications and relevant legislative standards (e.g., in Europe EU Reg 2073/2005 on 'microbiological criteria for foodstuffs').

Product shelf life must be assessed and validated during these development stages of a product using multiple samples from throughout a series of production runs.

Typically, when considering the microbiological safety of chilled foods an indicator organism is monitored to give assurance of the safety of the food. For chilled foods with a shelf life of less than 10 days *L. monocytogenes* is a pathogenic organism that is capable of survival and growth at the lower temperatures found in the manufacture and storage of chilled foods.

The following UK-based resources provide guidance on the review and consideration of the process (times and temperatures) and product criteria (including product pH, aw and shelf life) currently recommended to assure product safety with regard to *L. monocytogenes*.

Shelf life of ready-to-eat food in relation to *L. monocytogenes*: Guidance for food business operators – Chilled Foods Association (CFA), British Retail Consortium (BRC) and Food Standards Agency (FSA). https://www.chilledfood.org/wp-content/uploads/2015/08/Shelf-life-of-RTE-foods-in-relation-to-Lm-FINAL-v1.1.1-23-3-10-with-worked-examples.pdf

There is a wealth of microbiological guidance available on the setting of safe shelf lives for cook-chilled foods. The reader is recommended to review a wide range of resources which enable consideration of the particular product, process, intended supply chain and shelf life.

If considering extended chilled shelf lives then psychrotrophic *C. botulinum* may also require added consideration as this organism can potentially survive and produce toxin if anaerobic conditions develop in the products over time.

In addition to the potential risks with pathogens such as *L. monocytogenes* and *C. botulinum*, there are other spoilage organisms which should be considered when setting the shelf life of chilled pasteurised foods. Organisms of particular concern with respect to potential spoilage of pasteurised chilled products include psychrotrophic strains of *Bacillus* and *Clostridium* species. These organisms produce spores which may survive the various cooking steps and have the ability to grow at chilled storage temperatures.

For the reasons outlined above it is of vital importance to cook-chill operations that their technical management/HACCP team clearly understand and are capable of analysing and managing related microbiological safety/shelf life challenges. Where there is any doubt then specialist support must be engaged to ensure that all matters are thoroughly addressed.

It can sometimes be the product organoleptic performance that limits the total product shelf life rather than the overall product microbiological performance. Therefore, it is important to ensure that the organoleptic assessment of the product during shelf life testing evaluates product samples that have been held in the same storage conditions (e.g., times and temperatures) as the microbiological test samples.

It is prudent to assess product shelf life performance to a point which is past the length of shelf life that will be typically applied to the product. Not only is this good practice in demonstrating due diligence that the shelf life applied also allows for a margin of safety, but also by conducting such tests the business will gain an understanding of the product characteristics/features which are exhibited by the product as it deteriorates.

21.2.3.1 End of life testing once in full 'production and supply mode'

On a routine basis many businesses review the organoleptic performance of their products at the very end of the product shelf life as part of their quality assurance programme. These tests can involve comparison against newer stock of the product and focus on comparison against the agreed/specified sensory profile (which should define the target and limits of acceptability on aspects including product appearance, aroma, taste and texture).

Typically such assessments will involve the use of a multidisciplinary panel including members of the quality, NPD, process development, production and sales teams.

To best reflect the actual conditions that the product batches will encounter during distribution and consumer storage, businesses should seek to hold their end of life samples in temperature conditions which reflect the end customer shelf life testing criteria, and this is most easily achieved via the use of storage incubators. However, many businesses choose to simplify this approach by purchasing samples of their products from the relevant retail/food service outlets, storing them in domestic fridges (set to temperatures recognised by industry research to reflect typical consumer fridge conditions) for the remaining days of their shelf life – followed by taste panel evaluation on the last day of the shelf life to confirm that the organoleptic shelf life appears to be set correctly (i.e., the product quality has not yet deteriorated to a point of being unacceptable to the end consumer).

As with the routine daily site taste panels the businesses may choose to score the attendee's assessment of each 'end of life' sample whereas others choose a simple 'pass/fail' approach. The results can then be trend analysed over time and reacted to in the event of a problem issue being noted.

Similar to the sensory evaluation processes described, the product microbiological performance should also be routinely subjected to review at the end of the product shelf life. Such microbiological assessment is typically driven by customer demands as to frequency and testing suites/regimes.

21.3 The production process

21.3.1 Production process sensory evaluation techniques

To ensure absolute clarity on the sensory attributes required of the end product, a typical approach is for the development or process technologist to draft a 'product sensory profile' capturing the organoleptic expectations. This is usually agreed with the end customer during the development or product scale-up phase. A visual standard can also be achieved by the use of a photograph of the product which can be affixed or embedded in the sensory profile.

From the agreed sensory profile a 'product sensory analysis check-sheet' can be drafted for use when conducting quality checks on the end product (e.g., at the completion of cooking processes and during sensory assessment panels). Typical content of this check sheet will include:

product name,
product code,
batch code,
time of assessment,
date of assessment and
assessor name.

Followed by descriptions of the required product attributes. For example (leek and potato soup):

Appearance: A thick leek and potato soup with evident particulates of 10 mm diced potato and 20 mm sliced green leeks.
Colour: The soup has an off-white background with specks of black pepper, the diced potato should be bright white and the leek particulates should be a 50:50 mixture of green and white.
Aroma: A leek aroma dominates.
Flavour: Leek dominates followed by potato with a back note of pepper.
Aftertaste: Aftertaste of leek with a slight warmth from the pepper.
Texture: Product has a smooth and relatively thick background. The potato should still be evidently 10 mm diced (not broken down) and should be firm (but not crunchy) to the bite. Leek texture should be soft to the bite (not rubbery/hard).
Viscosity: This is a thick soup. Bostwick viscometer 14–16 cm over 30 s (measure at end of cooking at 75°C with product sieved to remove the inconsistency of particulates).

By each 'product attribute' descriptor on the 'product sensory analysis check-sheet' there is then space provided to state 'pass'/'fail' and any comments/further actions taken.

Especially with regard to ready meals, there may also be a number of intermediary stages in the production process where a sensory check is beneficial. For example, if cooking chicken fillets in an oven in readiness to chill, slice and eventually utilise in a lemon chicken ready meal, then the chicken fillets on exit from the cooker would benefit from a sensory check to confirm that the cook has been successful and the product can move on to the next stage. For this reason 'intermediary' of 'WIP (work in progress)' product sensory analysis check-sheets can also be drafted.

21.3.1.1 Sensory evaluation in the factory

When businesses own the product brand then they will tend to select particular sensory assessments and check frequencies based upon their staff and time resources available, the potential for product variability and the scale of the financial/business consequences of failure to supply to the quality specification required. However, when producing for a customer's brand, the customer will tend to stipulate what quality checks need to be conducted and how frequently.

As factory product evaluation relies on the sensory assessment skills of the production operatives and supervisory staff it is vital that all staff who are to be placed within factory operation roles involving sensory assessment of the product are screened to ensure that they can recognise to a sufficient extent the key flavours of sweet, sour, bitter, acid and umami.

Staff engaged in product assessment activities should also be screened for colour blindness and trained on how to adequately assess the sensory performance of a food product. Such training can include understanding the standards to be attained (and where to find these sensory descriptions within the business operating system), the importance of not allowing personal preference to influence the tests completed/test results and also review of the corrective actions to be taken in the event of a nonconformance.

There are International Organization for Standardization (ISO) guidelines and methodologies for sensory analysis of food available to support these processes. Many businesses choose to engage a specialist sensory evaluation trainer in order to train all of the staff required to formally assess the product.

While the use of a trained operator's pallet and visual assessment skills provides an excellent resource for quality assurance, such checks upon appearance, aroma, taste and texture can by their nature be quite subjective. Therefore it is also of great benefit to a manufacturer's quality assurance programme to incorporate equipment which can provide objective, measured tests upon the acceptability of the food products.

Quality assessment equipment typically used during the production processes of soups, sauces and ready meals include:

Viscosity meters: Equipment such as a Bostwick Consistometer (http://www.christison.co.uk/food/bostwick.html) or a Brookfield Viscometer (https://www.brookfieldengineering.com). For products such as soups and sauces these viscosity measurement devices provide an accurate and repeatable method of product viscosity assessment.

As viscosity varies with product temperature, for consistency of assessment businesses must stipulate at what temperature the product should be measured at. Usually manufacturers will choose a set temperature close to the temperature that the product is likely to be at that point of assessment. For example, a thermally processed sauce to be assessed at the stage of cooked batch completion may by assessed at 75°C, whereas a cold blend sauce to be assessed at the point of batch completion may be assessed at 4°C. To ensure accuracy a calibrated hand probe should be used to confirm that the product meets the required temperature at the point of assessment.

Another point of potential variation in results is that when assessing soups and sauces which contain particulates, the amount of particulates in each small test sample will affect the flow rate/viscosity of the product. It is therefore common business practice, when assessing the viscosity of particulate sauces/soups, to always conduct a 'sieved sample' where the sample is sieved through a set size sieve to remove the particulates before viscosity assessment in order to eliminate the 'particulate variable'.

Colour reference charts: Colours can be described, but the use of colour charts enables the assessor to work back to a consistent standard. Colour reference charts available include RHS (Royal Horticultural Society) (http://www.rhsshop.co.uk) and Pantone (https://www.pantone.com/color-finder). Many food companies now rely on photographic standards for colour assessments and often these standards will be reviewed on the computer screen at the relevant food manufacturing workstation.

Colorimeters (automated colour assessment) are also becoming increasingly prevalent in product evaluation where companies wish to move away from subjective visual grading scales.

In addition to supporting in-house process control, the use of specific test equipment also enables the manufacturing operation to communicate and work to a common standard with suppliers/customers, helping to ensure that quality standards for raw material supply and end product acceptance are clearly and objectively defined.

21.3.2 Ingredient and packaging considerations

Ingredient consistency is vital to achieving end product consistency. To help achieve a sufficiently consistent supply of ingredients to the food manufacturing operation the following steps can be helpful:

- The use of 'approved suppliers' (with approval confirming that the supplier is able to consistently control product quality, safety and legality, prior to being authorised to supply).
- Purchase to an agreed ingredient specification which clearly defines the attributes/performance requirements of the ingredient. By having a clear understanding and definition of the intended end product KSPs, checks can then focus on ensuring that the ingredient KSPs are aligned with the end product specification.

Definition of ingredient KSPs will include written description of the sensory aspects of the ingredient, including its appearance, aroma, taste and texture. These specified KSPs can then be checked against to confirm conformance to requirements at point of delivery.

- Incoming goods sensory assessment should be conducted by a member of staff who has been trained on such checks and confirmed to be capable of reviewing each ingredient against its specified quality criteria.

Over time a business may wish to vary its frequency of checks on each ingredient, with the extent of assessment dependent upon the supplier track history of consistency of supply, the potential for major product/business impact in the event of a fault, and some ingredients may require extra focus during particular 'at-risk' times of year with regard to consistency/seasonality of supply. It is often useful to build up a catalogue of these potential supply issues which can be used within a food production business for staff training and advanced warning purposes.

Laboratory assessment of ingredients can include:

- microbiological analysis to monitor supplier performance against specified limits;
- microbiological assessment of raw materials when seeking the source of high/'out of specification' end product micro results;
- nutritional analysis to monitor supplier performance against specified limits; and
- nutritional assessment of raw materials when seeking the source of 'out of specified limits' product nutritional results (e.g., perhaps when making a 'low-fat' claim on a product and finding during end product testing that the product is not achieving the required low-fat levels;)
- allergens, e.g., checking gluten free assurances on raw materials;
- GM free materials;
- illegal dyes, e.g., Sudan 1.

21.3.2.1 *Ingredient changes*

Circumstances can occur which drive a business to consider replacing a currently supplied ingredient with another. These circumstances include:

- Standard supplier is unable to achieve the consistency of supply required.
- Business initiative to improve the nutritional status of the end product (e.g., reduced salt or reduced fat projects).
- 'Cost control' or 'margin protection' – a cheaper supply is therefore being sought.
- The usual manufacturing site unexpectedly does not have the ingredient available and therefore the business needs to use a substitute ingredient in order to avoid significant disruption in their production/delivery plan.

The technical team must be involved in all such circumstances when considering options on adjustments to the ingredients used as the business must ensure that it is in control of key factors such as ingredient declaration changes, any food safety/allergen status changes and impacts upon end product specifications/customer approval.

Organoleptic assessment of the proposed new ingredient against the required ingredient specification and against a sample drawn from the current supply is a typical approach used by businesses when evaluating the adequacy of a proposed new ingredient. Such checks are usually conducted by a company panel, consisting of staff who have confirmed their competence in sensory evaluation (often quality and development staff), and involve the review (uncooked or cooked as appropriate) of aspects such as ingredient appearance, aroma, taste and texture (and also viscosity assessment for liquid products).

Ubiquitous ingredients. Some ingredients within a business manufacturing multi-component food products may be widely used across a large number of products. With regard to sauce and soup manufacture such core ingredients include tomatoes, milks and creams which are often used as the base for soup and sauce products, diced/sliced vegetables are frequently used as a particulate or blended component and herbs/spices are typically used to add extra flavour.

Clearly any changes to an ingredient which is used within many different products has the potential to cause widespread problems if the change is not controlled and ensured to be appropriate. For these reasons it is important when considering such ingredient changes that 'worst-case' scenarios are considered, that is, conducting kitchen or factory production trials upon the products which contain the highest quantities or are most influenced by the ingredient that is being considered for change. Sometimes the result may be that a new ingredient is approved for use only in recipes where it is used at a low level (e.g., below a threshold of noticeability/influence upon the KSPs of the end product).

Knowing the end customer: During all testing and evaluation of ingredients and their subsequent performance within the end products, it is important to understand the end consumer and the aspects (KSPs) of the end product that they would miss/notice when influenced by a change in the characteristics of an ingredient. Will the end consumer notice the change? If so will the consumer feel that the change makes the product better or worse, or will the consumer feel that the change has made no difference at all to their enjoyment of the product? It is important to bear in mind whether or not proposed changes are going to be communicated to the end consumer. If consumers are told that there is a difference then they may expect and seek to find a difference.

Consumer panels can be used to answer such questions and assure a business that the changes it is considering will either enhance or at least avoid damaging the KSPs of their end products. Consumer panels can utilise sensory evaluation techniques including preference testing, focus groups, central location testing and in-home placement, or can simply request the panellist to taste the product and state what they like and what they dislike about the product, and ultimately would they purchase the product at the price that it is intended to be sold for?

21.3.2.2 Ingredient and end product packaging considerations

Ingredients can be delivered in a range of packaging/container formats which can serve many functions including protection from physical damage, microbiological spoilage, contamination and flavour taints over shelf life.

When agreeing the delivery packaging it should be ensured that the format facilitates a good range of ingredient assessment, including sensory evaluation, upon arrival. Factors such as ensuring that all parts of a delivery can be accessed for inspection need to always be considered.

In addition, it may be costly to open packaging formats such as vacuum packed meats for evaluation immediately upon delivery if the ingredient is not to be used until many days after delivery (as opening will allow air into the pack which will consequently reduce the shelf life of the amount of ingredient remaining in the pack following testing). For such reasons the manufacturing operation may decide to defer a full assessment of the ingredient quality until nearer the time of use (but not so near that there would be insufficient time to deal with any problems arising), or perhaps arrange for a smaller 'sample pack' to be sent alongside the main larger delivery packs.

Such arrangements send a clear message to the supplier that their customer is monitoring the quality of their supply upon delivery and therefore heighten the supplier's focus on ensuring full adherence to the specified quality standards.

With regard to the final products produced, the packaging of the end product (and in particular the direct contact 'primary packaging') must be selected to be 'suitable for food contact' and possess sufficient barrier properties to help assure product quality and safety over the shelf life of the product.

During the initial approval of packaging for product use it is vital that a full organoleptic analysis is conducted on the food product which has been held within the packaging in a manner which reflects the worst-case scenarios and timescales of the production process, storage, distribution (including transit testing to assess physical robustness) and consumer end use.

Such packaging/product tests could be conducted by a food manufacturer on a routine basis in order to form part of a packaging quality monitoring programme, and should certainly be conducted upon any proposed change of packaging specification or packaging supplier.

21.3.2.3 Ingredient storage

With ingredient quality often being critical to the achievement of end product quality and safety in cook-chill foods, it is vitally important that all product ingredients are stored in a manner which reflects the supplier's recommendations and good manufacturing practice on aspects such as temperature control, relative humidity and the avoidance of physical damage (e.g., stacking/compression of goods).

Many ingredients will typically be used within a few days of arrival on site, and other longer shelf life ingredients may be stored for weeks or months before being required for use. It is therefore advisable for a business to monitor these 'longer term' ingredients during their storage phase for factors including organoleptic performance, as such checks will provide advanced notice of any developing quality deterioration issue and therefore ensure sufficient time to rectify the matter without disruption to the business production plan.

Routine checks during storage can also include review of the condition of the ingredient packaging, as damage and poor seals can allow air ingress which may accelerate spoilage, drying or oxidative reactions. An ever-increasing amount of foods are now reliant upon vacuum packaging or MAP (modified atmosphere packaging) for the achievement of their shelf lives and therefore a small seal failure across a batch if unnoticed can soon lead to a major failure in organoleptic performance and possibly food safety issues.

Ensuring that all ingredients are held within appropriate storage conditions is a key factor in assuring their consistent organoleptic and microbiological performance over the course of their shelf life. Optimal storage conditions will benefit each ingredient and a multicomponent ready meal, soup or sauce manufacturing business will have chilled product stores running at typically below +4°C and frozen goods stores running at below −18°C. With regard to chilled and frozen products, high airflow conditions can significantly dry any exposed product (in frozen products this is known as 'freezer burn' and can be protected against via thorough containment within the primary packaging) and higher than ideal storage temperatures can encourage microbiological growth which in addition to the associated food safety issues can also cause flavour taints and deterioration in product texture.

Further assurance of storage performance can be gained by installation of recording and alarm systems on chills and freezers to confirm that the optimum running conditions are being consistently achieved. Relative humidity could also be monitored and controlled in dry goods stores, as too much moisture within the air can lead to clumping of hygroscopic powders and the potential for elevated levels of microbial spoilage.

Extending the shelf life of ingredients: Sometimes in food manufacturing operations there may be circumstances where the business has a surplus of a particular ingredient which is at risk of exceeding its best before date before being scheduled for use. Such circumstances can occur due to overordering (perhaps resulting from a mistake or due to planning to predicted orders which have turned out to be unrealistically high) or perhaps due to delays in the production plan caused by production line breakdowns.

If it is not possible to pull forward the next production date, in an attempt to save the cost of ingredient stock losses, food production businesses will sometimes seek to extend the shelf life of the ingredients at risk of going out of date via factors such as formal review and agreement of shelf life extensions with the particular ingredient supplier, or sometimes will consider the freezing of the particular 'at-risk' ingredient stocks which would otherwise have perished before the date of the next scheduled use if continued to be stored in chilled conditions.

In such circumstances the use of sensory evaluation plays an important role in confirming that the decision to extend the shelf life of the ingredient does not adversely affect to an unacceptable extent the ingredient organoleptic properties (and as a consequence pose a threat to the quality of the end product).

It should be noted that when considering the potential for ingredient shelf life extensions the primary concern must always be that of product safety (e.g., during the additional shelf life required is there a risk of the safety of the ingredient being compromised?). If through a combination of detailed product/process knowledge and liaison with the ingredient supplier it can be ascertained that the application of additional shelf life would not pose a threat to food safety, then it would be appropriate to conduct a thorough sensory assessment in order to also confirm that the extended shelf life ingredient will still deliver (and not threaten to damage) the required KSPs within the final product.

Shelf life extension checks should be conducted by experienced members of the technical, quality and development teams, who have first-hand experience of the usual organoleptic properties of the specific ingredient and are well aware of the likely signs of deterioration/spoilage. Such signs can include off aromas/flavours, colour deterioration and texture changes (e.g., a change to become slimy or dry).

Points for consideration during such organoleptic assessments to underpin shelf life extensions include ensuring that a representative sample size is being assessed from the ingredient stock in question, as the early signs of ingredient deterioration may be localised and not yet widespread across an ingredient batch. Businesses should also consider the balance of 'risk to reward'. If by applying an ingredient shelf life extension the business is only saving a small amount of ingredient/money and has plenty of 'within standard shelf life' material in stock, is it worth the time, trouble and end product quality performance risk to extend the shelf life of the ingredient stock?

It should also be noted that the end customer, be it a supermarket or food service business, may have a supplier policy upon whether they authorise (or need to be advised of) the procedure of controlled extensions to ingredient shelf lives. Therefore with regard to ingredient shelf life extensions it is always wise for the manufacturer to check their customer policies before considering how to proceed with the best interests of all parties in mind.

21.3.3 Preparing the factory batch recipe

Preparation of the factory batch recipe typically involves the removal of the ingredients from their primary packaging and measurement into their required recipe weights to await further processing. This is therefore usually the first point in the production process at which 100% of each ingredient can be handled and closely reviewed.

Quality checks at this stage are predominantly 'vigilance' by the factory preparation staff who should be trained to ensure that each ingredient being prepared consistently appears to meet the quality standards required.

Preparation staff should be made aware that each ingredient may have been stored on site for a significant amount of time and therefore could have deteriorated since delivery. Staff should also understand that while ingredients may have passed an inspection on intake, the checker at that stage is only likely to have viewed a small percentage sample of that ingredient, whereas at the preparation stage all of the batch of that ingredient can be inspected to at least some extent. It is important therefore that staff see themselves far more as key operators who are providing a vital quality assurance role in monitoring product quality and questioning any issues, rather than as team members who have a relatively narrow remit of only weighing ingredients.

Factory staff in all areas should always be encouraged to check with the technical/quality team if ever uncertain with regard to the quality of the ingredients/end products being handled. Such regular interaction is a good sign of 'food safety and quality culture' in an organisation.

21.3.4 Work in progress (WIP) storage

It is possible for ingredient storage methods to adversely impact on the organoleptic performance of the ingredient. To help protect the prepared ingredients in storage, manufacturers of multicomponent foods will often prepare quite close to the time of production to reduce waiting time and tend to ensure that the ingredient (or ingredient mix) is stored in lidded food grade plastic containers. The term 'food grade' expects that the supplier of the container has selected/tested the material to ensure that it does not impart any flavours, taints or chemical compounds detrimental to the safety or quality of the ingredients to be contained.

Clear labelling is important at this preparation WIP stage as the primary packaging is often no longer present and the ingredient will therefore usually have a reduced 'prepared shelf life' which should be recorded on the containers together with identification of the ingredient and the destination product/batch. Usually the prepared shelf life of each ingredient is kept as short as possible to help assure the ingredient quality at point of use.

While the quality monitoring of WIP is often confined to routine checks on the prepared ingredient storage areas to ensure that the required holding conditions are being maintained and that none of the ingredients have exceeded their 'prepared shelf life', sensory assessment will be required typically in the event of an ingredient quality query, or in the circumstances where the standard 'prepared shelf life' has been exceeded (perhaps due to production delays/breakdowns) and the factory therefore requires a decision upon whether the prepared ingredient is still acceptable for use. Where shelf life extension is to be considered the primary consideration must always be that of product safety. The consideration points documented within the 'ingredient shelf life extension' section 21.3.2.3 of this chapter are equally as relevant in these circumstances at the WIP production stage.

21.3.5 Processing (mixing and cooking operations)

Thorough thermal processing of cook-chill foods is usually a key CCP with temperature checks completed at the cooking stage to confirm that the required product temperature and holding times are achieved before release to the packing/cooling phases of the operation.

The cooking phase often coupled with mixing and blending of ingredients instigates a range of physical and chemical interactions (e.g., Maillard reactions, blending or softening of particulates, formation of oil-in-water emulsions, etc.). As most food production operations are dealing with natural ingredients which can vary in variety, source and season, and also vary in factors such as their temperature and age upon addition to the product mix, such variations in these ingredients will often lead to variations in the processing performance of the end product.

Food manufacturing sites can possess a variety of processing equipment which can be used to produce the end products required (e.g., dicers, slicers, mixers, blenders, homogenisers, ovens, liquid batch cookers, packing machinery, chillers, etc.), and yet depending upon which equipment is available/selected, there can be variability in terms of the end results achieved (e.g., mixers, agitators and pumps will all vary in their degree of damage caused to the food product).

Considering the potential for process and product variability detailed earlier, in order to ensure that the end product meets with the required sensory standards ideally the product sensory description will have been written and agreed in a form which allows for an acceptable range of natural product variability from one batch to the next. For example, with regard to soups and sauces the colour reference may allow a shade either side of the ideal colour, and the product viscosity may allow for a set amount of deviation from the standard target.

As variation is common in most processing operations which use natural ingredients and a variety of processing equipment, the processing team often have to assess the product at key stages within the production flow, and make adjustments to the processing parameters in order to ensure that the desired end result is achieved (similar to a chef checking the quality of the product while being cooked in a kitchen).

Such checks and corrective actions typically involve sensory assessment of the product/batch at the point of phase completion, and before progression to the next stage of the operation (typically the packing stage for soups/sauces or the chilling stage for cooked ingredients destined for packing into ready meals). A sufficiently sized sample should be drawn from the most appropriate points in the batch (if the production batch is known to vary at certain points then all such points should be assessed) and the product should then be evaluated by the production operator or nominated quality assurance operative against the appropriate product sensory description.

The selected assessor should have been trained and screened in advance to confirm that they are capable of end product sensory analysis. Such staff training would typically take the form of the measures outlined earlier in this chapter.

Upon sensory assessment, if the product meets all of the defined requirements then the product can be allowed to progress to the next process stage; however, if there is a nonconformance raised at this stage then corrective action will typically be required.

Corrective actions in soups and sauces may include:

- addition of water if the batch is too thick;
- extra cooking if the batch particulates are too firm;
- extra homogenisation if the batch texture is too coarse;
- addition of more thickening agent (e.g., starch) if the batch requires greater viscosity and
- addition of extra particulates if the product texture has broken down.

Many such corrective actions have an impact on the product recipe and consequently the ingredient declaration. It is therefore prudent to have agreed the appropriate and acceptable corrective actions with the relevant authorities and customers in advance. Any product adjustments should also be logged on the process records to ensure full traceability. These circumstances should be progressed "by exception" rather than being allowed to become routine. They should also always be overseen by a capable and senior member of the team. With regard to such matters it is important to watch for trends over time and respond accordingly.

A similar approach can also be taken when assessing the results of intermediate stage processes (e.g., prefrying minced beef before addition to a bolognaise sauce or ready meal, or perhaps preblending a starch/powder slurry mix before addition to a batch of soup). Such assessments will require specific 'intermediate or WIP' sensory standards to be written. At their simplest these standards could be one line reminders placed upon the process sheets (e.g., 'check that the slurry mix is lump free before addition to the batch'), or could be a more complex full sensory description (e.g., appearance, aroma, taste, and texture guidance for a batch of par-cooked pilau rice which will then complete its cook at a later processing stage).

Any batch recipe or process corrections/adjustments should be formally fed back to the team member responsible for setting the production process/production sheets (typically a 'process technologist') as the requirement to routinely have to adjust a particular product may suggest that the standard production process requires a permanent adjustment or retrialing in order to increase the chances of first time success, and avoid the requirement for time consuming corrective action every time that particular product is made.

It is very important that where an operator has highlighted that there has been an issue which required 'in-process correction', the person responsible for adjusting future productions to avoid a recurrence of the issue should always feed back to the operator what corrective measures have been taken. The production operator will then feel that their feedback has been valued and will be focussed upon highlighting other opportunities for improvement in the future. In circumstances where operators feel that their feedback is not listened to or that 'we always have problems on this product' there is a danger of acceptance of less than ideal quality standards and a loss of quality focus over time.

With regard to the physical provision of sensory descriptions to the factory operators for reference against when conducting their product sensory checks, processing operations take different approaches, each of which can be effective providing that they are managed and monitored carefully with close focus upon document control and issue of updates as and when required. Common approaches include:

- printing a summary of the product sensory description upon the relevant production process instruction sheets (e.g., the mixing/cooking sheet);
- holding copies of all product sensory descriptions in files within the relevant processing areas and
- maintaining an electronic database of sensory descriptions which can be accessed at a computer terminal or printed out as and when required.

All batch assessment checks should be recorded to maintain full traceability within the operation and help prove in the event of an issue/complaint that the batch was correct at that particular point in the operation.

The recording of such results and any further corrective actions can also be used to trend analyse the product over time for any routine problems or seasonal variances (e.g., perhaps variation with regard to colour, flavour or texture which could be linked to seasonally varying produce. If the extent of such seasonal variability is unacceptable then a solution may be to opt to use a frozen version of the variable ingredients, which has been harvested and frozen at one set point in the year, and is then available to be used consistently all year round).

In batch production processes, where a number of separate batches of the same product are to be produced sequentially, a technique that is often also used during organoleptic assessment is that of maintaining a reference sample of each of the previous batches for review against as each new batch is made. Such comparison enables a swift comparison check to ensure that the product performance is not gradually drifting away from the required standard.

21.3.5.1 Production flow

In cook-chill processing systems, often following operations such as cooking/mixing, and prior to cooling/packing phases, there is the potential for delay/holding time to be incurred due to the next stage not yet being ready to process the product.

In circumstances where the product sensory attributes are particularly sensitive in the event of a delay in processing, this amount of time needs to be very closely controlled to help ensure that the final end product quality will not be adversely affected. For example, in ready meal manufacture, rice which has been cooked and chilled will ideally be used immediately but may be stored in a refrigerated area for approximately 24 h before packing as a component into the final ready meal packs (e.g., 'sweet and sour chicken with egg fried rice'). In such a scenario there is the potential during this holding time for the rice to become dry/hard, especially if the rice has been stored uncovered and placed in a chill store area which has high airflow.

Similarly when producing cook-chill risotto, upon completion of cooking the batch needs to be rapidly cooled otherwise if held up at high temperatures the perfect defined rice texture can quickly break down and the product is then spoilt.

It is therefore important not only to set an optimised holding method and maximum amount of time before the component can no longer be used (in ready meal production such limiting of holding times can be as much related to product microbiological control as they are related to product organoleptic control), but also to ensure that there is a confirmatory organoleptic assessment upon the component appearance, aroma, taste and texture before further processing/packing. At this stage if the component quality is found to have deteriorated beyond an acceptable point, the business would incur further significant product, time and packaging cost losses by continuing to process the product, only to realise later at final product analysis stage that a specific component has caused the entire product to be organoleptically unacceptable. Well-timed sensory checks therefore help assure product quality and help avoid incurring unnecessary costs by highlighting faults as early as possible.

21.3.6 Filling/packing operations

As covered earlier in this chapter, when processing cook-chill soups and sauces, often the temperature of 'hot-filling' the product into its packaging in a segregated designated 'high-risk' area of the factory will be classed as a CCP.

In the previous section it was touched upon that the packing phase of the product following the recipe mix/cooking operation can have significant impacts on the product organoleptic performance. For example, when manufacturing soups and sauces the products will typically have been transferred from the recipe mix/cooking operation via a series of vessels, pipework and pumps (which may or may not have been designed to handle the product as gently as possible) and will often then be held in an agitated vessel during the packing process in order to ensure a consistent mix/blend. In addition, if the products are being hot-filled into the packaging then the products will be hot and therefore continuing to cook while waiting to be packed.

Product heat and agitation factors can lead to deteriorating organoleptic quality such as the softening of particulates, deterioration of starches/gels, colour and viscosity changes and therefore often the process technologist who has scaled-up the product will have set a maximum batch size related to the rate at which the product can be packed (this packing rate is usually dictated by the speed of the packing machines, and sometimes by the cooling/freezing capacity of the operation post packing). The technologist may calculate and set such maximum holding times by organoleptically assessing trial batches during the course of the batch packing operation until the point is reached where the product has deteriorated past a point of acceptability.

Delays in the packing operation (perhaps due to machinery breakdown) may incur further deterioration of the food. In order to reduce the adverse impacts of such delays upon the end product in soup and sauce operations, factories will often choose to turn off the agitation of the batch during the delay to reduce the physical impacts upon the product, and where batches are held hot a factory may choose to also reduce the holding temperature during the time of the delay (while ensuring that the temperature reduction does not fall below a point where microbiological growth may become an issue).

In preparation for hold-ups in product processing cook-chill food businesses will benefit from use of a 'delay procedure' (a summary of all of the actions required to be taken in the event of a process delay) which not only seeks to minimise the organoleptic impacts on the product, but also ensures that the product quality is monitored closely during the delay (typically via the requirement for routine product assessments conducted by key staff) in order that the factory is quickly aware of when the delayed batch has gone past the point of being acceptable to pack. Where particulate deterioration in soups and sauces is a key concern, samples of the product may be routinely sieved to enable closer visual examination of the particulates. Sometimes colour change may be the main concern, requiring routine comparison against colour charts during the course of the delay.

The packing process can have significant physical effects on the product as it passes through vessels, agitators, pumps, pistons and pipework, resulting in product quality impacts, which will be further magnified if a significant proportion of the product is being recycled within the process, perhaps due to product being reworked back into the batch prepacking/holding system (e.g., if the packing of the batch is encountering machine problems which are causing a high level of pack weight or seal integrity rejects which are then being reworked to save wastage).

The potential for product organoleptic variation at the packing phase usually leads to businesses placing a great deal of organoleptic scrutiny upon samples drawn from the final product at the start, middle and end of a batch run, in order that the extent of deterioration can be measured and possible 'batch cut-off points' defined between 'acceptable product' and the point at which the product deteriorated to an extent where it is no longer acceptable for sale and therefore needs to be segregated, quarantined and either reworked or disposed of.

Sometimes such checks are conducted even more frequently/throughout the packing operation if particular problems are being encountered, for example, excessive variability in particulate distribution which therefore requires further analysis to monitor and ascertain the root cause of the issue.

With regard to batch soup/sauce production, focus on a 'particulate distribution issue' – inconsistency in the amount of particulates in end product packs – may necessitate routine 'washouts' of packs to ascertain the consistency of distribution of particulates across the packed batch and to assess whether all of the product components are evenly present throughout the batch. Some particulates may particularly sink/float when the batch is being packed which may lead to them being present mostly at the start or end of the batch. Also if certain larger size particulates are becoming blocked in the feed pipes/dosing pistons this may lead to their sporadic distribution as every so often the pressure may build up until a burst of those stuck particulates are finally dosed into the packaging.

21.3.7 Product 'cooling/chilling' phase

For food safety, quality and shelf life assurance it is very important that the 'chilling'/'cooling' phase of cook-chill food products is conducted as thoroughly and as swiftly as possible. Moving the product from a 'cooked'/'hot' state down to below 4°C within 90 min is a typical CCP target for many cook-chill processes, with the temperatures confirmed by both destructive and nondestructive temperature probing of the product.

Deterioration of product 'KSPs' is also a consideration in the product cooling/chilling phases. For example, if the product is cooled in an open state (e.g., a tray of oven cooked rice transferred to a blast chiller), the chilled airflow that it is exposed to may cause some drying of the product, whereas if a product is cooled in its packaging, while this may protect the product from the drying effect of chilled airflows, the product can sometimes be found to take longer to cool due to the insulating effect of the packaging, which can also cause organoleptic changes. In addition, some cooling/chilling operations may require the product to be agitated, or may require the outer edges of the product to be crust frozen. Such aspects of cooling processes will also impact upon the organoleptic performance of the end product.

With all of the scope for impacts on product organoleptic quality through the process, it is only after the product has been packed and cooled that a business can start to be confident that their organoleptic quality tests are assessing the product in a form that the end consumer is likely to experience. Therefore the final product sensory evaluation of appearance, aroma, taste and texture is a very important stage of the business quality assurance system.

At this final product organoleptic assessment stage businesses will often evaluate a number of samples drawn from the start, middle and end areas of the production batch, the number of which should reflect the potential for variability within the production process. Food manufacturers will typically ensure that these checks are conducted immediately upon the completed end product by an experienced member of the cooling/packing teams or by a quality assurance technician.

Taste panel: Most food manufacturers will also place all recently completed products on a routine (often daily, sometimes even more frequent than this) business taste panel which is attended by a multidisciplinary cross section of the business staff and management.

End product taste panels will ideally include members of the NPD department, process development and sales teams, as these team members will often be able to closely remember the customer's expectations of the particular products and how these expectations should be reflected within the KSPs of each particular product. Also having members of the factory production and quality assurance teams present at the formal end product, taste panel will help ensure a good transfer of product sensory knowledge and provide the opportunity to discuss any particular processing issues/problems being encountered.

Finished product taste panels will usually assess a representative sample of each production batch against the sensory description set either internally or with the end customer. Some businesses choose to score the attendee's assessment of each batch, others choose a simple 'pass/fail' approach. The more data that can be captured at this stage, the greater the potential for trend analysis of results over time. Such results can be used to drive business focus upon product quality (e.g., trend analysis of end product taste panel results may show a gradual deterioration in a product's colour or texture over time which due to the gradual drift in standard may not have been picked up by the individuals regularly attending each taste panel).

Some businesses insist that during the taste panel product assessment phase, every panellist should evaluate each sample in silence or isolation from other panellists to ensure that their judgement cannot be influenced in any way by more dominant, opinionated or senior members of the panel. Potentially some panel members may be biased in their opinions on the acceptability of products due to pressures such as production throughputs, financial impacts or customer demands.

The end product taste panel will typically be the most thorough organoleptic evaluation that the product is going to receive on site, therefore businesses should seek to ensure that the taste panels take place before the particular batches of product are due to be despatched to the customer. This will ensure that if a product is found at the business taste panel to be unacceptable (or in requirement of further scrutiny), then the product will still be within the control of the business, rather than incurring the difficult situation of having to consider a withdrawal/recall from the distribution chain or customer.

21.3.8 Frozen product formats

In circumstances where the end product is intended to be sold in a frozen format (after being produced initially via 'cook-chill' processes) then there are often extra organoleptic factors to be considered and monitored via the use of sensory assessment during the product/process design phase and the subsequent production quality assurance phase.

Many operations tend to freeze the product (starting from chilled condition) utilising blast freezers or spiral freezers. Target temperatures to achieve before placing in frozen storage tend to be −15°C to −18°C as by this stage the produce has moved out of the ice crystal formation phase.

Considerations include the fact that the freezing of food products will form ice crystals within the food. Typically, the slower the freezing process, the larger the ice crystals formed and therefore the greater the potential to damage the food product structure (including deterioration of the physical texture of particulates and damage to product starch/gel suspensions, which can lead to excessive product syneresis upon defrost).

Organoleptic quality issues that can also arise at the product freezing phase are the potential for 'freezer burn' (quality deterioration typically caused by product dehydration and oxidation, often linked to the product not being sufficiently wrapped in protective packaging) and also the potential for product quality loss during storage, which is especially a concern for products which are particularly susceptible to deterioration (e.g., high-fat meat products can be prone to rancidity during frozen storage). Poorly maintained freezer stores can also increase the potential for product deterioration due to significant fluctuations in air temperature.

21.3.9 Microbiological and chemical analysis of the end product

It is usual to test at least one sample of each production run of each product to verify the microbiological safety of the product. (Shelf life will already have been assessed and validated during the product development stages of a product using multiple samples from throughout a series of production runs.)

In addition to testing for the microbiological safety of production runs it is also important to test for any chemical, nutritional, speciation or chain of custody claims made for the products. The frequency of testing these parameters will typically be based on customer requirements, codes of practice or in accordance with the requirements of any scheme owners.

Customers will usually dictate the range of product microbiological and nutritional tests required and also the frequency of testing. Especially so when the product is sold under their own brand. The range of tests will usually be related to the product being manufactured and the test requirements, targets and tolerances will be expected to be stated on the end product specification.

Alongside the routine product assessments (e.g., microbiological testing and nutritional testing) conducted by the manufacturer, when supplying the major supermarkets with their 'supermarket own brand' products (and also an increasing number of own brand food service operations), they will often expect to also conduct a range of routine independent tests, completed across the range of products supplied.

These sampled products will typically be expected to be purchased in-store (e.g., to give a true reflection of the product microbiological performance in the supply chain) and to then be analysed by an independent laboratory. It is also 'usual practice' in many business settings for the supplier to receive the bill for this in-store product/supplier surveillance testing).

Many supermarkets tend to expect all new launch 'first production runs' of their own branded products to be sent to their nominated independent laboratories for microbiological and chemical (typically nutritional) analyses. This way they will be immediately aware if any launched products are outside of their specified expected standards on both microbiological and chemical analysis grounds.

Where retail pack labels are stating the nutritional characteristics of the product (this will be the case in just about all retail products) food manufacturing suppliers will also often be asked to submit samples to the independent 'supermarket assigned' laboratory routine nutritional testing/monitoring. The frequency of this testing will vary, often at the discretion of the customer technologist, and usually related to the risk/variability expected (e.g., products may receive a more frequent assessment if the products are bearing a nutritional claim such as 'high fibre', 'low fat', 'low salt', etc.).

In all cases where a customer nominated independent laboratory is used for routine product assessment, often the assessment results will be sent to both the manufacturer and the customer, together with a summary report on whether or not the products performed as expected (to specification). When the testing results are found to be outside of the specified limits (often classified as 'out of specification', 'action' or 'red' results), the food manufacturer will be expected to take swift and appropriate corrective action, usually under the full focus/scrutiny of the customer technologist.

Outside of such surveillance testing, the routine production testing conducted by the food manufacturing operation will typically be providing a steady stream of microbiological and other results for the technical/quality manager to assess and action where necessary. This routine testing will also require reporting to both site management and senior review meetings in order to provide a measure of site compliance to required standards. Also when packing for the major retailer brands, monthly technical KPI (key performance indicator) reports will usually also expect a summary of microbiological and nutritional performance for the month together with any corrective actions taken upon results which were outside of the specified limits.

Therefore particularly when manufacturing and supplying food retailer own brands and food service business own brands, product analysis can be a significant cost to the business which needs to be set and monitored very carefully and worked back into the end product costing to reflect the frequencies/demands of each specific customer. It is worth noting that the technical/quality manager should always check whether the proposed or current range and frequency of microbiological testing is appropriate for the specific product.

21.3.10 Environmental control and monitoring

Cook-chill food products are particularly susceptible to microbiological contamination due to their formulation and the fact that the chilled food manufacturing environment can often harbour a range of microorganisms including pathogens.

For this reason, cook-chill food businesses will typically progress a range of environmental cleaning and monitoring measures including microbiological swabbing and the use of rapid hygiene assessment technologies such as those using ATP (adenosine triphosphate) to instantly give an indication of the level of residual debris.

Certain microorganisms are used as hygiene indicators in the chilled food sector as the presence of indicator bacteria in chilled ready-to-eat food, although not inherently a hazard, can be indicative of poor practice that may be linked to one or more of the following:

- poor quality of raw materials or food components,
- undercooking,
- cross-contamination,
- poor cleaning, and
- poor temperature and time control.

Indicator organisms are useful in the assessment of food product safety because they tend to be present in higher numbers than most pathogens and are relatively quick and easy to identify. Indicator organisms typically used for chilled food production areas include Enterobacteriaceae, *Escherichia coli* and *Listeria* species.

21.4 Storage and supply

21.4.1 End product storage

Business efficiency pressures to produce larger product batch runs less frequently can lead to end products being held in storage for a significant amount of time pre-despatch to the customer.

During this holding time most products organoleptic performance will typically be deteriorating. This is especially so if the product bears a 'use-by' code rather than a 'best before' code (e.g., most chilled products).

Most chilled foods will deteriorate during storage due to a combination of microbiological, physical and chemical factors. Frozen foods may deteriorate due to ice crystal formation and aspects such as oxidative reactions and freezer burn. Ambient products could be susceptible to the absorbance of moisture from the environment over time (especially if not packaged in robust gas and moisture barrier packaging materials).

External factors can increase the potential for such deterioration to occur, including higher than ideal chill store/frozen store temperatures. For such reasons site cold stores are typically fitted with temperature monitoring probes and associated alarms. Given the large amounts of end product value located in these stores and the very high food safety/quality sensitivity to any failure of the chill/frozen chain, the stores are also typically subjected to routine manual checks on the area temperatures and the running temperature records should be routinely assessed by the technical department to check for any indication of an emerging issue with the running temperatures of the chill store system.

Due to the potential for product deterioration during frozen storage, some businesses choose to conduct a sensory evaluation of the stored (stock holding) batches on a routine basis and at a frequency which reflects the potential for deterioration to occur. Such checks are important as it is better to be aware of the deterioration of a product batch early in order that a fresh batch run can be planned in time, rather than awaiting receipt of a customer order, only to then find during final quality checks at point of despatch that the product does not meet the quality criteria required due to deterioration which has occurred during storage.

21.4.2 Product despatch

Completing a final taste panel on the end product just before despatch to the customer provides one final assurance. At this stage there is little that can be done to rectify any product problems which may have occurred earlier in the production process; however, such checks do serve as a useful quality control checkpoint and provide the major benefit of avoiding any problem products being despatched to the end customer.

For these reasons the final assessment of product organoleptic performance (compared against the requirements defined in the agreed customer specification) is typically seen as the final 'quality safety-net' prior to despatch.

At point of outload onto the distribution wagon the cased stock will receive a temperature check to confirm that it is being handed over to the haulier at the required temperature. Checks will also be made to ensure that the distribution wagon is set to the required temperature for the journey and may also confirm that the temperature recording devices in the unit are working correctly.

21.4.3 'Depot checks' and store inspection

On delivery to the customer distribution depot/store the cased end product is likely to receive a nondestructive temperature check to confirm that the cold chain is being maintained. Inspection will also confirm the sound condition of the stock and will record traceability information from the product codes present.

Sensory checks on the product are likely to continue even after it has been despatched, usually in the form of 'depot checks' (typically conducted upon delivery as part of the customer acceptance checks) or in-store/stock inspections conducted by the customer. By having a clearly defined and agreed sensory description/specification for each food product, problems at depot and in-store can be reduced as without clearly agreed product standards the manufacturer risks the potential for their foods to be judged and possibly rejected purely on subjective assessments alone.

21.5 Summary

Cook-chill processes are subject to the same product hazard groups as all other foodstuffs, namely physical, chemical (including allergens) and microbiological. Yet in this product category the 'microbiological' aspects are often particularly challenging due to the relatively 'light' thermal processing used. Also, in the case of ready meals the relatively 'open' processing and packing methods can leave foods susceptible to microbiological contamination/growth during the process. For such reasons a high degree of hygiene and process control is required at all times, as is strict adherence to industry best practice on the shelf lives to be applied and on the selection and validation of the key food safety control measures required in the manufacturing processes.

With regard to sensory quality control in cook-chill manufacturing operations, it is important to ensure that all processes are well maintained, robust and consistent. The quality manager should continually seek to make sensory checks clearly defined and objective rather than subjective wherever possible and always ensure that the sensory criteria required of the product is realistic and consistently achievable before being formally agreed and committed to with the end customer.

Further reading

Anaphylaxis Campaign promotes awareness and the need for allergen control. More details on allergens and their control can be found at the Anaphylaxis Campaign website. http://www.anaphylaxis.org.uk

BRC Global Standard Food Safety Issue 8. https://www.brcglobalstandards.com

Creating New Foods, The Product Developers Guide. Earle, M. D., Earle R.L, 1999. CCFRA.

Chilled Foods, A Comprehensive Guide. Brown, M., 2008. Woodhead.

Chilled Foods Association (CFA) – https://www.chilledfood.org – was formed in 1989 to establish, continuously improve and promote best hygienic practice standards in the production of retailed chilled prepared food.

European Commission Guide to the Machinery Directive 2006/42/EC, second ed. http://ec.europa.eu/DocsRoom/documents/9483/attachments/1/translations/en/renditions/native

EU Reg 2073/2005 on microbiological criteria for foodstuffs. http://eur-lex.europa.eu/LexUriServ/LexUriServ.do?uri = OJ:L:2005:338:0001:0026:EN:PDF

Evaluation of Product Shelf-Life for Chilled Foods. Betts, G.D., Brown, H.M., Everis, L.K., 2004. CCFRA, Chipping Campden. Guide on establishing the shelf-life of chilled food products in terms of both safety and quality, on an informed and scientific basis.

Food Hygiene – Basic Texts (fourth edition) Codex Alimentarius Commission. The Codex basic texts on food hygiene promote understanding of how rules and regulations on food hygiene are developed and applied. The general principles of food hygiene cover hygiene practices from primary production through to final consumption, highlighting the key hygiene controls at each stage. This publication also contains the most internationally used description of the HACCP system and guidelines for its application. http://www.fao.org/docrep/012/a1552e/a1552e00.htm

Food Manufacturing: An Overview. Hutton, T., 2001. Campden & Chorleywood Food Research Association Group, Chipping Campden.

Food Processing Technology Principles and Practice. Fellows, P., 2000. CRC Woodhead Publishing.

Food Quality Assurance, Principles and Practices. Inteaz Alli, 2004. CRC Press.

Food Safety Culture – Creating a Behavior-Based Food Safety Management System. Yiannas, F., 2009. Springer.

Food Safety Behavior. Yiannas, F., 2015. Springer.

Food Taints and Off-Flavours, first ed. Saxby, M.J., 1993. Blackie Academic & Professional; Chapman & Hall, London, New York, NY.

Guidelines for the Validation of Food Safety Control Measures, Codex Alimentarius Commission, 2008. http://www.fao.org/input/download/standards/11022/CXG_069e.pdf

Guidelines for Assessing the Microbiological Safety of Ready-to-Eat Foods Placed on the Market, Health Protection Agency, 2009. https://www.gov.uk/government/publications/ready-to-eat-foods-microbiological-safety-assessment-guidelines

Hygiene for Management, nineteenth ed. Sprenger, R.A., 2017. Highfield International Ltd.

International Featured Standards (IFS). http://www.ifs-certification.com

Ishihara's Tests for Colour Deficiency. Ishihara, S., 2004. Kanehara Trading Inc.

ISO 9000:2015 Quality management systems – fundamentals and vocabulary. https://www.iso.org/obp/ui/#iso:std:iso:9000:ed-4:v1:en

ISO 6658:2017(en) Sensory analysis – methodology – general guidance. https://www.iso.org/obp/ui/#iso:std:iso:6658:ed-3:v1:en

ISO 8589:2007. Sensory analysis – general guidance for the design of test rooms. https://www.iso.org/standard/36385.html

Microbiological quality of chilled pasteurised food products – Campden BRI (2014). https://www.campdenbri.co.uk/white-papers/microbiology-safety-chilled-foods.php

Predictive Microbiology:

PMP (Pathogen Modelling Programme), USDA. https://pmp.errc.ars.usda.gov/PMPOnline.aspx

ComBase. http://www.combase.cc/index.php/en/

Product Development Guide for the Food and Drink Industry: Guideline No. 8. Campden BRI (1997).

Safer Food Better Business Initiative (UK Food Standards Agency). https://www.food.gov.uk/business-industry/sfbb

Sensory Evaluation: A Practical Handbook. Kemp, S., Hollowood, T., Hort, J., 2009. Wiley Blackwell.

Shelf life of ready-to-eat food in relation to *L. monocytogenes*: Guidance for food business operators – Chilled Foods Association (CFA), British Retail Consortium (BRC) and Food Standards Agency (FSA). https://www.chilledfood.org/wp-content/uploads/2015/08/Shelf-life-of-RTE-foods-in-relation-to-Lm-FINAL-v1.1.1-23-3-10-with-worked-examples.pdf

The Bad Bug Book – Foodborne Pathogenic Microorganisms and Natural Toxins Handbook (second ed., 2012). US Food and Drug Administration. Published online at http://www.fda.gov/Food/FoodborneIllnessContaminants/CausesOfIllnessBadBugBook/

The safety and shelf-life of vacuum and modified atmosphere packed chilled foods with respect to nonproteolytic *Clostridium botulinum*. Food Standards Agency (2017) http://www.foodstandards.gov.scot/downloads/vacpacguide.pdf

UK Food Standards Agency, http://www.food.gov.uk

'Food Handlers Fitness to Work' guidance.

'Safer Food Better Business' initiative.

General UK/EU Food Law.

Food Safety Guidance.

UK HSE (Health & Safety Executive) Information Source: The hygienic design of food machinery: Supply of Machinery (Safety) Regulations 2008 and Machinery Directive (2006/42/EC), http://www.hse.gov.uk/foi/internalops/ocs/500–599/520_22.htm

Vacuum packing and modified atmosphere packing of food training course, UK Food Standards Agency 2018. http://vacuumpackingtraining.food.gov.uk/introduction/

Index

Note: Page numbers followed by "*f*" and "*t*" refer to figures and tables, respectively.

FSC
www.fsc.org
MIX
Papier | Fördert gute Waldnutzung
FSC® C083411

Kolibri 360 GmbH
Ferdinand-Jühlke-Straße 7
99095 Erfurt, Deutschland
produktsicherheit@kolibri360.de